Electronegativity Values of the Elements (Pauling Scale)

Legend:

atomic number →	35
element symbol →	Fe
electronegativity →	1.83

Main Table

1 1A	2 2A	3 3B	4 4B	5 5B	6 6B	7 7B	8 8B	9 8B	10 8B	11 1B	12 2B	13 3A	14 4A	15 5A	16 6A	17 7A	18 8A
1 H 2.20																	2 He
3 Li 0.98	4 Be 1.57											5 B 2.04	6 C 2.55	7 N 3.04	8 O 3.44	9 F 3.98	10 Ne
11 Na 0.93	12 Mg 1.31											13 Al 1.61	14 Si 1.90	15 P 2.19	16 S 2.58	17 Cl 3.16	18 Ar
19 K 0.82	20 Ca 1.00	21 Sc 1.36	22 Ti 1.54	23 V 1.63	24 Cr 1.66	25 Mn 1.55	35 Fe 1.83	27 Co 1.88	28 Ni 1.91	29 Cu 1.90	30 Zn 1.65	31 Ga 1.81	32 Ge 2.01	33 As 2.18	34 Se 2.55	35 Br 2.96	36 Kr 3.00
37 Rb 0.82	38 Sr 0.95	39 Y 1.22	40 Zr 1.33	41 Nb 1.6	42 Mo 2.16	43 Tc 1.9	53 I 2.66	45 Rh 2.28	46 Pd 2.20	47 Ag 1.93	48 Cd 1.69	49 In 1.78	50 Sn 1.96	51 Sb 2.05	52 Te 2.1	53 I 2.66	54 Xe 2.60
55 Cs 0.79	56 Ba 0.89	57 *La 1.1	72 Hf 1.3	73 Ta 1.5	74 W 2.36	75 Re 1.9	85 At 2.2	77 Ir 2.20	78 Pt 1.28	79 Au 2.54	80 Hg 2.00	81 Tl 1.62	82 Pb 2.33	83 Bi 2.02	84 Po 2.0	85 At 2.2	86 Rn 2.2
87 Fr 0.7	88 Ra 0.9	89 #Ac 1.1	104 Rf	105 Db	106 Sg	107 Bh	108 Hs	109 Mt	110 Ds	111 Rg	112 Cn	113 Uut	114 Uuq	115 Uup	116 Uuh	117 Uus	118 Uuo

*Lanthanide series

58 Ce 1.12	59 Pr 1.13	60 Nd 1.14	61 Pm 1.13	62 Sm 1.17	63 Eu 1.2	64 Gd 1.2	65 Tb 1.1	66 Dy 1.22	67 Ho 1.23	68 Er 1.24	69 Tm 1.25	70 Yb 1.1	71 Lu 1.27

#Actinide series

90 Th 1.3	91 Pa 1.5	92 U 1.38	93 Np 1.36	94 Pu 1.28	95 Am 1.13	96 Cm 1.28	97 Bk 1.3	98 Cf 1.3	99 Es 1.3	100 Fm 1.3	101 Md 1.3	102 No 1.3	103 Lr 1.3

Notes:

1. Electronegativity values are below and atomic numbers are above the element symbol.
2. Electronegativities in the same row tend to increase from left to right.
3. Electronegativities in the same column tend to increase from bottom to top.
4. Where electronegativity values are not given, they have not been determined because of experimental difficulties (very short half-lives, very weak bonding energies, etc.).

Solubility Table for Ionic Compounds in Water

Compounds Containing	Very Soluble	Slightly Soluble	Insoluble
Lithium (Li^+)	All		
Sodium (Na^+)	All		
Potassium (K^+)	All		
Ammonium (NH_4^+)	All		
Chloride (Cl^-)	All except ⇒	Pb^{2+}	Ag^+, Hg_2^{2+}, Hg^{2+}
Bromide (Br^-)	All except ⇒	Hg^{2+}, Pb^{2+}	Ag^+, Hg_2^{2+}
Iodide (I^-)	All except ⇒		Ag^+, Hg_2^{2+}, Hg^{2+}, Pb^{2+} Bi^{3+} Cr^{3+}
Fluoride (F^-)	All except ⇒		Mg^{2+}, Ca^{2+}, Sr^{2+}, Ba^{2+}, Pb^{2+}
Nitrate (NO_3^-)	All		
Chlorate (ClO_3^-)	All		
Perchlorate (ClO_4^-)	All		
Acetate ($C_2H_3O_2^-$)	All except ⇒	Ag^+, Hg_2^{2+}	Bi^{3+}
Carbonate (CO_3^{2-})	Li^+, Na^+, K^+, NH_4^+		⇐ All except
Phosphate (PO_4^{3-})	Li^+, Na^+, K^+, NH_4^+		⇐ All except
Oxalate ($C_2O_4^{2-}$)	Li^+, Na^+, K^+, NH_4^+	Ca^{2+} Cu^{2+} Cr^{3+} Fe^{3+}	⇐ All except
Chromate (CrO_4^{2-})	Li^+, Na^+, K^+, NH_4^+	Mg^{2+} Ca^{2+} Cu^{2+} Fe^{2+}	⇐ All except
Hydroxide (OH^-)	Li^+, Na^+, K^+	Ca^{2+}, $Sr^{2=}$, Ba^+	⇐ All except
Oxide (O^{2-})		Ca^{2+}, $Sr^{2=}$, Ba^+	⇐ All except
Sulfides (S^{2-})	Li^+, Na^+, K^+, Mg^{2+}, NH_4^+	Ca^{2+}, Sr^{2+}, Ba^{2+},	⇐ All except
Sulfates (SO_4^{2-})	All except ⇒	Ca^{2+}, Ag^+	Ag^+ $Sr^{2=}$, Ba^{2+}, Pb^{2+}, Hg_2^{2+}, Hg^{2+}
Sulfites (SO_3^{2-})	Li^+, Na^+, K^+, NH_4^+	Mg^{2+}	⇐ All except

Alphabetical Table of the Elements with Standard Atomic Weights[++]

(Standard atomic weights abridged to 4 significant digits, from: M.E. Wieser and T.B. Coplen, "Atomic weights of the elements 2009 (IUPAC Technical Report)", *Pure Appl. Chem.*, ASAP Article doi: 10.1351/PAC-REP-10-09-14, Web publication date: 12 December 2010, http://iupac.org/publications/pac/pdf/asap/pdf/PAC-REP-10-09-14.pdf)

Element	Symbol	Atomic Number	Atomic Weight	Element	Symbol	Atomic Number	Atomic Weight
actinium	Ac	89	[227]*	neodymium	Nd	60	144.2
aluminum	Al	13	26.98	neon	Ne	10	20.18
americium	Am	95	[243]*	neptunium	Np	93	[237]*
antimony	Sb	51	121.8	nickel	Ni	28	58.69
argon	Ar	18	39.95	niobium	Nb	41	92.91
arsenic	As	33	74.92	nitrogen	N	7	14.01
astatine	At	85	[210]*	nobelium	No	102	[259]*
barium	Ba	56	137.3	osmium	Os	76	190.2
berkelium	Bk	97	[247]*	oxygen	O	8	16.00
beryllium	Be	4	9.012	palladium	Pd	46	106.4
bismuth	Bi	83	209.0	phosphorus	P	15	30.97
bohrium	Bh	107	[262]*	platinum	Pt	78	195.1
boron	B	5	10.81	plutonium	Pu	94	[244]*
bromine	Br	35	79.90	polonium	Po	84	[208]*
cadmium	Cd	48	112.4	potassium	K	19	39.10
calcium	Ca	20	40.08	praseodymium	Pr	59	140.9
californium	Cf	98	[251]*	promethium	Pm	61	[145]*
carbon	C	6	12.01	protactinium	Pa	91	231.0
cerium	Ce	58	140.1	radium	Ra	88	[226]*
cesium	Cs	55	132.9	radon	Rn	86	[222]*
chlorine	Cl	17	35.45	rhenium	Re	75	186.2
chromium	Cr	24	52.00	rhodium	Rh	45	102.9
cobalt	Co	27	58.93	rubidium	Rb	37	85.47
copernicum	Cn	112	[277]*	roentgenium	Rg	111	[281]*
copper	Cu	29	63.55	ruthenium	Ru	44	101.1
curium	Cm	96	[247]*	rutherfordium	Rf	104	[261]
darmstadtium	Dm	110	[281]*	samarium	Sm	62	150.4
dubnium	Db	105	[262]*	scandium	Sc	21	44.96
dysprosium	Dy	66	162.5	seaborgium	Sg	106	[263]*
einsteinium	Es	99	[252]*	selenium	Se	34	78.96
erbium	Er	68	167.3	silicon	Si	14	28.09
europium	Eu	63	152.0	silver	Ag	47	107.9
fermium	Fm	100	[257]*	sodium	Na	11	22.99
fluorine	F	9	19.00	strontium	Sr	38	87.61
francium	Fr	87	[223]*	sulfur	S	16	32.06
gadolinium	Gd	64	157.3	tantalum	Ta	73	180.9
gallium	Ga	31	69.72	technetium	Tc	43	[98]*
germanium	Ge	32	72.64	tellurium	Te	52	127.6
gold	Au	79	197.0	terbium	Tb	65	158.9
hafnium	Hf	72	178.5	thallium	Tl	81	204.4
hassium	Hs	108	[265]*	thorium	Th	90	232.0
helium	He	2	4.003	thulium	Tm	69	168.9
holmium	Ho	67	164.9	tin	Sn	50	118.7
hydrogen	H	1	1.008	titanium	Ti	22	47.87
indium	In	49	114.8	tungsten	W	74	183.9
iodine	I	53	126.9	uranium	U	92	238.0
iridium	Ir	77	192.2	vanadium	V	23	50.94
iron	Fe	26	55.85	xenon	Xe	54	131.3
krypton	Kr	36	83.80	ytterbium	Yb	70	173.1
lanthanum	La	57	138.9	yttrium	Y	39	88.91
lawrencium	Lr	103	[260]*	zinc	Zn	30	65.38
lead	Pb	82	207.2	zirconium	Zr	40	91.22
lithium	Li	3	6.941				
lutetium	Lu	71	175.0				
magnesium	Mg	12	24.31	ununtrium[#]	Uut	113	[284]*
manganese	Mn	25	54.94	ununquadium[#]	Uuq	114	[289]*
meitnerium	Mt	109	[266]*	ununpentium[#]	Uup	115	[288]*
mendelevium	Md	101	[258]*	ununhexium[#]	Uuh	116	[292]*
mercury	Hg	80	200.6	ununoctium[#]	Uuo	118	[294]*
molybdenum	Mo	42	95.96				

[++]The standard atomic weight of an element is a weighted average of its isotopic composition. Many standard atomic weights are not invariant but depend on the prior history of the material in which the elements are found. They are calculated from knowledge of their isotopic abundances in a particular sample and the known masses of the different isotopes of each element. Since isotopic abundances do not have uniform terrestrial distributions, there is always a degree of uncertainty in how to calculate a single value for a substance whose origin is unspecified To emphasize that standard atomic weights are not constants of nature, the International Union of Pure and Applied Chemistry (IUPAC) now recommends that the atomic weights of certain elements is represented by a range, rather than a single value (see reference above). For users needing an atomic-weight value for an element in a sample of unspecified origin, IUPAC has provided a table of "conventional" atomic weights, which is used here.

* Element has no stable isotopes and no characteristic terrestrial isotopic composition, so no average atomic weight can be calculated. The number in brackets is the mass number of the longest-lived isotope.

[#] Element name is provisional until a permanent name is assigned.

PERIODIC TABLE OF THE ELEMENTS

Legend:
- atomic number →
- element symbol →
- atomic weight →

26
Fe
55.85

1A (1)	2A (2)	3B (3)	4B (4)	5B (5)	6B (6)	7B (7)	8B (8)	8B (9)	8B (10)	1B (11)	2B (12)	3A (13)	4A (14)	5A (15)	6A (16)	7A (17)	8A (18)
1 **H** 1.008																	2 **He** 4.003
3 **Li** 6.941	4 **Be** 9.012											5 **B** 10.81	6 **C** 12.01	7 **N** 14.01	8 **O** 16.00	9 **F** 19.00	10 **Ne** 20.18
11 **Na** 22.99	12 **Mg** 24.31											13 **Al** 26.98	14 **Si** 28.09	15 **P** 30.97	16 **S** 32.07	17 **Cl** 35.45	18 **Ar** 39.95
19 **K** 39.10	20 **Ca** 40.08	21 **Sc** 44.96	22 **Ti** 47.87	23 **V** 50.94	24 **Cr** 52.00	25 **Mn** 54.94	26 **Fe** 55.85	27 **Co** 58.93	28 **Ni** 58.69	29 **Cu** 63.55	30 **Zn** 65.38	31 **Ga** 69.72	32 **Ge** 72.64	33 **As** 74.92	34 **Se** 78.96	35 **Br** 79.90	36 **Kr** 83.80
37 **Rb** 85.47	38 **Sr** 87.61	39 **Y** 88.91	40 **Zr** 91.22	41 **Nb** 92.91	42 **Mo** 95.96	43 **Tc** [98]	44 **Ru** 101.1	45 **Rh** 102.9	46 **Pd** 106.4	47 **Ag** 107.9	48 **Cd** 112.4	49 **In** 114.8	50 **Sn** 118.7	51 **Sb** 121.8	52 **Te** 127.6	53 **I** 126.9	54 **Xe** 131.3
55 **Cs** 132.9	56 **Ba** 137.3	57 **·La** 138.9	72 **Hf** 178.5	73 **Ta** 180.9	74 **W** 183.9	75 **Re** 186.2	76 **Os** 190.2	77 **Ir** 192.2	78 **Pt** 195.1	79 **Au** 197.0	80 **Hg** 200.6	81 **Tl** 204.4	82 **Pb** 207.2	83 **Bi** 209.0	84 **Po** [209]	85 **At** [210]	86 **Rn** [222]
87 **Fr** [223]	88 **Ra** [226]	89 **#Ac** [227]	104 **Rf** [265]	105 **Db** [268]	106 **Sg** [271]	107 **Bh** [272]	108 **Hs** [277]	109 **Mt** [276]	110 **Ds** [281]	111 **Rg** [280]	112 **Cp** [285]	113 **Uut** [284]	114 **Uuq** [289]	115 **Uup** [288]	116 **Uuh** [292]	117 **Uus** [292]	118 **Uuo** [294]

·Lanthanide series

58 **Ce** 141.1	59 **Pr** 140.9	60 **Nd** 144.2	61 **Pm** [145]	62 **Sm** 150.4	63 **Eu** 152.0	64 **Gd** 157.3	65 **Tb** 158.9	66 **Dy** 162.5	67 **Ho** 164.9	68 **Er** 167.3	69 **Tm** 168.9	70 **Yb** 173.1	71 **Lu** 175.0

#Actinide series

90 **Th** 232.0	91 **Pa** 231.0	92 **U** 238.0	93 **Np** [237]	94 **Pu** [244]	95 **Am** [243]	96 **Cm** [247]	97 **Bk** [247]	98 **Cf** [251]	99 **Es** [252]	100 **Fm** [257]	101 **Md** [258]	102 **No** [259]	103 **Lr** [262]

Note: Periodic Table boxes have the atomic number at the top, element symbol in the center and atomic weight (to 4 significant digits) or mass number of most stable isotope (in brackets) at the bottom. See also the notes at the bottom of the Alphabetical Table of the Elements on opposite page.

THIRD EDITION

Applications of
Environmental
Aquatic
Chemistry

A Practical Guide

THIRD EDITION

Applications of
Environmental
Aquatic
Chemistry

A Practical Guide

Eugene R. Weiner

CRC Press
Taylor & Francis Group
Boca Raton London New York

CRC Press is an imprint of the
Taylor & Francis Group, an **informa** business

Cover Design: Jim Carr.

Chapter Title Photographs: Gary Witt and Eugene Weiner.

CRC Press
Taylor & Francis Group
6000 Broken Sound Parkway NW, Suite 300
Boca Raton, FL 33487-2742

First issued in paperback 2021

© 2013 by Taylor & Francis Group, LLC
CRC Press is an imprint of Taylor & Francis Group, an Informa business

No claim to original U.S. Government works

Version Date: 20121015

ISBN 13: 978-1-03-209915-6 (pbk)
ISBN 13: 978-1-4398-5332-0 (hbk)

Library of Congress Cataloging-in-Publication Data

Weiner, Eugene R.
  [Applications of environmental chemistry]
  Applications of environmental aquatic chemistry : a practical guide / Eugene R. Weiner. -- Third edition.
    pages cm
  Includes bibliographical references and index.
  ISBN 978-1-4398-5332-0 (hardback)
  1. Environmental chemistry. 2. Water quality. I. Title. II. Title: Environmental aquatic chemistry.

  TD193.W45 2013
  628.1'68--dc23   2012030504

Visit the Taylor & Francis Web site at
http://www.taylorandfrancis.com

and the CRC Press Web site at
http://www.crcpress.com

Even without human intervention, pollutant concentrations in the environment have a tendency to diminish with time due to natural causes. Where natural processes are fast enough and contaminant concentrations small enough, the simplest approach to remediation is to wait until pollutant levels are no longer deemed hazardous.

Chapter 2, Section 2.3.1

Contents

(*Photo by Eugene Weiner*)

Once eutrophication has started, it is hard to reverse, because bottom sediments become a reservoir for stable phosphorus, which can cycle between stable and labile forms, providing a continuing source of nutrient P long after the initial sources of excess P have been eliminated.

Chapter 4, Section 4.4

Preface to the Third Edition

Many environmental chemistry books can be put into one of two groups:

1. Those that are very thorough and include nearly all of the basic chemical principles that underlie the topics covered
2. Those that rely on an existing understanding of the basic principles and begin from there

This text tries to follow a middle path, where the basic principles needed for an environmental topic are included only to the extent of contributing to an understanding of the topic being discussed and to guide the reader in applying the information to actual environmental problems.

The fields of environmental science and engineering advance rapidly. This third edition is substantially updated from the second edition. Besides extensive reorganization and updating, there is a new chapter, "Nutrients and Odors: Nitrogen, Phosphorus, and Sulfur," two new appendices, "Solubility of Slightly Soluble Metal Salts" and "Glossary of Acronyms and Abbreviations Used in this Book," and much new material, especially in Chapters 5 and 11 and in Appendix A.

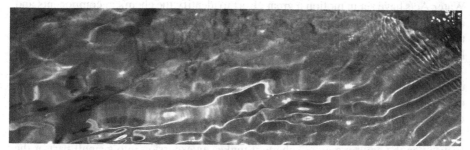

Generally, the most stringent water quality numerical standards are for drinking water and aquatic life classifications.

Chapter 1, Section 1.1.4

Preface to the Second Edition

Much new material has been added to this second edition. Besides a totally new chapter on radionuclides, the text has been reorganized and updated with separate chapters on metals, light nonaqueous phase liquids (LNAPLs), dense nonaqueous phase liquids (DNAPLs), and biodegradation. Also, some end-of-chapter exercises have been added. The dictionary of inorganic pollutants has been enlarged and some important organic pollutants added. The former appendices listing drinking water standards, water quality criteria, and sample collection protocols have been omitted because these data are continually changing and are readily available on the Internet. However, the goals of the book remain the same—to help nonchemist environmental professionals and students work with chemical information in their work and studies.

Preface to the First Edition

By sensible definition, any by-product of a chemical operation for which there is no profitable use is a waste. The most convenient, least expensive way of disposing of said waste—up the chimney or down the river—is the best.

From *American Chemical Industry—A History*, by
W. Haynes, Van Nostrand Publishers, 1954

The quote above describes the usual approach to waste disposal as it was practiced in the first half of the 1900s. Current disposal and cleanup regulations are aimed at correcting problems caused by such misguided advice and go further toward trying to maintain a nondegrading environment. Regulations such as federal and state Clean

Water Acts have set in motion a great effort to identify the chemical components and other characteristics that influence the quality of surface water and groundwater and the soils through which they flow. The number of drinking water contaminants regulated by the U.S. government has increased from about 5 in 1940 to over 150 in 1999.

There are two distinct spheres of interest for an environmental professional: the ever-changing, constructed sphere of regulations and the comparatively stable sphere of the natural environment. Much of the regulatory sphere is bounded by classifications and numerical standards for waters, soils, and wastes. The environmental sphere is bounded by the innate behavior of chemicals of concern. While this book focuses on the environmental sphere, it makes an excursion into a small part of the regulatory sphere in Chapter 1, where the rationale for stream classifications and standards and the regulatory definition of water quality are discussed.

This book is intended to be a guide and reference for professionals and students. It is structured to be especially useful for those who must use the concepts of environmental chemistry but are not chemists and do not have the time and/or inclination to learn all the relevant background material. Chemistry topics that are most important in environmental applications are succinctly summarized, with a genuine effort to walk the middle ground between too much and too little information. Frequently used reference materials are also included, such as water solubilities, partition coefficients, natural abundance of trace metals in soil, and federal drinking water standards. Particularly useful are the frequent Rules of Thumb lists, which conveniently offer ways to quickly estimate important aspects of the topic being discussed.

Although it is often true that "a little knowledge can be dangerous," it is also true that a little chemical knowledge of the "right sort" can be a great help to the busy nonchemist. Although no "practical guide" will please everyone with its choice of inclusions and omissions, I have based my choices on the most frequently asked questions from my colleagues and the material I find myself looking up over and over again. The main goal of this book is to offer nonchemist readers enough chemical insight to help them contend with those environmental chemistry problems that seem to arise most frequently in the work of an environmental professional. Environmental chemists and students of environmental chemistry should also find the book valuable as a general-purpose reference.

(Photo courtesy of Gary Witt)

The chemical makeup of a collected water sample reflects the history of its prior flow path, particularly the minerals it has contacted, its contact time with these minerals, and the values of temperature, pH, and redox potential along the flow path.

Chapter 11, Section 11.2.1

Acknowledgments

This third edition is enriched by contributions from three talented friends. The cover design is by Jim Carr, and Gary Witt took most of the title page photographs. Their artistry provides a welcome link between words on a page and the world they try to describe. Noah Greenberg, a wetland scientist, constructively reviewed parts of the book related to nutrients and agricultural water quality, Chapter 4 and Section 6.11. Their generous help is much appreciated.

Understanding intermolecular forces is the key to predicting how contaminants become distributed in the environment. When you can predict relative attractive forces between molecules, you can predict their relative solubility, volatility, and sorption behavior.

Chapter 2, Section 2.7.2

Author

Eugene R. Weiner, PhD, is professor emeritus of chemistry at the University of Denver, Colorado. He was a professor of chemistry at the University of Denver from 1965 to 1995. From 1967 to 1992, he worked with the U.S. Geological Survey, Water Resources Division in Denver as a consultant and has consulted on environmental issues for many other private, state, and federal entities. He served as a member of the Colorado Water Quality Control Commission from 1975 to 1981 and was chairman of the Standards and Classification Committee, responsible for developing regulations for classifying state waters and assigning appropriate numerical standards. After 30 years of research and teaching environmental and physical chemistry, he joined Wright Water Engineers Inc., an environmental and water resources engineering firm in Denver, as senior scientist.

He received a BS in mathematics from Ohio University, an MS in physics from the University of Illinois, and a PhD in chemistry from Johns Hopkins University. He has authored and coauthored more than 400 research articles, books, and technical reports. He has also conducted 16 short courses dealing with the movement and fate of contaminants in the environment in major cities around the United States for the continuing education program of the American Society of Civil Engineers.

The objective of this act is to restore and maintain the chemical, physical, and biological integrity of the Nation's waters.

Federal Water Pollution Control Act Amendments of 1972,
Public Law 92500 (Clean Water Act)

1 Water Quality

1.1 DEFINING ENVIRONMENTAL WATER QUALITY

Water quality means different things to different people, depending on their goals for the water. A chemist in a laboratory will regard high-quality laboratory water as water free from chemical impurities or suspended solids. High-quality environmental water has different criteria. The same chemist on a wilderness backpack trip might identify high-quality water as water in a pristine environment unaltered by human activity. If the chemist is also a fisherman, she or he might regard high-quality water as a good habitat for fish and other aquatic organisms. A farmer prefers water with low sodium or other dissolved salts that may degrade the soil or be toxic to crops, but may appreciate moderate levels of nutrients such as nitrogen and phosphorus. A drinking water treatment plant manager will define high-quality water as water with a minimum amount of substances that have to be removed or treated to produce safe and palatable drinking water. A broad view of high-quality water will take into consideration its suitability for particular uses.

The U.S. Congress recognized this when it enacted the Federal Water Pollution Control Act Amendments of 1972, Public Law 92500, also known as the Clean Water Act (CWA), where it is stated, "The objective of this act is to restore and maintain the chemical, physical, and biological integrity of the Nation's waters." In the law, water quality is a measure of its suitability for particular designated uses.

Implementing the law entails identifying these uses, setting standards that are protective of the designated uses, and providing enforcement procedures that require compliance with the standards.

1.1.1 WATER-USE CLASSIFICATIONS AND WATER QUALITY STANDARDS

In most parts of the world, the days are long gone when one could directly drink from many rivers, lakes, springs, and wells, and when untreated water supplies could readily meet almost all needs for high-quality water. Where such water remains, mostly in high mountain regions untouched by mining, grazing, or industrial fallout, it must be protected by strict regulations. In the United States, most states seek to preserve high-quality waters with antidegradation policies. But most of the water that is used for drinking water supplies, irrigation, and industry, not to mention serving as a supporting habitat for natural flora and fauna, is much-reused water that often needs treatment to become acceptable.

Whenever it is recognized that water treatment is required, new issues arise concerning the degree of water quality sought, the costs involved, and, perhaps, restrictions imposed on the uses of the water. Since it is economically impossible to make all waters suitable for all purposes, it becomes necessary to designate for which uses various waters are suitable.

In this context, a practical evaluation of water quality depends on what the water is used for, as well as its chemical makeup. The quality of water in a stream might be considered good if the water is used for irrigation but poor if it is used as a drinking water supply. To determine water quality, one must first identify the ways in which the water will be used and only then determine the appropriate numerical standards for important water quality parameters (maximum concentrations for chemical and biological constituents and values for physical parameters like turbidity and odor) that will support and protect the designated water uses.

Strictly speaking, a water impurity is any substance in the water that does not originate from a water molecule, and absolutely pure water is unattainable in any realistic water sample. High-quality water is not necessarily near-pure; it may contain significant levels of impurities that do not affect its intended uses, amounts of detrimental impurities too small to be harmful to its intended uses, or even impurities useful for its intended uses. Many impurities in water may be beneficial. For example, carbonate (CO_3^{2-}) and bicarbonate (HCO_3^{2-}) make water less sensitive to acid rain and acidic mine drainage; hardness and alkalinity decrease the solubility and toxicity of metals; nutrients, dissolved carbon dioxide (CO_2), and dissolved oxygen (O_2) are essential for aquatic life. Outside a chemical laboratory, extremely pure water generally is not desirable. Water with very low concentrations of certain dissolved impurities cannot sustain aquatic life, is not the best choice for crop irrigation, is more corrosive (aggressive) to metal pipes than water containing a measure of hardness, and certainly will not taste as good as natural water saturated with dissolved oxygen and containing a healthy mix of minerals. For most purposes, the quality of water is not judged by its purity but rather by its suitability for the different uses intended for it.

In this book, the following definitions are used:

- A water impurity is any substance in water that is not derived from a water molecule only,* regardless of whether it is considered harmful, beneficial, or neutral to the intended uses of the water.
- A water pollutant or contaminant is any substance in water that is not derived from a water molecule only, and is also considered, when present in sufficient concentration, to be harmful to the water's intended uses.

The water contaminant nitrate (NO_3^-) illustrates this point. In drinking water supplies, nitrate concentrations greater than 10 mg/L are considered a potential health hazard, particularly to young children. On the other hand, nitrate is a beneficial plant nutrient in agricultural water and is added as a fertilizer. Water containing more than 10 mg/L of nitrate is of poor quality if it is used for potable water but may be of good quality for agricultural irrigation.

Thus, water uses must be identified before water quality can be judged. Once the water uses are defined, numerical water quality standards for harmful impurities can be developed that will protect each use.

There are three different types of water quality standards set by state and federal regulations:

1. Surface and groundwater standards, for the ambient quality of natural waters (rivers, lakes, reservoirs, wetlands, and groundwater). These standards are chosen to protect the current and intended uses of natural waters, as discussed later.
2. Effluent standards, controlled by discharge permits under the National Pollutant Discharge Elimination System (NPDES). Effluent standards are chosen so that wastewaters discharging into natural waters do not cause the receiving waters to exceed their surface and groundwater standards. Effluent standards are affected by the ambient standards for the receiving water, the assimilation capacity of the receiving water, the total pollutant load contributed by all dischargers into that water, natural background concentrations, and so on.
3. Drinking water standards, which apply both to groundwater used as a public water supply and to water delivered to the public from drinking water treatment plants. Drinking water standards are chosen to protect public health.

---

* The species H^+, OH^-, H_3O^+, $H_5O_2^+$, and so on, can all be derived from water molecules only. Species such as Cl^-, Na^+, MnO_4^-, $Ca(OH)_2$, $PbOH^-$, Cd^{2+}, and so on, clearly cannot. Concentrations of H^+, OH^-, H_3O^+, $H_5O_2^+$, and so on that are high enough to be considered harmful (i.e., acid or basic conditions harmful to certain designated water uses) cannot occur from water alone; they require that acidic or basic substances be added to water in sufficient quantity to change the pH beyond its normal environmental range of about pH 6–9.

1.1.2 WATER QUALITY CLASSIFICATIONS AND STANDARDS FOR NATURAL WATERS

The following preliminary steps, taken by a state or federal agency, are a common approach to evaluating water quality in natural waters:

1. Define in general the basic purposes for which natural waters will potentially be used (water supply, aquatic life, recreation, agriculture, etc.). These will be the categories used for classifying uses for existing bodies of water.
2. Set numerical water quality standards for physical and chemical characteristics that will support and protect the different water-use categories. For certain parameters, like odors, color, or algal blooms, narrative rather than numerical standards may be appropriate.*
3. Compare the water quality standards with field measurements of existing bodies of water, and then assign appropriate use classifications to the water bodies according to whether their present or potential quality is suitable for the assigned water uses.
4. After a natural body of water is classified for one or more uses, compile an appropriate set of numerical standards to protect its assigned use classifications. Where different assigned classifications have different standards for the same parameter, the more stringent standard will always apply.
5. Review established water use classifications and associated quality standards on a regular basis and revise them when appropriate.

It is clear that measuring the chemical composition of a water sample collected in the field is just one step in determining water quality. The sample data must then be compared with the standards assigned to that water body. If no standards are exceeded, the water quality is defined as good within its classified uses. As new information is collected about environmental and health effects of individual water constituents, it may be necessary to revise the standards for different water uses. Federal and state regulations require that water quality standards be reviewed periodically and modified when appropriate.

In addition to the review process for existing standards, the Safe Drinking Water Act requires the U.S. Environmental Protection Agency (EPA) to periodically publish a Drinking Water Contaminant Candidate List (CCL), a list of contaminants that, "at the time of publication, are not subject to any proposed or promulgated national primary drinking water regulations, are known or anticipated to occur in public water systems, and may require regulations."†

---

* A narrative water quality standard is a statement that prohibits unacceptable conditions in or upon the water, such as floating solids, scum, visible oil film, or nuisance algae blooms. Narrative standards are sometimes called "free froms" because they help keep surface waters free from visible and chemical types of water pollution where numeric standards are not required by federal law. The association between a narrative standard and classified use is less well defined than it is for numeric standards; however, most narrative standards are used to protect, or provide additional protection for, aquatic life, recreation, and aesthetic classified uses. Because narrative standards are not quantitative, determining that one has been exceeded typically requires a "weight of evidence" approach to data analysis, showing a pattern of violations.
† More information may be found at www.epa.gov/safewater/ccl/index.html.

Contaminants on the list are studied until the EPA concludes that there are sufficient data and information to either propose appropriate regulations or conclude
that no action is currently necessary.

1.1.3 SETTING NUMERICAL WATER QUALITY STANDARDS

Water quality standards are intended to protect public health and welfare, enhance
the quality of natural waters, and serve the purposes of the CWA by

- Providing, wherever attainable, water quality suitable for the protection and
  propagation of fish, shellfish, and wildlife, as well as recreation in and on
  the water ("fishable/swimmable")
- Considering the use and value of state waters for public water supplies,
  propagation of fish and wildlife, recreation, agriculture and industrial purposes, and navigation

To do this, water quality standards consist of two parts:

1. Classifications that describe current and intended uses for a water body
2. Numerical standards that provide water of a suitable quality for attaining
   and protecting these uses

Numerical water quality standards for each water body must protect all the uses
for which the water is classified. In addition, site-specific or time-specific standards
may be established where special conditions exist, such as where aquatic life has
become acclimated to high levels of dissolved metals or during fish spawning periods. Each state has tables of numerical water quality standards for each classified
water body. In addition to numerical standards for environmental waters, there are
separate human health–based standards for groundwater used for public drinking
water supplies, and for treated drinking water as delivered from a water treatment
plant or, for some parameters such as lead and copper, which may be present in a
building's plumbing system, as delivered at the tap.

The states, not the EPA, have the primary responsibility for setting water quality standards. However, the EPA sets baseline standards for different use classifications that serve as minimum requirements for the state standards. In addition, the
EPA issues guidance and model regulations regarding standards. EPA approval is
required before standards can be adopted or changed by states.

Water quality numerical standards are determined for measurable water properties that are important for classified uses, such as

- Chemical composition: Concentrations of metals, organic compounds,
  chlorine, nitrates, ammonia, phosphorus, sulfate, and so on.
- General physical and chemical properties: Temperature, alkalinity, conductivity, pH, dissolved oxygen (DO), hardness, total dissolved solids (TDS),
  chemical oxygen demand (COD), and so on.

- Biological characteristics: Biological oxygen demand (BOD), *Escherichia coli*, fecal coliforms, whole effluent toxicity (WET), and so on.
- Radionuclides: Radium-226, radium-228, uranium, radon, gross alpha and gross beta emissions, and so on.

1.1.4 TYPICAL WATER-USE CLASSIFICATIONS

Consistent with the requirements of the CWA, states are free to develop and adopt any use classification system they find appropriate, except that using natural waters for waste transport and assimilation* is never an acceptable classification under any circumstances. All states classify surface waters and groundwaters according to their current and intended uses. In some cases, a future intended use serves as a goal for making water quality improvements.

Each state adopts those water use classifications (also called *beneficial uses*) that it considers necessary for its purposes, but the CWA describes certain uses of waters, listed below, that are always considered desirable, whether or not they are currently attained, and which always must be protected. States always must take these uses into consideration when classifying state waters, but are free to add other use classifications. Although multiple use classifications may be assigned to a particular water body, none are considered more important than others even though one classification may have more stringent numerical standards than another. Where one water body has several classifications with different numerical standards for the same parameter, the more stringent standard will always apply.

Some commonly used beneficial use water classifications are as follows:

- Public water supply
  - Surface waters or groundwaters that are suitable or intended to become suitable for potable water supplies, or in some cases, food processing. After receiving standard treatment—defined as coagulation, flocculation, sedimentation, filtration, and disinfection with chlorine or its equivalent—these waters will meet federal and state drinking water standards.
- Protection and propagation of fish, shellfish, and wildlife
  - May be divided into several more specific subcategories, including cold-water fish, warmwater fish, and shellfish. For example, some coastal states have a classification specifically for oyster propagation. The use may also include protection of aquatic flora. Wildlife protection should include waterfowl, shore birds, and other water-oriented wildlife.
- Recreation
  - Surface waters that are suitable or intended to become suitable for recreational activities. This classification is complicated by the fact that although the states and the EPA have an obligation to do as much as possible to protect the health of the public, people tend to use whatever water bodies are available for recreation, regardless of (or without realizing) unsafe sanitary or flow conditions. For this reason, the CWA

---

* This limitation is the origin of the slogan, "Dilution is not the solution to pollution."

requires that all state waters shall be presumed, by default, to be suitable or potentially suitable for a recreation classification of primary contact, Class 1a or Class 1b (see below), unless a use attainability analysis demonstrates that there is not a reasonable potential for primary contact uses to occur in the water segments in question within the next 20-year period.

- Recreation is traditionally divided into primary contact and secondary contact categories.
  - *Primary contact*: Requires numerical standards, mainly bacteriological, that will protect people from illness due to activities in or on the water when the ingestion of small quantities of water is likely to occur and prolonged and intimate contact with the body is expected. Primary contact uses include swimming, wading, skin-diving, surfing, rafting, kayaking, tubing, water-skiing, frequent water play by children, and other activities likely to result in immersion. Because of the CWA requirement that all state waters initially be classified for primary contact,* two subclassifications may be assigned:
  - *Primary Class 1a—Existing primary contact*: Class 1a waters are those in which primary contact uses have been documented or are presumed to be present. All State waters must be assigned a Class 1a classification by default, unless a reasonable level of inquiry has failed to identify any existing Class 1 uses of the water segment. Waters for which no use attainability analysis has been performed that demonstrates that a recreation Class 2 designation is appropriate, are also to be classified as Class 1a.
  - *Primary Class 1b—Potential primary contact*: Class 1b is assigned to water segments for which no use attainability analysis has been performed, demonstrating that a recreation Class 2 classification is appropriate, and, also, where a reasonable level of inquiry has failed to identify any existing Class 1 uses of the water segment.
  - *Secondary contact*: Requires numerical standards, mainly bacteriological, that are protective when immersion is unlikely. Examples are boating, wading, fishing, rowing, and other streamside or lakeside recreation activities. The secondary contact designation applies to surface waters that are not suitable for a primary contact classification but are suitable or intended to become suitable for recreational uses in or around the water that are not included in the primary contact categories, for example, wading, fishing, motor yachting, and so on.

---

* Before a state water can be assigned any recreational classification other than primary, the state must provide, and EPA approve, justification for concluding that the state water cannot now or in the foreseeable future be used for primary contact recreation. One reason for this stringent requirement is that public use of any accessible water body for recreational activities, prohibited or not, is almost impossible to control. Another reason is that a guiding principle of the CWA is that all natural waters be preserved or restored to a quality suitable for primary contact recreation wherever possible.

- *Class 2*—Class 2 recreational waters are those deemed unsuitable and not potentially attainable for recreation, such as wetlands that do not have sufficient water for recreation, at least seasonally. To assign this classification, states must conduct a use attainability analysis, demonstrating that recreational uses consistent with the CWA are not attainable. However, states are always encouraged to establish water quality standards that recognize and protect recreational uses that do not directly involve contact with water, including hiking, camping, and bird watching.
- Agriculture
  - The agricultural use classification defines waters that are suitable for irrigation of crops, consumption by livestock, support of vegetation for range grazing, and other uses in support of farming and ranching and for protecting livestock and crops from injury due to pollution of irrigation and other agricultural water uses.
- Groundwater
  - Subsurface waters in a zone of water-saturation that are at the ground surface or can be brought to the ground surface or brought to surface waters through wells, springs, seeps, or other discharge areas. Separate standards are applied to groundwater suitable for
    - *Domestic use*: Groundwaters that are used or are suitable for a potable water supply.
    - *Agricultural use*: Groundwaters that are used or are suitable for irrigating crops and livestock water supply.
    - *Surface water quality protection*: This classification is used for groundwaters that feed surface waters. It places restrictions on proposed or existing activities that could impact groundwaters in a way that water quality standards of classified surface water bodies could be exceeded.
    - *Potentially usable*: Groundwaters that are not used for domestic or agricultural purposes, where background levels are not known or do not meet human health and agricultural standards, where TDS levels are less than 10,000 mg/L, and where domestic or agricultural use can be reasonably expected in the future.
    - *Limited use*: Groundwaters where TDS levels are equal to or greater than 10,000 mg/L, where the groundwater has been specifically exempted by regulations of the state, or where the criteria for any of the above classifications are not met.
- Industry
  - The industrial use classification includes industrial cooling and process water supplies. This classification protects industrial equipment from damage from cooling and/or process waters. Specific criteria depend on the industry involved.
- Navigation
  - This use classification is designed to protect ships and their crews and to maintain water quality so as not to restrict or prevent navigation.

- Wetlands
  - Wetlands may be defined as areas that are inundated or saturated by surface or groundwater at a frequency and duration sufficient to support, and under normal circumstances do support, a prevalence of vegetation and organisms typically adapted for life under water-saturated soil conditions. Surface water and groundwater that supply wetlands may be subject to the same standards applied to wetlands.
  - A state may adopt a wetlands classification based on the functions of the wetlands in question. Wetland functions that may warrant site-specific protection include groundwater recharge or discharge, flood flow alteration, sediment stabilization, sediment or other pollutant retention, nutrient removal or transformation, biological diversity or uniqueness, wildlife diversity or abundance, aquatic life diversity or abundance, and recreation.
- Other uses
  - States may adopt other uses they consider to be necessary. Some examples include coral reef preservation, marinas, groundwater recharge, aquifer protection, and hydroelectric power. States may also establish criteria specifically designed to protect these uses.

RULE OF THUMB

Generally, the most stringent numerical standards are for drinking water and aquatic life classifications.

1.1.5 Staying Up-to-Date with Standards and Other Regulations

This is a daunting task and, in the opinion of some, an impossible one. Not only are the federal regulations continually changing, but also individual states may promulgate different rules because of local needs. The usual approach is to obtain the latest regulatory information as the need arises, always recognizing that your current knowledge may be outdated. Part of the problem is that few environmental professionals can find time to regularly read the Federal Register, where the EPA first publishes all proposed and final regulations.

Fortunately, most trade magazines and professional journals highlight important changes in standards and regulations that are of interest to their readers. If you stay abreast of this literature, you will be aware of the regulatory changes and their implications. For the greatest level of security, one often has to contact state and federal information centers to ensure you are working with the regulations that are currently being enforced. Among the most useful sources for staying abreast of the latest information is the EPA website (www.epa.gov). This website has links to information hotlines, laws and regulations, databases and software, available publications, and other information sources. Each state environmental agency also has its own website.

1.2 SOURCES OF WATER IMPURITIES

As discussed in Section 1.1.1, a water impurity is any substance other than water (H_2O) that is found in the water sample, whether harmful or beneficial. Thus, calcium carbonate ($CaCO_3$) is a water impurity even though it generally is not considered hazardous and is normally not regulated. Impurities can be divided into three classes: (1) regulated impurities (pollutants) considered harmful or aesthetically objectionable, (2) unregulated impurities not considered harmful, and (3) unregulated impurities not yet evaluated for their potential health risks.*

In water quality analysis, unregulated as well as regulated impurities may be measured. For example, hardness is a water quality parameter that results mainly from the presence of dissolved calcium and magnesium ions, which are unregulated impurities. However, hardness values may be of interest because high hardness levels can partially mitigate the toxicity of many dissolved metals to aquatic life. Numerical standards for metals in waters classified for aquatic life often contain an adjustment factor for the hardness level of the water. Hence, it is important to measure water hardness in order to evaluate the hazards to aquatic life of dissolved metals. Some reasons not related to health or environmental welfare for measuring hardness are that high hardness levels may cause objectionable deposits to form in plumbing and boilers and low hardness levels may increase the corrosivity of water to metals. Measuring other unregulated impurities can also help to identify the recharge sources of wells and springs, identify the mineral formations through which surface water or groundwater passes, and age-date water samples.

1.2.1 NATURAL SOURCES

Snow and rainwater contain dissolved and particulate minerals collected from atmospheric particulate matter, and small amounts of gases dissolved from atmospheric gases. Snow and rainwater have virtually no bacterial content until they reach the surface of the earth.

After precipitation reaches the surface of the earth and flows over and through the soil, there are innumerable opportunities for introduction of mineral, organic, and biological substances into the water. Water can dissolve at least a little of nearly anything it contacts from the atmosphere and soil. Because of its relatively high density, water can also transport suspended solids. Even under pristine conditions, surface and groundwater will usually contain various dissolved and suspended chemical substances.

1.2.2 HUMAN-CAUSED SOURCES

Many human activities cause additional possibilities for water contamination. Some important sources are

- Construction and mining where freshly exposed soils and minerals can contact flowing water
- Industrial waste discharges and spills

---

* See discussion of the EPA's Contaminant Candidate List in Section 1.1.2, last paragraph.

- Petroleum leaks and spills from storage tanks, pipelines, tankers, and trucks
- Agricultural applications of chemical fertilizers, herbicides, and pesticides
- Urban stormwater runoff, which may contact all the debris of a city, including spilled fuels, animal feces, dissolved metals, organic scraps, road salt, tire and brake particles, construction rubble, and so on
- Effluents from industries and waste treatment plants
- Leachate from landfills, septic tanks, treatment lagoons, and mine tailings
- Fallout from atmospheric pollution

Environmental professionals must remain alert to the possibility that natural impurity sources also may be contributing to pollution problems that at first appear to be solely the result of human-caused sources. Whenever possible, one should obtain background measurements that demonstrate what impurities are present in the absence of known human-caused contaminant sources. For instance, metal mines are developed where there are metal-rich ores. Surface and groundwater in an area impacted by mining often contains relatively high concentrations of dissolved metals. Before any remediation programs are initiated, it is important to determine what the surface and groundwater quality would have been if the mines had not been there. This generally requires finding, if possible, a location upgradient of the area influenced by mining, where surface and groundwater encounters mineral structures similar to those in the mined area.

1.3 MEASURING IMPURITIES

There are four characteristics of water impurities that are important for an initial assessment of water quality:

1. What kinds of impurities are present? Are they regulated compounds?
2. How much of each impurity is present? Are any standards exceeded for the water body being sampled?
3. How do the impurities influence water quality? Are they hazardous? Beneficial? Unaesthetic? Corrosive?
4. What is the fate of the impurities? How will their location, quantity, and chemical form change with time?

1.3.1 WHAT IMPURITIES ARE PRESENT?

The chemical content of a water body is estimated by qualitative chemical analysis of collected environmental samples. Qualitative analysis identifies the chemical species present in the sample but not the quantity (although qualitative and quantitative analyses are often combined in a single measurement). Some of the analytical methods used are gas and ion chromatography, mass spectroscopy, optical emission and absorption spectroscopy, electrochemical probes, so-called wet chemistry tests, and immunoassay testing.

1.3.2 How Do Impurities Influence Water Quality?

The effects of different impurities on water quality are found by research and experience. For example, concentrations of arsenic in drinking water greater than 0.01 mg/L are deemed hazardous to human health. This judgment is based on research and epidemiological studies. Frequently, regulations have to be based on an interpretation of studies that are not rigorously conclusive. Such regulations may be controversial, but until they are revised due to the emergence of new information, they serve as the legal definition of the concentration above which an impurity is deemed to have a harmful effect on water quality.

The EPA has a policy of publishing newly proposed regulatory rules before the rules are finalized, and explaining the rationale used to justify the rules, in order to receive feedback from interested parties. During the period dedicated for public comment, interested parties can support or take issue with the EPA's position. The public input is then added to the database used for establishing a final regulation. Examples of such regulations may be a numerical standard for a chemical not previously regulated, a revised standard for a chemical already regulated, or a new procedure for the analysis of a pollutant. The EPA publishes extensive documentation for all their standards, describing the data on which the numerical values are based.

1.3.3 How Much of Each Impurity Is Present?

The amount of impurity is found by quantitative chemical analysis of the water sample. The amount of impurity can be expressed in terms of total weight (e.g., "There are 15 tons of nitrate in the lake"), or in terms of concentration (e.g., "Nitrate is present at a concentration of 12 milligrams per liter (mg/L)"). Concentration is usually the measure of interest for predicting the effect of an impurity on the environment. It is used for defining environmental standards and is reported in most laboratory analyses. In addition to concentration standards, an additional limit of total weight for certain pollutants may be applied to some rivers in the form of total maximum daily loads (TMDLs). TMDLs are used in allocating the permissible discharges from all recognized sources of the regulated pollutants and setting limits for waste effluent discharges, so that the allowable total loads for a water body are not exceeded.

1.3.4 Working with Concentrations

Unfortunately, there is not one all-purpose method for expressing concentration. The best choice of concentration units depends in part on the medium (liquid, solid, or gas) and in part on the purpose of the measurement. The example problems in this chapter illustrate many applications of concentration calculations.

For regulatory compliance purposes, concentration is usually expressed either as weight of impurity per unit volume of sample or as weight of impurity per unit weight of sample.

- In water samples, impurity concentrations are typically reported as grams (g), milligrams (mg), micrograms (μg), nanograms (ng), picograms (pg), and so on of impurity per liter (L) of water sample. Although a ratio of weight per liter is actually comparing a weight to a volume, it is generally assumed that a liter of water sample weighs exactly 1000 grams, so that an impurity concentration of 1 mg/L is equivalent to 1 gram of impurity in 1 million grams of water, or one part per million (1 ppm).*

  1 g/L = 1 part per thousand (ppth)
  1 mg/L = 10^{-3} g/L = 1 part per million (ppm)
  1 μg/L = 10^{-6} g/L = 1 part per billion (ppb)
  1 ng/L = 10^{-9} g/L = 1 part per trillion (ppt)
  1 pg/L = 10^{-12} g/L = 1 part per quadrillion (ppq)

- In soil samples, impurity concentrations are typically reported as grams, milligrams, micrograms, nanograms, picograms, or femtograms of impurity per kilogram of soil sample.

  1 g/kg = 1 part per thousand (ppth)
  1 mg/kg = 10^{-3} g/kg = 1 part per million (ppm)
  1 μg/kg = 10^{-6} g/kg = 1 part per billion (ppb)[†]
  1 ng/kg = 10^{-9} g/kg = 1 part per trillion (ppt)[‡]
  1 pg/kg = 10^{-12} g/kg = 1 part per quadrillion (ppq)
  1 fg/kg = 10^{-15} g/kg = 1 part per quintillion

In gas samples (normally air samples), concentrations cannot be expressed as simply as in water or soils, because gas volumes and densities are strongly dependent on temperature and pressure, whereas volumes and densities of liquids and

---

* Note that the actual weight of the water sample includes the weight of the water plus the weight of the impurity. Since 1 L of pure water at 4°C and 1 atm pressure weighs 1000 g, there is an inherent assumption when equating 1 mg/L to 1 ppm that the weight of the impurity in the sample is negligible compared to the weight of water and that the density of the sample does not change significantly over the temperature and pressure ranges encountered in environmental sampling.

† The named numbers "billion," "trillion," and higher mean different values in different countries. The United States and England (after 1974) use the so-called short scale in which successive named numbers (ending in –*lion*) after million (10^6) increase by a factor of 10^3, for example, million = 10^6; billion = 10^9; trillion = 10^{12}; quadrillion = 10^{15}; quintillion = 10^{18}; sextillion = 10^{21}; and so on. However, most non-English-speaking countries use the so-called long scale, in which names ending in –*liard* are introduced. In continental Europe and many other countries million = 10^6; milliard = 10^9; billion = 10^{12}; billiard = 10^{15}; trillion = 10^{18}; trilliard = 10^{21}; and so on. The rationale for the two different scales and national usage may be found at http://en.wikipedia.org/wiki/Long_and_short_scales.

‡ To avoid misunderstanding, the use of ppb, ppt, and smaller should be avoided or defined numerically whenever used. Although the "parts-per-" notation is often used in chemistry and environmental literature, it is considered to be a pseudo unit and using it is fundamentally incorrect (see http://en.wikipedia.org/wiki/Parts-per_notation). Although "ppt" usually means "parts per trillion," it occasionally is used to mean "parts per thousand." Unless the meaning of "ppt" is defined explicitly, it has to be guessed from the context. Here, "parts per thousand" is abbreviated as "ppth" to avoid confusion. No "parts-per" abbreviation could be found for parts per quintillion (ppq conflicts with parts per quadrillion).

solids are not. In addition, the amount of some air pollutants (such as carbon monoxide or organic vapors) can be as large or greater than the oxygen and nitrogen levels in severely polluted air; thus, the approximation that the weight of the pollutant is negligible, used for water concentrations and discussed in footnote* on the previous page, does not apply to air concentrations.

For these reasons, parts per million for gases has a different meaning than does parts per million for liquids and solids. For liquids and solids, ppm is a ratio of two weights (sometimes written as "ppm (w/w)" or "ppmw"), whereas for gases, ppm means a ratio of two volumes* (sometimes written as "ppm (v/v)" or "ppmv"). Since air is approximately 1000 times less dense than water,[†] 1 $\mu g/m^3$ is roughly equivalent to parts per trillion for water, whereas for air, 1 $\mu g/m^3$ is roughly equivalent to parts per billion.

There is no consensus regarding the appropriate units by which to express concentrations of substances in air. Air pollution standards are usually promulgated as ppm, whereas air pollutant concentrations in technical reports and other literature may be expressed as

- Percent (parts per hundred) by volume (%v)
- Parts per million by volume (ppmv)
- Weight of pollutant per cubic meter of air (mg/m^3, $\mu g/m^3$, ng/m^3, etc.)
- Molecules of a pollutant gas per cubic centimeter of air (molecules/cc or molecules/cm^3)
- Moles of a pollutant gas per liter of air (mol/L)
- Partial pressure of pollutant gas
- Mole fraction of pollutant gas

With so many definitions of gas concentrations in common use, it clearly is useful to be able to convert gas concentrations from one set of units to another. The Rules of Thumb box on the next page illustrates how to convert between ppmv and mg/m^3, the two most commonly used concentration units for air pollutants in the environmental literature. Note that air pollutant standards generally assume reference conditions of 20°C and 1 atm of pressure. Then 1 ppmv of an air pollutant means one volume unit of pollutant gas per million volume units of air, assuming $T = 20$°C and $P = 1$ atm. For example,

$$1 \text{ ppmv of } H_2S = 1 \text{ L}_{H_2S}/10^6 \text{ L}_{air} = 10^{-6} \text{ L}_{H_2S}/\text{L}_{air} = 1 \text{ } \mu\text{L}_{H_2S}/\text{L}_{air}$$

---

* The ratio v/v means the ratio of the volume that a gaseous pollutant would have if it were isolated from the air, to the volume of air without the pollutant, both volumes having the same temperature and pressure. According to the ideal gas law, a v/v ratio is equivalent to a ratio of the number of pollutant gas molecules (n_{pol}) to the number of normal air molecules (n_{air}), or n_{pol}/n_{air} in the sample volume.

† Although one cubic meter of dry air at 1 atm pressure is 1.220 kg, air mass depends strongly on the actual pressure and humidity. However, 1 L of water is almost exactly 1 kg under all normal environmental conditions.

RULES OF THUMB

1. To convert a gaseous pollutant concentration (C_{pol}) from ppmv to mg/m³, use

$$C_{pol}(\text{mg/m}^3) = \left(\frac{C_{pol}(\text{ppmv}) \times MW_{pol} \times P_{atm} \times 12.188}{273.15 + T_{°C}} \right) \qquad (1.1)$$

2. To convert a gaseous pollutant concentration (C_{pol}) from mg/m³ to ppmv, use

$$C_{pol}(\text{ppmv}) = \left(\frac{C_{pol}(\text{mg/m}^3)(273.15 + T_{°C})}{MW_{pol} P_{atm} 12.188} \right) \qquad (1.2)$$

where

C_{pol} = pollutant concentration in the desired units
MW_{pol} = molecular weight of the pollutant in g/mol
P_{atm} = pressure of air (atm)
$T_{°C}$ = actual air temperature (°C)
12.188 (K mol)/(L atm) = R⁻¹, reciprocal of the ideal gas constant

Notes:

1. P(atm) = P(mm Hg)/760 = P(torr)/760 = P(Pa)/101,325 = P(bar)/1.01325 = P(dyne/cm²)/1.01325 × 10⁶ = P(psi)/14.696
2. See Example 2 for a sample calculation.
3. The pollution laws and regulations in the United States typically reference their air pollutant limits to an ambient temperature of 20–25°C and an ambient air pressure of 1 atm. In other nations, the reference ambient air conditions may be different, for example, $T_{°C}$ (reference) might be defined as 0°C.
4. One percent by volume = 10,000 ppmv.

Example 1: Calculating a Concentration in Water

A 45.6 mL water sample was found to contain 0.173 mg of sodium. What is the concentration in mg/L of sodium in the sample?

ANSWER

$$\frac{0.173 \text{ mg}}{45.6 \text{ mL}} \times \frac{1000 \text{ mL}}{L} = 3.79 \text{ mg/L or } 3.79 \text{ ppm}$$

Note that 3.79 ppm has all of the following meanings: 3.79 g of sodium in 10^6 g of solution, or 3.79×10^{-3} g (3.79 mg) of sodium in 10^3 g (1 L) of solution, or 3.79×10^{-6} g (3.79 µg) of sodium per gram of solution.

Example 2: Calculating a Concentration in Air

The equations used in these calculations are based on the ideal gas laws that are discussed in introductory general chemistry textbooks. The equations are used here without derivation.

The atmospheric ozone (O_3) level in Denver, Colorado, was reported to be 2.50 ppmv $\left(2.50\,\mu L_{O_3}/L_{air}\right)$. At the time of the ozone measurement, atmospheric temperature was 37°C and atmospheric pressure was 722 mm Hg. Express the ozone concentration in mg/m^3.

ANSWER

Step 1: Write the molecular formula for ozone and determine its molecular weight.

The molecular formula for ozone is O_3, which means that one molecule of ozone contains three atoms of oxygen. Obtain the molecular weight (MW) by adding the atomic weights (AWs)* of all the atoms in the molecule. Use the table of the elements inside the front cover of this book to find the AWs.

AW(O) = 15.9994; using three significant figures rounds it to AW(O) = 16.0

Therefore, the MW of O_3 is MW = (3×16.0) = 48.0 g/mol,† to three significant figures.

---

* AWs are dimensionless. They are relative values compared to the weight of a standard atom (C-12) and are determined by assigning the most abundant isotope of carbon (C-12) an exact AW of 12 and then experimentally measuring the relative weights of isotopes of all other elements as compared to C-12. The AW values in the table of the elements are weighted averages of the AWs of all the isotopes of each element. Hydrogen's AW of 1.008 means that the weight of an average H atom is 1.008/12 times the weight of a C-12 atom. Oxygen's AW of 15.9994 means that the weight of an average O atom is 15.9994/12 times the weight of a C-12 atom. Operationally, the AW of an element is the weight in grams of 6.022×10^{23} atoms (1 mol) of that element (see footnote † below).

† Because AWs are dimensionless, it is necessary to create a multiplier quantity to relate them to the measured weights of chemical samples. This quantity is based on the mole (see footnote * on the next page) and is called a *dalton*, abbreviated *Da*: 1 Da = 1 g/mol. By footnote * on this page, the AW of an oxygen atom is 15.9994. By footnote * on the next page, a quantity of 1 mol of oxygen atoms (6.022×10^{23} atoms) weighs 15.9994 grams, found by multiplying the AW of oxygen by the factor of 1 dalton (1g/mol) to give

Weight of 1 mol of O = AW(O) × 1 Da = 15.9994 × 1 g/mol = 15.9994 g/mol

In summary, 1 Da = 1 atomic mass unit (1/12 × mass of a C-12 atom or $1.660538921 \times 10^{-12}$ kg). 1 dalton is approximately equal to the molar mass of a proton or a neutron. The sum of AWs in any molecule multiplied by 1 dalton is its molar mass. For example, the sum of AWs in chloroform ($CHCl_3$) is 119. The weight of 1 mol of chloroform molecules is 119 Da or 119 g/mol.

Step 2: Use Equation 1.1:

$$C_{O_3}(\text{mg/m}^3) = \left(\frac{C_{O_3}(\text{ppmv}) \times MW_{O_3} \times P_{atm} \times 12.188}{273.15 + T_{°C}} \right)$$

$$MW_{O_3} = 48 \text{ g/mol}$$

$$P_{atm} = 722 \text{ mm Hg} \times \frac{1 \text{ atm}}{760 \text{ mm Hg}} = 0.950 \text{ atm}$$

$$T_{°C} = 37°C$$

$$C_{O_3}(\text{mg/m}^3) = \left(\frac{2.50 \text{ ppmv} \times 48 \text{ g/mol} \times 0.950 \text{ atm} \times 12.188}{273.15 + 37°C} \right)$$

$C_{O_3}(\text{mg/m}^3) = 4.4799 \text{ mg/m}^3$; or $C_{O_3} = 4.48 \text{ mg/m}^3$, to 3 significant figures

1.3.4 MOLES AND MOLAR CONCENTRATIONS

For chemical calculations (as opposed to regulatory compliance calculations), concentrations in any phase are usually expressed either as moles* of impurity per liter of sample (abbreviated as mol/L) or moles of impurity per kilogram of sample (mol/kg). Moles per liter or kilogram are related to the *number* of impurity molecules in a liter or kilogram of sample rather than the *weight* of impurity. Because chemical reactions involve one-on-one molecular interactions, regardless of the weight of the reacting molecules, moles are best for chemical calculations, such as balancing chemical reactions, predicting if a precipitate will form or a solid dissolve, and calculating reaction rates. A common chemical notation for expressing a concentration in mol/L is to enclose the constituent in square brackets. Thus, writing $[Na^+] = 16.4$, is the same as writing $Na^+ = 16.4$ mol/L.

---

* A *mole* is a unit of quantity. Operationally, 1 mol of any pure element or compound is that quantity of the substance that has a weight equal to the AW or MW, in grams, of that substance. Thus, 1 mol of pure sodium (Na) metal is the amount that weighs 23.00 g; 1 mol of sodium chloride (NaCl) is the amount that weighs 58.45 g. This arises from the definition of a mole (abbreviated mol in chemical notation, as in mol/L); the term "mole" indicates a particular number of things, just as a dozen indicates 12 things and a pair indicates 2 things. The number of things indicated by a mole is defined to be the number of carbon atoms found in exactly 12 g of the ^{12}C isotope. The number of atoms present in 12 g of ^{12}C has been determined experimentally to be 6.022×10^{23} atoms (given here to four significant figures). This large number is called Avogadro's number, after the first scientist to deduce its value. *The molecular (or atomic) weight of any molecule (or atom), expressed in grams, contains 1 mol, or 6.022×10^{23} molecules or atoms.* Thus, 1 mol of pure calcium metal (weighing 40.08 grams) contains 6.022×10^{23} calcium atoms. As an example of a molecule, 1 mol of pure calcium chloride ($CaCl_2$; weighing $40.08 + 2 \times 35.45 = 110.98$ g) contains 6.022×10^{23} calcium chloride molecules. When 1 mol of calcium chloride dissolves, it dissociates into 1 mol of Ca^{2+} ions and 2 mol of Cl^- ions.

A mole is the amount of a compound that has a weight in grams equal to its MW. MW is the sum of the AWs of all the atoms in the molecule. (See the table of the elements inside front cover for AWs.)

For example, the AW of oxygen is 16. An oxygen molecule (O_2) contains two oxygen atoms and, thus, has an MW of $2 \times 16 = 32$ g/mol. A mole of O_2 is the amount, or number of molecules, that weighs 32 g. Another example: the AWs of carbon and calcium are 12 and 40, respectively. The MW of $CaCO_3$ is $40 + 12 + 3 \times 16 = 100$ g. One mole of $CaCO_3$ is the quantity that weighs 100 g.

Example 3: Converting mg/l to mol/l

The concentration of benzene in a water sample was reported as 0.017 mg/L. Express this concentration as moles per liter (mol/L). One mole of a chemical compound is the amount of that compound that has a weight in grams equal to its MW. The MW of a compound is the sum of the AWs of all the atoms that make up the compound.

ANSWER

Step 1: Write the molecular formula for benzene and determine its MW.

The molecular formula for benzene is C_6H_6, which means that one molecule of benzene contains six atoms of carbon and six atoms of hydrogen.

Obtain the MW by adding the AWs of all the atoms in the molecule. Use the table of the elements inside the front cover of this book to find the AWs.

AW of C = 12.01; 6 atoms of C have a total AW of $6 \times 12.01 = 72.06$

AW of H = 1.008; 6 atoms of H have a total AW of $6 \times 1.008 = 6.048$

Therefore, its MW is $(12.06 + 6.048) = 78.054$ g/mol, or 78.1 g/mol to three significant figures.

Step 2: Convert the concentration of benzene from 0.017 mg/L of benzene to mol/L.

$$\frac{0.017\text{mg/L}}{1000\ \text{mg/g}} = 1.7 \times 10^{-5} \text{ g/L}$$

$$\frac{1.7 \times 10^{-5} \text{ g/L}}{78 \text{ g/mol}} = 2.18 \times 10^{-7} \text{ mol/L} = C_{\text{benzene}}$$

Example 4: Using Moles, ppm, and mg/L Together

The federal primary drinking water standard for nitrate is 10 mg of nitrate–nitrogen per liter of water (written as 10 mg NO_3–N/L). It is defined in terms of the weight of nitrogen that is contained in a sample containing nitrate anions (NO_3^-); the standard does not include the weight of the oxygen atoms that are also in the nitrate anion. It is common practice in environmental literature to neglect the charge sign of ions when the charge, which has insignificant weight, is not relevant to the topic.

If a laboratory analysis includes the weight of the oxygen atoms and reports the nitrate concentration in a water sample as 33 ppm NO_3/L (not NO_3–N/L, as the drinking water standard is defined), does the analysis indicate that the water source is in compliance with the federal drinking water standard?

ANSWER

$$33 \text{ ppm} = 33 \text{ mg/L} = 33 \times 10^{-3} \text{ g/L}$$

$$\text{Moles of } NO_3 \text{ in 1 L of sample} = \frac{\text{weight of } NO_3 \text{ in 1 L of sample}}{\text{molecular weight of } NO_3}$$

$$= \frac{33 \times 10^{-3} \text{ g/L}}{62.0 \text{ g/mol}}$$

$$= 0.53 \times 10^{-3} \text{ mol/L or } 0.53 \text{ mmol/L}$$

The drinking water standard is not based on the concentration of NO_3 but on the concentration of nitrogen that is in the form of NO_3. Since each anion of NO_3 contains one atom of nitrogen and three atoms of oxygen, each mole of NO_3 will contain 1 mol of N and 3 mol of O.

Therefore, 0.53 mmol/L of NO_3 contains 0.53 mmol/L of N.

$$0.53 \times 10^{-3} \text{ mol N/L} \times 14 \text{ g N/mol} = 7.4 \times 10^{-3} \text{ g } NO_3 - N/L$$

This sample does not exceed the federal standard of 10 mg NO_3–N/L, and the source is in compliance.

A simpler approach to this calculation, but one that obscures the concepts of MWs and moles, is as follows.

The MW of NO_3 is 62 g/mol and the AW of nitrogen is 14. Then, the weight fraction of N in NO_3 is 14/62 = 0.226, or 22.6%. Therefore, every sample of NO_3 contains 22.6% N by weight, and a sample that contains 33 ppm of NO_3 will contain

$$(33 \, ppm \, NO_3)(0.226 \, N/NO_3) = 7.4 \, ppm \, NO_3 - N = 7.4 \times 10^{-3} g \, NO_3 - N/L.$$

Example 5: Using Moles, ppm, and mg/l Together (*Continued*)

$Ca(OCl)_2$, a disinfectant sometimes used for treating hot tub water, dissociates in water by the reaction

$$Ca(OCl)_2 + 2H_2O \rightarrow 2HOCl + Ca^{2+} + 2OH^- \tag{1.3}$$

The HOCl product partially dissociates further by

$$HOCl + H_2O \rightleftharpoons OCl^- + H_3O^+ \tag{1.4}$$

a. If 2.00 g of the disinfectant calcium hypochlorite, $Ca(OCl)_2$, is added to a hot tub containing 1050 L of water, what would be the concentration of $Ca(OCl)_2$ in the hot tub water if $Ca(OCl)_2$ did not dissociate at all? In other words, what concentration of $Ca(OCl)_2$ is initially added to the hot tub water?

ANSWER

$$\text{Concentration of } Ca(OCl)_2 = \frac{2.00 \text{ g}}{1050 \text{ L}} = 0.0019 \text{ g/L} = 1.9 \text{ mg/L} = 1.9 \text{ ppm}$$

b. The active disinfecting species are $HOCl$ and OCl^-, which together are called the available free chlorine. What is the concentration of available free chlorine in ppm after the reaction of Equation 1.4 is complete? Assume that the 2.00 g of $Ca(OCl)_2$ of part a. dissociates in water completely, and that the available free chlorine is 50% $HOCl$ and 50% OCl^- (in terms of number of molecules, or moles, not by weight), as is the case at pH = 7.5.

ANSWER

(Use the periodic table or the table of the elements inside the front cover for AWs.):

1. *Determine the number of moles in 2.00 g of Ca(OCl)₂.*

$$\text{MW of } Ca(OCl)_2 = 40.1 + (2 \times 16.0) + (2 \times 35.4) = 142.9 \text{g / mol}$$

$$\text{Moles of } Ca(OCl)_2 \text{ in 2.00 g} = \frac{2.00 \text{ g}}{142.9 \text{ g/mol}} = 0.014 \text{ mol}$$

2. *Determine the moles of HOCl formed when 2.00 g of Ca(OCl)₂ dissociates by Equation 1.3, that is, if no further dissociation occurs.*
   Equation 1.3 indicates that 2 mol of $HOCl$ are formed from 1 mol of $Ca(OCl)_2$. Therefore,

$$2 \times 0.014 \text{ mol} = 0.028 \text{ moles of HOCl are formed from}$$
$$0.014 \text{ mol (2.00 g) of } Ca(OCl)_2.$$

3. *Determine the moles of HOCl and OCl⁻ in the water after one-half of the HOCl dissociates further by Equation 1.4.*
   This is the equilibrium state at pH = 7.5. After half of the $HOCl$ dissociates to OCl^-, the solution contains 0.028/2 = 0.014 mol of $HOCl$ and 0.014 mol of OCl^-.
4. *Determine the weight of HOCl and OCl⁻ in the water at equilibrium.*
   MW of $HOCl$ = 1.0 + 16.0 + 35.4 = 52.4 g/mol
   MW of OCl^- = 16.0 + 35.4 = 51.4 g/mol
   Weight of $HOCl$ in the water = 0.014 mol × 52.4 g/mol = 0.73 g

Weight of OCl$^-$ in the water $= 0.014$ mol $\times 51.4$ g/mol $= 0.72$ g
Total weight of HOCl $+$ OCl$^-$ = weight of free chlorine $= 1.45$ g
5. *Calculate the ppm of HOCl $+$ OCl$^-$ in 1050 L of water at equilibrium.*

$$\text{Concentration of free chlorine in hot tub} = \frac{1.45 \text{ g}}{1050 \text{ L}} = 0.0014 \text{ g/L}$$
$$= 1.4 \text{ ppm}$$

Example 6: Using Concentration Calculations to Predict a Precipitate

In this example, because the results of a chemical reaction must be determined, it is necessary to use concentration units of moles per liter (molarity).

A water sample is collected from a stream that passes through soils containing the metal salt calcium sulfate ($CaSO_4$), some of which dissolves. The stream already carries some Ca^{2+} and SO_4^{2-} dissolved from other mineral sources. Laboratory analysis of the water shows $SO_4^{2-} = 576$ mg/L; $Ca^{2+} = 244$ mg/L.

Will a precipitate of $CaSO_4$ develop in the stream?

ANSWER

The precipitation reaction is

$$Ca^{2+}(aq) + SO_4^{2-}(aq) \rightarrow CaSO_4(s),$$

where (aq) indicates an aqueous, or dissolved, species and (s) indicates an undissolved solid or precipitated species. It has been experimentally determined (see Appendix B for a discussion of the *solubility product*) that if the calculated value of the molar concentration of Ca^{2+} multiplied by the molar concentration of SO_4^{2-} exceeds the value 2.4×10^{-5} (written: $[Ca^{2+}] \times [SO_4^{2-}] > 2.4 \times 10^{-5}$), then a precipitate of solid $CaSO_4$ will form. The product of the in-stream dissolved calcium and sulfate ion concentrations (in mol/L) is called the *reaction quotient,* and their limiting value (2.4×10^{-5}) before precipitation will occur is called the *solubility product*. The solubility of metal salts is discussed further in Chapter 5.

If the product of the dissolved calcium and sulfate concentrations (in mol/L) exceeds 2.4×10^{-5}, then $CaSO_4$ will precipitate until the final dissolved concentrations of Ca^{2+} and SO_4^{2-} are such that their product equals 2.4×10^{-5} and the system has come to equilibrium.

PROCEDURE

1. Convert mg/L to mol/L:

$$Ca^{2+} : \frac{244 \times 10^{-3} \text{ g/L}}{40 \text{ g/mol}} = 6.1 \times 10^{-3} \text{ mol/L}$$

$$SO_4^{2-} : \frac{576 \times 10^{-3} \text{ g/L}}{96 \text{ g/mol}} = 6.0 \times 10^{-3} \text{ mol/L}$$

2. Calculate the product of the dissolved calcium and sulfate ion concentrations in moles per liter (the reaction quotient).

$$[Ca^{2+}] \times [SO_4^{2-}] = (6.1 \times 10^{-3} \text{ mol/L}) \times (6.0 \times 10^{-3} \text{ mol/L})$$
$$= 3.7 \times 10^{-5} \text{ mol}^2/L^2$$

The reaction quotient is larger than the solubility product of 2.4×10^{-5}.
Therefore, $CaSO_4$ may precipitate downstream when equilibrium is reached if the
stream is not diluted by additional water carrying less calcium or sulfate.

1.3.5 CASE STUDY

A shallow aquifer below an industrial park was contaminated with toxic halogenated hydrocarbons (in this case, hydrocarbons containing chlorine and bromine).
Although other pollutants were present, only the halogenated hydrocarbons were
found to threaten the municipal drinking water supply. The state environmental
authorities mandated a remediation program to be paid for by the responsible parties,
which were several industrial facilities in the park. In order to allocate an appropriate
share of the cleanup expenses to each responsible party, it was necessary to estimate
what percentage of the total pollution was caused by each party.

An automobile rental agency was cited as one of the responsible parties even
though it did not use halogenated chemicals in their business, because it had had a
leaking underground gasoline storage tank that released approximately 2500 gal of
leaded gasoline to the subsurface above the aquifer. The gasoline contained additives
with chlorine and bromine compounds.

During the time period between the late 1920s and the early 1990s, lead compounds, particularly tetraethyl lead (also called TEL), were added to automotive and
aviation gasoline as octane enhancers. During that time, it was common practice
to also add halogenated organic compounds, particularly 1,2-dichloroethane (also
called ethylene dichloride* or EDC) and 1,2-dibromoethane (also called ethylene
dibromide or EDB), to leaded gasoline to serve as lead scavengers, helping to prevent
lead deposits from accumulating in gasoline engines.

a. Use the data below to calculate the weight in grams of EDC and EDB that was
   potentially added to the aquifer from the spill of 2500 gal of leaded gasoline.
b. Measurements of the contaminant plume indicated that the gasoline spill
   impacted about 9×10^6 ft^3 of aquifer volume. Assume that 25% of this volume was occupied by water in the soil pore space and calculate the potential average concentrations of EDC and EDB in the aquifer water (ignoring
   biodegradation, evaporation, and other loss mechanisms).

---

* The use of these common trade names may be confusing because the name ethylene normally means
that there is a double bond between two carbons, whereas the compounds 1,2-dichloroethane and
1,2-dibromoethane contain only single bonds. The use of ethylene dichloride and ethylene dibromide
as trade names arose because 1,2-dichloroethane and 1,2-dibromoethane were often manufactured
from ethylene with chlorine or bromine.

Data:

Weight percentages in the gasoline additive package were 62% TEL, 18% EDC, 18% EDB, and 2% other nonrelevant compounds. The amount of additive used was sufficient to yield 2.0 g of lead per gallon of gasoline. Assume both EDC and EDB are fully dissolved in water.

Chemical formulas: TEL = $C_8H_{20}Pb$, EDC = $C_2H_4Cl_2$, and EDB = $C_2H_4Br_2$. Use the periodic table on the inside front cover to calculate MWs.

Conversion factors: 1 gal = 3.785412 L; 1 ft^3 = 28.31685 L

Calculation:

a. The formula for tetraethyl lead (TEL) is $Pb(C_2H_5)_4$ or PbC_8H_{20}:

MW TEL = 207.2 + (8 × 12.0) + (20 × 1.0) = 323.2 g/mol

wt% lead in TEL = %Pb = $\dfrac{207.2}{323.2} \times 100\% = 64\%$, to two significant figures

0.64 × (g of TEL per gallon of gasoline) = 2.0 g Pb/gal gasoline

$$g \text{ of TEL/gal} = \frac{2.0}{0.64} = 3.1 \text{ g TEL/gal}$$

TEL is 62% of the additive package.

$$\text{Therefore, total grams of additive} = \frac{3.1 \text{ g TEL}}{0.62} = 5.0 \text{ g additive/gal.}$$

EDC and EDB each equal 18% of the additive package.
EDC and EDB each = 0.18 (5.0 g additive/gal) = 0.9 g/gal.
Therefore, the 2500 gal spill contained about
    0.9 g/gal × 2500 gal = 2250 g each of EDC and EDB.

b. Volume of aquifer water receiving 2250 g each of EDC and EDB:

$$(9 \times 10^6 \text{ ft}^3 \text{ of aquifer}) \times (0.25) = 2.2 \times 10^6 \text{ ft}^3 \text{ of water} = 6.4 \times 10^7 \text{ L}$$

The potential concentration of each pollutant in mg/L is

$$\frac{2250 \text{ g}}{6.4 \times 10^7 \text{ L}} \times 1 \times 10^3 \text{ mg/g} = 0.035 \text{ mg/L, or 35 ppb}$$

For comparison, the drinking water MCLs (EPA maximum contaminant levels) are EDC = 0.005 ppm or 5 ppb and EDB = 0.00005 ppm or 0.05 ppb. Therefore, the gasoline releases had the potential to pollute the groundwater aquifer with EDC and EDB in excess of the drinking water standards.

1.3.6 EQUIVALENTS AND EQUIVALENT WEIGHTS

An equivalent is a unit of quantity related to the mole, which is especially useful for chemical calculations involving ions. For any dissolved ion, one equivalent of the ion is the amount that carries 1 mol of unit charge, independent of the sign of the charge. Thus, one equivalent of sodium ions (Na^+) is equal to 1 mol of sodium ions, because each sodium ion carries one unit of charge. But one equivalent of calcium ions (Ca^{2+}) is equal to ½ mol of calcium ions, because each calcium ion carries two units of charge and only ½ mol of calcium is needed to carry 1 mol of charge.

Calculations with equivalents are useful for ionic reactions because ionic reactions must always balance electrically, that is, with respect to ionic charge. One equivalent of sodium ions is equivalent to one equivalent of calcium ions in the sense that equal equivalents of any ions always carry the same total magnitude of electrical charge. Although it is never necessary to use equivalents, they are often easier to use than mg/L or moles for comparing the balance of positive and negative ions in a water sample or making cation exchange calculations. Since environmental waters normally contain many ionic species, equivalent weights are often useful in water quality calculations. Examples 7–9 illustrate how equivalent weights are calculated. See Table 1.1 for a convenient list of equivalent weights for some common water species.

Example 7

The equivalent weight (eq wt) of an ion is its weight in grams that carries 1 mol of charge. This will be its MW (for molecular ions such as PO_4^{3-}) or AW (for single atom ions such as Na^+ or Cl^-) divided by its magnitude of charge (without regard for the sign of the charge).

a. Atomic ion with one unit of charge:

$$1 \text{ eq wt of } Na^+ = \frac{\text{atomic wt. of } Na^+ \text{ (g/mol)}}{\text{magnitude of charge on } Na^+ \text{ (eq/mol)}}$$

$$= \frac{23.0 \text{ g/mol}}{1 \text{ eq/mol}} = 23.0 \text{ g/eq}$$

b. Atomic ion with two units of charge:

$$1 \text{ eq wt of } Ca^{2+} = \frac{\text{atomic wt. of } Ca^{2+} \text{ (g/mol)}}{\text{magnitude of charge on } Ca^{2+} \text{ (eq/mol)}}$$

$$= \frac{40.08 \text{ g/mol}}{2 \text{ eq/mol}} = 20.04 \text{ g/eq}$$

c. Molecular ion with three units of charge:

$$1 \text{ eq wt of } PO_4^{3-} = \frac{\text{molec. wt. of } PO_4^{3-} \text{ (g/mol)}}{\text{magnitude of charge on } PO_4^{3-} \text{ (eq/mol)}}$$

$$= \frac{95.0 \text{ g/mol}}{3 \text{ eq/mol}} = 31.7 \text{ g/eq}$$

TABLE 1.1

Atomic/Molecular Weights and Equivalent Weights of Some Common Water Species

Species	Molecular Weight	Absolute Charge	Equivalent Weight
Na^+	23.0	1	23.0
K^+	39.1	1	39.1
Li^+	6.9	1	6.9
Ca^{2+}	40.1	2	20.04
Mg^{2+}	24.3	2	12.2
Sr^{2+}	87.6	2	43.8
Ba^{2+}	137.3	2	68.7
Fe^{2+}	55.8	2	27.9
Mn^{2+}	54.9	2	27.5
Zn^{2+}	65.4	2	32.7
Al^{3+}	27.0	3	9.0
Cr^{3+}	52.0	3	17.3
NH_4^+	18.0	1	18.0
Cl^-	35.4	1	35.4
F^-	19.0	1	19.0
Br^-	79.9	1	79.9
NO^{3-}	62.0	1	62.0
NO^{2-}	46.0	1	46.0
HCO^{3-}	61.0	1	61.0
CO_3^{2-}	60.0	2	30.0
CrO_4^{2-}	116.0	2	58.0
SO_4^{2-}	96.1	2	48.03
S^{2-}	32.1	2	16.0
PO_4^{3-}	95.0	3	31.7
$CaCO_3$	100.1	2	50.04
$CaSO_4$	136.2	2	68.1

Example 8

The equivalent weight of a neutral (uncharged) compound that dissociates into cations and anions of equal absolute charge (like $CaSO_4$ or NaCl) is found by dividing its MW by what the ionic charge would be on either the cation or the anion if the molecules were dissolved (also called the oxidation number or valence).

a. NaCl → Na^+ + Cl^-: Each ion carries one unit of charge.

$$1 \text{ eq wt of NaCl} = \frac{\text{molec. wt. of NaCl (g/mol)}}{\text{charge number of either } Na^+ \text{ or } Cl^- \text{ (eq/mol)}}$$

$$= \frac{58.4 \text{ g/mol}}{1 \text{ eq/mol}} = 58.4 \text{ g/eq}$$

b. $CaSO_4 \rightarrow Ca^{2+} + SO_4^{2-}$: Each ion carries two units of charge.

$$1 \text{ eq wt of } CaSO_4 = \frac{\text{molec. wt. of } CaSO_4 \text{ (g/mol)}}{\text{charge number of either } Ca^{2+} \text{ or } SO_4^{2-} \text{ (eq/mol)}}$$

$$= \frac{136.2 \text{ g/mol}}{2 \text{ eq/mol}} = 68.1 \text{ g/eq}$$

Example 9

The equivalent weight of a neutral (uncharged) compound that dissociates into cations and anions of different absolute charge (such as Na_2SO_4, or $Ca_3(PO_4)_2$) must take into account the different quantities of ions that are formed.

a. $Na_2SO_4 \rightarrow 2 Na^+ + SO_4^{2-}$: The dissociation of 1 mol of Na_2SO_4 produces 2 mol of positive charge (2 mol of singly charged cations) and 2 mol of negative charge (1 mol of doubly charged anions).* Therefore, the equivalent weight of Na_2SO_4 is half of its MW because 1 mol of Na_2SO_4 dissociates into two equivalent weights of each ion. Either calculation below is correct.

$$1 \text{ eq wt of } Na_2SO_4 = \frac{\text{molec. wt. of } Na_2SO_4 \text{ (g/mol)}}{\text{total charge on } 2 Na^+ \text{ ions (eq/mol)}}$$

$$= \frac{142.04. \text{ g/mol}}{2 \text{ eq/mol}} = 71.02 \text{ g/eq,}$$

or

$$1 \text{ eq wt of } Na_2SO_4 = \frac{\text{molec. wt. of } Na_2SO_4 \text{ (g/mol)}}{\text{total charge on } 1 SO_4^{2-} \text{ ion (eq/mol)}}$$

$$= \frac{142.04 \text{ g/mol}}{2 \text{ eq/mol}} = 71.02 \text{ g/eq,}$$

b. $Ca_3(PO_4)_2 \rightarrow 3 Ca^{2+} + 2 PO_4^{3-}$: The dissociation of 1 mol of $Ca_3(PO_4)_2$ produces 6 mol of positive charge (3 mol of doubly charged cations) and 6 mol of negative charge (2 mol of triply charged anions). Therefore, the equivalent weight of $Ca_3(PO_4)_2$ is one-sixth of its MW, because 1 mol of $Ca_3(PO_4)_2$ dissociates into six equivalent weights of each ion. Either calculation below is correct.

$$1 \text{ eq wt of } Ca_3(PO_4)_2 = \frac{\text{molec. wt. of } Ca_3(PO_4)_2 \text{(g/mol)}}{\text{total charge on } 3 Ca^{2+} \text{ ions (eq/mol)}}$$

$$= \frac{310.23 \text{ g/mol}}{6 \text{ eq/mol}} = 51.71 \text{ g/eq}$$

---

* Because every uncharged compound is electrically neutral, if it dissociates into charged ions, the principle of charge conservation (electrical charge can neither be created nor destroyed) requires that it must necessarily produce equal quantities of positive and negative charge, so that the net change in charge is zero.

or

$$1 \text{ eq wt of } Ca_3(PO_4)_2 = \frac{\text{molec. wt. of } Ca_3(PO_4)_2 (g/mol)}{\text{total charge on 2 } PO_4^{3-} \text{ ions (eq/mol)}}$$

$$= \frac{310.23 \text{ g/mol}}{6 \text{ eq/mol}} = 51.71 \text{ g/eq}$$

RULES OF THUMB

1. How to find the equivalent weight of an ion (see Example 1):
    a. One equivalent weight of a singly charged ion is equal to its MW or AW, because 1 mol of the ions carries 1 mol of charge.
    b. One equivalent weight of a doubly charged ion is equal to one-half of its molecular or AW, because 1 mol of the ions carries 2 mol of charge.
    c. One equivalent weight of a triply charged ion is equal to one-third of its molecular or AW, because 1 mol of the ions carries 3 mol of charge.
    d. In general, one equivalent weight of any ion is equal to its MW or AW divided by its ionic charge (regardless of sign).
    e. Correspondingly, the number of equivalents in 1 mol of any ion is equal to the magnitude of its charge, for example, the number of equivalents in 1 mol of Ca^{2+} is two.
2. How to find the equivalent weight of a neutral compound. One equivalent weight of a neutral compound depends on its dissociation reaction (see Examples 2 and 3).
    a. If it dissociates into an equal number of cations and anions (the cations and anions will have equal absolute charges), divide the MW of the neutral compound by the magnitude of charge on either the cation or the anion.
    b. If it dissociates into cations and anions of different absolute charge (such as Na_2SO_4, or $Ca_3(PO_4)_2$), you must take into account the different quantities of ions that are formed.
3. To convert from mg/L to eq/L or meq/L, divide mg/L by the MW and multiply by the ionic charge.

$$150 \text{ mg/L of } Ca^{2+} = \frac{150 \text{mg/L}}{40.1 \text{ g/mol}} \times \frac{1 \text{ g}}{1000 \text{ mg}} \times 2 \text{ eq/mol}$$

$$= 0.00748 \text{ eq/L}$$

$$= 7.48 \text{ meq/L}$$

4. To convert from meq/L to mg/L, multiply meq/L by the MW and divide by the ionic charge.

$$15.2 \text{ meq/L of PO}_4^{3-} = \frac{15.2 \text{meq/L} \times 95.0 \text{ g/mol} \times 1000 \text{ mg/g}}{3 \text{ eq/mol} \times 1000 \text{ meq/eq}}$$

$$= 481 \text{ mg/L}$$

Example 10: Using mg/L, moles/L, and Equivalents Together

Using the dissolution reaction $CaCl_2 \rightarrow Ca^{2+} + 2Cl^-$

a. Calculate how many mol/L of $CaCl_2$ are needed to produce a solution with 500 mg/L of Ca^{2+}, with 500 mg/L of Cl^-.
b. Show that dissolving in water one equivalent weight of calcium chloride ($CaCl_2$), results in an electrically neutral solution containing 1 mol of positive ions and 1 mol of negative ions.

ANSWER

a. Calculate how many mol/L of Ca are in 500 mg/L of Ca^{2+}.

b. $\dfrac{500 \times 10^{-3} \text{ g/L}}{40.1 \text{ g/mol}} = 0.0125 \text{ mol/L}$

The dissolution reaction shows that the moles of Ca^{2+} produced are equal to the moles of $CaCl_2$ dissolved. Therefore, 0.0125 mol/L of $CaCl_2$ are needed. Calculate how many moles/L of Cl are in 500 mg/L of Cl.

$$\frac{500 \times 10^{-3} \text{ g/L}}{35.5 \text{ g/mol}} = 0.0141 \text{ mol/L}$$

The dissolution reaction shows that the moles of Cl^- produced are two times the moles of $CaCl_2$ dissolved. Therefore, $0.0141/2 = 0.00704$ mol/L of $CaCl_2$ are needed.

c. The exact amount of $CaCl_2$ dissolved is irrelevant. When any quantity of moles or equivalents are dissolved in any amount of water, the concentration of Cl^- produced is always twice the concentration of Ca^{2+} produced and the total negative charge is always equivalent to the total positive charge.

- When $CaCl_2$ dissolves, twice as much Cl^- is formed as Ca^{2+}.
- The total negative charge in the solution is $1 \times [Cl^-]$
- The total positive charge in the solution is $2 \times [Ca^{2+}]$
- The total negative charge must equal the total positive charge: $[Cl^-] = 2 \times [Ca^{2+}]$.

Then, negative charge $= 1 \times [Cl^-] = 1 \times (2 \times [Ca^{2+}]) =$ positive charge.
This result is general. Dissolving electrically neutral compounds in water always results in an electrically neutral solution.

EXERCISES

1. To determine the amount of dissolved salts contained in it, 475 mL of a water sample was evaporated. After evaporation, the dried precipitated salts weighed 1475 mg. What was the concentration in parts per million (ppm) of dissolved salts (also called total dissolved solids [TDS])?

2. The annual arithmetic mean ambient air quality standard for sulfur dioxide (SO_2) is 0.03 ppmv. What is this standard in $\mu g/m^3$?

3. The primary drinking water MCL for barium (Ba) is 2.0 mg/L. If the sole source of barium is barium sulfate ($BaSO_4$), how much $BaSO_4$ salt is present in 1 L of water that contains 2.0 mg/L of Ba? (Hint: The moles of Ba in 2.0 mg equal the moles of $BaSO_4$ in 1 L of sample.)

4. Most people can detect the odor of ozone in concentrations as low as 10 ppb. Could they detect the odor of ozone in samples with an ozone level of (a) 0.118 ppm, (b) 25 ppm, and (c) 0.001 ppm?

5. Determine the percentage by volume of the different gases in a mixture containing 0.3 L O_2, 1.6 L N_2, and 0.1 L CO_2. Assume that each separate gas is at 1 atm pressure before mixing and that the pressure of the combined gases after mixing is also at 1 atm.

6. What is the significance of the fact that the percentage of oxygen in air is 21% by numbers of molecules and 23% by mass?

7. Express the 0.9% argon content of air in ppm.

8. Express 400 ppm of CO_2 in cigarette smoke as a percentage of the smoke inhaled.

9. The permissible limit for ozone for a 1 h average is 0.12 ppm. If Little Rock, Arkansas, registers a reading of 0.15 ppm for 1 h, by what percent does Little Rock exceed the limit for atmospheric ozone?

10. A certain water-soluble pesticide is fatal to fish at 0.5 mg/L (ppm). Five kilograms of the pesticide are spilled into a stream. The stream flow rate was 10 L/s at 1 km/h. For what approximate distance downstream could fish potentially be killed?

11. Chromium(III) in a water sample is reported as 0.15 mg/L. Express the concentration as eq/L. (The Roman numeral III indicates that the oxidation number of chromium in the sample is +3. It also indicates that the dissolved ionic form would have a charge of +3.)

12. Alkalinity in a water sample is reported as 450 mg/L of $CaCO_3$. Convert this result to eq/L of $CaCO_3$. Alkalinity is a water quality parameter that results from more than one constituent. It is expressed as the amount of $CaCO_3$ that would produce the same analytical result as the actual sample (see Chapter 2).

It is directed to treat foul water by boiling it, exposing it to sunlight, and by dipping seven times into it a piece of hot copper, then to filter and cool in an earthen vessel. The direction is given by the God who is the incarnation of medical science.

From a Sanskrit record about 4000 years old*

2 Contaminant Behavior in the Environment
Basic Principles

2.1 BEHAVIOR OF CONTAMINANTS IN NATURAL WATERS

Every part of our world is continually changing, essential ecosystems as well as unwelcome contaminants. Some changes occur imperceptibly on a geological time scale; others are rapid, occurring within days, minutes, or less. Oil and coal are formed from dead animal and vegetable matter over millions of years. When oil and coal are burned, they can release their stored energy in fractions of a second. The products of their combustion cause further changes in the environment, often undesirable. While all technology is designed to create desired changes, it seems to be almost inevitable that undesired changes are also created, often in the form of dangerous waste pollutants. When this is the case, it is the role of environmental professionals to bring about still more changes that work to restore the environment to more acceptable conditions.

---

* This quote is one of the earliest known references to drinking water treatment and shows that water quality has been a concern from the time it was first connected with health problems.

Control of environmental contamination depends on learning how to bring about desired changes within a useful time scale, a task that requires an understanding of how pollutants are affected by environmental conditions. For example, metals that are dangerous to our health, such as lead, are often more soluble in water under acidic conditions than under basic conditions. Knowing this, one can plan to remove dissolved lead from drinking water by raising the pH and making the water basic. Under basic conditions, a large part of dissolved lead can be made to precipitate as a solid and can be removed from drinking water by settling out or filtering.

Contaminants in the environment are driven to change by

- Physical forces that move contaminants to new locations, often without significant change in their chemical properties. Contaminants released into the air, soil, and water can move into regions far from their origin under the forces of wind, gravity, and water flow. An increase in temperature will cause an increase in the rate at which gases and volatile substances evaporate from water or soil into the atmosphere. Electrostatic attractions can cause dissolved substances and small particles to adsorb to solid surfaces, where they may leave the water flow and become immobilized in soils or filters. Water flow can erode soils and transport sediments carrying sorbed pollutants over long distances.
- Chemical changes, such as oxidation and reduction, which break and make chemical bonds, allowing atoms to rearrange into new compounds with different properties. Chemical change often has the potential to destroy pollutants by converting them into less undesirable substances.
- Biological activity, whereby microbes in their constant search for survival energy break down many kinds of contaminant molecules and return their atoms to the environmental cycles that circulate carbon, oxygen, nitrogen, sulfur, phosphorus, and other elements repeatedly through our ecosystems. Biological processes are a special kind of chemical change.

Environmental workers are particularly interested in processes that

- Immobilize pollutants to prevent them from moving from locations where they pose little risk to locations where they may become unacceptable threats to public health or the environment.
- Mobilize pollutants to move them from locations where they endanger public health or the environment to less hazardous locations.
- Remove pollutants from the environment to permanent or temporary storage.
- Chemically change the nature of pollutants to less harmful forms.

These processes are the tools of environmental protection. Their effectiveness depends on the physical and chemical properties of the pollutant and its water and soil environment. It is often said that every remediation project is unique and site-specific. The reason for this is that, although each pollutant has its predictable and, frequently, tabulated chemical and physical properties, each project site has properties of water, soil, and environment that are always different from other sites to some extent, depending on climate, long-term geologic history, and more recent anthropo-morphic disturbances.

Important properties of pollutants can usually be found in handbooks or chemistry references. However, important properties of the water and soil in which the pollutant resides are always unique to the particular site and time of measurement and must be measured or estimated anew for every project.

2.1.1 IMPORTANT PROPERTIES OF POLLUTANTS

The six most important pollutant properties for predicting the environmental behavior of a pollutant are listed below. They are usually tabulated in handbooks and other chemistry references, to the extent of current knowledge:

1. Solubility in water
2. Volatility
3. Density
4. Chemical reactivity
5. Biodegradability
6. Strength of sorption* to solids

If measured or theoretical values are not readily found, these properties can often be estimated from the chemical structure of the pollutant. Whenever possible, this book will offer "rules of thumb" for estimating pollutant properties. The ability to "guesstimate" the environmental behavior of a pollutant is often an important first step in developing a remediation strategy.

2.1.2 IMPORTANT PROPERTIES OF WATER AND SOIL

The properties of water and soil that influence pollutant behavior can be expected to differ at every location and must be measured or estimated for each project. Since environmental conditions are so varied, it is difficult to generate a simple set of water and soil properties that should always be measured. The lists below include the most commonly needed properties. Discussions and examples throughout this book will illustrate how knowledge of important soil and water properties are used in protecting and restoring the natural environment.

Water properties:

- Temperature
- Water quality (chemical composition, pH, oxidation–reduction potential, alkalinity, hardness, turbidity, dissolved oxygen, biological oxygen demand, fecal coliforms, etc.)
- Flow rate and flow pattern

---

* Sorption is a general term that includes all the possible processes by which a molecule, originally in a gas or liquid phase, becomes bound to a solid. Sorption includes both adsorption (becoming bound to a solid surface) and absorption (becoming bound within pores and passages in the interior of a solid). It also includes all the variants of binding mechanisms, such as chemisorption (where chemical bonds are formed between a molecule and the surface), and physisorption (where physical attractions such as electrostatic, van der Waals, and London forces hold a molecule to a surface).

Properties of mineral solids and soils in contact with water:

- Mineral composition
- Percentage of organic matter
- Sorption coefficients for contaminants (attractive forces between solids and contaminants)
- Mobility of suspended solids (colloid and particulate movement)
- Particle size distribution
- Soil compaction, erodibility, cohesion, etc.
- Porosity
- Permeability to water and contaminant liquids
- Hydraulic conductivity

The properties of environmental waters and soils are always site-specific and must be estimated or measured in the field.

Weather conditions can be considered another important site-specific variable, even though weather is influenced by many factors completely unrelated to any particular site. Long periods of sunlight and warm weather promote algae growth in ponds and lakes. A tendency to severe rainfall events can cause significant soil erosion that mobilizes soil-bound contaminants. Wind patterns can carry airborne contaminants long distances to finally settle to earth as acid rain or radioactive dust. Spring snowmelt can mitigate pollutant concentrations in some streams by dilution or worsen them because of sudden flushing of pollutants such as metals from contaminated soils or mine tailings.

Environmental protection and remediation programs often must consider external properties such as seasonal weather changes, arid or humid climates, sun intensity and exposure, and wind patterns.

2.2   WHAT ARE THE FATES OF DIFFERENT POLLUTANTS?

There are three possible naturally occurring (rather than engineered) fates of pollutants:

1. All or a portion might remain unchanged in their present location.
2. All or a portion might be carried elsewhere by transport processes.
   a. Movement to other phases (air, water, or soil) by volatilization, dissolution, adsorption, and precipitation.
   b. Movement within a phase under gravity, diffusion, and advection.
3. All or a portion might be transformed into other chemical species by natural chemical and biological processes.
   a. Biodegradation (aerobic and anaerobic): Pollutants are altered structurally by biological processes, mainly the metabolism of microorganisms present in aquatic and soil environments.
   b. Bioaccumulation: Pollutants accumulate in plant and animal tissues to higher concentrations than in their original environmental locations.

c. Weathering: Pollutants undergo a series of environmentally induced nonbiological chemical changes, by processes such as oxidation–reduction, acid–base, hydration, hydrolysis, complexation, and photolysis reactions.

2.3 PROCESSES THAT REMOVE POLLUTANTS FROM WATER

2.3.1 NATURAL ATTENUATION

Even without human intervention, pollutant concentrations in the environment have a tendency to diminish with time due to natural causes. The rate of attenuation, however, depends strongly on the chemical and physical properties of the pollutants (e.g., solubility; biodegradability; chemical stability; whether solid, liquid, or gas; etc.) and on many characteristics of the polluted site (soil permeability, average rain and snow precipitation and temperature, geologic features, etc.). Where natural processes are fast enough and contaminant concentrations small enough, the simplest approach to remediation is to wait until pollutant levels are no longer deemed hazardous. The case study in Chapter 9, Section 9.3.1 is an example of when this approach may be the best choice.

Because every case is different and highly site-specific, the progress of natural attenuation generally must be closely monitored before considering it as the preferred remediation option. Monitored natural attenuation is a recognized approach to pollution remediation (OSWER 1999), described by the EPA as:

the reliance on natural attenuation processes (within the context of a carefully controlled and monitored site cleanup approach) to achieve site-specific remediation objectives within a time frame that is reasonable compared to that offered by other more active methods. The 'natural attenuation processes' that are at work in such a remediation approach include a variety of physical, chemical, or biological processes that, under favorable conditions, act without human intervention to reduce the mass, toxicity, mobility, volume, or concentration of contaminants in soil or groundwater. These in-situ processes include biodegradation; dispersion; dilution; sorption; volatilization; radioactive decay; and chemical or biological stabilization, transformation, or destruction of contaminants.

Natural attenuation processes are described more fully below and in later chapters.

2.3.2 TRANSPORT PROCESSES

Contaminants that are dissolved or suspended in water can move to other phases by the following three processes:

1. Volatilization: Dissolved and sorbed contaminants move from water and soil into air in the form of gases or vapors.
2. Sorption: Dissolved and airborne contaminants become bound to solids by attractive chemical, physical, and electrostatic forces.

3. Sedimentation: Small suspended solids in water grow large enough, or water flow slows enough, to settle solids out of water under gravity. There are two stages to sedimentation:

   a. Coagulation: Suspended solids generally carry an electrostatic charge that keeps them apart. Some fraction of particle collisions, depending on the temperature, will have enough energy to overcome the repulsive energy barrier between particles and allow two colliding particles to approach close enough that short-range attractive forces (van der Waals/London attractions) prevail. Then the particles can coalesce to form a larger, heavier particle. Chemicals may be added to lower the repulsive electrostatic energy barrier (destabilization), allowing higher rates of coagulation at lower temperatures.

   b. Flocculation: Coagulation allows suspended solids to collide and clump together because of short-range attractive forces, to form a floc. When floc particles aggregate sufficiently (flocculation), they can become heavy enough to settle out of the water.

2.3.3 Environmental Chemical Reactions

The following are brief descriptions of some important environmental chemical reactions that can remove pollutants from water. More detailed discussions are given throughout this book.

- *Photolysis*: In molecules that absorb solar radiation, exposure to sunlight can break chemical bonds and start chemical breakdown. Many natural and synthetic organic compounds are susceptible to photolysis.
- *Complexation and chelation*: Polar or charged dissolved species (such as metal ions) bind to electron-donor ligands* to form complex or coordination compounds. Complex compounds are often soluble and resist removal by precipitation because the ligands must be displaced by other anions (such as sulfide) before an insoluble species can be formed. Common ligands include hydroxyl, carbonate, carboxylate, phosphate, and cyanide anions, as well as water molecules, humic acids, and synthetic chelating agents such as nitrilotriacetate (NTA) and ethylenediaminetetraacetate (EDTA).
- *Acid–base*: Protons (H^+ ions) are transferred between chemical species. Acid–base reactions are part of many environmental processes and influence the chemical fate of many pollutants.
- *Oxidation–reduction (OR or redox)*: Electrons are transferred between chemical species, changing the oxidation states and the chemical properties of the electron donor and the electron acceptor. Water disinfection, electrochemical reactions such as metal corrosion, and most microbial reactions, including those causing biodegradation, are oxidation–reduction reactions.

---

* Ligands are polyatomic chemical species containing nonbonding electron pairs that are used to bond the ligand to a central atom. The ligand contributes both of the electrons that form the bond, instead of the more common case where each bonded atom contributes one electron.

- *Hydrolysis and hydration*: A compound reacts with water to form dissolved ions (hydrolysis) or incorporate water molecules into its chemical structure (hydration). In water, all ions and polar compounds develop a hydration shell of water molecules. When the attraction to water is strong enough, hydrolysis reactions may cause hydrolyzed metal ions to form hydroxides and oxides of low solubility; see Chapter 4. With organic compounds, a water molecule may replace an atom or group, a step that often breaks the organic compound into smaller fragments. Even dissolved gases can undergo hydration. Hydration of dissolved carbon dioxide (CO_2) and sulfur dioxide (SO_2) forms carbonic acid (H_2CO_3) and sulfurous acid (H_2SO_3), respectively, which then increase acidity and lower the pH of the water.
- *Chemical precipitation*: Two or more dissolved species may react to form an insoluble solid compound, or a change in pH, redox potential, temperature, or concentrations of dissolved species may result in the formation of a solid from dissolved species. For example, precipitation can occur if a solution of a salt becomes oversaturated (when the concentration of a salt is greater than its solubility limit). For example, the mineral gypsum ($CaSO_4 \cdot 2H_2O$) has a solubility in pure water at 20°C and pH 7 of about 2400 mg/L (Appendix B, Table B.2). In a water solution containing 1000 mg/L of gypsum, all the gypsum will be dissolved under these conditions of pH and temperature. If more gypsum is added or water is evaporated, the concentration of dissolved gypsum can increase only to about 2400 mg/L. Any gypsum in excess of the solubility limit will precipitate as solid gypsum. Precipitation of dissolved minerals is treated in more detail in Appendix B.

Chemical precipitation can also occur if two soluble salts react to form a different salt of low solubility. For example, silver nitrate ($AgNO_3$) and sodium chloride (NaCl) are both highly soluble. They react in solution to form the very slightly soluble salt silver chloride (AgCl) and the highly soluble salt sodium nitrate ($NaNO_3$). The almost insoluble silver chloride precipitates as a solid, while the soluble silver nitrate remains dissolved. Breaking this reaction into two separate conceptual steps (Equations 2.1 and 2.2) helps to visualize what happens. Refer to the solubility table inside the back cover, which gives qualitative solubilities for ionic compounds in water.

In the first conceptual step, the soluble salts silver nitrate and sodium chloride are added to water and dissolve as ions:

$$AgNO_3\,(s) \xrightarrow{\;H_2O\;} Ag^+\,(aq) \;+\; NO_3^-\,(aq) \qquad\qquad (2.1)*$$

$$NaCl\,(s) \xrightarrow{\;H_2O\;} Na^+\,(aq) \;+\; Cl^-\,(aq) \qquad\qquad (2.2)$$

---

* Placing the chemical formula for water, H_2O, above the reaction arrows means that the reaction requires the presence of water, even though water does not react chemically with the other reagents and does not appear in the overall reaction. The suffix (aq), abbreviation for aqueous, following a chemical species means that the species are dissolved in water. The suffix (s), abbreviation for solid, following a chemical species means that the species is in solid form.

After the dissolution steps, Equations 2.1 and 2.2, and before any further reactions, the solution contains Ag^+, Na^+, Cl^-, and NO_3^- ions.

While in solution, all ions move about freely. Ions with charges having opposite signs (positive/negative) are attracted to one another, while ions with charges having the same sign (positive/positive and negative/negative) are repelled from one another. Charged ions with opposite signs tend to pair up randomly, regardless of their chemical identity. Therefore, in the second conceptual step, the ions can combine in all possible ways that pair a positive ion with a negative ion. Besides the original Ag^+/NO_3^- and Na^+/Cl^- pairs, both of which are very soluble, Ag^+/Cl^- and Na^+/NOJ_3^- pairs are also possible. Since $NaNO_3$ also is a very soluble ionic compound, dissolving to form Na^+ and NO_3^-, the Na^+ and NO_3^- ions simply remain in solution. However, Ag^+ and Cl^- can combine to form $AgCl$, which has a very low solubility and will precipitate as a solid.

The overall reaction is written as

$$AgNO_3(aq) + NaCl(aq) \xrightarrow{H_2O} Na^+(aq) + NO_3^-(aq) + AgCl(s) \qquad (2.3)$$

Thus, adding the two soluble salts, $AgNO_3$ and $NaCl$ to water results in a solution containing Na^+ and NO_3^- ions and the precipitated solid compound $AgCl$. If equal moles of the two salts, $AgNO_3$ and $NaCl$, were mixed initially, only very small amounts of Ag^+ and Cl^- will remain unprecipitated, because the solubility of $AgCl$ is very small. Appendix B contains a more quantitative discussion of metal salt solubility.

2.3.4 ENVIRONMENTAL BIOLOGICAL PROCESSES

Microbes can degrade organic pollutants by facilitating oxidation–reduction reactions. Microbial metabolism (the biological reactions that convert organic compounds into energy and carbon for microbial growth) often involves the transfer of electrons from a pollutant molecule (the electron donor) to other compounds present (electron acceptors) in the soil or water environment. The electron acceptors most commonly available in the environment are molecular oxygen (O_2), carbon dioxide (CO_2), nitrate (NO_3^-), sulfate (SO_4^{2-}), manganese (Mn^{2+}), and iron (Fe^{3+}). When molecular oxygen (O_2) is available, it is always the preferred electron acceptor and the process is called aerobic biodegradation. In the absence of O_2, the process is called anaerobic biodegradation. Aerobic and anaerobic biodegradation are examples of oxidation–reduction reactions, discussed in Chapter 3, Section 3.5.

Organic pollutants are generally toxic because of their chemical structure. Changing their structure in any way will often change the way living organisms react to them and may make them nontoxic or, in a few cases, more toxic. Eventually, usually after many reaction steps, in a process called mineralization, biodegradation converts organic pollutants into carbon dioxide, water, and mineral salts. Although these final products represent the destruction of the original pollutant, some of the intermediate steps may temporarily produce compounds that are also pollutants, sometimes more toxic than the original. Biodegradation is discussed in more detail in Chapter 9.

2.4 SOME MAJOR CONTAMINANT GROUPS AND NATURAL PATHWAYS FOR THEIR REMOVAL FROM WATER

Only brief introductory descriptions of some major contaminant groups and natural removal processes are given here as an introduction to the discussion of the intermolecular forces that are the basis for their removal processes. Other important contaminant groups, such as polychlorinated biphenyls (PCBs), dioxins, polycyclic aromatic hydrocarbons (PAHs), radioisotopes, and so on are discussed elsewhere in this book. There also are less common removal pathways not discussed here, such as photolysis and radiolysis, which can become important or even dominant under special conditions.

2.4.1 METALS

Dissolved metals such as iron, lead, copper, cadmium, mercury, and so on are removed from water mainly by sorption, precipitation, and filtration processes. Some metals—particularly As, Cd, Hg, Ni, Pb, Se, Te, Sn, and Zn—can form volatile metal-organic compounds in the natural environment by microbial reactions. For these, volatilization can be an important removal mechanism. Bioaccumulation of metals in animals usually is not very significant as a removal process, although it can have very toxic effects on the affected animals. Bioaccumulation in plants on the other hand has been developed into a useful remediation technique called phytoremediation. Biotransformation of metals, in which redox reactions involving bacteria can cause some metals to precipitate, has also shown promise as a removal method. The aqueous chemistry of metals is discussed in Chapter 5.

2.4.2 CHLORINATED PESTICIDES

Chlorinated pesticides, such as atrazine, chlordane, DDT, dicamba, endrin, heptachlor, lindane, 2,4-D, etc., are removed from water mainly by sorption (sometimes followed by catalytic decomposition), volatilization, filtration, and biotransformation. Chemical processes like oxidation, hydrolysis, and photolysis usually play a minor role.

2.4.3 HALOGENATED ALIPHATIC HYDROCARBONS

Halogenated aliphatic* hydrocarbons in the environment arise mostly from industrial and household solvents. Compounds such as 1,2-dichloropropane, 1,1,2-trichlorethane, tetrachloroethylene, carbon tetrachloride, chloroform, and so on are removed mainly by volatilization and sometimes by membrane filtration and catalytic decomposition. Under natural conditions, aerobic biotransformation and biodegradation processes are usually very slow, with half-lives of tens to hundreds of years. However, natural and engineered anaerobic biodegradation processes have been developed with short enough half-lives to be useful remediation techniques.

---

* Aliphatic hydrocarbons are organic compounds that do not contain aromatic ring structures. Aromatic ring structures contain delocalized electrons that impart extra structural stability (resistance to bond-breaking forces) to aromatic ring compounds like benzene, naphthalene, pyrene, etc.

2.4.4　Fuel Hydrocarbons

Gasoline, diesel fuel, and heating oils are mainly distillation products of petroleum and are mixtures of hundreds of different organic hydrocarbons. Gasoline is mainly the lighter weight compounds such as benzene, toluene, ethylbenzene, xylenes, naphthalene, trimethylbenzenes, the smaller alkanes, and so on, which are removed mainly by sorption, volatilization, and biotransformation, and sometimes by incineration, membrane filtration, and catalytic decomposition. Diesel and heating fuels are mostly the heavier distillation compounds, which include polycyclic aromatic hydrocarbons (PAHs) such as fluorene, benzo(a)pyrene, anthracene, phenanthrene, and so on, and are not very volatile; they are removed mainly by sorption, sedimentation, incineration, and biodegradation.

2.4.5　Inorganic Nonmetal Species

These include ammonia, chloride, bromide, fluoride, cyanide, nitrite, nitrate, phosphate, sulfate, sulfide, and so on. They are removed mainly by sorption, volatilization, chemical transformation, and biotransformation.

It should be noted that many normally minor removal pathways, such as photolysis, can become important, or even dominant, in special circumstances. For example, low volatility pesticides in a clear, shallow stream with little organic matter might be degraded primarily by photolysis.

2.5　CHEMICAL AND PHYSICAL REACTIONS IN THE WATER ENVIRONMENT

Chemical and physical reactions in water can be classified either as *homogeneous* (occurring entirely among dissolved species) or *heterogeneous* (occurring at the liquid–solid–gas interfaces). Most environmental water reactions are heterogeneous. Purely homogeneous reactions are rare in natural waters and wastewaters. Among the most important heterogeneous reactions are those that move pollutants from one phase to another: volatilization, dissolution, and sorption.

- *Volatilization* moves pollutants into the vapor phase and depends mostly on the intermolecular forces holding a liquid or solid together and the temperature; a substance volatilizes when its thermal energy is high enough to overcome attractive forces within the substance. At the liquid–air and solid–air interfaces, volatilization transfers volatile contaminants from water and solid surfaces into the atmosphere and into air contained in soil pore spaces. It is most important for compounds with high vapor pressures. Contaminants in the vapor phase travel by diffusion and wind transport and are the most mobile in the environment.
- *Dissolution* moves pollutants into the liquid phase. At the solid–liquid and air–liquid interfaces, dissolution transfers contaminants from air and solids to water. It is most important for contaminants of significant water solubility. A solid substance dissolves when water molecules adjacent to a solid surface attract solid molecules more strongly than they are attracted to other

solid molecules. Gaseous molecules have little attraction to other molecules in the gas phase; they dissolve in water when water molecules attract them more strongly than their thermal energy can overcome. Contaminants dissolved in water travel by diffusion and with the water flow. The environmental mobility of contaminants dissolved in water is generally intermediate between volatilized and sorbed contaminants.

*   *Sorption* binds pollutants to solid surfaces; a substance sorbs to a solid surface when pollutant attractions to surfaces become dominant relative to their attractions to water or their thermal energy. At the liquid–solid and air–solid interfaces, sorption transfers contaminants from water and air to soils and sediments. It is most important for compounds of low solubility and low volatility. Sorbed compounds undergo chemical and biological transformations at different rates and by different pathways than dissolved or airborne compounds. The binding strength with which different contaminants become sorbed depends on the nature of the solid surface (sand, clays, organic particles, etc.) and on the properties of the contaminant. Contaminants sorbed to solids move significantly only if the solid moves, for example, during soil erosion, with blowing dust, as suspended sediment in streams, and so on, and usually are the least mobile in the environment (except for solid colloids; see Chapter 6, Section 6.8.

2.6   PARTITIONING BEHAVIOR OF POLLUTANTS

Eventually, a portion of every pollutant released to the environment becomes distributed by heterogeneous reactions into all the liquid, gas, and solid phases with which it comes into contact. The most important factor for predicting the partitioning behavior of contaminants in the environment is an understanding of the intermolecular attractive forces between contaminants and the water and soil materials in which they are found. Predicting the amount of pollutant that will enter different phases is an important subject that recurs frequently throughout this text.

A pollutant in contact with water, soil, and air will partially dissolve into the water, partially volatilize into the air, and partially sorb to the soil surfaces, as illustrated in Figure 2.1. The relative amounts of pollutant that are found in each phase with which it is in contact depend on intermolecular attractive forces existing between pollutant, water, and soil molecules.

2.6.1   Partitioning from A Diesel Oil Spill

Consider, for example, what happens when diesel oil is spilled at the soil surface. Some of the liquid diesel oil (often called *free product**) flows downward under gravity through the soil toward the groundwater table. Before the spill, the soil pore spaces above the water table (called the soil *unsaturated zone* or *vadose zone*) were

---

* *Free product* is a common term for the original physical form of a pollutant before being transformed by physical, biological, or chemical processes, such as a pool of industrial solvent formed by a leaking storage tank or a surface layer of oil floating down a river; sometimes called the bulk pollutant.

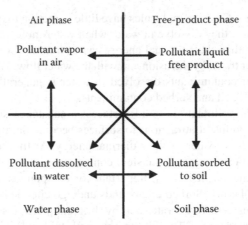

Air phase | Free-product phase

Pollutant vapor in air ←→ Pollutant liquid free product

Pollutant dissolved in water ←→ Pollutant sorbed to soil

Water phase | Soil phase

FIGURE 2.1 Partitioning of a pollutant among air, water, soil, and free product phases. Arrows indicate all possible phase change pathways.

filled with air and capillary-held water, and the soil surfaces were partially covered with sorbed water.

As diesel oil, which is a mixture of hundreds of different compounds, passes downward through the soil, a portion of its different components continually partitions from the liquid diesel oil to the air and water within the soil pore spaces, and to the soil particle surfaces. After the spill, the soil pore spaces contain diesel fuel that has partitioned into all the phases of Figure 2.1:

- Air phase, with vapors of the volatile diesel components
- Water phase, carrying dissolved soluble diesel components
- Soil phase, where diesel components of low solubility and low volatility have sorbed, along with sorbed diesel-free product and sorbed water containing dissolved diesel components
- Liquid diesel oil (free product) phase, that has changed in composition by losing some of its more volatile and soluble components to other phases

The many different compounds in diesel oil each have a unique partitioning, or distribution pattern. The pore space air will contain mainly the most volatile components, the pore space water will contain mainly the most-soluble components, and the soil particles will sorb mainly the most strongly sorbed, least-volatile, and least-soluble components. The quantity of free product remaining diminishes continually as it moves downward through the soil because a significant portion is lost to other phases. The composition of the free product also changes continually because the most volatile, soluble, and strongly sorbed compounds are lost preferentially.

The distribution of the various diesel compounds among different phases attains quasi-equilibrium, with compounds continually passing back and forth across each phase interface, as indicated in Figure 2.1. As the remaining liquid diesel free product flows downward into new soil, it continues to change in composition by losing components to other phases (part of the "weathering" process), and becomes increasingly resistant to further change. Since the lightest weight components tend to be the most volatile and soluble, they are the first to be lost to other phases, and the remaining liquid diesel becomes increasingly more viscous and less mobile. Severely weathered liquid diesel is very resistant to further change, and can persist in the soil for decades. It only disappears by biodegradation or by actively engineered removal.

Depending on the amount of diesel oil spilled and the distance from the ground surface to groundwater, it is possible that all of the liquid diesel becomes "immobilized" in the soil before it can reach the water table. This occurs when the mass of liquid diesel free product diminishes and its viscosity increases to the point where capillary forces in the soil pore spaces can hold the remaining liquid diesel in place against the force of gravity. There is still pollutant movement, however, mainly in the nonliquid diesel phases. The volatile components in the vapor state usually diffuse rapidly through the soil, moving mostly upward toward the soil surface and along any high permeability pathways through the soil, such as a sewer line backfill. New water percolating downward, from precipitation or other sources, can dissolve additional diesel compounds from the sorbed phase and carry them downward. The weight of downward percolating water can also displace some liquid diesel held by capillary forces, as well as soil pore water already containing dissolved pollutants, forcing them to move farther downward. Although the liquid diesel is never truly immobilized, its downward movement can become imperceptible.

However, if the spill is large enough, liquid diesel free product may reach the water table before becoming immobilized. If this occurs, liquid diesel free product, being less dense than water, cannot enter the water-saturated zone but remains above it, effectively floating on top of the water table. There, the liquid diesel free product spreads horizontally on the groundwater surface, as a growing subsurface contamination "plume," continuing to partition into groundwater, soil pore space air, and soil particle surfaces. In other words, wherever it moves, a portion of the liquid diesel free product continues to become distributed among all the solid, liquid, and gas phases with which it comes into contact. This behavior is governed by intermolecular forces that exist between molecules.

RULES OF THUMB

When a pollutant consisting of a mixture of different compounds, such as diesel fuel or gasoline, is released to the environment, its composition and physical properties change as time passes.

1. The most volatile components tend to leave the free product and pass into the atmosphere or into air in the soil pore space.
2. The most water-soluble components tend to dissolve into any water they contact.
3. The least volatile and soluble components tend to sorb to soil and sediment surfaces as the pollutant is moved by gravity and water flow forces.

The remaining free product progressively becomes denser, more viscous, less mobile, and more resistant to further change. Sometimes, comparing the composition of "weathered" free product with the known (or estimated) composition of fresh free product can be useful for estimating the "age" of the weathered free product (see Age-Dating Fuel Spills, Chapter 7, Section 7.6.3).

2.7 INTERMOLECULAR FORCES

Volatilization, dissolution, and sorption processes all result from the interplay between intermolecular forces. All molecules have attractive forces acting between them. The attractive forces are electrostatic in nature, created by a nonuniform distribution of outer orbit electrons surrounding a molecule. This occurs when different nuclei in a molecule have differing strengths of electrostatic attraction for the electrons, preferentially drawing electrons closer to the more electropositive nuclei.

When electrons are not uniformly distributed, the molecule is said to be *polar* or *ionic*. A nonuniform electron distribution results in a molecule having locations with net positive and/or negative charges. A charged region on one molecule is attracted to oppositely charged regions on adjacent molecules, resulting in the so-called polar and ionic attractive forces. Even nonpolar molecules have attractive forces between them (although much weaker than polar or ionic molecules) because of temporary distortions of the electron clouds that occur during molecular collisions; see the definition of dispersion (London) forces in Section 2.8.6.

2.7.1 TEMPERATURE-DEPENDENT PHASE CHANGES

Electrostatic attractive forces always work to bring order to molecular configurations, in opposition to thermal energy, which always works to randomize configurations. Gases are always the higher-temperature form of any substance and are the most randomized state of matter. If the temperature of a gas is lowered enough, every gas will condense to a liquid, a more ordered state. Condensation is a manifestation of intermolecular attractive forces. As the temperature falls, the thermal energy of gas molecules decreases, eventually reaching a point where there is insufficient thermal kinetic energy to keep the molecules separated against their intermolecular attractive forces. The temperature at which condensation occurs is called the boiling point (bp); it is dependent on environmental pressure as well as temperature.

If the temperature of a liquid is lowered further, it eventually freezes to a solid at the freezing point (fp), when its thermal energy becomes low enough for intermolecular attractions to pull the molecules into a rigid solid arrangement. Solids are the most highly ordered state of matter. Whenever lowering the temperature causes a change of phase (gas to liquid or liquid to solid), the thermal energy has decreased sufficiently to allow the always-present attractive forces to overcome molecular kinetic energy and pull gas and liquid molecules closer together into more-ordered liquid or solid phases.

2.7.2 VOLATILITY, SOLUBILITY, AND SORPTION

The model of intermolecular forces that describes how phase changes result from attractive forces that always work to bring increased order to molecular distributions, against the randomizing effects of thermal energy, also explains the volatility, solubility, and sorption behavior of molecules.

- Molecules of volatile liquids have relatively weak attractions to one another. Thermal energy at ordinary environmental temperatures is sufficient to allow the most energetic of the weakly held liquid molecules to escape from electrostatic attractions to their slower liquid neighbors and fly into the gas phase.
- Molecules in water-soluble solids are attracted to water more strongly than they are attracted to themselves. If a water-soluble solid is placed in water, its surface molecules are drawn from the solid phase into the liquid phase by the stronger attractions to water molecules.
- Dissolved molecules that become sorbed to sediment surfaces are held to the sediment particle by stronger attractive forces that pull them away from water molecules.

Understanding intermolecular forces is the key to predicting how contaminants become distributed in the environment.

2.7.3 PREDICTING RELATIVE ATTRACTIVE FORCES

When you can predict relative attractive forces between molecules, you can predict their relative solubility, volatility, and sorption behavior. For example, the volatility of a substance is closely related to its freezing and boiling temperatures, which in turn are related to the attractive forces between molecules of that substance. The water solubility of a compound is related to the relative strengths of the attractive forces between molecules of the compound to other molecules within the compound, compared to its attraction to adjacent molecules of water. The soil–water partition coefficient of a compound indicates the relative strengths of that compound's attraction to water and to soil.

For any compound, the temperature at which it changes phase is an indicator of the strength of the intermolecular attractive forces existing between its molecules:

- Boiling a liquid means that it has been heated to the point where thermal energy imparts sufficient kinetic energy to its molecules to allow them to overcome their attractive forces and move apart from one another into the gas phase. A higher boiling temperature indicates stronger intermolecular attractive forces between the liquid molecules because they must attain higher kinetic energy to pull apart. The thermal energy has to be higher in order to overcome the stronger attractions and allow liquid molecules to escape into the gas phase. Thus, the fact that methanol boils at a lower temperature than water means that methanol molecules are attracted to one another more weakly than are water molecules. The reasons for this are discussed in Section 2.8.
- Freezing a liquid means that its thermal energy has been reduced by cooling to the point where attractive forces can overcome the randomizing effects of

thermal motion and pull freely moving liquid molecules into fixed positions in a solid phase. A lower freezing point indicates weaker attractive forces. The thermal energy has to be reduced to lower values so that the weaker attractive forces can pull the molecules into fixed positions in a solid phase. The fact that methanol freezes at a lower temperature than water is another indicator that attractive forces are weaker between methanol molecules than between water molecules.

- Wax, a mixture of hydrocarbons consisting mainly of carbon and hydrogen atoms, is solid at room temperature (20°C), whereas diesel fuel, also a mixture of hydrocarbons, is liquid. The freezing temperature of diesel fuel is well below room temperature. This indicates that the attractive forces between wax molecules are stronger than between molecules in diesel fuel. At the same temperature where diesel molecules can still move about randomly in the liquid phase, wax molecules are held by their stronger attractive forces in fixed positions in the solid phase. Understanding the reasons for differences in intermolecular attractive forces, discussed in Section 2.8, is important for remediation strategies.
- Compounds that are highly soluble in water have strong attractions to water molecules. When a soluble solid substance is added to water, water molecules are attracted to the solid surface, where they literally pull the solid molecules apart from one another and carry them into solution.
- When compounds released to the environment are mostly found sorbed to soils and sediments, rather than dissolved in water or vaporized into the air, it indicates that they have stronger attractions to soil than to water or to molecules of their own kind.
- Compounds found in the environment as a gas, because they volatilize readily at environmental temperatures from water, soil, and their own molecules, must have relatively weak attractions to water, soil, and molecules of their own kind.

2.8 ORIGINS OF INTERMOLECULAR FORCES: ELECTRONEGATIVITIES, CHEMICAL BONDS, AND MOLECULAR GEOMETRY

Intermolecular forces are electrostatic in nature. Molecules are composed of electrically charged particles (electrons and protons), and it is common to have regions within a single molecule that are predominantly charged positive or negative. Attractive forces between molecules arise when electrostatic forces attract positive regions on one molecule to negative regions on another. The strength of the attractions between different molecules depends on the polarities of chemical bonds within the molecules, the geometrical shapes of the molecules, and their relative sizes.*

2.8.1 CHEMICAL BONDS

At the simplest level, the chemical bonds that hold atoms together in a molecule are of two types: ionic and covalent.

---

* Large molecules are more *polarizable*, a property that allows them to develop temporary electrostatic charges, resulting in attractive *dispersion forces*; see Section 2.8.6.

1. *Ionic* bonds occur when one atom in a molecule attracts an electron away from another atom to form a positive ion (from the atom losing an electron) and a negative ion (from the atom receiving an electron). These ions are then bound together by electrostatic attraction. The electron transfer occurs because the electron-receiving atom has a much stronger attraction for electrons in its vicinity than does the electron-losing atom.

2. *Covalent* bonds are formed when two atoms share electrons, called bonding electrons, in the space between their nuclei. The electron-attracting properties of two covalent bonded atoms are not different enough to allow one atom to pull an electron entirely away from the other. However, unless both atoms attract the bonding electrons equally, the average position of the bonding electrons will be closer to one of the atoms. The atoms are held together because their positive nuclei are attracted to the negative charge of the shared electrons in the space between them.

When two covalent bonded atoms are identical, as in Cl_2, the bonding electrons are always equally attracted to each atom and the electron charge is uniformly distributed between the atoms. Such a bond is called a *nonpolar covalent bond*, meaning that it has no polarity, that is, no regions with net positive or negative charge.

When two covalent bonded atoms are of different kinds, as in HCl, one atom can attract the bonding electrons more strongly than the other. This produces a nonuniform distribution of electron charge between the atoms such that one end of the bond is more negative than the other, resulting in a polar covalent bond.

Figure 2.2 illustrates the electron distributions in nonpolar and polar covalent bonds. The strength with which an atom attracts bonding electrons to itself is indicated by a quantity called *electronegativity*, abbreviated EN. Electronegativities of the elements, shown in Table 2.1, are relative numbers with an arbitrary maximum

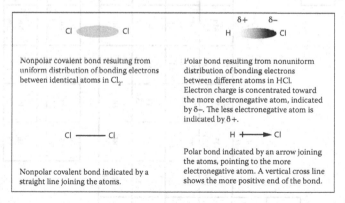

FIGURE 2.2 Uniform and nonuniform electron distributions, resulting in nonpolar and polar covalent chemical bonds. The oval between the atoms in the top two figures indicates the electron distribution in the chemical bond. Shading of the oval indicates variation in the electron density within the bond; light regions are low density and dark regions are high density. In the right-hand top figure, the use of a delta (δ) in front of the + and – signs signifies that the charges are partial, arising from a nonuniform electron charge distribution rather than the transfer of a complete electron.

TABLE 2.1
Electronegativity Values of the Elements (Pauling Scale)

Legend (example):

Atomic number → 26
Element symbol → Fe
Electronegativity → 1.83

1 (1A)	2 (2A)	3 (3B)	4 (4B)	5 (5B)	6 (6B)	7 (7B)	8 (8B)	9 (8B)	10 (8B)	11 (1B)	12 (2B)	13 (3A)	14 (4A)	15 (5A)	16 (6A)	17 (7A)	18 (8A)
1 H 2.20																	2 He
3 Li 0.98	4 Be 1.57											5 B 2.04	6 C 2.55	7 N 3.04	8 O 3.44	9 F 3.98	10 Ne
11 Na 0.93	12 Mg 1.31											13 Al 1.61	14 Si 1.90	15 P 2.19	16 S 2.58	17 Cl 3.16	18 Ar
19 K 0.82	20 Ca 1.00	21 Sc 1.36	22 Ti 1.54	23 V 1.63	24 Cr 1.66	25 Mn 1.55	26 Fe 1.83	27 Co 1.88	28 Ni 1.91	29 Cu 1.90	30 Zn 1.65	31 Ga 1.81	32 Ge 2.01	33 As 2.18	34 Se 2.55	35 Br 2.96	36 Kr 3.00
37 Rb 0.82	38 Sr 0.95	39 Y 1.22	40 Zr 1.33	41 Nb 1.6	42 Mo 2.16	43 Tc 1.9	44 Ru 2.2	45 Rh 2.28	46 Pd 2.20	47 Ag 1.93	48 Cd 1.69	49 In 1.78	50 Sn 1.96	51 Sb 2.05	52 Te 2.1	53 I 2.66	54 Xe 2.60
55 Cs 0.79	56 Ba 0.89	57 *La 1.1	72 Hf 1.3	73 Ta 1.5	74 W 2.36	75 Re 1.9	76 Os 2.2	77 Ir 2.20	78 Pt 1.28	79 Au 2.54	80 Hg 2.00	81 Tl 1.62	82 Pb 2.33	83 Bi 2.02	84 Po 2.0	85 At 2.2	86 Rn 2.2
87 Fr 0.7	88 Ra 0.9	89 *Ac 1.1	104 Rf	105 Db	106 Sg	107 Bh	108 Hs	109 Mt	110 Ds	111 Rg	112 Cn	113 Uut	114 Uuq	115 Uup	116 Uuh	117 Uus	118 Uuo

*Lanthanide series

58 Ce 1.12	59 Pr 1.13	60 Nd 1.14	61 Pm 1.13	62 Sm 1.17	63 Eu 1.2	64 Gd 1.2	65 Tb 1.1	66 Dy 1.22	67 Ho 1.23	68 Er 1.24	69 Tm 1.25	70 Yb 1.1	71 Lu 1.27

*Actinide series

90 Th 1.3	91 Pa 1.5	92 U 1.38	93 Np 1.36	94 Pu 1.28	95 Am 1.13	96 Cm 1.28	97 Bk 1.3	98 Cf 1.3	99 Es 1.3	100 Fm 1.3	101 Md 1.3	102 No 1.3	103

Notes:
1. Electronegativity values are below and atomic numbers are above the element symbol.
2. Electronegativities in the same row tend to increase from left to right.
3. Electronegativities in the same column tend to increase from bottom to top.
4. Where electronegativity values are not given, they have not been determined because of experimental difficulties (very short half-lives, very weak bonding energies).

value of 3.98 for fluorine,* the most electronegative element. Electronegativity values are dimensionless numbers, to be used primarily for predicting the relative polarities of covalent bonds and relative bond strengths. The electronegativity difference, ΔEN, between two atoms indicates what kind of bond they will form. The greater the difference in electronegativities of two bonded atoms, the more strongly the bonding electrons are attracted to the more electronegative atom, and the more polar is the resulting bond. The following rules of thumb usually apply, with very few exceptions.

Because electronegativity differences can vary continuously between zero and 3.98, bond character also can vary continuously between nonpolar covalent and ionic, as illustrated in Figure 2.3.

RULES OF THUMB (USE TABLE 2.1 FOR ELECTRONEGATIVITY VALUES)

1. If the electronegativity difference between two bonded atoms is zero, they will form a nonpolar covalent bond. Examples are O_2, H_2, and N_2.
2. If the electronegativity difference between two atoms is greater than zero and less than 1.7, they will generally form a polar covalent bond. Examples are HCl, NO, and CO.
3. If the electronegativity difference between two atoms is 1.7 or greater, they will generally form an ionic bond, typical of salt crystals. Examples are NaCl, HF, and KBr.
4. Relative electronegativities of the elements can be estimated by an element's position in the periodic table. Ignoring the inert gases (Column 18):
   a. The most electronegative element (F) is at the upper right corner of the periodic table.
   b. The least electronegative element (Fr) is at the lower left corner of the periodic table.
   c. In general, electronegativities increase diagonally up and to the right in the periodic table. Within a given period (or row), electronegativities tend to increase in going from left to right; within a given group (or column), electronegativities tend to increase in going from bottom to top.
   d. The farther apart two elements are in the periodic table, the more different are their electronegativities, and the more polar will be a bond between them.

---

* Linus Pauling first proposed the concept of electronegativity in 1932 to explain why covalent bonds between two different kinds of atoms in a molecule were stronger than predicted by existing theories. Because his approach only determined electronegativity differences between different atoms, it was necessary to choose an arbitrary reference atom to construct a scale. Hydrogen is chosen as the reference atom, and given a convenient EN value of 2.20, because it forms covalent bonds with very many different elements. Other definitions of electronegativity, with slightly different scales, have since been developed (see http://en.wikipedia.org/wiki/Electronegativity), but the Pauling scale remains the most commonly used.

5. The solubility in water of a pure compound is roughly proportional to the polarity of its molecules.
   a. Molecules with no, or small, polarity are generally insoluble or only slightly soluble in water.
   b. Molecules that are ionic or have large polarity are generally soluble in water.*

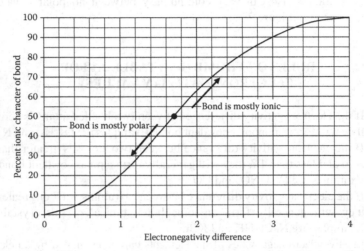

FIGURE 2.3   Bond character as a function of the electronegativity difference between two bonded atoms.

2.8.2   CHEMICAL BOND DIPOLE MOMENTS

For polar bonds, a property called the dipole moment has been defined that serves as a measure of the nonuniform charge separation. The dipole moment is a vector quantity that measures the magnitude and direction of the bond polarity; the more polar the bond, the larger is its dipole moment. In Figure 2.4, the dipole moment, μ, is multiplied by the distance, d, between the charges.

Polarity arrows, as shown in Figure 2.4, are vector quantities. They show both the magnitude and direction of the bond dipole moment. The length of the arrow indicates the magnitude of the dipole moment, and the direction of the arrow points from the positive region toward the negative region of the separated charges.

2.8.3   MOLECULAR GEOMETRY AND MOLECULAR POLARITY

Depending on its geometry, a molecule that contains polar bonds may or may not be a polar molecule. A molecule with polar bonds will not be a polar molecule if

---

* There are some exceptions to this principle, as when a crystalline substance, such as LiF, has a higher than common lattice energy, indicating that the atoms are held in place by stronger forces. LiF is ionic ($\Delta EN = 3.0$) but is only slightly soluble in water.

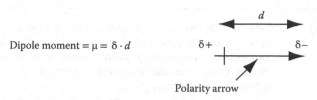

Dipole moment $= \mu = \delta \cdot d$

FIGURE 2.4 Bond dipole moment as indicated by a polarity arrow.

the polar bonds are oriented in a way that the polarity vectors cancel each other (see Section 2.8.4). A molecule with polar bonds will be polar if the polarity vectors of all its bonds add up to give a net polarity vector to the molecule, as in Section 2.8.5. The polarity of a molecule is the vector sum of all its bond polarity vectors. A polar molecule can be experimentally detected by observing whether an electric field exerts a force that makes the molecule align its charged regions with the direction of the field. Polar molecules will point their negative ends toward the positive source of the field, and their positive ends toward the negative source.

When a molecule containing polar bonds is itself polar, its polarity will always contribute to its strength of attraction to other molecules. When we know whether a molecule is polar or not, we can estimate its relative water solubility and several other properties. The presence of polar bonds in a molecule is necessary, but not sufficient, for the molecule also to be polar. The geometric symmetry of the molecule also is important.

RULES OF THUMB

To predict if a molecule is polar, we need to answer two questions:

1. Does the molecule contain polar bonds?
   a. If it does, then, depending on its symmetry, it might be polar.
   b. If it does not, it cannot be polar.
2. If the molecule contains polar bonds, do all the bond polarity vectors add to give a resultant molecular polarity?
   a. If the molecule is symmetrical in a way that the bond polarity vectors add to zero, then the molecule is not polar, even though it contains polar bonds.
   b. If the molecule is asymmetric and/or bond polarity vectors add to give a resultant polarity vector, the resultant vector indicates the molecular polarity.

2.8.4 EXAMPLES OF NONPOLAR MOLECULES WITH POLAR BONDS

Nonpolar molecules invariably have low water solubility because of weak attractive forces to polar water molecules. A molecule with no polar bonds can never be a polar molecule. Thus, all diatomic molecules having two identical atoms, such as H_2, O_2, N_2, and Cl_2, are nonpolar (and have low water solubility) because there is no electronegativity difference across the bond. Similarly, monatomic crystalline solids

O ◄──┼ C ┼──► O   Carbon dioxide: Oxygen is more electronegative ($EN(O_2) = 3.5$) than carbon
($EN(C) = 2.5$). The polarity of each bond is equal in magnitude but opposite
in direction, with an oxygen atom at the negative end of each dipole.
Because CO_2 is linear with carbon in the center, the polarity vectors cancel
each other and CO_2 is nonpolar.

Carbon tetrachloride: $EN(C) = 2.5$, $EN(Cl) = 3.2$, C ┼──► Cl. Although the
polarity of each bond has the same magnitude, the tetrahedral symmetry of
the bonds results in no net dipole moment, so that CCl_4 is nonpolar.

Hexachlorobenzene: The bond polarities are the same as in CCl_4 above. C_6Cl_6
is planar with hexagonal symmetry. All the bond polarities cancel one
another and the molecule is nonpolar.

p-Dichlorobenzene: This molecule also is planar. It has polar bonds of two
magnitudes, the H ┼──► C bond with the smaller polarity and the C ┼──► Cl
bond with the larger polarity. The H and Cl atoms are positioned so that all
polarity vectors cancel and the molecule is nonpolar. If the positions of one
Cl and one H were exchanged, the molecule would become asymmetrical
and polar. Check the electronegativity values in Table 2.1.

Boron tribromide: $EN(B) = 2.0$, $EN(Br) = 2.8$, B ┼──► Br. BBr_3 has trigonal
planar symmetry, with 120° between adjacent bonds. All the polarity vectors
cancel and the molecule is nonpolar.

like diamond or pure metals are not water-soluble.* On the other hand, a molecule
with polar bonds whose dipole moments add to zero because of molecular symmetry
is also not a polar molecule. Carbon dioxide, carbon tetrachloride, hexachlorobenzene, p-dichlorobenzene, and boron tribromide are all symmetrical and nonpolar,
although all contain polar bonds.

2.8.5 EXAMPLES OF POLAR MOLECULES

Polar molecules are generally more water-soluble than nonpolar molecules of similar
molecular weight. Any molecule with polar bonds whose dipole moments do not add
to zero is a polar molecule. Carbon monoxide, carbon trichloride, pentachlorobenzene, o-dichlorobenzene, boron dibromochloride, and water are all polar.

---

* The terms *insoluble* and *not soluble* should be interpreted as allowing a very small degree of water
solubility caused by the statistical likelihood that, at any temperature, there is always some small probability that a few molecules of a solid at a solid–water interface may acquire sufficient thermal energy to
break free from the solid and enter the liquid phase. "Thermal" solubility has different causes than does
"chemical" solubility. This explains why "insoluble" diatomic gases, which have very weak attractions to
other molecules in the gas phase, are nevertheless more soluble in water than are "insoluble" pure metals.

C ←→ O

Carbon monoxide: EN(O) = 3.5, EN(C) = 2.5. Oxygen is more electronegative than carbon. Every diatomic molecule with a polar bond must be a polar molecule.

Carbon trichloride: EN(C) = 2.5, EN(Cl) = 3.2, EN(H) = 2.2. Carbon trichloride has polar bonds of two magnitudes, the smaller polarity H ←→ C bond and the larger polarity C ←→ Cl bond. The asymmetry of the molecule results in a net dipole moment, so that $CHCl_3$ is polar.

Pentachlorobenzene: The bond polarities are the same as in $CHCl_3$ above. All bond polarities do not cancel one another and the molecule is polar.

o Dichlorobenzene: This molecule is planar and has two kinds of polar bonds: H ←→ C and C ←→ Cl. The bond polarity vectors do not cancel, making the molecule polar. Compare with *p*-dichlorobenzene in Section 2.8.4.

Boron dibromochloride: EN(B) = 2.0, EN(Br) = 2.8, EN(Cl) = 3.2. In BBr_2Cl, the polarity vectors of the polar bonds, B ←→ Br and B ←→ Cl, do not quite cancel and the molecule is slightly polar.

Resultant molecule
polarity vector

Bond polarity vectors

Water is a particularly important polar molecule. Its bond polarity vectors add to give the water molecule a high polarity (i.e., dipole moment). The dipole–dipole attractions between water molecules are greatly strengthened by hydrogen bonding (see discussion below), which contributes to many of the unique characteristics of water, such as relatively high boiling point and viscosity, low vapor pressure, and high heat capacity.

2.8.6 THE NATURE OF INTERMOLECULAR ATTRACTIONS

Ionic and dipole forces: All molecules are attracted to one another because of electrostatic forces. Polar molecules are attracted to one another because negative parts of one molecule are attracted to positive parts of other molecules, and vice versa. Attractions between polar molecules are called dipole–dipole forces. Similarly, positive ions are attracted to negative ions. Attractions between ions are called ion–ion forces. If ions and polar molecules are present together, as when sodium chloride is dissolved in water, there can be ion–dipole forces, where positive and negative ions (e.g., Na^+ and Cl^-) are attracted to the oppositely charged ends of polar molecules (e.g., H_2O).

Dispersion (London) forces: Nonpolar molecules also are attracted to one another, even though they do not have ionic charges or dipole moments. Evidence of attractions between nonpolar molecules is demonstrated by the fact that nonpolar gases, such as methane (CH_4), oxygen (O_2), nitrogen (N_2), ethane (CH_3CH_3), and carbon tetrachloride (CCl_4), condense to liquids and solids when the temperature is lowered sufficiently. Knowing that positive and negative charges attract one another makes it easy to understand the existence of attractive forces among polar molecules and ions. But how can the attractions among nonpolar molecules be explained?

In nonpolar molecules, the valence electrons are distributed about the nuclei so that, on average, the electron distribution is uniform and there is no net dipole moment. However, molecules are in constant motion, often colliding and approaching one another closely. When two molecules approach closely, their electron clouds interact by electrostatically repelling one another. These repulsive forces momentarily distort the electron distributions within molecules and create transitory dipole moments in molecules that would be nonpolar if isolated from neighbors. A transitory dipole moment in one molecule induces electron charge distortions and transitory dipole moments in all nearby molecules. At any instant in an assemblage of molecules, nearly every molecule will have a nonuniform charge distribution and an instantaneous dipole moment. An instant later, these dipole moments will have changed direction or disappeared so that, averaged over time, nonpolar molecules have no net dipole moment.

However, at any instant, the temporary dipoles produce temporary regions of net positive and negative charge that appear here and there over a molecule's surface. This causes every collection of molecules, both polar and nonpolar, to experience momentary electrostatic repulsive and attractive forces, in addition to any permanent polar and ionic attractions. On average, however, molecules tend to move away from the repulsive forces and toward the attractive forces. Because the magnitude of electrostatic forces decreases when molecules move farther apart and increases when molecules approach closer, the temporary attractive electrostatic forces always dominate over the temporary repulsive forces. Thus, the effect of these transitory dipole moments is to create another net attractive force in addition to the polar and ionic attractions. Even in a substance of only nonpolar molecules, there will be attractive forces that induce phase changes with changing temperature.

Attractions between nonpolar molecules are called dispersion forces or London forces (after Professor Fritz London, who gave a theoretical explanation for them in 1928). The magnitude of dispersion force attractions depends on how easily a molecule's electron cloud can be distorted in a collision, a property called *polarizability* (see Section 2.8.7). The electron cloud of large molecules having many electrons at longer distances from their nuclei is more easily distorted (high polarizability) than the electron cloud of small molecules, where the charge cloud is closer and held more tightly to the nuclei (low polarizability). Thus, the large nonpolar hydrocarbon molecules of wax have strong enough dispersion attractive forces to be solid at room temperature, while the smaller nonpolar molecules of gasoline are liquid and the still smaller nonpolar molecules of methane are gaseous.

Hydrogen bonding: An especially strong type of dipole–dipole attraction, called hydrogen bonding, occurs among molecules containing a hydrogen atom covalently bonded to a small, highly electronegative atom that contains at least one valence shell nonbonding electron pair. An examination of Table 2.1 shows that fluorine, oxygen, and nitrogen are the smallest (implied by their position at the top of their columns in the periodic table)* and the most electronegative elements that also contain nonbonding valence electron pairs. Although chlorine and sulfur have similarly high electronegativities and contain nonbonding valence electron pairs, they are too large to consistently form hydrogen bonds (H-bonds). Because hydrogen bonds are both strong and common, they influence many substances in important ways.

Hydrogen bonds are very strong (10–40 kJ/mole) compared to other dipole–dipole forces (from less than 1 to 5 kJ/mole). The very small size of the hydrogen atom makes hydrogen bonding uniquely strong. Hydrogen has only one electron. When hydrogen is covalently bonded to a small, highly electronegative atom, the shift of bonding electrons toward the more electronegative atom leaves the hydrogen nucleus nearly bare. With no inner core electrons to shield it, the partially positive hydrogen can approach very closely to a nonbonding electron pair on nearby small polar molecules. Since electrostatic forces increase quadratically with decreasing distance between the attracting charges, the very close approach results in stronger attractions than with other dipole–dipole forces.

Because of the strong intermolecular attractions, hydrogen bonds markedly influence the properties of the substances in which they occur. Compared with nonhydrogen-bonded compounds of similar size, hydrogen-bonded substances have relatively high boiling and melting points, low volatilities, high heats of vaporization, and high specific heats. Molecules that can H-bond with water are highly soluble in water; thus, all the substances in Figure 2.5 are water-soluble.

2.8.7 COMPARATIVE STRENGTHS OF NONPOLAR INTERMOLECULAR ATTRACTIONS

The strength of nonpolar intermolecular attractions arising from dispersion forces depends on the polarizability of the nonpolar molecules. While polarizability is a measurable property, it may be most useful to be able to estimate the relative magnitudes of the intermolecular attractions from knowledge of molecular structure and geometry.

Polarizability is a measure of how easily the electron distribution can be distorted by an electric field—that is, how easily a dipole moment can be induced in an atom or a molecule by the electric fields surrounding nearby atoms and molecules. Large atoms and molecules have more electrons and larger electron clouds than small ones. In large atoms and molecules, the outer shell electrons are farther from the nuclei and, consequently, are more loosely bound. Their electron distributions can more easily be distorted by external electric fields. In small atoms and molecules, the outer electrons are closer to the nuclei and are more tightly held. Electron charge distributions in small atoms and molecules are less easily distorted.

---

* The atomic sizes of atoms in the same column of the periodic table tend to decrease from the bottom of the column to the top. For example, the diameter of F < Cl < Br < I < At. Similarly, O and N are the smallest atoms in their respective columns.

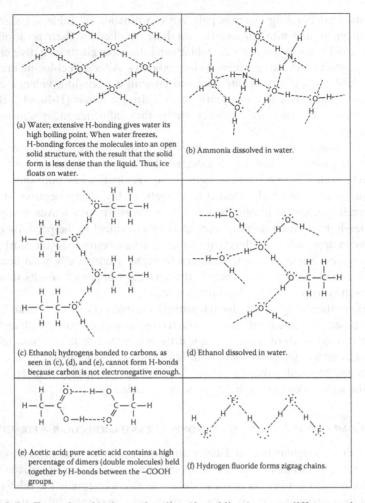

(a) Water; extensive H-bonding gives water its high boiling point. When water freezes, H-bonding forces the molecules into an open solid structure, with the result that the solid form is less dense than the liquid. Thus, ice floats on water.

(b) Ammonia dissolved in water.

(c) Ethanol; hydrogens bonded to carbons, as seen in (c), (d), and (e), cannot form H-bonds because carbon is not electronegative enough.

(d) Ethanol dissolved in water.

(e) Acetic acid; pure acetic acid contains a high percentage of dimers (double molecules) held together by H-bonds between the –COOH groups.

(f) Hydrogen fluoride forms zigzag chains.

FIGURE 2.5   Examples of hydrogen bonding (dotted lines) among different molecules.

Therefore, large atoms and molecules are generally more polarizable than small ones. Since atomic and molecular sizes are closely related to atomic and molecular weights, we can generalize that polarizability increases with increasing atomic and molecular weights. The greater the polarizability of atoms and molecules, the stronger the attractive intermolecular dispersion forces between them. Molecular shape also affects polarizability. Elongated molecules are more polarizable than compact molecules. Thus, a linear alkane is more polarizable than a branched alkane of the same molecular weight.

All atoms and molecules have some degree of polarizability. Therefore, all atoms and molecules experience attractive dispersion forces, whether or not they also have dipole moments, ionic charges, or can hydrogen bond. Small polar molecules are dominated by dipole–dipole forces since the contribution to attractions from dispersion forces is small. However, dispersion forces may dominate in very large molecules, including those that are polar.

RULES OF THUMB

1. The higher the atomic or molecular weights of molecules, polar or nonpolar, the stronger are the attractive dispersion forces between them.
2. For different molecules with the same molecular weight, those with a linear shape have stronger attractive dispersion forces than do those that are branched and more compact.
3. For polar and nonpolar molecules alike, the stronger the attractive forces, the higher the boiling point and freezing point, and the lower the volatility of the substance.

Example 1

Consider the halogen gases fluorine (F_2, MW = 38), chlorine (Cl_2, MW = 71), bromine (Br_2, MW = 160), and iodine (I_2, MW = 254). All are nonpolar, with progressively greater molecular weights and correspondingly stronger attractive dispersion forces as you go from F_2 to I_2. Accordingly, their boiling and melting points increase with their molecular weights. At room temperature, F_2 is a gas (bp = –188°C), Cl_2 is also a gas but with a higher boiling point (bp = –34°C), Br_2 is a liquid (bp = 58.8°C), and I_2 is a solid (mp = 184°C).

Example 2

Alkane hydrocarbons are compounds of singly bonded carbon and hydrogen atoms. Although C–H bonds are slightly polar (EN(C) = 2.5; EN(H) = 2.1), all alkanes are nonpolar because of their bond geometry. In the straight-chain alkanes (called normal alkanes), as the alkane carbon chain becomes longer, the molecular weights and, consequently, the attractive dispersion forces become greater. Consequently, melting points and boiling points become progressively higher. The physical properties of the normal alkanes in Table 2.2 reflect this trend.

Example 3

Normal-pentane (n-C_5H_{12}) and 2,2-dimethylpropane ($CH_3C(CH_3)_2CH_3$) are both nonpolar and have the same molecular weights (MW = 72). However, n-C_5H_{12} is a straight-chain alkane, whereas $CH_3C(CH_3)_2CH_3$ is branched. Thus, n-C_5H_{12} has stronger dispersion attractive forces than $CH_3C(CH_3)_2CH_3$ and a correspondingly higher boiling point (36°C for n-pentane compared to 9.5°C for 2,2-dimethylpropane).

2.9 SOLUBILITY AND INTERMOLECULAR ATTRACTIONS

In liquids and gases, the molecules are in constant, random, thermal motion, colliding and intermingling with one another. Even in solids, the molecules are in constant, although more limited, motion. If different kinds of molecules are present, random

TABLE 2.2

Some Properties of the First Twelve Straight-Chain Alkanes

Alkane	Formula	Molecular Weight	Melting Point[a] (°C)	Boiling Point (°C)
Methane	CH_4	16	−183	−161
Ethane	C_2H_6	30	−172	−88
Propane	C_3H_8	44	−188	−42
n-Butane	C_4H_{10}	58	−138	−0.5
n-Pentane	C_5H_{12}	72	−130	36
n-Hexane	C_6H_{14}	86	−95	69
n-Heptane	C_7H_{16}	100	−91	98
n-Octane	C_8H_{18}	114	−57	126
n-Nonane	C_9H_{20}	128	−51	151
n-Decane	$C_{10}H_{22}$	142	−29	174
n-Dodecane	$C_{12}H_{26}$	170	−10	216

[a] Deviations from the general trend in melting points occur because melting points for the smallest alkanes are more strongly influenced by differences in crystal structure and lattice energy of the solid.

movement tends to mix them uniformly. If there were no other considerations, random motion would cause all substances to eventually dissolve completely into one another. Gases and liquids would dissolve more quickly and solids more slowly.

However, intermolecular attractions must also be considered. Attractive forces between molecules tend to hold them together. Consider two different substances A and B, where A molecules are attracted strongly to other A molecules and B molecules are attracted strongly to other B molecules, but A and B molecules are attracted only weakly to one another. Then, A and B molecules tend to stay separated from each other. A molecules try to stay together and B molecules try to stay together, each excluding entry from the other. In this case, A and B are not very soluble (immiscible) in one another.

As an example of this situation, let A be a nonpolar, straight-chain liquid hydrocarbon, such as n-octane (C_8H_{18}), and let B be water (H_2O). Octane molecules are attracted to one another by strong dispersion forces, and water molecules are attracted strongly to one another by dipole–dipole forces and H-bonding. Dispersion attractions are weak between the small water molecules. Because the small water molecules have low polarizability, octane cannot induce a strong dispersion force attraction to water. Because octane is nonpolar, there are no dipole–dipole attractions to water. When water and octane are placed in the same container, they remain separate, forming two liquid layers, with the less dense octane floating on top of the water.

However, if there were strong attractive forces between A and B molecules, it would help them to mix. The solubility of one substance (the solute) in another (the solvent) depends mostly on intermolecular forces and, to a much lesser extent, on conditions such as temperature and pressure. Substances are more soluble in one another when intermolecular attractions between solute and solvent molecules are similar (or stronger) in magnitude to the intermolecular attractions between molecules of the pure substances. This principle is the origin of the rules of thumb that say, "like dissolves like" or "oil and water do not mix." "Like" molecules have similar polar or nonpolar properties and, consequently, similar intermolecular attractions. Oil and water do not mix because water molecules are attracted strongly to one another, and oil molecules are attracted strongly to one another; but water molecules and oil molecules are attracted only weakly to one another.

RULES OF THUMB

1. The more symmetrical the structure of a molecule containing polar bonds, the less polar and the less soluble it is in water.
2. Molecules with OH, NO, or NH groups can form hydrogen bonds to water molecules. They are the most water-soluble nonionic compounds, even if geometrical symmetry makes them nonpolar.
3. The next most water-soluble compounds contain O, N, and F atoms. All have high electronegativities and allow water molecules to H-bond with them.
4. Charged regions in ionic solids (like sodium chloride) are attracted to polar water molecules. This makes them more soluble.
5. Most compounds in oil and gasoline mixtures are nonpolar. They are attracted to water very weakly and have very low solubilities.
6. All molecules, including nonpolar molecules, are attracted to one another by dispersion forces. The larger the molecule, the stronger the dispersion force.
7. Nonpolar molecules, large or small (except for those that can H-bond, see 2 and 3 above), have low solubilities in water because the small-sized water molecules have weak dispersion forces, and nonpolar molecules have no dipole moments. Thus, there are neither dispersion nor polar attractions to encourage solubility.

Example 4

Alcohols of low molecular weight are very soluble in water because of hydrogen bonding. However, their water solubilities decrease as the number of carbons increase. The –OH group on alcohols is hydrophilic (attracted to water by hydrogen bonding), whereas the hydrocarbon part is hydrophobic (repelled* from water) because it is affected mainly by dispersion forces. If the hydrocarbon part of an alcohol is large enough, the hydrophobic behavior overcomes the hydrophilic behavior of the –OH group and the alcohol has low solubility. Solubilities for alcohols with increasingly larger hydrocarbon chains are given in Table 2.3.

---

* "Repelled" in this case actually means "expelled," as the strong attractions between water molecules "squeeze out" the hydrocarbon parts of the alcohol molecules.

TABLE 2.3
Solubilities and Boiling Points of Some Straight-Chain Alcohols

Name	Formula	Molecular Weight	Melting Point[a] (°C)	Boiling Point (°C)	Aqueous Solubility at 25°C (mol/L)
Methanol	CH_3OH	32	−98	65	∞(miscible)
Ethanol	C_2H_5OH	46	−130	78	∞(miscible)
1-Propanol	C_3H_7OH	60	−127	97	∞(miscible)
1-Butanol	C_4H_9OH	74	−90	117	0.95
1-Pentanol	$C_5H_{11}OH$	88	−79	138	0.25
1,5-Pentanediol[b]	$C_5H_{10}(OH)_2$	104	−18	239	∞(miscible)
1-Hexanol	$C_6H_{13}OH$	102	−47	158	0.059
1-Octanol	$C_8H_{17}OH$	130	−17	194	0.0085
1-Nonanol	$C_9H_{19}OH$	144	−6	214	0.00074
1-Decanol	$C_{10}H_{21}OH$	158	+6	233	0.00024
1-Dodecanol	$C_{12}H_{25}OH$	186	+24	259	0.000019

[a] Deviations from the general trend in melting points occur because melting points for the smallest alcohols are more strongly influenced by differences in crystal structure and lattice energy of the solid.

[b] The properties of 1,5-pentanediol deviate from the trends of the other alcohols because it is a diol and has two –OH groups available for hydrogen bonding, making it more soluble; see Example 5.

Example 5

For compounds of comparable molecular weight, the more hydrogen bonds to water a compound can form, the more water-soluble the compound; the more hydrogen bonds made to adjacent molecules in the pure compound, the higher the boiling and melting points of the pure compound. In Table 2.3, notice the effect of adding another –OH group to an alcohol. The double alcohol 1,5-pentanediol is more water-soluble and has a higher boiling point than single alcohols of comparable molecular weights, because it has two –OH groups capable of hydrogen bonding. This effect is general. Double alcohols (diols) are more water-soluble and have higher boiling and melting points than single alcohols of comparable molecular weight. Triple alcohols (triols) are still more water-soluble and have higher boiling and melting points.

2.9.1 SOLUBILITY AND OTHER PHASE CHANGES RESULT WHEN UNBALANCED INTERMOLECULAR FORCES SEEK AN EQUILIBRIUM

When intermolecular forces are in balance, the different phases of a chemical system (solid, liquid, and gas) and the distribution of different chemical species within a phase (e.g., H_2O, OH^-, H^+, H_3O^+, etc.) are in equilibrium and no changes in the relative amounts of existing phases or species distributions are observable. When intermolecular forces are unbalanced, the system seeks a balance by causing the atoms and molecules to change phase and/or the distribution of chemical species:

- Solids may dissolve, melt, or vaporize.
- Liquids may freeze, precipitate some components, dissociate some species, sorb to surfaces, or evaporate.
- Nonpolar liquids may separate from polar liquids, and vice versa.
- Gases may condense, dissolve, or sorb onto surfaces.
- Different species in any phase may chemically react.

These tendencies to reach equilibrium can be described quantitatively with various equations, all of which are some form of ratios of species concentrations, generally known as *equilibrium constants*. Depending on the particular reaction of interest, a particular equilibrium constant may be called a *dissociation constant, partition coefficient, distribution coefficient, solubility product, Henry's Law Constant*, or even *equilibrium constant*. Equilibrium constants are important tools for predicting the environmental behavior of metal species. Their uses for environmental calculations are developed in subsequent parts of this book, particularly Chapters 5 and 6 and Appendix B.

EXERCISES

1. a. Describe briefly what physical changes will condense a gas to a liquid and a liquid to a solid. Explain, in terms of forces between molecules, why these changes have the effect described.
   b. In light of your answer to 1a, what conclusions can you draw from the fact that water boils at a higher temperature than does ammonia?
2. Describe briefly the relation between temperature and molecular motion.
3. a. Balance the equation: $H_2 + N_2 \rightarrow NH_3$.
   b. When 3 g of hydrogen (H_2) react with 14 g of nitrogen (N_2), 17 g of a compound called ammonia (NH_3) are made: 3.0 g H_2 + 14 g $N_2 \rightarrow$ 17 g NH_3. How many grams of ammonia will be made if 6.0 g of hydrogen react with 14 g of nitrogen?
4. Balance the following equation: $C_5H_{10} + O_2 \rightarrow CO_2 + H_2O$.
5. Various compounds and some of their properties are tabulated as follows:

Compound	Boiling Point (°C)	Water Solubility (g/100 mL)	Dipole Moment (D)
H_2	−253	2×10^{-4}	0
HCl	−84.9	82	1.08
HBr	−67	221	0.82
CO_2	−78	0.15	0
CH_4	−164	2×10^{-3}	0
NH_3	−33.5	90	1.3
H_2O	100	∞	1.85
HF	19.5	∞	1.82
LiF	1676	0.27	6.33

a. Which compounds are nonpolar?
b. Which compounds are polar?
c. Make a rough plot of boiling point versus dipole moment. What conclusions may be inferred from the graph?
d. Discuss briefly the trends in water solubility. Why is solubility not related to dipole moment in the same manner as boiling point? Note that zero dipole moment indicates low solubility and higher dipole moments up to at least about D = 2 tend to indicate higher solubility. However, LiF, with a very high dipole moment of D = 6.3, has a low solubility, more characteristic of compounds with dipole moments around D = 0.
e. The ionic compound LiF appears to be unique. Try to suggest a reason for its low solubility.

6. The chemical structures of several compounds are shown below. Fill in the table with estimated properties based only on their chemical structures. Do not look up reference data for the answers.

Compound	Structure	Physical State at Room Temperature (Gas, Liquid, Solid)	Water Solubility (Very Low, Low, Moderate, High)	Solubility in Nonpolar Liquids (Very Low, Low, Moderate, High)
Ethane	H_3C-CH_3			
Benzene				
Citric acid				
Anthracene				
Glyphosate				

REFERENCE

OSWER. 1999. USEPA Office of Solid Waste and Emergency Response, Directive 9200. 4-17P. Use of Monitored Natural Attenuation at Superfund, RCRA Corrective Action, and Underground Storage Tank Sites, April 21, 1999. http://www.epa.gov/swerust1/directiv/d9200417.htm.

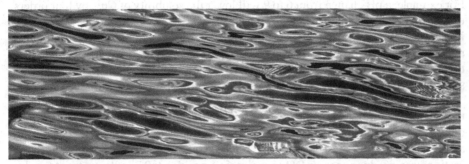

(Photo courtesy of Gary Witt)

Alkalinity in environmental waters is beneficial because it minimizes (buffers) both natural and human-induced pH changes, reduces the toxicity of many metals by forming complexes with them, and provides nutrient carbon for aquatic plants. Its pH-buffering action is important to the well-being of fish and other aquatic life.

Chapter 3, Section 3.4.4

3 Major Water Quality Parameters and Applications

3.1 INTERACTIONS AMONG WATER QUALITY PARAMETERS

This chapter deals with important water quality parameters that serve as controlling variables; controlling variables are those that strongly influence the behavior of many other constituents present in the water. The major controlling variables are pH, alkalinity and acidity, oxidation–reduction potential (ORP),* temperature, and total dissolved solids (TDS). This chapter also discusses several other important parameters that are strongly affected by changes in the controlling variables, such as carbonate, dissolved oxygen (DO), and biochemical oxygen demand (BOD).

Chemical constituents in natural water bodies react in an environment in far more complicated ways than if they were isolated in water and surrounded only by a large number of water molecules. Various impurities in water interact in ways that can affect their chemical behavior markedly. The water quality parameters defined above as controlling variables have an especially strong effect on water chemistry.

---

* Also called redox potential.

63

For example, a pH change from 6 to 9 will lower the solubility of Cu^{2+} by four orders of magnitude. The solubility of Cu^{2+} in water at pH 6 is about 40 mg/L, while at pH 9 it is about 4×10^{-3} mg/L, or 10,000 times smaller. If, for example, a water solution at pH 6 contained 40 mg/L of Cu^{2+} and its pH was raised to 9, all but 4×10^{-3} mg/L of the Cu^{2+} would precipitate as solid $Cu(OH)_2$, a behavior that is utilized in treatment systems for removing copper (as well as other metals) from industrial waste streams.

As another example, consider a shallow lake with algae and other aquatic vegetation growing in it. Its suspended and bottom sediments contain high concentrations of decaying organic matter. The lake is fed by surface and groundwaters containing a high concentration of sulfate. During daytime, photosynthesis of underwater vegetation* can produce enough DO to maintain a positive ORP in the water. At night, photosynthesis stops and biodegradation of suspended and lake-bottom organic sediments, along with underwater plant respiration, consumes nearly all the DO in the lake. This changes the water from oxidizing (aerobic) to reducing (anaerobic) conditions and causes the ORP to change from positive to negative values. At negative ORP (reducing conditions), aquatic life is no longer supported, and dissolved sulfate in the lake is reduced to sulfide, producing hydrogen sulfide (H_2S) gas, which smells like rotten eggs. If there are residences around the lake, there may be an odor problem at night that generally dissipates during the day. A remedy for these problems entails finding a way to maintain a positive ORP for longer periods.

RULE OF THUMB

Because they strongly influence other water quality parameters, the controlling variables listed here are usually included among the parameters that are routinely measured in water quality sampling programs.

- pH
- Temperature
- Alkalinity and acidity
- Total dissolved solids or conductivity
- ORP (redox potential)

3.2 PH

3.2.1 BACKGROUND

Acidity, basicity, and neutrality of water solutions are properties closely related to a quantity called pH. pH is a measure of the hydrogen ion (H^+) concentration in water solutions. Hydrogen ions arise in water from dissociation of the water molecules themselves or from dissociation of other molecules dissolved in water that contain ionizable hydrogen atoms. Pure water always contains a small number of

---

* See Section 3.3.2 and Chapter 6, Section 6.4.1.

water molecules that have dissociated because of thermal energy into hydrogen ions $(H^+)^*$ and hydroxyl ions (OH^-), as illustrated by Equation 3.1:

$$H_2O \rightleftarrows H^+ + OH^- \tag{3.1}$$

Equation 3.1 is a reversible reaction (indicated by double arrows); after they are formed, H^+ and OH^- ions can recombine to uncharged water molecules. At any given temperature, collisions between the more energetic water molecules initiate the forward dissociation reaction of Equation 3.1, and collisions between less energetic H^+ and OH^- ions initiate the reverse recombination reaction. The condition of equilibrium, when the rates of dissociation and recombination are equal, determines the equilibrium concentrations of H^+ and OH^- ions.

The equilibrium constant (see Chapter 2, Section 2.9.1) for Equation 3.1 is called the water dissociation constant, K_w, defined as the product of the equilibrium concentrations of H^+ and OH^- ions, expressed as moles per liter:

$$K_w = [H^+][OH^-] \tag{3.2}$$

Enclosing the symbol of a chemical species in square brackets is chemical symbolism that represents the species concentration, expressed as moles per liter. For example, if $[Na^+] = 10^{-3}$ mol/L, the concentration of Na^+ in the solution is 0.001 mol/L, and, since the atomic weight of Na is 23, this is equivalent to 0.023 g/L (or 23 ppm) (see Chapter 1, Section 1.3.4).

Because the degree of dissociation increases with temperature, K_w is temperature dependent. At 25°C

$$K_{w,25°C} = [H^+][OH^-] = 1.0 \times 10^{-14} (mol/L)^2 \tag{3.3}$$

whereas at 50°C

$$K_{w,50°C} = [H^+][OH^-] = 1.83 \times 10^{-13} (mol/L)^2 \tag{3.4}$$

If, for example, an acid is added to water at 25°C (often considered the standard default temperature when no other is specified), it dissociates, releasing additional H^+ into the water. The H^+ concentration increases, but the product $[H^+] \times [OH^-]$ in Equation 3.3 must remain equal to 1.0×10^{-14} $(mol/L)^2$. This means that if $[H^+]$ increases, $[OH^-]$ must decrease by a corresponding amount. Similarly, adding a base causes $[OH^-]$ to increase; therefore, $[H^+]$ must decrease correspondingly.

---

* Note that a hydrogen atom consists of a nucleus with just one proton, surrounded by an electron cloud containing just one electron. Its positive ion is formed by losing its only electron, leaving a bare nucleus of one proton. Without an electron cloud, H^+ has the dimensions of a proton, very much smaller than any other ionic species. Thus, the hydrogen ion is a unique chemical entity, lying between subatomic and atomic domains. It is structurally identical to the subatomic positive proton particle found in all atomic nuclei, but is not within a nucleus because it is formed in a chemical reaction and is available for further chemical interactions. Its very large ratio of charge-to-diameter makes it extremely reactive (see discussion of Equations 3.5a through 3.5d).

In pure water, or in water with no sources or sinks for H^+ or OH^- other than the dissociation reaction of water, Equation 3.1 predicts that there would always be equal numbers of H^+ and OH^- species. By definition, pure water is neither acidic nor basic; it is defined as neutral with respect to its acid–base properties. Thus, the neutral condition in water is when there are equal concentrations of H^+ and OH^- ions. When $[H^+]$ is greater than $[OH]$, the water solution is acidic; when $[H^+]$ is less than $[OH^-]$, it is basic.

In pure water at 25°C, the values of $[H^+]$ and $[OH^-]$ must be equal to each other; therefore, each must be equal to 1.0×10^{-7} mol/L, since by Equation 3.3,

$$K_{w,25°C} = [H^+][OH^-] = (1.0 \times 10^{-7} \text{mol/L})(1.0 \times 10^{-7} \text{mol/L})$$
$$= 1.0 \times 10^{-14} (\text{mol/L})^2$$

Since pure water defines the condition of acid–base neutrality, neutral water always has equal concentrations of H^+ and OH^-, or $[H^+] = [OH^-]$.

Whatever their separate values, the product of hydrogen ion and hydroxyl ion concentrations must be equal to 1×10^{-14} at 25°C, as in Equation 3.3. If, for example, an acid, such as HCl, were added to pure water in an amount that caused $[H^+]$ in the solution to increase from 10^{-7} mol/L to 10^{-5} mol/L, then it is necessary that $[OH^-]$ decreases from 10^{-7} mol/L to 10^{-9} mol/L, so that their product remains 10^{-14} $(\text{mol/L})^2$. In this example, because $[H^+] > [OH^-]$, the solution is acidic.

Many compounds dissociate in water to form ions or react with ions already present in water. Those that form hydrogen ions (H^+) or consume hydroxyl ions (OH^-) are called acids because, when dissolved in pure water, they produce the condition $[H^+] > [OH^-]$; examples are nitric acid (HNO_3), sulfuric acid (H_2SO_4), hydrochloric acid (HCl), and carbon dioxide (CO_2; see Section 3.3.2). Compounds that produce the condition $[H^+] < [OH^-]$ when added to pure water are called bases; examples are sodium hydroxide (NaOH), calcium carbonate ($CaCO_3$), and ammonia (NH_3).*

Although not very relevant to our further discussion, certain details about the properties of the unique cation H^+ should be understood. As pointed out in Footnote p. 65, H^+ is an unusually reactive species. Under common environmental conditions, it is far too reactive to ever exist alone and is always attached to other molecular species. In water solutions, some of the H^+ cations are always chemically bound to water molecules by strong ion-polar attraction, most commonly to one or two water molecules. Equation 3.5a is almost instantaneous; Equations 3.5b through 3.5d occur progressively more slowly. Thus, H^+ by itself does not exist in water long enough to be observed as an individual species.

---

* Sodium hydroxide is a base because it introduces additional OH^- when it dissociates in water by:

$$NaOH \xrightarrow{H_2O} Na^+ + OH^-$$

Calcium carbonate and ammonia are bases because they cause the condition $[H^+] < [OH^-]$ by combining with H^+ when they dissolve in water, thus reducing the $[H^+]/[OH^-]$ ratio. The relevant reactions are

$$CaCO_3 \xrightarrow{H_2O} Ca^{2+} + CO_3^{2-}, \text{ followed by: } CO_3^{2-} + H^+ + OH^- \rightarrow HCO_3^- (bicarbonate\ anion) + OH^-$$

and

$$NH_3 + H^+ + OH^- \rightarrow NH_4^+ + OH^-$$

$$H^+ + H_2O \rightarrow H_3O^+ \text{(very fast)} \tag{3.5a}$$

$$H_3O^+ + H_2O \rightarrow H_5O_2^+ \tag{3.5b}$$

$$H_5O_2^+ + H_2O \rightarrow H_7O_3^+, \text{ etc.} \tag{3.5c}$$

$$H^+ + nH_2O \rightarrow H_{2n+1}O_n^+, \text{(May also be written } (H_2O)_n H^+) \tag{3.5d}$$

It is the hydrogen ion–water molecule complex $(H_2O)_n H^+$ (also written $H_{2n+1}O_n^+$), called the hydrated proton, that gives acidic characteristics to water solutions. The species corresponding to $n = 1$, H_3O^+ is by far the most common, distantly followed by $H_5O_2^+$ (for $n = 2$); reactions yielding values for $n > 2$ rapidly become less and less likely.

Since the chemical structure of the hydrated proton is not precisely defined, a convention must be adopted to chemically describe the hydrated proton. Although H_3O^+ and $H_5O_2^+$ are the two most common varieties of hydrated protons, it does not make any difference to the meaning of a chemical equation whether the presence of an acid is indicated by H^+, H_3O^+, or $H_5O_2^+$. All hydrated protons consist of some number of water molecules attached to a hydrogen ion (H^+); H_3O^+ or $H_5O_2^+$ could equally well be written as $(H_2O)H^+$ or $(H_2O)_2 H^+$. The attached water molecules frequently exchange with other unattached water molecules and play no role in the chemical reactions involving the H^+ ion.*

For example, the addition of nitric acid (HNO_3) to water can be written in a manner that emphasizes the hydration process and identifies the hydrated protons:

$$HNO_3 + H_2O \rightarrow H_3O^+ + NO_3^- \tag{3.6a}$$

$$HNO_3 + 2H_2O \rightarrow H_5O_2^+ + NO_3^- \tag{3.6b}$$

and so on.

Equation 3.7 is an equivalent, but more generic, equation that emphasizes the dissociation process:

$$HNO_3 \xrightarrow{\text{H}_2\text{O}} H^+ + NO_3^- \tag{3.7}$$

All the equations are read, "HNO_3 added to water forms H^+ (or $H_3O^+/H_5O_2^+$) and NO_3^- ions." In other words, 1 mole of HNO_3 dissociates in water to form 1 mole of hydrated protons, no matter whether the reaction is with 1, 2, 3, or more water molecules, or whether we choose to write the hydrated proton as H^+, H_3O^+, or $H_5O_2^+$. For almost all environmental chemistry purposes, the dissociation process is more important than the degree of hydration, and this text will discuss acidic solutions in terms of their hydrogen ion (H^+) concentration.

---

* In water solutions, the water molecules attached to a hydrated proton (as H_3O^+, $H_5O_2^+$, etc.) readily exchange with other unattached water molecules and are indistinguishable from them. It makes no difference in quantitative calculations when the water molecules bound to hydrated protons are simply included with those of the water solvent.

3.2.2 DEFINING pH

The concentration of H^+ in water solutions commonly ranges from about 1 mol/L for extremely acidic water to about 10^{-14} mol/L (10^{-14} g/L or 10^{-11} ppm) for extremely basic water. Under special circumstances, the range can be even wider.* Since the molecular weight of H^+ is 1.0 g/mol,[†] this is equivalent to a concentration range of 1 to 10^{-14} g/L. Rather than working with such a wide numerical range (~15 orders of magnitude) for a measurement that is so common, chemists have developed a way to use logarithmic units for expressing $[H^+]$ as a positive decimal number whose value normally lies between 0 and 14. This number is called the pH, and is defined in Equation 3.8 as the negative of the base 10 logarithm of the hydrogen ion concentration, expressed in moles per liter:

$$pH = -\log_{10}[H^+] \qquad (3.8)$$

An alternate and useful form of Equation 3.8 is

$$[H^+] = 10^{-pH} \qquad (3.9)$$

Note that since logarithms are dimensionless, pH value has no dimensions or units. Frequently, pH is (unnecessarily) assigned units called standard units (SU), even though pH is unitless. This mainly serves to avoid blank spaces in a table that contains a column for units, or to satisfy a computer database that requires an entry in a units field. Also note that if $[H^+] = 10^{-7}$, then $pH = -\log_{10}(10^{-7}) = -(-7) = 7$. A higher concentration of H^+ such as $[H^+] = 10^{-5}$ yields a lower value for pH, that is, $pH = -\log_{10}(10^{-5}) = 5$. Thus, if pH is less than 7 in a solution at 25°C, the solution contains more H^+ than OH^- and is acidic; if pH is greater than 7, the solution is basic.

RULES OF THUMB

1. Because the pH scale is logarithmic, a change of one pH unit, say from pH 5 to pH 6, indicates a change in H^+ concentration by a factor of 10.
   a. At $pH = 5$, $H^+ = 10^{-5}$ g/L.
   b. At $pH = 6$, $H^+ = 10^{-6}$ g/L, one-tenth of its concentration at pH 5.

---

* It is possible, although not very common, to have H^+ concentrations greater than 1 ppm or less than 10^{-14} ppm. This, of course, would result in pH values less than 0 (negative) or greater than 14. For example, if $[H^+] = 2$ g/L, then $pH = -\log_{10}(2) = -(+0.301) \approx -0.3$. (see Nordstrom 2000).
[†] By the reasoning of footnote † on the previous page, the molecular weight of the hydrated proton may be taken as 1 g/mol, the same as for H^+.

2. The extent of water self-dissociation increases with temperature. For example,
   a. In neutral water at 25°C, $[H^+] = [OH^-] = 1 \times 10^{-7}$ mol/L, and pH 7.0 (by Equation 3.8).
   b. In neutral water at 50°C, $[H^+] = [OH^-] = 4.3 \times 10^{-7}$ mol/L, and pH 6.4 (by Equation 3.8).
   c. Although their pH values are different, both (a) and (b) represent neutral solutions because $[H^+] = [OH^-]$.
3. A solution is acidic, basic, or neutral depending on its relative concentrations of $[H^+]$ and $[OH]$:
   a. If $[H^+] > [OH^-]$, the water solution is acidic.
   b. If $[H^+] < [OH^-]$, the water solution is basic.
   c. If $[H^+] = [OH^-]$, the water solution is neutral.
4. In an acid–base reaction, H^+ ions are exchanged between chemical species. The species that donates the H^+ is the acid. The species that accepts the H^+ is the base.
5. The concentration of H^+ in water solutions is an indication of how many hydrogen ions are available, at the time of measurement, for exchange between chemical species.
6. The exchange of hydrogen ions changes the chemical properties of the species between which the exchange occurs; the donor, which was an acid, becomes a base because it now can accept a hydrogen ion, and the acceptor, which was a base, becomes an acid, because it now can donate a hydrogen ion.
7. By definition, $pH = -\log_{10}[H^+]$ and serves as a measure of $[H^+]$, the hydrogen ion concentration. The value of pH describes the acidic or basic quality of water solutions. At 25°C,
   a. When pH < 7, $[H^+] > [OH^-]$ and a water solution is acidic.
   b. When pH = 7, $[H^+] = [OH^-]$ and a water solution is neutral.
   c. When pH > 7, $[H^+] < [OH^-]$ and a water solution is basic.

Example 1

The $[H^+]$ of water in a stream is 3.5×10^{-6} mol/L.

a. What is the pH?
b. What is $[OH^-]$?
c. Express $[H^+]$ and $[OH^-]$ in ppb.

ANSWER

a. $pH = -\log_{10}[H^+] = -\log_{10}(3.5 \times 10^{-6}) = -(-5.46) = 5.46$.
b. Assuming a default temperature of 25°C, Equation 3.3 is valid:

$$K_{w,25°C} = [H^+][OH^-] = 1.0 \times 10^{-14} (mol/L)^2 = (3.5 \times 10^{-6}\ mol/L) \times [OH^-]$$

$$[OH^-] = \frac{1.0 \times 10^{-14} \text{ (mol/L)}^2}{3.5 \times 10^{-6} \text{ mol/L}} = 2.86 \times 10^{-9} \text{ mol/L}^*$$

c. The molecular weights of H^+ and OH^- are 1 g/mol and 17 g/mol, respectively.
$[H^+] = (3.5 \times 10^{-6} \text{ mol/L})(1 \text{ g/mol}) = 3.5 \times 10^{-6} \text{ g/L} = 3.5 \text{ μg/L} = 3.5 \text{ ppb}.$
$[OH^-] = (2.86 \times 10^{-9} \text{ mol/L})(17 \text{ g/mol}) = 4.86 \times 10^{-8} \text{ g/L} = 0.0486 \times 10^{-6} \text{ g/L}.$
$[OH^-] = 0.049 \text{ ppb}.$

This example illustrates that, in spite of their strong influence on water quality, H^+ and OH^- normally occur in very low concentrations.

Example 2

The pH of water in a stream is 6.65. What is the hydrogen ion concentration in ppb?

ANSWER

From Equation 3.9, $[H^+] = 10^{-pH} = 10^{-6.65} = 2.24 \times 10^{-7}$ mol/L.
$[H^+] = (2.24 \times 10^{-7} \text{ mol/L})(1 \text{ g/mol}) = 2.24 \times 10^{-7} \text{ g/L} = 0.224 \text{ μg/L} = 0.224 \text{ ppb}.$

3.2.3 IMPORTANCE OF pH

Measurement of pH is one of the most important and frequently used tests in water chemistry. pH is an important factor for determining the chemical and biological properties of water. It affects the chemical forms and environmental impacts of many chemical substances in water. For example, many metals dissolve as ions at lower pH values, precipitate as hydroxides and oxides at higher pH, and redissolve again at still higher pH (see Chapter 5). Figure 3.1 shows the pH scale and typical pH values of some common substances.

pH also influences the degree of ionization, volatility, and toxicity to aquatic life of certain dissolved substances, such as ammonia, hydrogen sulfide, and hydrogen cyanide. The ionized form of ammonia, which predominates at low pH, is the less toxic ammonium ion (NH_4^+). At higher pH, NH_4^+ loses H^+ to become the more toxic unionized form of ammonia, NH_3 (see Chapter 4, Section 4.2.3). Both hydrogen sulfide (H_2S) and hydrogen cyanide (HCN) behave conversely to ammonia; the less toxic ionized forms, S^{2-} and CN^-, are predominant at high pH, and the more toxic unionized forms, H_2S and HCN, are predominant at low pH (see Chapter 4,

---

* An alternate approach involves defining a quantity pOH $= -\log_{10}[OH^-]$, analogous to pH $= -\log_{10}[H^+]$. Then, analogous to Equation 3.9, we also have $[OH^-] = 10^{-pOH}$. Since $[H^+][OH^-] = 10^{-14} = 10^{-(pH+pOH)}$, it follows that pH + pOH = 14. Given that pH = 5.46, pOH = 14 − 5.46 = 8.54 and $[OH^-] = 10^{-8.54} = 2.88 \times 10^{-9}$ mol/L. The difference between this and the answer obtained above is due to rounding errors.

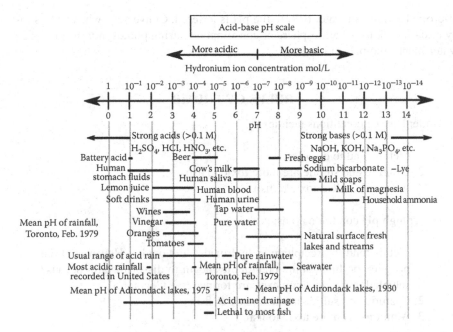

FIGURE 3.1 pH scale and typical pH values of some common substances. Note that for hydronium ion, the concentration in mol/L equals concentration in g/L (see footnote † on page 69 and Example 1).

Section 4.2.3 and Appendix A, Sections A.2.26 and A.2.38). The pH value is an indicator of the chemical state in which these compounds will be found and must be considered when establishing water quality standards.

3.2.4 MEASURING pH

The pH value of environmental waters is most commonly measured with electronic pH meters or by wetting special papers with the sample being measured. These pH-indicator papers are impregnated with dyes that change color to indicate a pH value. Battery-operated field pH meters are common.

The pH value of a water sample is altered by many processes that can occur after the sample is collected, such as loss or gain of dissolved CO_2, microbial reactions, or the oxidation of dissolved iron. Therefore, a pH measurement of surface or groundwater is valid only when made in the field or very shortly after sampling. A laboratory determination of pH made hours or days after sampling may be more than one full pH unit (a factor of 10 in H^+ concentration) different from the value at the time of sampling.

Loss or gain of dissolved CO_2 is one of the most common causes for pH changes in water bodies or water samples collected for analysis. When additional CO_2 dissolves into water, by diffusion from the atmosphere, respiration of aquatic vegetation, or microbial activity in water or soil, the pH is lowered. Conversely, when CO_2 is lost, by release back to the atmosphere or consumption during photosynthesis of algae or water plants, the pH is raised.

RULES OF THUMB

Under low pH conditions (acidic):

1. Metals tend to dissolve.
2. Cyanide and sulfide are more toxic to fish.
3. Ammonia is less toxic to fish.

Under high pH conditions (basic):

1. Metals tend to precipitate as hydroxides and oxides. However, if the pH gets too high, some precipitates begin to dissolve again because soluble hydroxide complexes are formed (see Chapter 5).
2. Cyanide and sulfide are less toxic to fish.
3. Ammonia is more toxic to fish.

3.2.5 WATER QUALITY CRITERIA AND STANDARDS FOR pH

The pH of pure water at 25°C is 7.0, but the pH of environmental waters is affected by dissolved CO_2 and exposure to minerals and other dissolved substances. Most unpolluted surface water in the United States has pH values between about 6 and 9, depending on the substances dissolved from minerals, soils, and other materials contacted in the water flow path; pH of unpolluted groundwaters tends to range between 5.5 and 8.5. Unpolluted groundwaters tend to have lower pH because biodegradation of organic matter in the subsurface, both aerobic and anaerobic, tends to increase the concentration of CO_2 in groundwater (see Chapter 6, Section 6.4.1 and Chapter 9, Section 9.3), which is slow to equilibrate with atmospheric CO_2. However, higher and lower values can occur in groundwater because of special conditions, such as sulfide oxidation, which lowers the pH, or contact with alkaline minerals, which raises the pH.

During daylight, photosynthesis in surface waters by aquatic organisms may consume dissolved CO_2 more rapidly than it is replenished from the atmosphere and released from underwater biodegradation, causing pH to rise because of a net loss of CO_2. At night, after photosynthesis has ceased, CO_2 dissolution from biodegradation and diffusion from the atmosphere continues, lowering the pH again. In this manner, photosynthesis can cause diurnal pH fluctuations, the magnitude of which depends on the alkalinity buffering capacity of the water. In poorly buffered lakes or rivers, the daytime pH may reach 9.0–12.0.

The permissible pH range for protecting aquatic life depends on factors such as DO, temperature, and concentrations of dissolved anions and cations. A pH range of 6.5–9.0, with no short-term change greater than 0.5 units beyond the normal seasonal maximum or minimum, is deemed protective of freshwater aquatic life and considered harmless to fish. In irrigation waters, the pH should not fall outside the range of 4.5–9.0 to protect plants. The most common Environmental Protection Agency (EPA) criteria for pH are

- Domestic water supplies: 5.0–9.0
- Freshwater aquatic life: 6.5–9.0

RULES OF THUMB

1. The pH of natural unpolluted river water is generally between 6 and 9.
2. The pH of natural unpolluted groundwater is generally between 5.5 and 8.5.
3. Clean rainwater has a pH of about 5.6 because of CO_2 dissolved from the atmosphere.
4. After reaching the surface of the earth, rainwater usually acquires alkalinity (see Section 3.4.3) from carbonate minerals while moving over and through the earth. This typically raises the pH above 5.7 and buffers the water against severe pH changes.
5. The pH of drinking water supplies should be between 5.0 and 9.0.
6. Fish can acclimate to ambient pH conditions. To protect aquatic life, pH should be between 6.5 and 9.0, and not vary more than 0.5 units beyond the normal seasonal maximum or minimum because of human activities like industrial discharges.

3.3 CARBON DIOXIDE, BICARBONATE, AND CARBONATE

3.3.1 BACKGROUND

Both the inorganic and organic forms of carbon have important and interrelated roles in environmental processes. The main reactive inorganic forms of environmental carbon are CO_2, bicarbonate (HCO_3^-), and carbonate (CO_3^{2-}). Organic carbon substances, such as cellulose and starch, are made by plants from CO_2 and water during photosynthesis. Carbon dioxide is present in the atmosphere and in soil pore space as a gas, and in surface waters and groundwaters as a dissolved gas. The carbon cycle is based on the mobility of CO_2, which is distributed readily through the environment as a gas in the atmosphere and dissolved in rainwater, surface water, and groundwater. Most of the earth's carbon, however, is relatively immobile, being contained in ocean sediments and on continents as minerals.

The atmosphere, with about 360 ppmv (parts per million by volume) of mobile CO_2, is the second smallest of the Earth's global carbon reservoirs, just above the living biomass reservoir, which is the smallest. Sedimentary carbonates and kerogen are the largest carbon reservoirs. Between are marine dissolved inorganic carbon (DIC; CO_2, H_2CO_3, HCO_3^-, and CO_3^{2-}), soils, and terrestrial sediments. Forests store about 86% of the Earth's above-ground carbon and about 76% of Earth's soil carbon. On land, solid forms of carbon are mobilized as particulates, mainly by weathering of carbonate minerals, biodegradation and burning of organic carbon, and burning of fossil fuels.

3.3.2 SOLUBILITY OF CO_2 IN WATER

Carbon dioxide plays a fundamental role in determining the pH of natural waters. Although CO_2 itself is not an acid, it reacts in water (reversibly) to make an acidic solution by forming carbonic acid (H_2CO_3), as shown in Equation 3.10. Carbonic acid can subsequently dissociate in two steps to release hydrogen ions (Equations 3.11 and 3.12):

$$CO_2 + H_2O \rightleftharpoons H_2CO_3 \tag{3.10}$$

$$H_2CO_3 \rightleftharpoons H^+ + HCO_3^- \tag{3.11}$$

$$HCO_3^- \rightleftharpoons H^+ + CO_3^{2-} \tag{3.12}$$

As a result, pure water exposed to air is not acid–base neutral with a pH near 7.0 because CO_2 dissolved from the atmosphere makes it acidic, with a pH around 5.6. The pH dependence of Equations 3.10 through 3.12 is shown in Figure 3.2 and Table 3.1.

Observations from Figure 3.2 and Table 3.1 are as follows:

- As pH increases, all equilibria in Equations 3.10 through 3.12 shift to the right.
- As pH decreases, all equilibria shift to the left.

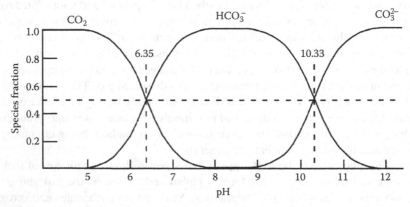

FIGURE 3.2 Distribution diagram showing how the mole fraction of carbonate species in water depends on pH.

TABLE 3.1

pH Dependence of Different Carbonate Species Mole Fractions

pH	Mole Fraction as CO_2	Mole Fraction as HCO_3^-	Mole Fraction as CO_3^{2-}
$\ll 6.35$	Essentially 1.00	Essentially 0	Essentially 0
6.35	0.50	0.50	Essentially 0
$\frac{1}{2}(6.35+10.33) = 8.34$	0.01	0.98	0.01
10.33	Essentially 0	0.50	0.50
$\gg 10.33$	Essentially 0	Essentially 0	Essentially 1.00

Note: Data corresponds to Figure 3.2.

- Above pH 10.3, carbonate ion (CO_3^{2-}) is the dominant species.
- Below pH 6.3, dissolved CO_2 is the dominant species.
- Between pH 6.3 and 10.3, a range common to most environmental waters, bicarbonate ion (HCO_3^-) is the dominant species.

The equilibria among only the carbon species (omitting the H^+ species) are

$$CO_2(gas, atm) \rightleftarrows CO_2(aq) \rightleftarrows H_2CO_3(aq) \rightleftarrows HCO_3^-(aq) \rightleftarrows CO_3^{2-}(aq) \quad (3.13)$$

These dissolved carbon species are sometimes referred to as DIC.

3.3.3 Soil CO_2

Processes such as biodegradation of organic matter and respiration of plants and organisms, which commonly occur in the root zone of the soil subsurface, consume O_2 and produce CO_2. Since air in the pore spaces of subsurface soils cannot readily equilibrate with the atmosphere, subsurface pore space air becomes depleted in O_2 and enriched in CO_2 relative to atmospheric air.

- Oxygen may decrease from about 21% (210,000 ppmv) in the atmosphere to between 15% and 0% (150,000–0 ppmv) in the soil.
- Carbon dioxide may increase from about 0.04% (~360 ppmv) in the atmosphere to between 0.1% and 10% (1,000–100,000 ppmv) in the soil.

When water moves through the subsurface, it equilibrates with soil gases and may become more acidic because of a higher soil concentration of dissolved CO_2. Acidic groundwater has an increased capacity for dissolving minerals. The higher the CO_2 concentration in soil air, the lower is the pH of groundwater. Acidic groundwater more easily dissolves soil minerals, particularly calcium carbonate ($CaCO_3$, limestone), which may increase the pH into the 6–8 range and, at the same time, may buffer the water so that it becomes more resistant to further pH changes. Limestone caves are formed when low-pH groundwater moves through limestone deposits and dissolves the limestone minerals.

RULES OF THUMB

1. Unpolluted rainwater is acidic, around pH 5.6, because of dissolved CO_2 from the atmosphere. Rainwater acidity appears to be increasing as atmospheric CO_2 levels increase, mainly due to increased burning of fossil fuels.

2. Acid rain has still lower pH values, reaching pH 2.0 or lower, because of dissolved sulfuric, nitric, and hydrochloric acids, which result mainly from industrial air emissions.

3. CO_2 transferred from the atmosphere causes dissolved carbonate species ($CO_{2,(aq)}$ [equivalent to H_2CO_3], HCO_3^-, and CO_3^{2-}) to be present in all natural water systems at the surface and in the near-subsurface of the earth. The relative proportions depend on pH.

4. At pH values between 7.0 and 10.0, bicarbonate is the dominant DIC species in water. Between pH 7.8 and 9.2, bicarbonate is close to 100%, while carbonate and dissolved CO_2 concentrations are essentially zero.

5. In subsurface soil pore space, oxygen is depleted and CO_2 is increased compared to the atmosphere. Oxygen typically decreases from 21% in atmospheric air to 15% or less in soil pore space air, and CO_2 typically increases from ~360 ppmv in atmospheric air to between 1,000 and 100,000 ppmv in soil pore space air. Thus, unpolluted groundwaters tend to be more acidic than unpolluted surface waters because of higher dissolved concentrations of CO_2.

3.4   ACIDITY AND ALKALINITY

3.4.1   BACKGROUND

The acidity of water is its base-neutralizing capacity (not its pH, which indicates whether the water exhibits acidic or basic properties). The alkalinity of water is its acid-neutralizing capacity. Both parameters are related to the buffering capacity of water (the ability to resist changes in pH when an acid or base is added). Water with high acidity can neutralize a large quantity of base without large changes in pH*; the same amount of base will cause a greater pH increase in water with less acidity. Similarly, water with high alkalinity can neutralize a large quantity of acid without large changes in pH, while the same amount of acid would cause a greater decrease of pH in water with less alkalinity.

---

* To *neutralize* in this statement means that, when a given quantity of a basic substance (like calcium carbonate or sodium hydroxide) is added to water, the pH change is less in water of higher acidity than in water of lower acidity. The degree of pH buffering (resistance to pH change) is proportional to the magnitude of the acidity of the water.

3.4.2 ACIDITY

Acidity of water is its quantitative capacity to react with a strong base to a designated pH. Acidity is a measure of an aggregate property of water and can be interpreted in terms of specific substances only when the chemical composition of the sample is known. Acids contribute to corrosiveness and influence chemical reaction rates, chemical speciation, and biological processes (*Standard Methods*, Eaton et al. 1995).

Acidity is determined by measuring how much standard base must be added to a water sample to raise the pH to a specified value. Acidity is the net effect of the presence of several dissolved species, including dissolved CO_2, dissolved multivalent metal ions, strong mineral acids such as sulfuric, nitric, and hydrochloric acids, and weak organic acids such as acetic acid; however, the analytical measurement does not specifically identify them. Dissolved CO_2 is the main source of acidity in unpolluted waters. Acidity from sources other than dissolved CO_2 is not commonly encountered in unpolluted natural waters and is often an indicator of pollution.

- Titrating any water sample with base to pH 8.3 measures phenolphthalein* acidity, also called total acidity. Total acidity measures the neutralizing effects of essentially all the acid species present, both strong and weak.
- Titrating an acidic water sample (having pH lower than 3.7) with base to pH 3.7 measures methyl orange* acidity. Methyl orange acidity primarily measures acidity due to the presence of strong mineral acids, such as sulfuric acid (H_2SO_4), hydrochloric acid (HCl), and nitric acid (HNO_3).

3.4.3 ALKALINITY

Alkalinity of water is its acid-neutralizing capacity. Alkalinity is a measure of an aggregate property of water and can be interpreted in terms of specific substances only when the chemical composition of the sample is known. Because the alkalinity of many surface waters is primarily a function of carbonate, bicarbonate, and hydroxide content, it (often) is taken as an indication of the concentration of these constituents. The measured values may also include contributions from borates, phosphates, silicates, or other bases if these are present (Eaton et al. 1995).

In natural waters that are not highly polluted, alkalinity is more commonly found than acidity, because the basic carbonate, bicarbonate, and hydroxide minerals are more commonly encountered than acidic minerals. Alkalinity is often a good indicator of the total DIC (usually bicarbonate between pH 6.3 and 10.3, and carbonate above pH 10.3; see Section 3.3.2). Alkalinity is determined by measuring how much standard acid must be added to a given amount of water to lower the

---

* Phenolphthalein and methyl orange are pH-indicator dyes that change color at pH 8.3 and 3.7, respectively.

pH to a specified value. Like acidity, alkalinity is the net effect of the presence of several constituents that are not identified by the titration method of analysis. Usually, however, the most important are bicarbonate (HCO_3^-), carbonate (CO_3^{2-}), hydroxyl (OH^-), and phosphate* (PO_4^{3-}) anions; alkalinity is often taken as an indicator for the concentration of these constituents. There are other, usually minor, contributors to alkalinity, such as ammonia, borates, silicates, and other basic substances.

- Titrating a basic water sample (having pH greater than 8.3) with acid down to pH 8.3 measures phenolphthalein alkalinity. Phenolphthalein alkalinity primarily measures the amount of carbonate ion (CO_3^{2-}) present, called the carbonate alkalinity.
- Titrating any water sample with acid to pH 3.7 measures methyl orange alkalinity or total alkalinity. Total alkalinity measures the neutralizing effects of essentially all the bases present.

Because alkalinity is more often measured than acidity and is a property caused by several constituents, some convention must be used for reporting it quantitatively as a concentration. The usual convention is to express alkalinity as the ppm or mg/L of calcium carbonate ($CaCO_3$) that, if it were the only source of alkalinity, would produce the same alkalinity as measured in the sample. This is done by calculating how much $CaCO_3$ would be neutralized by the same amount of acid as was used in titrating the water sample when measuring either phenolphthalein or methyl orange alkalinity. Calcium carbonate neutralizes acids and generates alkalinity by Equation 3.14, in which acidity (H^+) is consumed and water and CO_2 are produced.

$$CaCO_3 + 2\,H^+ \rightleftarrows Ca^{2+} + CO_2(g) + H_2O \tag{3.14}$$

Whether it is present or not, $CaCO_3$ is used as a proxy for all the base species that are actually present in the water. For example, if the measured total alkalinity of a water sample were reported as 1200 mg/L, it means that titration of a given volume of a solution containing 1200 mg/L of $CaCO_3$ to pH 3.7 would consume the same amount of standardized acid as would titration of an equal volume of the water sample. This result would be reported as follows: total alkalinity = 1200 mg/L as $CaCO_3$.

All unpolluted natural waters can be expected to have some degree of alkalinity. Since all natural waters contain dissolved CO_2, they all will have some alkalinity contributed by carbonate species,[†] unless acidic pollutants have consumed the alkalinity. It is not unusual for alkalinity to range from 0 to 750 mg/L as $CaCO_3$. For surface waters, alkalinity levels less than 30 mg/L are considered low, and levels greater than 250 mg/L are considered high. Average values for rivers are around 100–150 mg/L.

---

* The presence of significant phosphate concentrations is often an indication of nutrient pollution from fertilizers.
† Although adding dissolved CO_2 to water lowers its pH, see Figure 3.3 and accompanying text.

3.4.4 IMPORTANCE OF ALKALINITY

Alkalinity in environmental waters is beneficial because it minimizes (buffers) both natural and human-induced pH changes, reduces the toxicity of many metals by forming complexes with them, and provides nutrient carbon for aquatic plants. Its pH-buffering action is important for the well-being of fish and other aquatic life. The chemical species that cause alkalinity, such as carbonate, bicarbonate, hydroxyl, and phosphate ions, can form chemical complexes with many toxic heavy metal ions, often reducing their toxicity. Water with high alkalinity generally has a high concentration of DIC in the forms of HCO_3^- and CO_3^{2-}, which can be converted to biomass by photosynthesis. A minimum alkalinity of 20 mg/L as $CaCO_3$ is recommended for environmental waters and levels between 25 and 400 mg/L are generally beneficial for aquatic life. More productive waterfowl habitats correlate with increased alkalinity above 25 mg/L as $CaCO_3$. A range between 100 and 250 mg/L is considered normal for surface waters.

3.4.5 WATER QUALITY CRITERIA AND STANDARDS FOR ALKALINITY

Naturally occurring levels of alkalinity up to at least 400 mg/L as $CaCO_3$ are not considered a health hazard. An upper limit of 500 mg/L is considered safe for livestock. EPA guidelines recommend a minimum alkalinity of 20 mg/L as $CaCO_3$, also specifying that changes from natural background alkalinity be kept to a minimum, but not reduced by more than 25% by any discharge. For waters where the natural level is less than 20 mg/L, alkalinity should not be further reduced at all.

RULES OF THUMB

1. Alkalinity is the mg/L of $CaCO_3$ that would neutralize the same amount of standard acid as does the actual water sample.
2. Phenolphthalein alkalinity (titration with standard acid to pH 8.3) measures the amount of carbonate ion (CO_3^{2-}) present.
3. Total or methyl orange alkalinity (titration with standard acid to pH 3.7) measures the neutralizing effects of essentially all the bases present.
4. Surface waters and groundwaters draining carbonate mineral formations become more alkaline due to dissolved minerals.
5. High alkalinity can partially mitigate the toxic effects of heavy metals to aquatic life.
6. Alkalinity greater than 25 mg/L $CaCO_3$ is beneficial to water quality.
7. Surface waters without carbonate buffering may be more acidic than pH 5.7 (the value established by equilibration of dissolved CO_2 with atmospheric CO_2) because of water reactions with minerals and organic substances, biochemical reactions, and acid rain.

3.4.6 CALCULATING ALKALINITY

Although measured alkalinity is a resultant property of multiple water species, it is reported as the calcium carbonate equivalent. Example 3 is a sample calculation of how to do this for a case where alkalinity was not measured but the bicarbonate concentration and pH were available.

Example 3

A groundwater sample contains 300 mg/L of bicarbonate at pH 10.0. The carbonate concentration was not reported. Assuming all alkalinity in the sample is due to DIC, calculate the total carbonate alkalinity as $CaCO_3$. Total carbonate alkalinity is equal to the sum of the concentrations of bicarbonate and carbonate ions, expressed as the equivalent concentration of $CaCO_3$.

ANSWER

1. Use the measured values of bicarbonate and pH, with Figure 3.2, to determine the carbonate concentration. Figure 3.2 shows that at pH 10.0, total carbonate is roughly 73% bicarbonate ion and 27% carbonate ion. Although these percentages are related to mol/L rather than mg/L, the molecular weights of bicarbonate and carbonate ions differ by only about 1.7%; therefore, mole fractions of bicarbonate and carbonate are close to weight fractions based on mg/L, so that mg/L can be used in the percentage calculation without significant error.

$$\text{Total carbonate } (HCO_3^- + CO_3^{2-}) = \frac{300 \text{ mg/L}}{0.73} = 411 \text{ mg/L}$$

$$CO_3^{2-} = 0.27 \times 411 = 111 \text{ mg/L, or alternatively, } 411 - 300 = 111 \text{ mg/L.}$$

2. Determine the equivalent weights of HCO_3^-, CO_3^{2-}, and $CaCO_3$ (or use Table 1.1, Chapter 1, Section 1.3.6).

$$\text{Eq. wt.} = \frac{\text{Molecular or atomic weight}}{\text{Magnitude of ionic charge or oxidation number}}$$

$$\text{Eq. wt. of } HCO_3^- = \frac{61.0}{1} = 61.0 \text{ g/eq}$$

$$\text{Eq. wt. of } CO_3^{2-} = \frac{60.0}{2} = 30.0 \text{ g/eq}$$

$$\text{Eq. wt. of } CaCO_3 = \frac{100.1}{2} = 50.0 \text{ g/eq}$$

3. Determine the multiplying factors to obtain the equivalent concentrations of $CaCO_3$.

$$\text{Multiplying factor for } HCO_3^- \text{ as } CaCO_3 = \frac{\text{Eq. wt. of } CaCO_3}{\text{Eq. wt. of } HCO_3^-} = \frac{50.0 \text{ g/eq}}{61.0 \text{ g/eq}} = 0.820$$

$$\text{Multiplying factor for } CO_3^{2-} \text{ as } CaCO_3 = \frac{\text{Eq. wt. of } CaCO_3}{\text{Eq. wt. of } CO_3^{2-}} = \frac{50.0 \text{ g/eq}}{30.0 \text{ g/eq}} = 1.667$$

4. Use the multiplying factors and concentrations to calculate the carbonate alkalinity, expressed as mg/L of $CaCO_3$.

$$\text{Carbonate alk. (as } CaCO_3)$$
$$= 0.820 \, (HCO_3^-, \text{mg/L}) + 1.667 \, (CO_3^{2-}, \text{mg/L}) \qquad (3.15)$$

$$\text{Carbonate alk.} = 0.820(300 \text{ mg/L}) + 1.667(111 \text{ mg/L}) = 431 \text{ mg/L } CaCO_3$$

Equation 3.15 and Figure 3.2 may be used to calculate carbonate alkalinity whenever pH and the total carbonate concentration are known.

3.4.7 CALCULATING CHANGES IN ALKALINITY, CARBONATE, AND pH

A detailed calculation of how pH, total carbonate, and total alkalinity are related to one another is moderately complicated because of the three simultaneous carbonate equilibria reactions (Equations 3.10 through 3.12). However, the relations can be conveniently plotted on a total alkalinity/pH/total carbonate graph, also called a Deffeyes diagram, or capacity diagram (see Figures 3.3 and 3.4). Details of the construction of these diagrams may be found in Stumm and Morgan (1996) and Deffeyes (1965).

In a total alkalinity/pH/total carbonate graph, shown in Figure 3.3, a vertical line represents adding strong base or acid without changing the total carbonate. The added base or acid changes the pH and, therefore, shifts the carbonate equilibrium, but does not add or remove any carbonate. The amount of strong base or acid in meq/L equals the vertical distance on the graph. You can see from Figure 3.3 that if the total carbonate is small, the system is poorly buffered; so, a little base or acid causes large changes in pH. As total carbonate increases, the system buffering capacity also increases, and it takes more base or acid for the same pH change.

In Figure 3.3, a horizontal line represents changing total carbonate, generally by adding or losing CO_2, without changing alkalinity. For alkalinity to remain constant when total carbonate changes, the pH must also change. Changes caused by adding bicarbonate are indicated with a line of increasing 1:1 slope; changes from simple dilution are indicated by a line of decreasing 1:1 slope; changes from adding carbonate are indicated with a line of increasing 2:1 slope.

Figure 3.4 is a total acidity/pH/total carbonate graph. Note that changes in composition, caused by adding or removing CO_2, bicarbonate, and carbonate, are indicated by different movement vectors in the acidity and alkalinity graphs. The examples below illustrate the uses of the diagrams.

Example 4

Designers of a wastewater treatment facility for a meat-rendering plant planned to control ammonia concentrations in the wastewater by raising its pH to 11, in order to convert about 90% of the ammonia to the volatile form. The wastewater would then be passed through an air-stripping tower to transfer volatile ammonia to the atmosphere. Average initial conditions for alkalinity and pH in the wastewater were expected to be about 0.5 meq/L and 6.0, respectively.

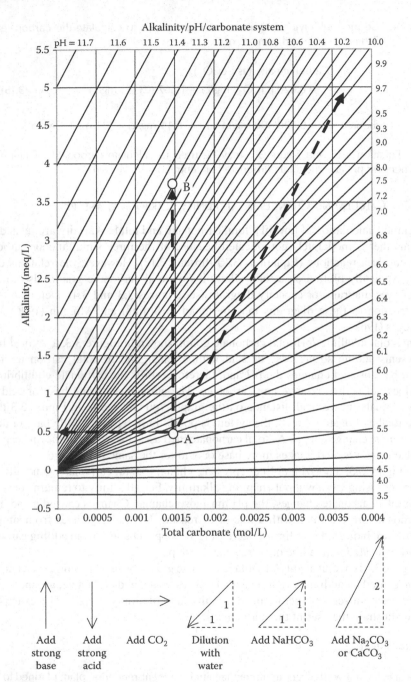

FIGURE 3.3 Total alkalinity/pH/total carbonate diagram (Deffeyes diagram). Relationships among total alkalinity, pH, and total carbonate are shown here. If any two of these quantities are known, the third may be determined from the plot. The composition changes indicated refer to Example 4.

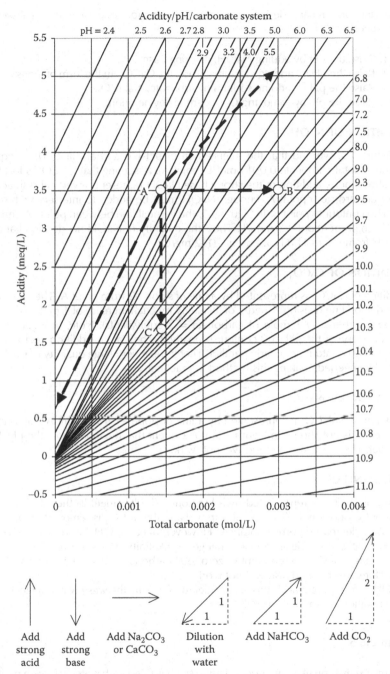

FIGURE 3.4 Total acidity/pH/total carbonate diagram (Deffeyes diagram). Relationships among total acidity, pH, and total carbonate are shown here. If any two of these quantities are known, the third may be determined from the plot. The composition changes indicated refer to Example 5.

In the preliminary design plan, four options for increasing the pH were considered:

1. Raise the pH by adding NaOH, a strong base.
2. Raise the pH by adding calcium carbonate, $CaCO_3$, in the form of limestone.
3. Raise the pH by adding sodium bicarbonate, $NaHCO_3$.
4. Raise the pH by removing CO_2, perhaps by aeration.

ADDITION OF NaOH

In Figure 3.3, we find that the intersection of pH 6.0 and alkalinity 0.5 meq/L occurs at total carbonate 0.0015 mol/L (point A). Assuming that no CO_2 is lost to the atmosphere, addition of the strong base NaOH represents a vertical displacement upward from point A. Enough NaOH must be added to intersect with the pH 11.0 contour at point B. In Figure 3.3, the vertical line between points A and B has a length of about 3.3 meq/L. Thus, the quantity of NaOH needed to change the pH from 6.0 to 11.0 is 3.3 meq/L (132 mg/L).

ADDITION OF CaCO₃

Addition of $CaCO_3$ is represented by a line of slope +2 from point A. The carbonate addition line rises by 2 meq/L of alkalinity for each increase of 1 mol/L of total carbonate (because 1 mole of carbonate = 2 equivalents). Note, in Figure 3.3, that the slope of the pH 11.0 contour is nearly 2. The $CaCO_3$ addition vector and the pH 11.0 contour are nearly parallel. Therefore, a very large quantity of $CaCO_3$ would be needed, making this method impractical.

ADDITION OF NaHCO₃

Addition of $NaHCO_3$ is represented by a line of slope +1 from point A. Although this vector is not shown in Figure 3.3, it is evident that it cannot cross the pH 11 contour. Therefore, this method will not work.

REMOVING CO₂

Removal of CO_2 is represented by a horizontal displacement to the left. Loss or gain of CO_2 does not affect the alkalinity. Note that if CO_2 is removed, total carbonate is decreased correspondingly. However, pH and $[OH^-]$ also increase correspondingly, resulting in no net change in alkalinity. We see from Figure 3.3 that removal of CO_2 to the point of zero total carbonate cannot achieve pH 11.0. Therefore, this method also will not work.

Of the four potential methods considered for raising the wastewater pH to 11.0, only addition of NaOH is useful.

Example 5

A large excavation at an abandoned mine site has filled with water. Because pyrite minerals were exposed in the pit, the water is acidic, with pH 3.2. The acidity was measured at 3.5 meq/L. Because the pit overflows into a stream during heavy rains, the managers of the site must meet the conditions of a discharge permit, which include a requirement that pH of the overflow water is between 6.0 and

9.0. The site manager decides to treat the water to pH 7.0 to provide a safety margin. Use Figure 3.4 to evaluate the same options for raising the pH as considered in Example 4.

ADDITION OF NaOH

In Figure 3.4, we find that the intersection of pH 3.2 and acidity 3.5 meq/L occurs at about total carbonate 0.0014 mol/L, point A. Assuming that no CO_2 is lost to the atmosphere, addition of the strong base NaOH represents a vertical displacement downward from point A to point C. Enough NaOH must be added to intersect with the pH 7.0 contour. The vertical line between points A and C has a length of about 1.8 meq/L. Thus, the quantity of NaOH needed to change the pH from 3.0 to 7.0 is 1.8 meq/L (72 mg/L).

ADDITION OF CaCO₃

In the acidity diagram, addition of $CaCO_3$ is represented by a horizontal line to the right. In Figure 3.4, the $CaCO_3$ addition line intersects the pH 7.0 contour at point B, where total carbonate is 0.0030 mol/L. Therefore, the quantity of $CaCO_3$ required to reach pH 7.0 is $0.0030 - 0.0014 = 0.0016$ mol/L (160 mg/L).

ADDITION OF NaHCO₃

The addition of $NaHCO_3$ is represented by a line of slope +1 (the vector upward to the right from point A in Figure 3.4). Note that the slope of the pH 7.0 contour is just a little greater than +1. The $NaHCO_3$ addition vector and the pH 7.0 contour are nearly parallel. Therefore, a very large quantity of $NaHCO_3$ would be needed, making this method impractical.

REMOVING CO₂

In the acidity diagram, the removal of CO_2 is represented by a line downward to the left with slope +2. We see from Figure 3.4 that removal of CO_2 to the point of zero total carbonate cannot achieve pH 7.0. Therefore, this method will not work.

Of the four potential methods considered for raising the wastewater pH to 7.0, addition of either NaOH or $CaCO_3$ will work. The choice will be based on other considerations, such as costs or availability.

Example 6

Figures 3.3 and 3.4 can also be used for finding the properties of mixed waters. As a simple example, consider an industry that requires its process water to be at pH 7.0. It uses two different water sources for filling a storage tank for its process water. Let water A be the remaining water in the half-full tank that needs replenishing. Water A has pH = 7.0 and alkalinity = 2.5 meq/L. Water B, used to fill the tank, has pH = 8.0 and alkalinity = 3.5 meq/L.

  a. What is the pH of the mixture after water B fills the tank? Assume no CO_2 is lost to the atmosphere.
  b. What would be a reasonable way to bring the mixture back to pH 7.0?

ANSWER

a. From Figure 3.3, for water A, at the intersection of pH 7.0 and alkalinity 2.5 meq/L, total carbonate is 0.0030 mol/L, and for water B, at the intersection of pH 8.0 and alkalinity 3.5, total carbonate is 0.0036. Because the volumes of waters A and B are equal and it is assumed that no CO_2 is lost, the mixture will have alkalinity and total carbonate values that are the average of waters A and B; that is, for the mixture, alkalinity is 3.0 meq/L and total carbonate is 0.0033 meq/L. From Figure 3.3, the mixture's pH is very close to 7.3.

b. Adding 0.2 meq/L of strong acid will lower the pH to 7.0.

Note: If unequal volumes of water are mixed, as is usually the case, the only change needed in the above approach is to use weighted averages of alkalinity and total carbon for the mixture.

3.5   OXIDATION–REDUCTION POTENTIAL (ORP)

3.5.1   BACKGROUND

Oxidizing or reducing conditions in water are indicated by a measured quantity known as the oxidation–reduction potential (ORP; also known as redox potential). The ORP measures the tendency for electrons to be exchanged between chemical species. This may be viewed as analogous to pH, which measures the tendency for protons (H^+ ions) to be exchanged between chemical species. When H^+ ions are exchanged, the acid or base properties of the species are changed, resulting in acid–base reactions. When electrons are exchanged, the oxidation states of the species and their chemical properties are changed, resulting in ORP reactions.

The electron donor is said to be oxidized; the electron acceptor is said to be reduced. In other words, from a chemical's point of view, to lose an electron is to be oxidized and to gain an electron is to be reduced. For every electron donor, there must be an electron acceptor; there cannot be any free electron accumulation in electrically neutral matter. This implies that whenever one substance is oxidized, another must be reduced. Because receiving an electron from the donor causes the acceptor to be reduced, the donor is called a reducing agent. Conversely, the acceptor, by providing a means for the donor to lose an electron, is called an oxidizing agent.

The tendency for electron exchange to occur is increased when the system contains at least two chemical species with very different strengths of attraction for their outer shell electrons, that is, with very different electronegativities. In an ORP reaction, the species with the lesser electronegativity (the donor) loses outer-shell electrons that become part of the outer-shell electron structure of the species with the stronger electronegativity (the acceptor).

The ORP is measured as a positive or negative voltage, generally in millivolts (mV). Each aqueous solution at equilibrium has its own intrinsic electrical ORP, which varies under different conditions of chemical composition, concentrations of species, and temperature. When the ORP of a solution is positive, it means that the solution predominantly contains species with high electron affinities; a solution with a negative ORP predominantly contains species with low electron affinities. The magnitude of a solution's ORP indicates how the solution will react to the introduction of a new species. A solution with an ORP more positive than that of a newly added species will gain electrons from the new species, oxidizing the new species while becoming itself reduced; a solution with an ORP more negative than that of the new species will lose electrons to the new species, reducing the new species while becoming itself oxidized.

ORP reactions are important because the reactivity of atoms and molecules is controlled mostly by their outer-shell electron structure. The gain or loss of electrons in ORP reactions changes this structure, thereby affecting reactivity, often resulting in significant changes in properties like solubility and toxicity.

The most common oxidizing agent (electron acceptor) in the natural environment is molecular oxygen (O_2). When the measured ORP of a natural water body is a positive value, indicating oxidizing (or aerobic) conditions, there is sufficient DO present to allow many dissolved metals and organic species to be oxidized, thereby undergoing chemical changes that influence their toxicity and solubility. Stronger oxidizing agents (although less environmentally common) than oxygen, such as ozone (O_3), chlorine (Cl_2), or permanganate ion (MnO_4^-), are those that take electrons from many substances even more readily than molecular oxygen. As noted above, the electron donor is oxidized, and, by accepting electrons, the oxidizing agents are themselves reduced. Strong oxidizing agents are those that strongly attract electrons to their outer shell; that is, they are easily reduced. Similarly, strong reducing agents are those that more weakly hold outer-shell electrons; that is, they lose electrons easily to become oxidized.

For example, chlorine is a strong oxidizing agent widely used to treat water and sewage. Chlorine oxidizes many pollutants to less objectionable forms more readily than does oxygen. When chlorine reacts with hydrogen sulfide (H_2S)—a common sewage pollutant that smells like rotten eggs—it oxidizes the sulfur in the H_2S to insoluble elemental sulfur (S_8), which is easily removed by settling or filtering. The reaction is

$$8Cl_2(g) + 8H_2S(aq) \rightarrow S_8(s) + 16HCl(aq) \qquad (3.16)$$

Since the sulfur in H_2S is more electronegative than the hydrogens, the sulfur atom in H_2S effectively carries two extra electrons that it has attracted from the hydrogens, giving it an oxidation number of -2 and a redox structure of S^{2-}. Because chlorine is more electronegative than sulfur, it can attract the extra two electrons from S^{2-}, leaving the sulfur with oxidation number zero,* at the same time forming two chloride ions in oxidation state -1, as in Equation 3.17, which shows only the species from Equation 3.16 that undergo a change in oxidation number.

$$Cl_2 + S^{2-} \rightarrow 2Cl^- + S^0 \qquad (3.17)$$

---

* Every atom in its elemental state has the same number of electrons around its nucleus as there are protons in its nucleus. Hence, the net charge on an elemental atom, and its oxidation number (also called oxidation state), equals zero. When an elemental atom loses an electron (is oxidized) its net charge and its oxidation number becomes $+1$; when an elemental atom gains an electron (is reduced) its net charge and its oxidation number becomes -1. These changes in its outer-shell electron structure always change its chemical behavior.

In Equation 3.17, the sulfur in H_2S is oxidized because it donates two electrons that are accepted by the chlorine atoms in Cl_2. Chlorine is reduced because it accepts the electrons. Since chlorine is the agent that causes the oxidation of H_2S, chlorine is called an oxidizing agent. Because H_2S is the agent that causes the reduction of chlorine, H_2S is called a reducing agent.

The class of redox reactions is very large. It includes all combustion processes, for example, the burning of gasoline or wood; most microbial reactions, such as those that occur in biodegradation; and all electrochemical reactions, such as those that occur in batteries and metal corrosion. Wastewater treatment relies heavily on redox reactions, many of them based on biological processes. Remediating contaminated subsurface groundwater by using permeable barrier walls containing finely divided iron is based on the reducing properties of iron. Such treatment walls are placed in the path of groundwater-contaminant plumes. The iron donates electrons to pollutants as they pass through the permeable barrier. Thus, the iron is oxidized and the pollutant reduced. This often causes the pollutant to decompose into less harmful or inert fragments.

RULES OF THUMB

1. Oxygen gas (O_2) is always an oxidizing agent in its reactions with metals and most nonmetals. If a compound has combined with O_2, it has been oxidized and the O_2 has been reduced. By accepting electrons, O_2 is either changed to the oxide ion (O^{2-}) or combined in compounds such as CO_2 or H_2O.
2. Like O_2, halogen gases (F_2, Cl_2, Br_2, and I_2) are always oxidizing agents in reactions with metals and most nonmetals. They accept electrons to become halide ions (F^-, Cl^-, Br^-, and I^-) or are combined in compounds such as HCl or $CHBrCl_2$.
3. If an elemental metal (Fe, Al, Zn, etc.) reacts with a compound, the metal acts as a reducing agent by donating electrons, usually forming a soluble positive ion such as Fe^{2+}, Al^{3+}, or Zn^{2+}.

3.5.2 CASE STUDY: TREATING FOUL ODORS FROM AN URBAN MARSH

A shallow natural marsh, with no surface inflow or outflow, in a Western U.S. town was valued for its scenic beauty and its use as a wetland complex providing sanctuary for birds and other wetland wildlife. Before the region was urbanized, groundwater inflow to the marsh was modest and climate fluctuations produced periods of high and low water levels. It was estimated that the marsh would have contained ponded water only for a few months each year; in dry years, the marsh would remain dried up. However, urbanization created new inflows to the marsh from surface runoff and seepage from an upgradient irrigation ditch. The marsh became a wetland that was always more or less full of water. With no surface water flow through the marsh, it became hypertrophic,* with accumulating bottom sediments, anaerobic water conditions, excessive algal growth, and high levels of BOD (see Section 3.8). Investigations showed that the hypertrophic conditions were caused by excessive nutrient inputs due to stormwater runoff from lawns and gardens surrounding the marsh, inadequate water column oxygenation, and continually high water levels that maintained anaerobic conditions in the bottom sediments. These bottom sediments were black in color, indicating sulfides of iron and other metals (see Chapter 4, Section 4.5.3 and Chapter 5, Section 5.4.3). Residents around the marsh complained to the city health department about persistent foul odors like rotten eggs emanating from the marsh, particularly on summer mornings.

Analysis of air samples confirmed the presence of hydrogen sulfide (H_2S) gas (see Chapter 4, Sections 4.5.3 through 4.5.5) at atmospheric concentrations that slightly exceeded 1 ppmv in the early morning, but subsided during the day. Analysis of the marsh water showed concentrations of sulfate (SO_4^-) > 5000 mg/L, pH = 6–10, and BOD_5 = 40–50 mg/L. Most importantly, mid-day measurements showed DO = 9–10 mg/L and ORP = +110 to +175 mV, but after dark, DO < 1 mg/L and ORP < −200 mV. Where bottom sediments around the marsh shore had been exposed to the atmosphere due to a receding waterline, the dominant sediment color had changed from black to light gray, indicating that sulfide metal salts had been oxidized to sulfate metal salts. These oxidized sediments were free of foul odors.

These observations led to the following explanation of why the foul odors occurred as they did and suggested a method for treating the marsh to reduce the odors to an acceptable level.

- The odor problems were a symptom of the hypertrophic conditions that prevailed at the marsh, combined with high levels of dissolved sulfate.
- Under anaerobic conditions created by biodegradation of organic debris from dead plants and algae, sulfate was reduced by bacterial activity to toxic hydrogen sulfide gas having the disagreeable odor of rotten eggs (see Chapter 4, Section 4.5.1).
- The strongest odors came from the layer of continuously anaerobic bottom sediments, which were about 1 to 1.5 feet thick. The water column also contributed to odor generation at night when the water became anaerobic due to algae and plant respiration and BOD.
- When exposed to the atmosphere, the sulfide in the bottom sediments was rapidly oxidized to odor-free sulfate.

The overall goal for treating the odor problem at the marsh was to reduce the magnitude and extent of anaerobic conditions causing excessive H_2S gas generation. Multiple approaches were implemented, involving both short- and long-term management:

- Water-level management was accomplished by constructing an outlet structure and pipeline discharging to a nearby stream. The outlet structure was designed to allow almost complete drainage of the marsh. This allowed

---

* Hypertrophic status means that the lake/marsh was slowly filling in with organic sediments because of accelerated biomass production and anaerobic conditions.

TABLE 3.2
Optimal Oxidation–Reduction Potential (ORP) Values for Biochemical Reactions Used in Wastewater Treatment

Biochemical Reaction	ORP (mV)	More Details of Process
Nitrification	+100 to +350	Chapter 4, Section 4.2.6.2
Denitrification	+50 to −50	Chapter 4, Section 4.2.6.2
Carbonaceous biochemical oxygen demand biodegradation using molecular oxygen	+50 to +250	Chapter 3: 3.8.2
Biological phosphorus removal	+25 to +250	Chapter 4, Section 4.3.6.1
Sulfide (H_2S) formation	−50 to −250	Chapter 4, Section 4.5
Acid formation (fermentation)	−100 to −225	
Biological phosphorus release	−100 to −250	
Methane production	−175 to −400	

both short-term (for rapid response to an odor situation) and long-term control of the marsh water level for drying exposed bottom sediments and air-oxidizing troublesome sulfides.

- Surface drainage from fertilized lawns and gardens was intercepted and diverted to a point downgradient from the marsh.
- Limited engineered aeration of parts of the marsh with a air compressor attached to perforated plastic pipes was considered, but not found to be necessary.
- Public education concerning excessive fertilization and watering of lawns and gardens was implemented through homeowners associations and by circulating informational flyers.

3.5.3 OXIDATION–REDUCTION POTENTIAL AS A CONTROL INDICATOR IN WASTEWATER TREATMENT

Wastewater treatments that involve ORP reactions include all biological processes, such as removal of nutrients (nitrogen and phosphorus), odor control, and the removal of BOD. The ORP of the waste stream must be continually controlled to maintain the values that promote desired reactions and inhibit undesired reactions. Table 3.2 lists some optimal ORP ranges used for various wastewater treatment procedures and includes references to the sections in this text where further details of the processes may be found.

3.6 HARDNESS

3.6.1 BACKGROUND

Hardness is frequently used as an assessment of the quality of water supplies. The hardness of water is governed by the content of calcium and magnesium salts (temporary hardness), largely combined with bicarbonate and carbonate and with sulfates, chlorides, and other anions of mineral acids (permanent hardness).

Originally, water hardness was a measure of the ability of water to precipitate soap. It was measured by the amount of soap needed for adequate lathering, which only occurs after most of the dissolved calcium and magnesium have been precipitated as solids. Hardness also served as an indicator of the rate of scale formation in hot water heaters and boilers. Soap is precipitated as a gray bathtub ring deposit mainly by reacting with the calcium and magnesium cations (Ca^{2+} and Mg^{2+}) present, although other doubly and triply charged cations may play a minor role. Now hardness is usually determined by measuring the dissolved concentrations of calcium and magnesium and calculating a value for hardness as described in Section 3.6.3.

Hardness has some similarities to alkalinity. Like alkalinity, hardness is a property of water that is not attributable to a single constituent; therefore, some convention must be adopted to express hardness quantitatively as a concentration. As with alkalinity, hardness is usually expressed as an equivalent concentration of $CaCO_3$. However, hardness is a property of cations (mainly Ca^{2+} and Mg^{2+}), whereas alkalinity is a property of anions (mainly HCO_3^- and CO_3^{2-}).

3.6.2 IMPORTANCE OF HARDNESS

Hardness is sometimes useful as an indicator that is approximately proportional to the TDS present, since Ca^{2+}, Mg^{2+}, CO_3^{2-}, and HCO_3^- often represent the largest part of the TDS. No human health effects due to hardness have been proven; however, an inverse relation with cardiovascular disease has been reported. Higher levels of drinking water hardness correlate with lower incidence of cardiovascular disease. High levels of water hardness may limit the growth of fish; on the other hand, low hardness (soft water) may increase fish sensitivity to toxic metals. In general, higher hardness is beneficial by reducing metal toxicity to fish. Aquatic life water quality standards for many metals are calculated by using an equation that includes water hardness as a variable.

The main advantages in limiting the level of hardness (by softening water) are economical: less soap requirements in domestic and industrial cleaning, and less scale formation in pipes and boilers. Water treatment by reverse osmosis (RO) often requires a water-softening pretreatment to prevent scale formation on RO membranes. Increased use of detergents, which do not form precipitates with Ca^{2+} and Mg^{2+}, has lessened the importance of hardness with respect to soap consumption.

On the other hand, a drawback to soft water is that it is more corrosive or aggressive than hard water. In this context, corrosive means that soft water more readily dissolves metal from plumbing systems than does hard water. Thus, in plumbing systems where brass, copper, galvanized iron pipe, or lead solders are used, a soft water system will carry higher levels of dissolved copper, zinc, lead, and iron than will a hard water system. Table 3.3 describes different degrees of hard and soft water.

TABLE 3.3

Degrees of Hardness and Hardness Effects

Degree of Hardness	mg/L as CaCO$_3$	Grains	Effects
Soft	<75	<4.5	May increase toxicity to fish of dissolved metals
			May increase corrosivity of water to metals
			No scale deposits
			Efficient use of soap
Moderately hard	75–120	4.5–7.0	Not objectionable for most purposes
			Requires somewhat more soap for cleaning
			Above 100 mg/L may deposit significant scale in boilers
Hard	120–200	7.0–11.5	Considerable scale buildup and staining. Generally softened if >200 mg/L
Very hard	>200	>11.5	Requires softening for household or commercial use

RULES OF THUMB

1. The higher the hardness, the more tolerant are many surface water metal standards for aquatic life.
2. Hardness above 100 mg/L can cause significant scale deposits to form in boilers.
3. The softer the water, the greater the tendency to dissolve metals from the pipes of water distribution systems.
4. An ideal quality goal for total hardness is about 70–90 mg/L. Municipal treatment sometimes allows up to 150 mg/L of total hardness in order to reduce costs of water softening.

Water will be hard wherever groundwater passes through calcium and magnesium carbonate mineral deposits. Such deposits are very widespread, and hard to moderately hard groundwater is more common than soft groundwater. Very hard groundwater occurs frequently. Calcium and magnesium carbonates are the most common carbonate minerals in the earth's crust and are the main sources of hard water. A geologic map showing the distribution of carbonate minerals serves also as an approximate map of the distribution of hard groundwater. The most common sources of soft water are where rainwater is used directly, or where surface waters are fed more by precipitation than by groundwater.

In industrial usage, hardness is sometimes expressed as grains per gallon (gpg). The conversion between gpg and mg/L is shown in Figure 3.5.

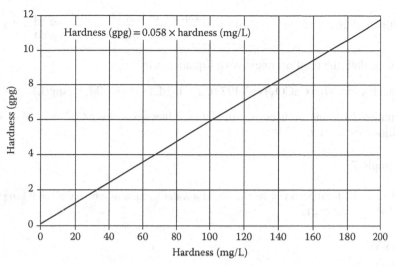

FIGURE 3.5 Relationship between hardness expressed as milligrams per liter as $CaCO_3$ (mg/L $CaCO_3$) and grains per gallon as $CaCO_3$ (gpg $CaCO_3$).

3.6.3 CALCULATING HARDNESS

Current practice is to define total hardness as the sum of the calcium and magnesium ion concentrations in mg/L, both expressed as a calcium carbonate equivalent. Hardness is usually calculated from separate measurements of calcium and magnesium, rather than measured directly by colorimetric titration.

Calcium and magnesium ion concentrations are converted to equivalent concentrations of $CaCO_3$ as follows:

1. Find the equivalent weights of Ca^{2+}, Mg^{2+}, and $CaCO_3$ (also, see Table 1.1, Chapter 1, Section 1.3.6).

2. $$Eq.\ wt. = \frac{Molar\ weight}{Magnitude\ of\ ionic\ charge\ or\ oxidation\ number}$$

$$Eq.\ wt.\ of\ Ca^{2+} = \frac{40.08\ g/mol}{2\ eq/mol} = 20.04\ g/eq$$

$$Eq.\ wt.\ of\ Mg^{2+} = \frac{24.31\ g/mol}{2\ eq/mol} = 12.15\ g/eq$$

$$Eq.\ wt.\ of\ CaCO_3 = \frac{100.09\ g/mol}{2\ eq/mol} = 50.04\ g/eq$$

3. Determine the multiplying factors to obtain the equivalent concentration of $CaCO_3$.

$$Multiplying\ factor\ for\ Ca^{2+}\ as\ CaCO_3 = \frac{Eq.\ wt.\ of\ CaCO_3}{Eq.\ wt.\ of\ Ca^{2+}} = \frac{50.04\ g/eq}{20.04\ g/eq} = 2.497$$

Multiplying factor of Mg^{2+} as $CaCO_3 = \dfrac{Eq.\ wt.\ of\ CaCO_3}{Eq.\ wt.\ of\ Mg^{2+}} = \dfrac{50.04\ g/eq}{12.15\ g/eq} = 4.119$

4. Calculate the total hardness with Equation 3.18.

Total hardness (as $CaCO_3$) = 2.497 (Ca^{2+}, mg/L) + 4.118 (Mg^{2+}, mg/L)   (3.18)

Equation 3.18 may be used to calculate hardness whenever Ca^{2+} and Mg^{2+} concentrations are known.

Example 7

Calculate the total hardness as $CaCO_3$ of a water sample in which Ca^{2+} = 98 mg/L and Mg^{2+} = 22 mg/L.

ANSWER

Use Equation 3.18:

Total hardness = 2.497(98 mg/L) + 4.118 (22 mg/L) = 335 mg/L $CaCO_3$

Both alkalinity and hardness are expressed in terms of an equivalent concentration of calcium carbonate. As noted before, alkalinity is a property that arises from reactions of the anions, CO_3^{2-} and HCO_3^-, whereas hardness arises from reactions of the cations, Ca^{2+} and Mg^{2+}.

It is possible for total hardness as $CaCO_3$ to exceed the total alkalinity as $CaCO_3$. When this occurs,

- The portion of total hardness equal to the alkalinity is referred to as temporary hardness or carbonate hardness.
- Temporary hardness measures the contribution to total hardness from bicarbonate and carbonate salts of calcium and magnesium. When total hardness is equal to alkalinity, essentially the only sources of Ca^{2+} and Mg^{2+} cations are the bicarbonate and carbonate salts.
- The portion of total hardness in excess of alkalinity is referred to as permanent hardness or noncarbonate hardness.
- Permanent hardness measures the contribution to total hardness from other salts of calcium and magnesium, mainly sulfate, chloride, fluoride, phosphate, and nitrate salts.

3.7   DISSOLVED OXYGEN

3.7.1   BACKGROUND

Sufficient DO is important for high-quality water. DO is crucial for the survival of fish and most other aquatic life forms. It oxidizes many sources of objectionable tastes and odors. Oxygen becomes dissolved in surface waters by diffusion from the atmosphere and from aquatic-plant photosynthesis.

On average, most DO is transferred directly from the atmosphere; only a little net DO is produced by aquatic-plant photosynthesis. Although water plants produce oxygen during the day, they consume oxygen at night as an energy source. When they die and decay, dead plant matter serves as an energy source for microbes, which consume additional oxygen. The net change in DO from aquatic-plant photosynthesis is small during the life cycle of aquatic plants.

Environmental factors that affect DO in lakes and streams are:

- Temperature: The solubility of oxygen, like all gases, decreases as temperature increases.
- Flow: DO concentrations vary with the volume and velocity of water flowing in a stream. Faster-flowing turbulent water exposes more surface area to the atmosphere than slow-moving smoother water, facilitating oxygen diffusion from the atmosphere.
- Aquatic plants: Aquatic plants release oxygen (also consuming CO_2) into the water during daytime photosynthesis. At night, plants consume oxygen (also releasing CO2) by respiration like animals. Thus, DO levels are higher during the day than at night.
- Altitude: At lower altitudes, the diffusion rate from the atmosphere to water is greater because of the higher atmospheric concentration of oxygen in the atmosphere.
- Dissolved and suspended solids: Oxygen is more soluble in water with lower concentrations of dissolved and suspended solids.
- Human activities
  - Removal of riparian vegetation may lower oxygen concentrations due to increased water temperature resulting from a lack of canopy shade, as well as increased suspended solids resulting from erosion of bare soil.
  - Storm runoff from impervious surfaces such as roads, parking lots, and building roofs carry salts, sediments, and other pollutants, increasing the amount of suspended and dissolved solids in stream water.
  - Organic wastes and other nutrient inputs from sewage and industrial discharges, septic tanks, and agricultural and urban runoff can result in decreased oxygen levels. Nutrient input can lead to excessive algal growth. When algae die, bacterial decomposition consumes oxygen (see Section 3.8).
  - Ammonia discharged from industrial and wastewater treatment plants is oxidized by DO to nitrite and nitrate forms, consuming about 4.3 mg of DO per milligram of ammonia nitrogen oxidized (Chapter 4, Section 4.2.2).
  - Dams may cause an oxygen deficiency in receiving waters if water from the bottom of the reservoir is released. Although water at the bottom of a dammed reservoir is cooler than the water on top, it will be lower in oxygen if the bottom sediments have a high BOD (see Section 3.8), because deep water bodies tend to be temperature-stratified and DO dissolved from the atmosphere tends to remain near the surface.

Streams with DO >8 mg/L are normally considered to be healthy streams; they carry cold, clear water with enough riffles to provide sufficient mixing of atmospheric oxygen into the water and support a diversity of aquatic organisms. Streams with DO <3 mg/L are stressful to most aquatic organisms. Most fish die at 1–2 mg/L. However, fish can move away from low DO areas. Water with low DO in the range 2–0.5 mg/L are considered hypoxic; waters with <0.5 mg/L are anoxic.

Summer can be a crucial season for low levels of DO in water bodies because stream flows tend to decrease, water temperatures to increase, dissolved and solid organic matter to increase, and microbial activity consuming DO to increase. Water quality in a stream or lake can be estimated by comparing its measured DO concentration with its theoretical saturated DO concentration* at the temperature of the water body (Table 3.2). The percent DO saturation, calculated by Equation 3.19, is a useful indicator of water quality with respect to aquatic life support (see Table 3.4).

TABLE 3.4
Maximum Dissolved Oxygen (DO) Concentration (Saturation Concentration) and Temperature

Temperature (°C/°F)	Saturated DO (mg/L)	Temperature (°C/°F)	Saturated DO (mg/L)
0/32.0	14.60	23/73.4	8.56
1/33.8	14.19	24/75.2	8.40
2/35.6	13.81	25/77.0	8.24
3/37.4	13.44	26/78.8	8.09
4/39.2	13.09	27/80.6	7.95
5/41.0	12.75	28/82.4	7.81
6/42.8	12.43	29/84.2	7.67
7/44.6	12.12	30/86.0	7.54
8/46.4	11.83	31/87.8	7.41
9/48.2	11.55	32/89.6	7.28
10/50.0	11.27	33/91.4	7.16
11/51.8	11.01	34/93.2	7.05
12/53.6	10.76	35/95.0	6.93
13/55.4	10.52	36/96.8	6.82
14/57.2	10.29	37/98.6	6.71
15/59.0	10.07	38/100.4	6.61
16/60.8	9.85	39/102.2	6.51
17/62.6	9.65	40/104.0	6.41
18/64.4	9.45	41/105.8	6.31
19/66.2	9.26	42/107.6	6.22
20/68.0	9.07	43/109.4	6.13
21/69.8	8.90	44/111.2	6.04
22/71.6	8.72	45/113.0	5.95

---

* The saturation DO concentration is another way of saying "solubility of oxygen gas in water at equilibrium." It is defined as the maximum level of DO that can be present in water at a specific temperature, in the absence of other influences (BOD, COD, TDS, etc.).

$$\text{Percent DO saturation} = \frac{\text{measured DO concentration}}{\text{saturated DO concentration}} \times 100 \qquad (3.19)$$

Table 3.4 illustrates that higher water temperatures require a higher percent DO saturation to maintain good water quality for aquatic life.

DO is consumed by the degradation (oxidation) of organic matter in water. Because the concentration of DO is never very large, oxygen-depleting processes can rapidly reduce it to near zero in the absence of efficient aeration mechanisms. Fish need at least 5–6 ppm DO to grow and thrive. They stop feeding if the level drops to around 3–4 ppm and die if DO falls to 1 ppm. Many fish kills are not caused by the direct toxicity of contaminants, but instead by a deficiency of oxygen caused by the biodegradation of organic contaminants (see Table 3.5).

Example 8

A stream temperature is measured at 5°C. From Table 3.4, the maximum saturation value is 12.75 mg/L. If DO in the stream is measured at 5.6 mg/L, the percent saturation would be (5.6/12.75) × 100 = 44.4%. Table 3.3 indicates that a stream with good water quality at 5°C should have a minimum percent saturation of DO somewhere between 55% and 71%. The value of 44.4% for 5°C water indicates that the stream is moderately polluted and that something else besides temperature is affecting oxygen levels adversely (e.g., suspended or dissolved solids, or bacteria decomposition).

Typical state standards for DO in surface waters classified for aquatic life are

- 7.0 ppm for cold-water spawning periods
- 6.0 ppm for class 1 cold-water biota
- 5.0 ppm for class 1 warm-water biota

TABLE 3.5
Dissolved Oxygen and Surface Water Quality for Aquatic Life

Water Quality	Dissolved Oxygen (mg/L)	% Saturation		
		0°C	10°C	20°C
Good	>8.0	>55	>71	>88
Slightly polluted	6.5–8.0	45–55	58–71	72–88
Moderately polluted	4.5–6.5	31–45	40–58	50–72
Heavily polluted	4.0–4.5	27–31	35–40	44–50
Severely polluted	<4.0	<27	<35	<44

RULES OF THUMB

1. The solubility in water of oxygen (and other gases) decreases as water temperature increases.
2. The percent saturation of DO required to maintain good water quality for aquatic life increases as water temperature increases, from >55% at 0°C to >88% at 20°C.
3. DO > 8 mg/L is considered good-quality water.
   DO = 5–6 mg/L at a minimum is required for healthy fish.
   DO < 3 mg/L is stressful for most aquatic life.
   DO < 1 mg/L initiates fish kills.

3.8 BIOCHEMICAL OXYGEN DEMAND AND CHEMICAL OXYGEN DEMAND

3.8.1 BIOCHEMICAL OXYGEN DEMAND

BOD refers to the amount of DO potentially consumed if all the biochemically degradable organic matter in a given volume of water were aerobically biodegraded. BOD is determined by splitting the collected sample into two portions and measuring the initial DO concentration in one portion and comparing it to the DO in the second portion after it has been incubated under specific conditions of temperature, nutrient and microbe concentrations for a certain number of days (see Section 3.8.4). The difference between the two DO values gives the amount of DO consumed by biodegradation under the conditions of the analysis.

BOD is an indicator of the potential for a water body to become depleted in oxygen and possibly become anaerobic because of biodegradation. BOD measurements do not take into account reoxygenation of water by naturally occurring diffusion from the atmosphere or mechanical aeration, nor do they exclude the consumption of DO when various inorganic chemicals in the water, such as sulfides, ferrous iron, and ammonia, are also oxidized. Water with a high BOD and an active microbial population may become depleted in oxygen to the extent of not supporting aquatic life, unless there is a means for rapidly replenishing DO. While a DO analysis tells you how much oxygen is available, a BOD analysis tells you how much oxygen can potentially be consumed by biodegradation of organic matter. Unpolluted, natural waters are expected to have a BOD of 5 mg/L or less. Raw sewage may have BOD levels ranging from 150 to 300 mg/L.

3.8.2 CARBONACEOUS BIOCHEMICAL OXYGEN DEMAND

Carbonaceous biochemical oxygen demand (CBOD) is a subset of BOD. CBOD is the fraction of BOD that measures only the oxygen demand from biodegradable hydrocarbon matter, omitting the oxygen demand from nitrogenous matter. This is accomplished by chemically suppressing the activity of nitrifying bacteria. The measure of CBOD is widely used as an indication of the pollutant removal efficiency of wastewater treatment plants.

3.8.3 Nitrogenous Biochemical Oxygen Demand

Nitrogenous biochemical oxygen demand (NBOD) is the amount of DO consumed by bioconversion of inorganic and organic nitrogen compounds. Organic nitrogen matter consists mainly of urea, protein, and amines, which are converted by microorganisms to ammonia (NH_3), nitrites (NO_2^-), and nitrates (NO_3^-), consuming DO in the process. Inorganic nitrogen compounds are mainly ammonia, nitrate, and nitrite. The chief inorganic sources of NBOD are direct releases of ammonia and nitrites from wastewater treatment plants and agricultural sources. The oxidation of ammonia to nitrites and nitrites to nitrates consumes DO as indicated above.

3.8.4 BOD$_5$ and CBOD$_5$

BOD$_5$ and CBOD$_5$ refer to particular empirical BOD analyses, accepted as standard procedures, in which a specified volume of sample water is seeded with bacteria and nutrients (nitrogen and phosphorus) and then incubated for 5 days at 20°C in the dark. BOD$_5$ and CBOD$_5$ are measured as the decrease in DO (in mg/L) after 5 days of incubation. They may or may not have degraded all the biodegradable organic matter in the sample by the end of the 5 days, but the amount of oxygen consumed is accepted as a reliable indicator of whether or not the amount of BOD present is an environmental hazard. These 5-day tests are said to have originated in England and Ireland, where biodegradable contaminants in any river that are not decomposed within 5 days will reach the ocean.

Water surface turbulence helps to dissolve oxygen from the atmosphere by increasing the water surface area. A BOD$_5$ of 5 mg/L in a slow-moving stream might be enough to produce anaerobic conditions, while a turbulent mountain stream might be able to assimilate a BOD$_5$ of 50 mg/L without appreciable oxygen depletion (see Figure 3.6 and Table 3.6).

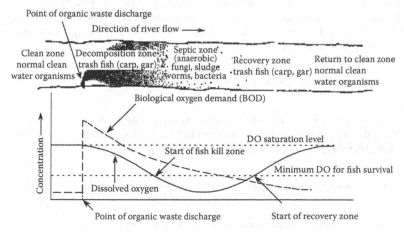

FIGURE 3.6   Dissolved oxygen sag curve caused by discharge of organic wastes into a river.

TABLE 3.6

Typical BOD$_5$ Concentrations

Type of Water Flow	Characteristic BOD$_5$ Concentrations (mg/L)
Pristine streams	<1
Unpolluted, natural streams	<5
Moderately polluted streams	5–8
Raw untreated sewage in the United States	150–300
Raw untreated sewage in Europe	~600
Adequately treated municipal sewage	<20

Note: The much lower concentration of BOD$_5$ in raw untreated sewage in the United States as compared to Europe is due to dilution in the sewer lines caused by the much greater water use per capita in the United States than in Europe.

3.8.5   CBOD CALCULATION

Example 9

While collecting a 1-L water sample for analysis, an insect weighing 50.0 mg was accidentally trapped in the bottle. The initial water sample DO was 7.0 mg/L. Assume that 15% of the insect's fresh weight is readily biodegradable and has the approximate unit formula CH$_2$O.* Also, assume that microbes are present that will metabolize the insect. If the laboratory does not analyze the sample until biodegradation is complete, what DO will they measure?

ANSWER

The chemical reaction for oxidation of hydrocarbon organic matter is

$$CH_2O + O_2 \rightarrow CO_2 + H_2O \qquad (3.20)$$

Equation 3.20 shows that 1 mole of O$_2$ oxidizes 1 mole of CH$_2$O. Therefore, the moles of CH$_2$O biodegraded will equal the moles of O$_2$ consumed during biodegradation. First, find the moles of O$_2$ initially present and the biodegradable moles of organic matter in the insect:

$$\text{Moles } O_2 \text{ initially present as DO} = \frac{7.0 \times 10^{-3} \text{ g/L}}{32 \text{ g/mol}} = 2.19 \times 10^{-4} \text{ mol/L}$$

Initial weight of insect organic matter (CH$_2$O) present = 50.0 mg.
Weight of insect matter that is biodegradable = 0.15 × (50.0 mg) = 7.50 mg.
Molecular weight of CH$_2$O = 12 + 2 + 16 = 30 g/mol.

$$\text{Moles biodegradable organic matter} = \frac{7.50 \times 10^{-3} \text{ g/L}}{30 \text{ g/mol}} = 2.50 \times 10^{-4} \text{ mol}$$

---

* Carbonaceous organic biomass contains carbon, hydrogen, and oxygen atoms approximately in the ratio of 1:2:1, so that CH$_2$O serves as a convenient unit molecule of organic matter.

The moles of biodegradable organic matter exceed the moles of DO in the 1-L sample. Therefore, biodegradation of the insect will consume all the DO in the sample, with some biodegradable organic matter left over.

Moles of organic matter biodegraded = moles of O_2 consumed = 2.19×10^{-4} mol/L

Mass of organic matter biodegraded = $(2.19 \times 10^{-4}$ mol/L$) \times (30$ g/mol$) = 6.57 \times 10^{-3}$ g

Mass of biodegradable organic matter remaining = $7.50 - 6.57$ mg = 0.093 mg

Since all the sample DO has been consumed, the laboratory will find the water anaerobic with no DO remaining. There will be $(50.0 - 6.57$ mg$) = 43.4$ mg of insect remaining, of which 0.093 mg is still biodegradable. Note that the concentration of CBOD degraded (6.6 mg/L) is similar to the concentration of DO consumed (7.0 mg/L), being about 6% less—a point that implies the following Rule of Thumb.

RULE OF THUMB

The mg/L of CBOD biodegraded in a $CBOD_5$ test $\approx$ mg/L of O_2 consumed by microbes in the sample.

Note: This convenient approximation is valid because, by Equation 3.20, the moles of O_2 consumed in a $CBOD_5$ test equal the moles of CBOD consumed, and the molecular weight of O_2 (32 g/mol) is close to that of the unit molecule, CH_2O, of organic matter (molecular weight of $CH_2O = 30$ g/mol). This means that the weight–concentration ratio of CBOD to DO consumed in a $CBOD_5$ test is 30/32, or 0.9375, very close to 1/1.

3.8.6 CHEMICAL OXYGEN DEMAND

Chemical oxygen demand (COD) refers to the amount of oxygen consumed when 95–100% of all the oxidizable organic matter (volatile organics and certain oxidation-resistant organics may not be completely oxidized) in a given volume of water is chemically oxidized to CO_2 and H_2O by a strong chemical oxidant such as chlorine, permanganate, or dichromate. COD is sometimes used as a measure of general pollution. For example, in an industrial area built on landfill dirt, COD in the groundwater might be used as an indicator of organic materials leached from the landfill materials. Leachate from landfills often has high levels of COD.

BOD is a subset of COD. The COD analysis oxidizes organic matter that is both chemically and biologically oxidizable. If a reliable correlation between COD and BOD can be established at a particular site, the simpler COD test may be used in place of the more complicated BOD analysis.

3.8.7 COD CALCULATION

Example 10

COD levels of 60 mg/L were measured in groundwater. It is suspected that diesel fuel contamination is the main cause. What concentration of hydrocarbons from diesel fuel is necessary to account for all of the COD observed?

ANSWER

For simplicity, assume fuel hydrocarbons (mostly alkanes, C_nH_{2n+2}, with $n \approx 6$–24) to have an average unit formula of CH_2. The oxidation reaction is

$$CH_2 + 1.5\ O_2 \rightarrow CO_2 + H_2O \qquad\qquad (3.21)$$

For each carbon atom in the fuel, 1.5 oxygen molecules are consumed.

Weight of 1 mole of $CH_2 = 12 + 2 = 14$ g.
Weight of 1.5 mole of $O_2 = 1.5 \times 32 = 48$ g.

Weight ratio of oxygen to fuel consumed during oxidation is $\dfrac{48\ g}{14\ g} = 3{:}4$.

A COD of 60 mg/L requires $\dfrac{60\ mg/L}{3.4} = 18$ mg/L fuel hydrocarbons.

If dissolved fuel hydrocarbons in the groundwater are around 18 mg/L or greater, the fuel alone could account for all the measured COD.

If dissolved fuel hydrocarbons in the groundwater are much less than 18 mg/L, then fuels could account for only a small part of the COD; other organic substances, such as pesticides, fertilizers, solvents, and PCBs, must account for the rest.

3.9 SOLIDS (TOTAL, SUSPENDED, AND DISSOLVED)

3.9.1 BACKGROUND

The general term "solids" refers to matter that is suspended (insoluble solids) or dissolved (soluble solids) in water. Solids can affect water quality in several ways. Drinking water with high dissolved solids may not taste good and may have a laxative effect. Boiler water with high dissolved solids requires pretreatment to prevent scale formation. Water high in suspended solids may harm aquatic life by causing abrasion damage, clogging fish gills, harming spawning beds, and reducing photosynthesis by blocking sunlight penetration, among other consequences. On the other hand, hard water (caused mainly by dissolved calcium and magnesium compounds) reduces the toxicity of metals to aquatic life and is less corrosive to metals.

Total solids (sometimes called residue) are the solids remaining after evaporating the water from an unfiltered sample. It includes two subclasses of solids that are separated by filtering (generally with a filter having a nominal 0.45 μm or smaller pore size):

1. Total suspended solids (TSS; sometimes called filterable solids) in water are organic and mineral particulate matter that do not pass through a 0.45-μm

filter. They include undissolved material such as silt, clay, metal oxides, sulfides, algae, bacteria, and fungi. TSS contributes to turbidity, which limits light penetration for photosynthesis and visibility in recreational waters. TSS is generally removed by flocculation and filtering. TSS in treated discharges is sometimes used as a preliminary screening measure of pollutant removal; effluent with TSS concentrations below 45 mg/L generally complies with effluent limits. If effluent TSS concentrations exceed 45 mg/L, the discharge should be analyzed for other constituents that have discharge limits.

2. Total dissolved solids (TDS; sometimes called nonfilterable solids) are substances that will pass through a 0.45-μm filter. If the water passed through the filter is evaporated, the TDS will remain behind as a solid residue. TDS includes dissolved metals, minerals and salts, humic acids, tannin, and pyrogens. TDS is removed by precipitation, ion exchange, and RO. In natural waters, the major contributors to TDS are carbonate, bicarbonate, chloride, sulfate, phosphate, and nitrate salts. Taste problems in water often arise from the presence of high TDS levels with certain metals present, particularly iron, copper, manganese, and zinc.

The difference between suspended and dissolved solids is a matter of definition based on the filtering procedure. Solids are always measured as the dry weight, and careful attention must be paid to the drying procedure to avoid errors caused by retained moisture or a loss or gain in mass by volatilization or oxidation.

3.9.2 TOTAL SUSPENDED SOLIDS AND TURBIDITY

Turbidity measures the cloudiness of water, defined as the extent that light is either scattered or absorbed by suspended and colloidal solids (e.g., clays, silt, and finely divided organic matter) and microscopic organisms (e.g., algae and plankton) in water, instead of being transmitted. Turbidity, measured in nephelometric turbidity units (NTU), may be used to estimate the concentration of TSS, in mg/L (see Table 3.7). Turbidity is commonly measured in a laboratory with an optical nephelometer, which determines the intensity of light scattered at 90 degrees to a collimated light beam. In the field, the transparency of water (inversely related to turbidity) is generally measured with a Secchi disc, a black-and-white patterned metal or plastic disk mounted on a line or pole. The disc is lowered slowly into the water until the pattern on the disc is no longer discernable. The depth at which this occurs is called the Secchi depth and is correlated to water turbidity.

TABLE 3.7

Total Suspended Solids (TSS) Concentration Estimated from Turbidity Measurements

Turbidity (NTU)	2	5	10	20	50
Corresponding TSS due to soil sediment (mg/L)	2.2	6.3	12	24	64
Corresponding TSS due to algae (μg chlorophyll/L)	2.2	4.7	10	36	54

Although turbidity has no specific health effects, it is used as an indicator of potential problems:

- Turbidity can indicate excessive soil erosion from earth-disturbing activities such as construction, agriculture, and logging.
- Turbidity often increases sharply during heavy rainfall in developed watersheds, especially those having a relatively high proportion of impervious surfaces (rooftops, roads, parking lots, etc.).
- Turbidity is used to monitor the erosion rates of stream banks and channels that may require additional erosion control practices.
- Not only does turbidity interfere with drinking water disinfection, but also suspended particles can provide a habitat for pathogens. The World Health Organization (WHO 2010) recommends that the turbidity of drinking water be less than 1 NTU, but never more than 5 NTU.
- Moderately low levels of turbidity may indicate a healthy, well-functioning ecosystem, with moderate amounts of plankton present to fuel the food chain.
- At higher levels of turbidity:
  - Suspended particles absorb energy from sunlight, making water bodies warmer and potentially reducing the concentration of DO to stressful levels for aquatic life (see Section 3.7).
  - Photosynthesis of aquatic plants and algae is diminished, also decreasing DO.
  - In slow-moving waters, settling particles may blanket and suffocate fish eggs, benthic animals, and insect larvae.
  - Shallow lakes fill in faster to become marshland.
  - Suspended solids can transport sorbed metals, pesticides, nutrients, and other low-solubility pollutants throughout a stream system.
  - Suspended solids can clog fish gills, reducing resistance to disease in fish, lowering growth rates, and affecting egg and larval development.

RULES OF THUMB

1. Turbidity < 0.1 NTU is required for effective disinfection.
   Turbidity ≈ 5 NTU in drinking water is visible but generally acceptable to consumers.
   Turbidity < 10 NTU is generally regarded as low turbidity.
   Turbidity > 10 NTU is generally regarded as turbid.
   Turbidity > 50 NTU is generally regarded as high turbidity.
2. TSS is detrimental to fish health by decreasing growth, disease resistance, and egg development.
   a. TSS should not reduce the maximum depth of photosynthetic activity by more than 10% from the seasonally established norm.
3. At low concentrations, TSS (in mg/L for soil erosion, or μg-chlorophyll/L for algae) is roughly equal to turbidity in NTU (see Table 3.7).

4. TSS in treated discharges may be used as a preliminary screening measure of pollutant removal.
   a. If effluent TSS < 45 mg/L, discharge generally complies with effluent limits.
   b. If effluent TSS > 45 mg/L, analyze discharge for other constituents that may exceed discharge limits.
5. Water with TDS < 1200 mg/L generally has an acceptable taste. Higher TDS can adversely influence the taste of drinking water and may have a laxative effect.
6. In water to be treated for domestic potable supply, TDS < 650 mg/L is a preferred goal.
7. For drinking water, recommended TDS is <500 mg/L; the upper limit is 1000 mg/L.

3.9.3 TOTAL DISSOLVED SOLIDS AND SALINITY

The terms *total dissolved solids* and *salinity* both indicate dissolved salts. Table 3.8 offers a qualitative comparison between the terms. Sometimes salinity is defined by the quantitative definition of weight percentage of dissolved salts in a water solution. More commonly, the salinity of irrigation water is considered to be the total dissolved salt concentration in mg/L, numerically the same as TDS.

3.9.4 ELECTRICAL CONDUCTIVITY AND TDS

Electrical conductivity (EC), also called specific conductivity or just conductivity, is a measure of the ability of a water sample to carry an electrical current, a property that depends mainly on the kinds and concentrations of dissolved salts, and often serves as an indirect measure of TDS (mainly ionic dissolved salts) concentration. Ions such as chloride, nitrate, sulfate, phosphate, sodium, magnesium, calcium, iron, and aluminum will carry an electrical current. Most organic substances, like oil, alcohol, and sugar, do not conduct electricity very well and thus have a low conductivity in water.

TABLE 3.8

Comparison of Total Dissolved Solids (TDS) and Qualitative Degree of Salinity

TDS (mg/L)	Qualitative Degree of Salinity
<1,000	Fresh water
1,000–3,000	Slightly saline
3,000–10,000	Moderately saline
10,000–35,000	Very saline
35,000–100,000	Briny
>100,000	Brackish

Inorganic dissolved solids are essential ingredients for aquatic life. They regulate the flow of water in and out of organisms' cells and are building blocks of the molecules necessary for life. A high concentration of dissolved solids, however, can cause chemical balance problems for aquatic organisms. Also, because ions from the dissolved salts are attracted much more strongly to the polar water molecules than are nonpolar oxygen molecules, cations and anions tend to displace DO from water and lower DO levels. This effect depends on the temperature and kinds of ions present, as well as their concentrations; as an example, the solubility of oxygen in seawater is 21% less than that of freshwater at 0°C.

EC is directly related to TDS and can serve as a check on TDS measurements, although it must always be remembered that EC is an indirect measurement of dissolved ions while TDS is a direct measurement of all dissolved solids, ionic or not. The internationally accepted standard unit for reporting EC is decisiemens per meter (dS/m).* An older name still encountered for the same electrical quantity is millimhos per centimeter (mmho/cm); 1 dS/m = 1 mmho/cm. EC-measuring instruments are often calibrated in micromhos per cm (μmhos/cm); 1 dS/m = 1 mmho/cm = 1000 μmhos/cm.

RULES OF THUMB

1. Common units for EC are:

   micromho/cm (μmho/cm)

   microsiemens/cm (μS/cm)

   millisiemens/cm (mS/cm)

   decisiemens/m (dS/m)

   For conversions, note that: 1 mho/cm = 1 S/cm, and 1 mS/cm = 1 dS/m

   This leads to: 1 μmho/cm = 1 μS/cm = 10^{-3} mS/cm = 10^{-3} dS/m,

   and

   1 dS/m = 1 mS/cm = 1000 μS/cm = 1000 μmho/cm

2. TDS in mg/L can be estimated from a measurement of electrical conductivity.

   a. For seawater (NaCl-based):

      TDS (mg/L) ≈ 0.5 × (EC in μS/cm) ≈ 500 × (EC in dS/m)

   b. For surface and groundwaters (carbonate or sulfate-based):[†]

      TDS (mg/L) ≈ 0.67 × (EC in μS/cm) ≈ 670 × (EC in dS/m)

---

* Because electrical conductivity is defined as the reciprocal of electrical resistance, it was assigned the unit of mho, which is the unit of electrical resistance (ohm) written backward. It was renamed the siemens as the SI unit in 1881, for German electrical engineer Ernst Werner von Siemens. Because siemens (often capitalized as Siemens and using the capitalized symbol S without regard for the official SI practice of writing units derived from proper names with a lower case first letter), is the official SI unit, it is generally used in science and often in electrical applications. The unit mho is still used primarily in electronic applications. However, mho is sometimes preferred over S because it is less likely to be confused with other variables that might use the letter S, or even the lower case s where time units of seconds are intended.

† The multiplying factor for converting EC measurements to TDS depends on the particular salts dissolved in a water solution as well as their concentrations and can range between about 0.6 and 1.0. A factor of 0.67 is commonly used for an approximation when the actual factor is not known.

Note: The factor of 0.67 in the above equation is generally suitable for surface and groundwaters with EC <2000 μS/cm (2 dS/m). The factor increases as EC increases. For surface and groundwaters with EC >2000 μS/cm or for greater accuracy in general, use Figure 3.7 to obtain the multiplying factor.

If the ratio $\frac{TDS_{means}}{Sp.Cond.}$ is demonstrated to be consistent, the simpler specific conductivity measurement may sometimes be substituted for TDS analysis (Table 3.9).

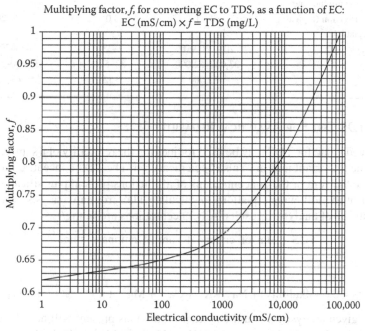

Multiplying factor, f, for converting EC to TDS, as a function of EC: EC (mS/cm) $\times f$ = TDS (mg/L)

FIGURE 3.7 Multiplying factor, f, for converting EC to TDS, as a function of EC. EC (μS/cm) $\times f$ = TDS (mg/L). Data were obtained from measurements on standard solutions containing a mixture of salts characteristic of fresh surface and groundwaters.

TABLE 3.9

Typical Electrical Conductivities and Calculated Total Dissolved Solids (TDS) of Some Aqueous Solutions

Solution	μS/cm	mS/cm = dS/m	TDS (mg/L) (using Figure 3.7)
Pure water	0.055	0.000055	0.034
Deionized (DI) water	0.1	0.0001	0.061
Distilled water	0.5	0.0005	0.31
Reverse osmosis (RO) water	50–100	0.05–0.1	32–65
Rainwater	<32	<0.032	<20

(Continued)

TABLE 3.9 (*Continued*)

Typical Electrical Conductivities and Calculated Total Dissolved Solids (TDS) of Some Aqueous Solutions

Solution	μS/cm	mS/cm = dS/m	TDS (mg/L) (using Figure 3.7)
Domestic tap water	150–800	0.15–0.80	98–544
Potable water (max. acceptable)	1350	1.35	1026
Environmental freshwaters: lakes, rivers, shallow groundwater	32–1600	0.032–1.6	20–1136
Agriculture irrigation (typical max.≈acceptable)	3000	3.0	2220
Brackish water	700–33,000	7–33	475–30,000
Seawater	33,000–52,000	33–52	30,000–50,000,
Brine	>52,000	>52	>50,000

3.9.5 TDS TEST FOR ANALYTICAL RELIABILITY

A calculated value for TDS may be used for judging the reliability of a sample analysis if all the important ions have also been measured. The TDS concentration should be equal to the sum of the concentrations of all the ions present plus silica. You can use either of the following equations to calculate TDS from an analysis or to check on the validity of analytical results. All concentrations are in mg/L.

$$TDS = \text{sum of cations} + \text{sum of anions} + \text{silica} \tag{3.22}$$

or

$$TDS = 0.6 \,(\text{alkalinity}) + Na^+ + K^+ + Ca^{2+} + Mg^{2+} + Cl^- + SO_4^{2-} + SiO_3 \tag{3.23}$$

In any given analysis, it is unlikely that all the ions present will have been measured. Frequently, only the major ions (Na^+, K^+, Ca^{2+}, Mg^{2+}, Cl^-, HCO_3^-, and SO_4^{2-}) are necessary for the calculations, as other ion concentrations are likely to be insignificant by comparison.

Use the following guidelines for checking the accuracy of a TDS analysis:

1. TDS_{meas} should always be equal to or somewhat larger than TDS_{calc} because a significant ion contributor might not have been included in the calculation.
2. An analysis is acceptable if the ratio of measured-to-calculated TDS is in the range

$$1.0 < \frac{\text{measured TDS}}{\text{calculated TDS}} < 1.2$$

3. If $TDS_{meas} < TDS_{calc}$ within the limits of analytical error (usually around 5–10%; the errors for each analyte are additive) the sample should be reanalyzed.
4. If $TDS_{meas} > 1.2 \times TDS_{calc}$, the sample should be reanalyzed, perhaps with a more complete set of ions.

3.10   TEMPERATURE

Temperature affects all water uses.

- The solubility of gases such as oxygen and CO_2 decreases as water temperature increases (Section 3.7).
- Biodegradation of organic material in water and sediments is accelerated with increased temperatures, increasing the demand on DO (Section 3.8).
- Fish and plant metabolism depends on temperature. If stream temperatures increase, decrease, or fluctuate too widely, metabolic activities can speed up, slow down, malfunction, or stop altogether.
- Most chemical equilibria are temperature dependent. Important environmental examples are the equilibria between ionized and unionized forms of ammonia, hydrogen cyanide, and hydrogen sulfide.

Temperature regulatory limits are set to maintain a normal pattern of diurnal and seasonal fluctuations, with no changes deleterious to aquatic life. Maximum-induced change is limited to a 3°C increase over a 4-hour period, lasting for 12 hours maximum.

Many factors can influence water temperatures:

- Water temperatures fluctuate with the seasons.
- On a shorter time scale, they undergo day–night variations.
- An overhanging canopy of stream vegetation provides shade and helps to lessen the effects of high air temperatures.
- Water temperatures are influenced by the quantity and velocity of stream flow. The sun has less effect in warming the waters of streams with greater and swifter flows than of streams with smaller, slower flows.
- Municipal and industrial discharges may raise water temperatures

3.11   DRINKING WATER TREATMENT

Until the germ theory of disease was generally accepted a little more than one century ago, people mostly assumed that good-tasting water was healthy and safe to drink. Although writings from as early as 4000 B.C. record the use of various forms of treatment to improve the taste and odor of drinking water, such as filtration, boiling, and exposure to sunlight, the fact that disease could be spread through drinking water was not commonly recognized until the 1800s.*

Disinfecting water with chlorine was first proposed around the year 1800 in Europe. It was first used regularly in Belgium and Great Britain in the early 1900s, resulting in a sharp decline in typhoid deaths. In 1908, chlorine disinfection of drinking water was used for the first time in the United States in Jersey City, New

---

* In 1855, epidemiologist Dr. John Snow linked an outbreak of illness in London to a public well that was contaminated by sewage, proving that cholera was a waterborne disease. He used chlorine to disinfect the well water and stopped the epidemic. In 1864, Louis Pasteur lectured on the germ theory of disease at a scientific meeting, and in 1881 he announced a successful vaccine for anthrax.

Jersey. Shortly after, the use of chlorine disinfection in the United States became widespread, bringing the virtual elimination of waterborne diseases such as cholera, typhoid, dysentery, and hepatitis A. Other disinfectants such as ozone were also used in Europe around this time, but were not routinely employed in the United States until several decades later.

Federal regulation of drinking water quality began in 1914, when the U.S. Public Health Service set standards for the bacteriological quality of drinking water. The standards applied only to water systems that provided drinking water to interstate carriers like ships and trains, and only applied to contaminants capable of causing contagious disease. It was not until the Safe Drinking Water Act of 1974 that all U.S. states adopted health-based standards for all public water systems.

Clean drinking water still remains the most important public health issue. Over 2 billion people worldwide do not have adequate supplies of safe drinking water. Worldwide, between 15 and 20 million babies die every year due to waterborne diarrhea diseases such as typhoid fever, dysentery, and cholera. Contaminated water supplies and poor sanitation cause 80% of the diseases that afflict people in the poorest countries. The development of municipal water purification in the last century has allowed cities in the developed countries to be essentially free of waterborne diseases.

It was discovered around 1974 that water disinfectants themselves react with organic compounds naturally occurring in water to form unintended disinfection byproducts (DBPs) that may pose health risks and the EPA began regulation of trihalomethane DBPs in drinking water. Since then, several DBPs (bromodichloromethane, bromoform, chloroform, dichloroacetic acid, and bromate) have been shown to be carcinogenic in laboratory animals at high doses. Some DBPs (bromodichloromethane, chlorite, and certain haloacetic acids) also caused adverse reproductive or developmental effects in laboratory animals. The goal of EPA disinfectant and DBP regulations is to balance the health risks of pathogen contamination against the health risks of DPB formation.

3.11.1 WATER SOURCES

Drinking water supplies are derived from both surface water and groundwater. In the United States, groundwater sources (wells) supply about 53% of all drinking water, and surface water sources (reservoirs, rivers, and lakes) supply the remaining 47%. Groundwater comes from underground aquifers, where wells are bored to recover the water. Wells may be from tens to hundreds of meters deep. In general, water in deep aquifers is replaced by percolation from the surface very slowly, over hundreds to thousands of years. Groundwater tends to be less contaminated than surface water. It is normally more protected from surface contamination; because it moves more slowly, organic matter has time to be decomposed by soil bacteria. The soil itself acts as a filter so that less suspended matter is present.

Surface water comes from lakes, rivers, and reservoirs. Surface water usually contains more suspended materials than groundwater, and so requires more processing to make it safe to drink. Surface waters are used for other purposes besides drinking and often become polluted by sewage, industrial, and recreational activities. On most rivers, the fraction of new water diminishes with distance from the headwaters, as the water becomes more used and reused. On the Rhine River in Europe, for example, communities near the mouth of the river receive as little as 40% new water in the river. All the other water has been discharged previously by an upstream city, or originates as nonpoint source return flow from agricultural activities. Water treatment must make this quality of river water fit to drink. Filtration through sand was the first successful method of municipal water treatment, used in London in the mid-1800s. It led to an immediate decline in the amount of waterborne disease.

3.11.2 WATER TREATMENT

Major changes are occurring in the water treatment field, driven by increasingly tighter water quality standards, a steady increase in the number of regulated drinking water contaminants (from about 5 in 1940 to around 100 in 2011), and new regulations affecting disinfection and DBPs. Municipalities constantly seek to refine their water treatment and provide high-quality water by more economical means. A recent development in water treatment is the application of membrane filtration to drinking water treatment. Membrane filters have been refined to the point where, in certain cases, they are suitable as stand-alone treatment for small systems. More often, they are used in conjunction with other treatment methods to economically improve the overall quality of finished drinking water.

3.11.3 BASIC DRINKING WATER TREATMENT

The purpose of water treatment is, first, to make water safe to drink by ensuring that it is free of pathogens and toxic substances. The second goal is to make it a desirable drink by removing offensive turbidity, tastes, colors, and odors.

Conventional drinking water treatment addresses both of these goals. It consists of four steps:

1. Primary settling
2. Aeration
3. Coagulation
4. Disinfection

Not all four of the basic steps are needed in every treatment plant. Groundwaters, in particular, usually need much less treatment than surface waters. Groundwaters may need no settling, aeration, or coagulation. For clean groundwaters, only a little chlorine (≈ 0.16 ppm) is added to protect the water while in the distribution system. The relatively new treatment technology of membrane filtration is increasingly being used in conjunction with the more traditional treatments and as a stand-alone treatment.

3.11.3.1 Primary Settling

Water, which has been coarsely screened to remove large particulate matter, is brought into a large holding basin to allow finer particulates to settle. Chemical coagulants may be added to form floc. Lime may be added at this point to help clarification if pH <6.5. The floc settles by gravity, removing solids larger than 25 μm.

3.11.3.2 Aeration

The clarified water is agitated with air. This promotes oxidation of any easily oxidizable substances, that is, those which are strong reducing agents. Chlorine will be added later. If chlorine was added at this point and reducing agents were still in the water, they would reduce the chlorine and make it ineffective as a disinfectant.

Ferrous iron (Fe^{2+}) is a particularly troublesome reducing agent. It may arise from the water passing through iron pyrite (FeS_2) or iron carbonate ($FeCO_3$) minerals. If DO is present, Fe^{2+} is oxidized to Fe^{3+}, which precipitates as ferric hydroxide, $Fe(OH)_3$, at pH greater than 3.5. $Fe(OH)_3$ gives a metallic taste to the water and causes the ugly red-brown stain commonly found in sinks and toilets in iron-rich regions. The stain usually is easily removed with weak acid solutions, such as vinegar.

3.11.3.3 Coagulation

The finest sediments, such as pollen, spores, bacteria, and colloidal minerals, do not settle out in the primary settling step. For the finished water to look clear and sparkling, these fine sediments must be removed. Hydrated aluminum sulfate, $Al_2(SO_4)_3 \cdot 18H_2O$, sometimes called alum or filter alum, applied with lime, $Ca(OH)_2$, is the most common filtering agent used for secondary settling:

$$Al_2(SO_4)_3 + Ca(OH)_2 \rightarrow Al(OH)_3(s) + CaSO_4 \qquad (3.24)$$

At pH 6–8, $Al(OH)_3(s)$ is formed as a light, fluffy, gelatinous flocculant having an extremely large surface area that attracts and traps small suspended particles, carrying them to the bottom of the tank as the precipitate slowly settles. In this pH range, $Al(OH)_3$ is near its minimum solubility and very little Al^{3+} is left in solution.

3.11.3.4 Disinfection

Killing bacteria and viruses is the most important part of water treatment. Proper disinfection provides a residual disinfectant level that persists throughout the distribution system. This not only kills organisms that pass through filtration and coagulation at the treatment plant, but it also prevents reinfection during the time the water is in the distribution system. In a large city, water may remain in the system for 5 days or more before it is used. Five days is long enough for any missed microorganisms to multiply. Leaks and breaks in the water mains can permit recontamination, especially at the extremities of the system where the pressure is low. High pressure always causes the flow at leaks to be from inside to outside, but at low pressure, bacteria can seep into a water distribution system.

3.11.4 DISINFECTION PROCEDURES

Most disinfectants are strong oxidizing agents that react with organic and inorganic oxidizable compounds in water. In some cases, the oxidant is produced as a reaction byproduct; hydroxyl radical is formed in this way. In addition to destroying pathogens, disinfectants are also used for removal of disagreeable tastes, odors, and color.

They also can assist in the oxidation of dissolved iron and manganese, prevention of algal growth, improvement of coagulation and filtration efficiency, and control of nuisance water organisms such as Asiatic clams and zebra mussels.

The most commonly used water treatment disinfectant is chlorine. Other disinfectants sometimes used are ozone, chlorine dioxide, and ultraviolet (UV) radiation. Of these, only chlorine and chlorine dioxide have residual disinfectant capability. With chlorine or chlorine dioxide, addition of a small excess of disinfectant maintains protection of the drinking water throughout the distribution system. Normally, residual chlorine or chlorine dioxide concentration of about 0.2–0.5 mg/L is sought. Disinfectants that do not have residual protection are normally followed by a low dose of chlorine in order to preserve the disinfection capability throughout the distribution system.

3.11.4.1 Chlorine Disinfection

Chlorine is a corrosive and toxic yellow-green gas at room temperature, with a strong irritating odor. It is stored and shipped as liquefied gas. Chlorine is the most widely used water treatment disinfectant because of its many attractive features:

* It is effective against a wide range of pathogens commonly found in water, particularly bacteria and viruses.
* It leaves a residual that stabilizes water in distribution systems against reinfection.
* It is economical and easily measured and controlled.
* It has been used for a long time and represents a well-understood treatment technology. It maintains an excellent safety record despite the hazards of handling chlorine gas.
* Chlorine disinfection is available from sodium and calcium hypochlorite salts, as well as from chlorine gas. Hypochlorite solutions may be more economical and convenient than chlorine gas for small treatment systems.

In addition to disinfection, chlorination is used for

* Taste and odor control, including destruction of hydrogen sulfide
* Color bleaching
* Controlling algal growth
* Precipitation of soluble iron and manganese
* Sterilizing and maintaining wells, water mains, distribution pipelines, and filter systems
* Improving some coagulation processes

Problems with chlorine usage include the following:

* Not effective against *Cryptosporidium* and limited effectiveness against *Giardia lamblia* protozoa.
* Reactions with naturally occurring organic matter (NOM) can result in the formation of undesirable DBPs.
* The hazards of handling chlorine gas require special equipment and safety programs.
* If site conditions require high chlorine doses, taste and odor problems may arise.

Chlorine dissolves in water by the following equilibrium reactions:

$$Cl_2(g) \rightleftarrows Cl_2(aq) \tag{3.25}$$

$$Cl_2(aq) + H_2O \rightleftarrows H^+(aq) + Cl^-(aq) + HOCl(aq) \tag{3.26}$$

$$HOCl(aq) \rightleftarrows H^+(aq) + OCl^-(aq) \tag{3.27}$$

At pH values below 7.5, hypochlorous acid (HOCl) is the dominant dissolved chlorine species (see Figure 3.8). Above pH 7.5, chlorite anion (OCl⁻) is dominant. The formation of H^+ indicates that chlorination reduces total alkalinity.

The active disinfection species, Cl_2, HOCl, and OCl⁻, are called the total free available chlorine, although Cl_2 is insignificant above pH 2. All these species are oxidizing agents, but chloride ion (Cl⁻) is not. HOCl is about 100 times more effective as a disinfectant than OCl⁻. Thus, the amount of chlorine required for a given level of disinfection depends on the pH. Higher doses are needed at higher pH. For the same amount of disinfection, the required chlorine dose at pH 8.5 is 7.6 times larger than at pH 7.0. HOCl is more effective than OCl⁻ because, as a neutral molecule, it can penetrate into the cell membranes of microorganisms more easily than can OCl⁻.

When chlorine gas is added to a water system, it dissolves according to Equations 3.25 through 3.27. The expression *chlorine demand* means the amount of chlorine required to oxidize all substances present in the water that are oxidizable by chlorine (see Figure 3.9). Until oxidation of these substances is complete, all the added chlorine is consumed and the net dissolved chlorine concentration remains zero as more chlorine is added. When no chlorine-oxidizable matter is left, that is, when the chlorine demand has been met, the dissolved chlorine concentration (chlorine residual) increases in direct proportion to the additional dose.

If chlorine demand is zero, residual always equals the dose, and the plot is a straight line with slope = 1, passing through the origin of Figure 3.9. Because the boiling point of molecular chlorine is −35°C at 1 atm pressure, chlorine is supplied and stored as the bulk liquid under pressure. The total time of water in the chlorine

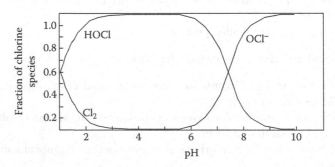

FIGURE 3.8 Distribution diagram for dissolved chlorine species. Free chlorine molecules, Cl_2, exist only below about pH 1. At pH 7.5, [HOCl] = [OCl⁻].

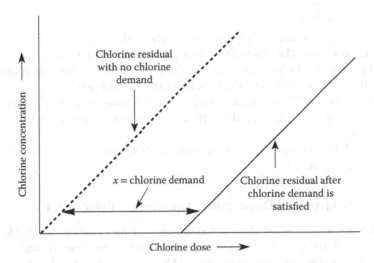

FIGURE 3.9 Relations among chlorine dose, chlorine demand, and chlorine residual.

disinfection tank is generally about 20–60 minutes. A typical concentration of residual chlorine in the finished water is 1 ppm or less.

3.11.4.2 Hypochlorite

In addition to chlorine gas, the active disinfecting species HOCl and OCl⁻ can be obtained from hypochlorite salts, chiefly sodium hypochlorite (NaOCl) and calcium hypochlorite (Ca(OCl)$_2$). The salts react in water according to Equations 3.28 and 3.29.

$$NaOCl + H_2O \rightarrow HOCl + Na^+ + OH^- \qquad (3.28)$$

$$Ca(OCl)_2 + 2H_2O \rightarrow 2HOCl + Ca^{2+} + 2OH^- \qquad (3.29)$$

Note that adding chlorine gas to water lowers the pH (Equations 3.25 through 3.27), while hypochlorite salts raise the pH.

Sodium hypochlorite salts are available as a dry salt or in aqueous solution. The solution is corrosive, with a pH of about 12. One gallon of 12.5% sodium hypochlorite solution is equivalent to about 1 lb of chlorine gas as a disinfectant. Unfortunately, sodium hypochlorite presents storage problems; after 1 month of storage under the best of conditions (low temperature, dark, and no metal contact), a 12.5% solution will have degraded to about 10%. On-site generation of sodium hypochlorite is accomplished by passing low-voltage electrical current through an NaCl solution. On-site generation allows smaller quantities to be stored and makes the use of more stable dilute solutions (0.8%) feasible.

Calcium hypochlorite is commonly available as dry salt, which contains about 65% available chlorine. Almost 1.5 lb of calcium hypochlorite is equivalent to about 1 lb of chlorine gas as a disinfectant. Storage is less of a problem with calcium hypochlorite; normal storage conditions result in a 3–5% loss of its available chlorine per year.

3.11.4.3 Definitions

Chlorine dose: The amount of chlorine originally used.

Chlorine residual: The amount remaining at the time of analysis.

Maximum residual disinfection level (MRDL): The maximum concentration of residual disinfectant allowed by the EPA in finished drinking water.

Chlorine demand: The amount used up in oxidizing organic substances and pathogens in the water, that is, the difference between the chlorine dose and the chlorine residual.

Free available chlorine: The total amount of HOCl and ClO⁻ in solution. (Cl_2 is not present above pH 2.)

3.11.5 Side Effects of Disinfection: Disinfection Byproducts

Most drinking water must be disinfected to make it reliably safe to drink. However, chemicals used for killing pathogens are necessarily very reactive and have the potential for reacting in undesirable ways. Most drinking water disinfectants are oxidizing agents and tend to oxidize everything else in the water that is susceptible. When disinfectants such as chlorine, chlorine dioxide, or ozone react with organic and inorganic substances present in the raw water, DBPs can form, some of which are harmful and must be controlled.

The formation of DBPs is a greater concern for water systems that use surface water, such as rivers, lakes, and streams, because surface water sources are more likely than groundwater to contain the organic and inorganic materials (called DBP precursors) that react with disinfectants to form DBPs. The primary organic DBP precursor is naturally occurring organic matter (NOM), derived from terrestrial and aquatic plants. The main inorganic DBP precursor is bromide ion.

NOM is usually measured as total organic carbon (TOC) or dissolved organic carbon (DOC). Typically in surface waters, about 90% of TOC is in the form of DOC. DOC is defined as the part of TOC that passes through a 0.45-µm filter. Halogenated organic byproducts are formed in water when NOM reacts with free chlorine (Cl_2) or free bromine (Br_2). Free chlorine may be introduced when chlorine gas, chlorine dioxide, or chloramines are added for disinfection. Free bromine is a product of the oxidation by bromide ion already present in the source water. Bromide ion (Br^-) may be present, especially where geothermal waters impact surface and groundwaters, and in coastal areas where saltwater incursion is occurring. Ozone or free chlorine oxidizes Br^- to form brominated DBPs, such as bromate, bromoform, cyanogen bromide, bromopicrin, and brominated acetic acid.

Reactions of TOC with strong nonhalogen oxidants, such as ozone, peroxone, permanganate, and peroxide, will also form nonhalogenated DBPs. Common nonhalogenated DBPs include aldehydes, ketones, organic acids, ammonia, and hydrogen peroxide. Because all water disinfectants will react with both particulate and dissolved organic matter, the EPA regulates TOC in its program for controlling the levels of DBPs.

Because of concerns about DBPs, the EPA and the water treatment industry are placing more emphasis on the use of disinfectants other than chlorine, which at present is the most commonly used water disinfectant. However, use of alternative disinfectants has also been found to produce DBPs and current regulations try to balance the risks between microbial pathogens and DBPs. Another approach to reducing the probability of DBP formation is by removing DBP precursors (TOC) from water before disinfection. DBPs include the following, not all of which pose health risks:

- Halogenated organic compounds, such as trihalomethanes (THMs), haloacetic acids (HAAs), haloketones, and other halogenated compounds that are formed primarily when chlorine or ozone (in the presence of bromide ion) are used for disinfection.
- Organic oxidation byproducts, such as aldehydes, ketones, assimilable organic carbon (AOC), and biodegradable organic carbon (BDOC). The latter two DBPs result from large organic molecules being oxidized to smaller molecules, which are more available to microbes, plant, and aquatic life as a nutrient source. Oxidized organics are formed when strong oxidizing agents (ozone, permanganate, chlorine dioxide, or hydroxyl radical) are used.
- Inorganic compounds, such as chlorate, chlorite, and bromate ions, formed when chlorine dioxide and ozone disinfectants are used.

3.11.5.1 Trihalomethanes and Haloacetic Acids

Chlorine, in gas, liquid, or solid formulations, is the most common oxidant and disinfectant used for the treatment of drinking water. A major advantage over other disinfectants of using chlorine is that it can be made to persist in treated water long enough to provide continuous disinfection throughout the entire distribution system, from the treatment plant to the household water tap. The major problem with the use of chlorine as a disinfectant is the formation of chlorination byproducts, particularly THMs and HAAs. THMs and HAAs can continue to form within a drinking water distribution system if residual chlorine contacts dissolved organic matter.

- It was once thought that THMs were formed by chlorination of dissolved methane. It is now known that they come from the reaction of HOCl with acetyl groups in TOC, chiefly humic acids. Humic acids are breakdown products of plant materials like lignin. There is no evidence that chlorine itself is carcinogenic.
- At least 600 halogenated DBPs have been identified. The low levels of many of these DBPs, coupled with the analytical costs in testing water samples for them, means that in practice, only a few DBPs are monitored. The most important halogenated DBPs are listed in Table 3.10. Four THMs (together regulated as total THMs or TTHM) and five HAAs (regulated as HAA5) serve as indicators of the bulk of DBPs that are not readily measurable; any treatment that reduces TTHM and HAA5 will also reduce the formation of other unregulated DBPs.

TABLE 3.10
Important Halogenated Disinfection Byproducts

Compound	Formula	Chemical Class	Cancer Classification[a,b]
Bromate	BrO_3^-	Inorganic	B2
Bromochloroacetic acid	CHBrClCOOH	HAA	—
Bromodichloroacetic acid	$CBrCl_2COOH$	HAA	—
Bromodichloromethane	$CHBrCl_2$	THM	L
Bromoform	$CHBr_3$	THM	L
Chlorate	ClO_3^-	Inorganic	—
Chlorite	ClO_2^-	Inorganic	D
Chloroform	$CHCl_3$	THM	L/N
Dibromoacetic acid	$CHBr_2COOH$	HAA	—
Dibromochloroacetic acid	$CBr_2ClCOOH$	HAA	—
Dibromochloromethane	$CHBr_2Cl$	THM	S
Dichloroacetic acid	$CHCl_2COOH$	HAA	L
Monobromoacetic acid	$CH_2BrCOOH$	HAA	—
Monochloroacetic acid	$CH_2ClCOOH$	HAA	I
Tribromoacetic acid	CBr_2COOH	HAA	—
Trichloroacetic acid	CCl_3COOH	HAA	S

Sources: U.S. Environmental Protection Agency (USEPA), *Alternative Disinfectants and Oxidants Guidance Manual*, EPA 815-R-99–014, Environmental Protection Agency, Washington, DC, 1999; U.S. Environmental Protection Agency (USEPA), *2011 Edition of the Drinking Water Standards and Health Advisories*, EPA 820-R-11-002. Environmental Protection Agency, Washington, DC, 2011.

[a] The EPA classifications for the carcinogenic potential of chemicals are a descriptive, weight-of-evidence judgment as to the likelihood that an agent is a human carcinogen. Under the 2005 EPA *Guidelines for Carcinogen Risk Assessment*, descriptive terms for carcinogenicity replace the earlier alphanumeric Cancer Group designations (USEPA 1986 guidelines). In the table above, both the new and the old classification symbols are given: the new symbols for chemicals that have been reevaluated under the new guidelines and the old symbols for those that have not.

[b] Carcinogenicity based on inhalation exposure.

Old Cancer Classifications (EPA 1986 guidelines):

A = Human carcinogen: Sufficient evidence in epidemiologic studies to support causal association between exposure and cancer.

B = Probable human carcinogen.

B1 = Limited evidence in epidemiologic studies and/or sufficient evidence from animal studies.

B2 = Sufficient evidence in animals and inadequate or no evidence in humans.

C = Possible human carcinogen: Limited evidence from animal studies and inadequate or no data in humans.

D = Not classifiable: Inadequate or no animal and human evidence of carcinogenicity.

E = No evidence of carcinogenicity for humans: No evidence of carcinogenicity in at least two adequate animal tests or in adequate epidemiologic and animal studies.

New Cancer Classifications (EPA 2005 guidelines):

H = Carcinogenic to humans.

L = Likely to be carcinogenic to humans.

L/N = Likely to be carcinogenic above a specified dose but not likely to be carcinogenic below that dose because a key event in tumor formation does not occur below that dose.

S = Suggestive evidence of carcinogenic potential.

I = Inadequate information to assess carcinogenic potential.

N = Not likely to be carcinogenic to humans.

3.11.5.2   Strategies for Controlling Disinfection Byproducts

Once formed, DBPs are difficult to remove from a water supply. Therefore, DBP control is focused on preventing their formation. The chief control measures for DBPs are as follows:

- Lowering TOC concentrations in source water by coagulation and settling, filtering, and oxidation.
- Using sorption on granulated activated carbon (GAC) to remove DOC.
- Moving the disinfection step later in the treatment train, so that it comes after all processes that decrease TOC.
- Using chloramines as a residual disinfectant in distribution systems or limiting chlorine to providing residual disinfection. Residual disinfection protects the distribution system from pathogens and follows a primary disinfection step that uses disinfectants less likely than chlorine to form DBPs, such as ozone, chlorine dioxide, chloramines, or UV radiation. These alternate primary disinfectants are discussed in Sections 3.11.6 through 3.11.11
- Protection of source water from bromide ion.

3.11.5.3 Regulation of Disinfection Byproducts

The EPA's program for regulating DBPs has the challenge of minimizing health risks from microbial pathogens by requiring adequate disinfection of drinking water while, at the same time, not allowing disinfection byproducts to become health risks in themselves. At present, it appears that DBPs are best managed by using treatment techniques that reduce the formation of DBPs, rather than trying to remove them after formation. Therefore, the Disinfectants/Disinfection Byproducts Rule (D/DBPR), with which the EPA regulates disinfectants and DBPs, includes removing TOC by coagulation, water softening, or filtration, as part of the disinfection procedure.

Any TOC concentration greater than 2.0 mg/L, a value that indicates a high potential for DBP formation, requires removal by either enhanced coagulation or enhanced softening. Enhanced coagulation is determined, in part, as the coagulant dose where an incremental addition of 10 mg/L of alum (or an equivalent amount of ferric salt) results in a TOC removal to below 0.3 mg/L. Typical required reduction percentages of TOC for conventional treatment plants are given in Table 3.11.

In addition to regulating the removal of TOC, the EPA regulates THMs and HAAs. The D/DBPR was designed to be implemented in three stages, the first of which was promulgated in December 1998. In 2012, the D/DBPR is in Stage 2. In Stage 1, THMs and HAAs were not regulated as individual compounds but as indicators that other unmeasured DBPs may be present. The EPA set MCLs (maximum contaminant levels) and MCLGs (maximum contaminant level goals) for total THMs (TTHMs, defined as four specific THMs: chloroform, bromoform, bromodichloromethane, and dibromochloromethane) and haloacetic acids (HAA5, defined as five specific acetic acids: monobromoacetic acid, dibromoacetic acid, monochloroacetic acid, dichloroacetic acid, and trichloroacetic acid).

In the current Stage 2 D/DBPR, MCLs for TTHMs and HAA5 were retained, but, in addition, some THMs and HAAs were also given individual MCLs and MCLGs due to new information concerning their carcinogen hazards. MCLs for Stage 1 and Stage 2 D/DBPR are given in Table 3.12.

TABLE 3.11
Required Percentage Removal of Total Organic Carbon (TOC) by Enhanced Coagulation for Conventional Water Treatment Systems[a]

	Source Water Alkalinity (mg/L as CaCO$_3$)		
Source Water TOC (mg/L)	0–60	>60–120	>120
>2.0–4.0	35.0%	25.0%	15.0%
>4.0–8.0	45.0%	35.0%	25.0%
>8.0	50.0%	40.0%	30.0%

Note: Enhanced coagulation is determined, in part, as the coagulant dose where an incremental addition of 10 mg/L of alum (or an equivalent amount of ferric salt) results in a TOC removal to below 0.3 mg/L.

[a] Conventional water treatment systems here apply to utilities using surface water and groundwater impacted by surface water.

TABLE 3.12
Disinfectants and Disinfection Byproducts Rule (D/DBPR) Drinking Water Standards[a]

Contaminant	MCL/MCLG (mg/L, unless noted)
Stage 1 D/DBPR: Disinfectants	
Chlorine	4.0 (as Cl$_2$) MRDL/4.0 MRDLG
Chloramines	4.0 (as Cl$_2$) MRDL/4.0 MRDLG
Chlorine dioxide	0.8 (as ClO$_2$) MRDL/0.8 MRDLG
Stage 1 D/DBPR: Disinfection Byproducts	
Total trihalomethanes (TTHMs)	0.080
Haloacetic acids (HAA5)	0.060
Chlorite	1.0/0.8
Bromate	0.010/0
Total organic carbon (TOC)	Best available treatment technique
Stage 2 D/DBPR: Disinfection Byproducts	
Total trihalomethanes (TTHMs)	0.080
Chloroform	0.07
Bromodichloromethane (BCDM)	0.06/0
Bromoform	0
Haloacetic acids (HAA5)	0.060
Monochloroacetic acid (MCAA)	0.07
Dichloroacetic acid (DCAA)	0
Trichloroacetic acid (TCAA)	0.02

[a] MRDL/MRDLG = maximum residual detection limit/maximum residual detection limit goal, applicable to chlorine residuals, Section 3.11.4.1

3.11.5.4 Chlorinated Phenols

If phenol or its derivatives from industrial activities are in the water, taste and color can be a problem. Phenols are easily chlorinated, forming compounds with extremely penetrating antiseptic odors. The most common chlorinated phenols arising from chlorine disinfection are shown in Table 3.13, with their odor thresholds, which are in the ppb (µg/L) range. At the ppm level, chlorinated phenols make water completely unfit for drinking or cooking. If phenol is present in the intake water, treatment choices are to employ additional nonchlorine oxidation for removing phenol,

---

TABLE 3.13

Odor Thresholds of Phenol and Chlorinated Derivatives from Drinking Water Disinfection with Chlorine

Phenol Compound	Chemical Structure	Odor Threshold in Water (ppb)
Phenol		>1000
2-Chlorophenol		2
4-Chlorophenol		250
2,4-Chlorophenol		2
2,6-Chlorophenol		3
2,4,6-Chlorophenol		>1000

---

remove phenol with activated charcoal, or use a different disinfectant. The activated charcoal treatment is expensive and few communities use it.

Example 11

Water started to seep into the basement of a home. The home's foundation is well above the water table and this problem had not been experienced before. The house is located about 50 ft downgradient from a main water line and one possibility is that a leak has occurred in the pipeline. The water utility company tested water entering the basement for the presence of chlorine, reasoning that if the water source was the pipeline, the chlorine residual should be detected. When no chlorine was found, the utility concluded that they were not responsible for the seep. Was this conclusion justified? Note that the MRDL mandated by the EPA is 4.0 mg/L.

ANSWER

No. Water would have to travel at least 50 ft through soil from the pipeline to the house. The chlorine residual in treated domestic use should not exceed 4 mg/L and would most likely contact enough oxidizable organic and inorganic matter in the soil to be depleted below detection. A better water source marker would be fluoride, assuming that the water supply is fluoridated. Although fluoride might react with calcium and magnesium in the soil to form solid precipitates, it is more likely to be detectable at the house than chlorine. However, neither test is conclusive. The simplest and best test would be to turn off the water in the pipeline for long enough to observe any change in water flow into the house. This, however, might not be possible. Another approach would be to examine the water line for leaks, using a video camera probe or soil conductivity measuring equipment.

3.11.5.5 Chloramines

Many utilities use chlorine for disinfection and chloramines for residual maintenance. Chloramines are formed in the reaction of ammonia with HOCl from chlorine, a process that is inexpensive and easy to control. The reactions are described in the discussion of breakpoint chlorination in Chapter 4, Section 4.2.6.3. Although the reaction of chlorine with ammonia can be used for the purpose of destroying ammonia, it also serves to generate chloramines, which are useful disinfectants that are more stable and long lasting in a water distribution system than free chlorine. Thus, chloramines are effective for controlling bacterial regrowth in water systems, although they are not very effective against viruses and protozoa.

The primary role of chloramine use in the United States is as a secondary disinfectant to provide residual treatment. Being weaker oxidizers than chlorine, chloramines also form far fewer DBPs. However, they are not useful for oxidizing iron and manganese. When chloramine disinfection is the goal, ammonia is added in the final chlorination step. Chloramines are always generated on-site.

Optimal chloramine disinfection occurs when the chlorine:ammonia–nitrogen (Cl_2:N) ratio by weight is around 4, before the chlorination breakpoint occurs. Under these conditions, monochloramine (NH_2Cl) and dichloramine ($NHCl_2$) are the main reaction products and these are the effective disinfectant species. The normal dose of chloramines is between 1 and 4 mg/L. Residual concentrations are usually maintained between 0.5 and 1 mg/L. The MRDL mandated by the EPA is 4.0 mg/L.

3.11.5.6 Chlorine Dioxide Disinfection Treatment

Chlorine dioxide (ClO_2) is a gas above 12°C, with high water solubility. Unlike chlorine, it reacts quite slowly with water, remaining mostly dissolved as a neutral molecule. It is a very good disinfectant, about twice as effective as HOCl from Cl_2, but also about twice as expensive. ClO_2 was first used as a municipal water disinfectant in Niagara Falls, New York, in 1944. In 1977, about 100 municipalities in the United States were using it and thousands in Europe. The main drawback to its use is that it is unstable and cannot be stored. It must be made and used on-site, whereas chlorine can be delivered in tank cars.

Much of its reactivity is because it is a free radical. ClO_2 cannot be compressed for storage because it is explosive when pressurized or at concentrations above 10% by volume in air. It decomposes in storage and can decompose explosively in sunlight, when heated or agitated suddenly. Therefore, it is never shipped and is always prepared on-site and used immediately. Typical dose rates are 0.1–1.0 ppm.

Sodium chlorite is used to make ClO_2 by one of three methods:

$$5NaClO_2 + 4HCl \leftrightarrow 4ClO_2(g) + 5NaCl + 2H_2O \qquad (3.30)$$

$$2NaClO_2 + Cl_2(g) \rightarrow 2ClO_2(g) + NaCl \qquad (3.31)$$

$$2NaClO_2 + HOCl \rightarrow 2ClO_2(g) + NaCl + NaOH \qquad (3.32)$$

Sodium chlorite is extremely reactive, especially in the dry form, and it must be handled with care to prevent potentially explosive conditions. If chlorine dioxide generator conditions are not carefully controlled (pH, feedstock ratios, low feedstock concentrations, etc.), the undesirable byproducts chlorite (ClO_2^-) and chlorate (ClO_3^-) may be formed.

Chlorine dioxide solutions below about 10 g/L will not have sufficiently high vapor pressures to create an explosive hazard under normal environmental conditions of temperature and pressure. For drinking water treatment, ClO_2 solutions are generally less than 4 g/L and treatment levels are generally between 0.07 and 2.0 mg/L.

As ClO_2 is an oxidizer and not a chlorinating agent, it does not form THMs or chlorinated phenols. Hence, it does not have taste or odor problems. Common applications for ClO_2 have been to control taste and odor problems associated with algae and decaying vegetation, reducing the concentrations of phenolic compounds, and oxidizing iron and manganese to insoluble forms. Chlorine dioxide can maintain a residual disinfection concentration in distribution systems. The toxicity of ClO_2 restricts the maximum dose. At 50 ppm, ClO_2 can lead to breakdown of red corpuscles with the release of hemoglobin. Therefore, the dose of ClO_2 is limited to 1 ppm.

3.11.6 NON-CHLORINE-BASED WATER TREATMENTS THAT ALSO PROVIDE DISINFECTION

A major advantage of chlorine-based disinfectants is that they can persist long enough to provide residual disinfection between the water treatment plant and the point of consumer use. When this is not an important consideration, as in swimming pools, hot tubs, and the like, other procedures such as those that follow may be preferable.

3.11.6.1 Ozone Disinfection Treatment

Ozone (O_3) is a colorless, highly corrosive gas at room temperature, with a pungent odor, easily detectable at concentrations as low as 0.02 ppmv, well below a hazardous level. It is one of the strongest chemical oxidizing agents available, second only to hydroxyl free radical ($HO\cdot$),* among disinfectants commonly used in water treatment. Ozone use for water disinfection started in 1893 in the Netherlands and in 1901 in Germany. Significant use in the United States did not occur until the 1980s. Ozone is one of the most potent disinfectants used in water treatment today. Ozone disinfection is effective against bacteria, viruses, and protozoan cysts, including *Cryptosporidium* and *G. lamblia*.

Ozone is made by passing a high-voltage electric discharge of about 20,000 V through dry, pressurized air.

$$3O_2(g) + energy \rightarrow 2O_3(g) \tag{3.33}$$

Equation 3.33 is endothermic and requires a large input of electrical energy. The ozone gas is transferred to water through bubble diffusers, injectors, or turbine mixers. Once dissolved in water, ozone reacts with pathogens and oxidizable organic and inorganic compounds. Gas not dissolved is released to the surroundings as off-gas and must be collected and destroyed by conversion back to oxygen before release to the atmosphere. Ozonator off-gas may contain as much as 3000 ppmv of ozone, well above a fatal level. Ozone is readily converted to oxygen by heating to above 350°C or by passing it through a catalyst held above 100°C. The Occupational Safety and Health Administration (OSHA) currently requires released gases to contain no more than 0.1 ppmv ozone for worker exposure. Typical dissolved ozone concentrations in water near the ozonator are around 1 mg/L.

Dissolved ozone gas decomposes spontaneously in water by a complex mechanism that includes the formation of hydroxyl free radical, which is the strongest oxidizing agent available for water treatment. Hydroxyl radical essentially reacts at every molecular collision with many organic compounds. The very high reaction rate of hydroxyl radicals limits their half-life in water to the order of microseconds and their concentration to less than about 10^{-12} mol/L. Both ozone molecules and hydroxyl free radicals play prominent oxidant roles in water treatment by ozonation.

Because ozone is unstable and highly reactive, it cannot be stored and must be generated as needed. Ozone concentrations of about 4–6% are achieved in municipal and industrial ozonators. Ozone reacts quickly and completely in water, leaving no active residual concentration. Decomposition of ozone in water produces DO in addition to hydroxyl radical (a very reactive short-lived oxidant) both of which further aid in disinfection and reducing BOD, COD, color, and odor problems. The air–ozone mixture is typically bubbled through water for a 10–15-minute contact time. The main drawbacks to ozone use have been its high capital and operating costs and the that it leaves no residual disinfection concentration. Since it offers no residual

---

* The dot following OH represents an unpaired orbital electron, which is a feature of free radicals and gives them their characteristic very high degree of reactivity.

protection, ozone can only be used as a primary disinfectant. It must be followed by a light dose of secondary disinfectant, such as chlorine, chloramine, or chlorine dioxide, for a complete disinfection system.

Several ways to assist ozonation are by adding hydrogen peroxide (H_2O_2), using UV radiation, and raising the pH to around 10–11. Hydrogen peroxide decomposes to form the very reactive hydroxyl radical, greatly increasing the hydroxyl radical concentration above that generated by simple ozone reaction with water. Reactions of hydroxyl radicals with organic matter cause structural changes that make organic matter still more susceptible to ozone attack. Adding hydrogen peroxide to ozonation is known as the advanced oxidation process (AOP) or PEROXONE process. UV radiation dissociates peroxide, forming hydroxyl radicals at a more rapid rate. Raising the pH allows ozone to react with hydroxyl ions (OH^-, not the radical $HO\cdot$) to form additional hydrogen peroxide. In addition to increasing the effectiveness of ozone oxidation, peroxide and UV radiation are effective as disinfectants in themselves. The use of these ozonation enhancers is known as the AOP.

The equipment for ozonation is expensive, but the cost per gallon decreases with large-scale operations. In general, only large cities use ozone. ClO_2 is not as problem-free as ozone, but it is cheaper to use for small systems.

In addition to disinfection, ozone is used for

- DBP precursor control
- Protection against *Cryptosporidium* and *Giardi*
- Taste and odor control, including destruction of hydrogen sulfide
- Color bleaching
- Precipitation of soluble iron and manganese
- Sterilizing and maintaining wells, water mains, distribution pipelines, and filter systems
- Improving some coagulation processes

3.11.6.2 Ozone Disinfection Byproducts

Although it does not form the chlorinated DBPs that are of concern with chlorine use, ozone can react to form its own set of oxidation byproducts. When bromide ion (Br^-) is present, as where geothermal waters impact surface and groundwaters or in coastal areas where saltwater incursion is occurring, ozonation can produce bromate ion (BrO_3^-), a suspected carcinogen, as well as brominated THMs and other brominated DBPs. Controlling the formation of unwanted ozonation byproducts is accomplished by pretreatment to remove organic matter (activated carbon filters and membrane filtration) and scavenge BrO_3^- (pH lowering and hydrogen peroxide addition).

When bromide is present, addition of ammonia with ozone forms bromamines, by reactions analogous to the formation of chloramines with ammonia and chlorine, and lessens the formation of bromate ion and organic DBPs.

3.11.6.3 Potassium Permanganate

Potassium permanganate salt ($KMnO_4$) dissolves to form the permanganate anion (MnO_4^-), a strong oxidant effective at oxidizing a wide variety of organic and inorganic substances. In the process, manganese is reduced to manganese dioxide

(MnO_2), an insoluble solid that precipitates from solution. Permanganate imparts a pink to purple color to water and is therefore unsuitable as a residual disinfectant.

Although easy to transport, store, and apply, permanganate is generally too expensive for use as a primary or secondary disinfectant. It is used in drinking water treatment primarily as an alternative to chlorine for taste and odor control, iron and manganese oxidation, oxidation of DPB precursors, control of algae, and control of nuisance organisms such as zebra mussels and the Asiatic clam. It contains no chlorine and does not contribute to the formation of THMs. When used to oxidize TOC early in a water treatment train that includes post-treatment chlorination, permanganate can reduce the formation of THMs.

3.11.6.4   Peroxone (Ozone Plus Hydrogen Peroxide)

The peroxone process is among those called Advanced Oxidation Processes (AOPs), which employ highly reactive hydroxyl radicals (OH·) as a major oxidizing species. Hydroxyl radicals are produced when ozone decomposes spontaneously. Accelerating ozone decomposition using, for example, UV radiation or addition of hydrogen peroxide, elevates the hydroxyl radical concentration and increases the rate of contaminant oxidation. When hydrogen peroxide is used, the process is called peroxone.

Like ozonation, the peroxone process does not provide a lasting disinfectant residual. Oxidation is more complete and much faster with peroxone than with ozone. Peroxone is the treatment of choice for oxidizing many chlorinated hydrocarbons, difficult to treat by any other oxidant. It is also used for inactivating pathogens and destroying pesticides, herbicides, and volatile organic compounds (VOCs). It can be more effective than ozone for removing taste and odor causing compounds such as geosmin and 2-methyliosborneol (MIB). However, it is less effective than ozone for oxidizing iron and manganese. Because hydroxyl radicals react readily with carbonate, it may be necessary to lower the alkalinity in water with a high carbonate level to maintain a useful level of radicals. Peroxone treatment produces similar DBPs as does ozonation. In general, it forms more bromate than does ozone under similar water conditions and bromine concentrations.

3.11.6.5   Ultraviolet Disinfection Treatment

Ultraviolet radiation at wavelengths below 300 nm is very damaging to life forms, including microorganisms. Low-pressure mercury lamps, known as germicidal lamps, have their maximum energy output at 254 nm. They are very efficient, with about 40% of their electrical input being converted to 254-nm radiation. Protein and DNA in microorganisms absorb radiation at 254 nm, leading to photochemical reactions that destroy the ability to reproduce. UV doses required to inactivate bacteria and viruses are relatively low, of the order of 20–40 (mW·s)/cm^2. Much higher doses, 200 (mW·s)/cm^2 or higher, are needed to inactivate *Cryptosporidium* and *G. lamblia*.

Color or high levels of suspended solids can interfere with transmission of UV through the treatment cell and UV absorption by iron species diminishes the UV energy absorbed by microorganisms. Such problems may necessitate higher UV dose rates or pretreatment filtration. UV reaction cells are designed to induce turbulent flow, have long water flow paths and short light paths (around 3 in.), and provide for cleaning of residues from the lamp housings to minimize these problems. Where used, usually in small water treatment systems, UV irradiation is generally the last step in the water treatment process, just after final filtration and before entering the distribution system. UV systems are normally easy to operate and maintain, although severe site conditions, such as high levels of dissolved iron or hardness, may require pretreatment.

UV does not introduce any chemicals into the water and causes little, if any, chemical change in water. Therefore, overdosing does not cause water quality problems. UV is used mostly for inactivating pathogens to regulated levels. Since it leaves no residual, it can serve only as a primary disinfectant and must be followed by some form of chemical secondary disinfection, generally chlorine or chloramine. UV water treatment is used more in Europe than in the United States. Small-scale units are available for individuals who have wells with high microbial levels.

3.11.6.6 Characteristics of UV Treatment

* Short contact time of 1–10 seconds. Ozone and chlorine require 10–50 minutes, necessitating large reaction tanks. Ozonation can be run on a flow-through basis.
* Destroys most viruses and bacteria without chemical additives. The destruction of *G. lamblia*, however, requires prefiltration. Leaves no residual disinfection potential in the water so that, for water entering a distribution system, light chlorination is still needed to provide prolonged disinfection.
* Low overall installation costs. Ozone generators are expensive. Chlorine-metering systems are not especially expensive, but large reaction tanks and safety systems are high-cost items.
* Not influenced by pH or temperature. Chlorination and ozonation work best at lower pH (chlorine because it is in the HOCl form; ozone because it decomposes more rapidly at higher pH). Chlorination and ozonation both require longer contact time at lower temperatures.
* No toxic residues. It adds nothing to the water unless some organics are present that photoreact to form toxic compounds. The formation of THMs or other DBPs is minimal.

3.11.6.7 Membrane Filtration Water Treatment

Membrane filters are being used to treat groundwater, surface water, and reclaimed wastewater. Membrane filtration is a physical separation process and removes unwanted substances from water without utilizing chemical reactions that can lead to undesirable byproducts. The range of membrane filters available is shown in Figure 3.10, along with common substances that can be removed by filtering. Although membranes sometimes serve as a stand-alone treatment, they are more often combined with other treatment technologies. For example, currently available microfiltration (MF) and ultrafiltration (UF) membranes are not very effective in removing DOC, some synthetic organic compounds, or THM precursors. Their performance in these respects is improved by adding powdered sorbent material to the wastewater flow. Contaminants that might pass through the filters are sorbed to the larger sorbent particles and rejected by the filters.

Organic membrane filters are made from several different organic polymer films, normally formed as a thin film on a supporting woven or nonwoven fabric. Inorganic membrane filters are made from ceramics, glass, or carbon. They generally consist

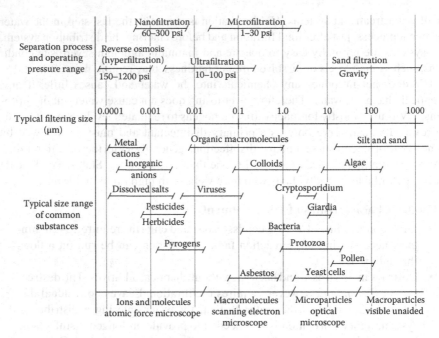

FIGURE 3.10   Comparison of filter processes and size ranges.

of a more porous supporting layer on which a thin microporous layer is chemically deposited. Inorganic membranes resist higher pressures, a wide pH range, and more extreme temperatures than do organic membranes. Their main disadvantages are greater weight and expense.

Filters can be fabricated to remove substances as small as dissolved ions, so they are useful for removing TDS, nitrate sulfate, radium, iron, manganese, DBP precursors, bacteria, viruses, and other pathogens from water without adding chemicals. It must also be recognized that there will be imperfections in manufactured membrane filters through which contaminants may pass. This is of particular concern with pathogens. Therefore, filters must never be regarded as having 100% rejection for any size range. Furthermore, filters do not protect water from reinfection after it has entered a distribution system, so it is common to add chlorine or another residual disinfectant at the end of the treatment chain for this purpose. As most organic matter has already been removed, end-of-treatment chlorination does not generate significant DBPs.

Unlike coarser filters operating in a normal mode, where all of the water passes through the filter surface, membrane filters operate in a crossflow mode. In crossflow filtration, the feed, or influent, stream is separated into two separate effluent streams. The pressurized feed water flows parallel to the membrane filter surface and some of the water diffuses through the filter. The remaining feed stream continues parallel to the membrane to exit the system without passing through the membrane surface. Filtered contaminants remain in the feed stream water, increasing in concentration until the feed stream exits the filter unit. The filtered water is called the permeate effluent, and the exiting feed stream water is called the concentrate effluent. Crossflow filtration provides a self-cleaning effect that allows continuous flushing away of contaminants, which, in normal filtration, would plug filters of small pore size very quickly.

Depending on the nominal size of the pores engineered into the membrane, crossflow filters are used in filtering applications classified as reverse osmosis (RO), nanofiltration (NF), ultrafiltration (UF), and microfiltration (MF), listed in order of increasing pore size range. The pore sizes in these membranes are so small that significant pressure is required to force water through them; the smaller the pore size, the higher the required pressure. Figure 3.10 illustrates some of the uses for the different membrane filter types.

3.11.6.8 Reverse Osmosis

Reverse osmosis, sometimes called hyperfiltration, was the first crossflow membrane separation process to be widely used for water treatment. It requires operating pressures of 150–1200 pounds per square inch (psi). It removes up to 99% of ions and most dissolved organic compounds. RO can meet most drinking water standards with a single-pass system. Although it might not be the most economical approach, using RO in multiple-pass systems allows the most stringent drinking water standards to be met. For example, rejection of 99.9% of viruses, bacteria, and pyrogens is achievable with a double-pass system.

3.11.6.9 Nanofiltration

Nanofiltration membranes can separate organic compounds with molecular weights as small as 250 Da.* It will also separate most divalent ions, and is effective for softening water (removing Ca^{2+} and Mg^{2+}). It allows greater water flow through at lower operating pressure (60–300 psi) than RO.

3.11.6.10 Ultrafiltration

Ultrafiltration membranes do not remove ions. They reject compounds greater than about 700 Da. The larger pore size permits lower operating pressures, in the range of 10–100 psi. UF is useful for separating larger organic compounds, colloids, bacteria, pyrogens, *G. lamblia*, and *Cryptosporidium*. Most ions and smaller molecules such as chloroform and sucrose will pass through UF membranes.

3.11.6.11 Microfiltration

Microfiltration membranes reject contaminants in the 0.05–3.0-μm range and operate at pressures of 1–30 psi. They are available in polymer, metal, and ceramic materials. MF is sometimes used as a pretreatment for RO, increasing the efficiency and duty cycle of RO membranes significantly. MF will remove *G. lamblia and Cryptosporidium*, whose spores range between 3 and 18 μm in diameter, but not viruses and most bacteria. MF-RO treatment trains are reported to be economical, easy to operate, and very reliable.

---

* See Chapter 1, footnote † on page 16 for a definition of Da, which essentially is the molecular weight in grams of a compound. For example, the molecular weight of chloroform ($CHCl_3$) is 119 Da or 119 g/mol. Chloroform will not be separated by nanofiltration, which does not reject molecules smaller than 250 Da.

EXERCISES

1. a. The measured pH of a seawater sample is 8.30. What is the hydrogen ion (H^+) concentration in mol/L and in mg/L?
   b. What is the pH of a water sample in which $[H^+] = 1.5 \times 10^{-10}$ M?
2. a. Water in a stream had $[H^+] = 6.1 \times 10^{-8}$ mol/L. What is the pH?
   b. The pH of water in a stream is 9.3. What is the hydrogen ion concentration?
3. An engineer requested a water sample analysis that included the parameters pH, carbonate ion, and bicarbonate ion. Explain why she probably is wasting money.
4. a. Water sample analysis indicated a total carbonate concentration of 0.003 mol/L and a pH of 6.6. What is the alkalinity in meq/L and in mg/L?
   b. If 1 L of the sample is diluted with enough pure water to change the total carbonate concentration to 0.0025 mol/L, what would be the new pH and alkalinity? Use Figure 3.3.
5. A water sample contains 150 mg/L of Ca^{2+} and 33 mg/L of Mg^{2+}. Calculate the total hardness of the water.
6. A water sample from a lake has a measured alkalinity of 0.8 eq/L.
   In the early morning, a monitoring team measures the lake's pH as part of an acid rain study and finds pH 6.0. The survey team returns after lunch to recheck their data. By this time, algae and other aquatic plants have consumed enough dissolved CO_2 to reduce the lake's total carbonate (C_T) to one-half of its morning value.
   a. What was the morning value for C_T?
   b. What was the pH after lunch?
7. A groundwater has the following analysis at pH 7.6:

Analyte	Concentration (mg/L)
Calcium	75
Magnesium	40
Sodium	10
Bicarbonate	300
Chloride	10
Sulfate	112

Calculate alkalinity, total hardness, carbonate (temporary) hardness, and noncarbonate (permanent) hardness.
8. A stream has a measured DO of 6.0 mg/L at 12°C. What can you say about its ability to support aquatic life?

REFERENCES

Deffeyes, K. S. 1965. "Carbonate Equilibria: A Graphic and Algebraic Approach," *Limnology and Oceanography* 10 (3): 412–26.

Nordstrom, D. K., et al. 2000. "Negative pH and Extremely Acid Mine Waters from Iron Mountain, California," *Environmental Science & Technology* 34 (2): 254–58.

Standard Methods, 1995, *Standard Methods for the Examination of Water and Wastewater*, 19th Ed., Edited by Eaton A., et al. Washington, DC: APHA, AWWA, WEF.

Stumm, W., and Morgan, J. J. 1996. *Aquatic Chemistry, Chemical Equilibria and Rates in Natural Waters*, 3rd ed. New York: John Wiley & Sons, Inc., pp 1022.

U.S. Environmental Protection Agency (USEPA). 1999. *Alternative Disinfectants and Oxidants Guidance Manual*, EPA 815-R-99–014. Washington, DC: Environmental Protection Agency.

U.S. Environmental Protection Agency (USEPA). 2011. *2011 Edition of the Drinking Water Standards and Health Advisories*, EPA 820-R-11-002. Washington, DC: Environmental Protection Agency.

World Health Organization (WHO). 2010. "Water for Health: WHO Guidelines for Drinking Water Quality." www.who.int/entity/water_sanitation_health/publications/2011/dwq_ guidelines/en/. Accessed September 17, 2012.

(Photo by Eugene Weiner)

Eutrophication is a naturally occurring process that is often accelerated with escalation of agriculture or other practices that result in increased flow of nutrients to water bodies.

Chapter 4, Section 4.3.5.4

4 Nutrients and Odors
Nitrogen, Phosphorus, and Sulfur

4.1 NUTRIENTS AND ODORS

Of the 16 or so essential plant nutrients,* nitrogen (N), phosphorus (P), and sulfur (S) are the three nutrients that currently are of greatest environmental concern because of elevated concentrations in environmental waters. According to the U.S. Environmental Protection Agency (EPA), nutrient enrichment ranks as one of the top causes of water resource impairment. In environmental waters that failed to meet their designated use criteria, nutrients (mainly nitrogen and phosphorus) were listed as a primary cause of impairment in 40% of rivers, 51% of lakes, and 57% of estuaries (EPA 2000).

Excessive concentrations of nitrogen and phosphorus are of concern because they are the nutrients that most frequently cause algal blooms, eutrophication, and diminished dissolved oxygen (DO). They also facilitate several serious health-related issues, including trihalomethane formation in chlorinated drinking water, non-Hodgkin lymphoma, and methemoglobonemia in water with elevated nitrate/nitrite.

* Other plant nutrients are discussed in Chapter 6, Section 6.11, Agricultural Water Quality.

Although a minor nutrient, sulfur is also of concern, because of the toxicity of many waterborne sulfur species to humans and aquatic life and the unpleasant taste and odor sulfur compounds can impart to water. These problems with sulfur species arise almost entirely in anaerobic waters, a condition that connects sulfur to the concerns with nitrogen and phosphorus nutrients, because anaerobic waters are found where algal blooms, eutrophication, and low DO occur.

4.2 NITROGEN: AMMONIA, NITRITE, AND NITRATE

4.2.1 BACKGROUND

Nitrogen compounds of greatest interest to water quality are those that are biologically available as nutrients to plants or exhibit toxicity to humans or aquatic life. Atmospheric nitrogen (N_2) is the primary source of all nitrogen species found in water and soil, but N_2 is not directly available to plants or animals as a nutrient because the $N\equiv N$ triple bond in the gaseous nitrogen molecule is too strong to be broken by photosynthesis or animal metabolic reactions. Atmospheric nitrogen must be converted to other nitrogen compounds before it can become available to life forms.*

The conversion of atmospheric nitrogen to other chemical forms is called nitrogen fixation and is accomplished in nature by certain bacteria that are present in water, soil, and root nodules of alfalfa, clover, peas, beans, and other legumes. Atmospheric lightning is another significant source of nitrogen fixation because the high temperatures generated in lightning strikes are sufficient to break the chemical bonds within atmospheric nitrogen (N_2), oxygen (O_2), and water vapor (H_2O), making possible the formation in the atmosphere of, mainly, nitrates (NO_3^-), with much lesser amounts of nitrites (NO_2^-) and ammonia (NH_3). These soluble nitrogen compounds created within lightning discharges dissolve in atmospheric precipitation and can be absorbed by plant roots, thus entering the nitrogen nutrient subcycles (see Figure 4.1).

The rate at which atmospheric nitrogen can enter the nitrogen cycle by natural processes is too low to support today's intensive agricultural production. The shortage of bioavailable nitrogen is made up with fertilizers containing nitrogen fixed by industrial processes, which are dependent on petroleum fuel energy. Modern large-scale farming has been called a method for converting petroleum into food.

4.2.2 NITROGEN CYCLE

In the nitrogen cycle (Figure 4.1), plants take up ammonia and nitrogen oxides dissolved in soil pore water and convert them into proteins, DNA, and other nitrogen-containing compounds. Animals get their nitrogen by eating plants or other animals; similar to photosynthesis, biological reactions are not energetic enough to process atmospheric N_2. Once in terrestrial ecosystems, the bioavailable nitrogen forms are

---

* There are specialized microorganism life forms in soil called diazotrophs that can break the $N\equiv N$ triple bond in atmospheric nitrogen molecules to form ammonia, providing bioavailable nitrogen. Common diazotrophs include cyanobacteria, green sulfur bacteria, Azotobacteraceae, Rhizobia, and Frankia.

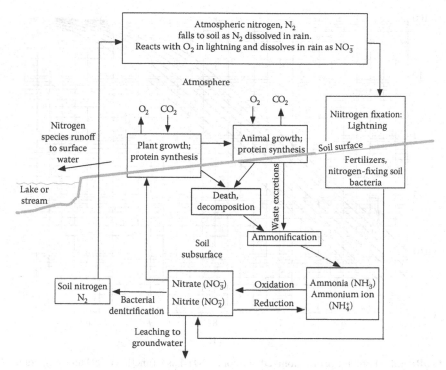

FIGURE 4.1 Nitrogen cycle.

recycled through repeated biological birth, growth, death, and decay steps. There is a continual and relatively small loss of fixed nitrogen from the nonatmospheric nutrient subcycles, when specialized soil bacteria convert fixed nitrogen back in to N_2 gas (denitrification), which is ultimately released back into the atmosphere, from which it can eventually reenter the nutrient subcycles again.

When nitrogen is circulating in the nutrient subcycles, it undergoes a series of reversible oxidation–reduction reactions (in degradation processes) that convert it from nitrogenous organic molecules, such as proteins, urea, and amines, to ammonia (NH_3), nitrite (NO_2^-), and nitrate (NO_3^-). Ammonia is the first product in the oxidative decay of nitrogenous organic compounds, whether from the presence of DO or nitrifying bacteria (see Section 4.2.6.2); further oxidation leads to nitrite and then to nitrate. Ammonia is naturally present in most surface waters and wastewaters. Under aerobic conditions, ammonia in environmental waters is then oxidized to nitrites and nitrates. These oxidation processes consume approximately 4.3 mg of DO per milligram of ammonia nitrogen oxidized (Equation 4.1) and can be a significant contribution to nitrogen biochemical oxygen demand (NBOD; Chapter 3, Section 3.8.3).

$$\text{Organic N} \xrightarrow{\text{biodegradation}} NH_3 \underset{\text{reduction}}{\overset{\text{oxidation}}{\rightleftharpoons}} NO_2^- \underset{\text{reduction}}{\overset{\text{oxidation}}{\rightleftharpoons}} NO_3^- \quad (4.1)$$

Under anaerobic conditions, a part of Equation 4.1 is reversed; nitrate is reduced to nitrite (see Equation 4.3 below), which in turn can be reduced to ammonia.

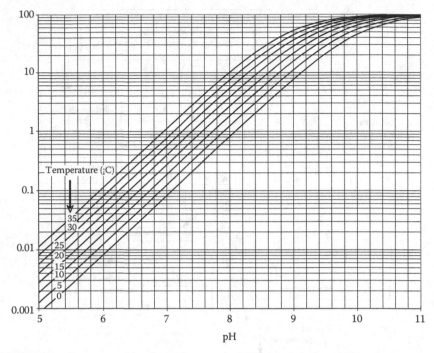

FIGURE 4.2 Percentage of unionized ammonia (NH_3) as a function of pH and temperature.

4.2.3 AMMONIA/AMMONIUM ION

Dissolved ammonia reacts with water as a base, raising the pH by generating OH^- ions (Equation 4.2).

$$NH_3 + H_2O \rightleftarrows NH_4^+ + OH^- \tag{4.2}$$

The equilibrium of Equation 4.2 depends on pH and temperature (see Figure 4.2). In a laboratory analysis for total ammonia nitrogen ($NH_3 + NH_4^+$), pH is adjusted to 11 or greater to assure all ammonia nitrogen is in the NH_3 form. For reliable measurement of ammonium ion (NH_4^+), the sample must be analyzed soon after being collected to avoid changes in pH and temperature during storage. If analysis at the sampling site is not feasible, the laboratory can use field-measured pH and temperature to calculate the original distribution of NH_3 and NH_4^+ in the sample from an analysis of total ammonia nitrogen. Since the unionized form is far more toxic to aquatic life than the ionized form,* field measurements of water pH and temperature at the sampling site are very important.

The two forms of ammonia have different mobilities in the environment. Although ammonium cations are highly soluble, a portion will sorb strongly to negatively

---

* Because it is not electrically charged, the NH_3 molecule passes more easily through fish respiratory membranes than ammonium, nitrite, or nitrate ions (NH_4^+, NO_2^-, and NO_3^-).

charged sites on mineral surfaces, where it is effectively immobilized. In contrast, unionized ammonia is also soluble because it can form hydrogen bonds to water molecules, but is not sorbed strongly to minerals and, therefore, is transported readily by water movement. However, unionized ammonia is also volatile and a portion will partition from water into the atmosphere. If suspended sediment carrying sorbed NH_4^+ is carried by a stream into a region with a higher pH, a portion will be converted to unionized NH_3, which can then desorb and become available to aquatic life forms as a toxic pollutant and to volatilization into the atmosphere.

In Figure 4.1, nitrogen passes through several different chemical forms within the nutrient subcycles. To allow quantities of these different forms to be directly compared with one another, analytical results often report their concentrations in terms of their elemental nitrogen content. For example, 10.0 mg/L of unionized ammonia may be reported as 8.22 mg/L NH_3–N (nitrogen is 82.2% of the molecular weight of NH_3); 10.0 mg/L of nitrate may be reported as 2.26 mg/L NO_3–N (nitrogen is 22.6% of the molecular weight of NO_3^-).

RULES OF THUMB

1. The percentage of dissolved total ammonia $(NH_3 + NH_4^+)$ present in the unionized form (NH_3) increases as either pH or temperature increases (see Figure 4.2).
2. Ammonia toxicity to aquatic life is mainly due to the unionized NH_3 form.
3. An increase in either pH or temperature shifts the equilibrium of Equation 4.2, $NH_3 + H_2O \rightleftarrows NH_4^+ + OH^-$, to the left, favoring NH_3, the toxic form.
4. At 20°C and pH > 9.3, the equilibrium of Equation 4.2 is mainly to the left, with NH_3 (the toxic form) >50%.
5. At 20°C and pH < 9.3, the equilibrium of Equation 4.2 is mainly to the right, with NH_4^+ (the nontoxic form) >50%.
6. NH_3 concentrations >0.5 mg NH_3–N/L cause significant toxicity to fish.
7. The unionized form is volatile (air-strippable), and the ionized form is nonvolatile.

Changes in environmental conditions that raise stream temperature or pH can cause an initially acceptable concentration of total ammonia to become unacceptable and in violation of a stream standard because of conversion of ammonia from the ionized to the unionized form. For example, consider a wastewater treatment plant that discharges its effluent into a detention pond that, in turn, periodically releases its water into a stream. The treatment plant meets its discharge limit for unionized ammonia when its effluent is measured at the end of its discharge pipe. However, suppose that the detention pond receiving the effluent supports algal growth. In such a situation, it is common for algae to grow to a level that influences the pond's pH. During daytime photosynthesis, algae may remove enough dissolved CO_2 from the pond to raise the pH and shift the equilibrium of Equation 4.2 to the left far enough that the pond concentration of NH_3 becomes higher than the discharge permit limit. In this case, discharges from the pond could exceed the stream standard for unionized ammonia even though the total ammonia concentration is unchanged. For this reason, discharge limits are usually written in terms of total ammonia limits.

Example 1

Ammonia is removed from an industrial wastewater stream by an air-stripping tower. To meet the effluent discharge limit of 5-ppm total ammonia, the wastewater stream entering the air-stripping tower must be adjusted so that 60% of the total ammonia is in the volatile form. To what pH must the wastewater stream be adjusted if the wastewater in the stripping tower is at 10°C? Use Figure 4.2.

ANSWER

In Figure 4.2, the 60% unionized ammonia gridline crosses the 10°C curve between pH 9.7 and 9.8. Thus, the influent must be adjusted to pH 9.8 or higher to meet the discharge limit.

4.2.4 CASE STUDY 1: UNIONIZED AMMONIA IN AN AQUACULTURE POND

In an aquaculture pond where fish are farmed, the main sources of ammonia are from fish excretions and bottom sediment decay of nitrogenous organic matter, particularly dead algae. If allowed to accumulate, ammonia concentrations may become sufficiently toxic to fish to have negative affects that do not always result in death, but cause reduced growth rates, poor feed conversion, and reduced disease resistance. Ammonia concentrations are reduced by nutrient uptake of algae and other plants during photosynthesis and by bacterial nitrification (Section 4.2.6.2). For adequate control of average total ammonia concentrations in aquaculture systems, it is important to consider diurnal variations in the percentage of total ammonia that are in the unionized form. Diurnal variations in the percentage of NH_3 are caused by the pH and temperature changes (Figure 4.2) that accompany day–night changes in the photosynthesis activity of pond vegetation, especially algae. Consumption of dissolved CO_2 by aquatic vegetation during sunlight hours raises the pond pH, whereas release of CO_2 into the water at night during plant respiration lowers the pH. (There are corresponding diurnal changes in DO; see Chapter 3, Section 3.5.2 and 3.7.1.)

During the late afternoon, when photosynthesis has reached its peak, pond pH may approach 9 or higher due to loss of dissolved CO_2. Using Figure 4.2, pH = 9 can result in around 10% of total ammonia being in the unionized form, depending on the temperature. To avoid detrimental stress to many types of fish, the unionized ammonia concentration should be maintained below about 0.02 mg/L. Assuming that the daytime peak pH can be maintained below pH = 9, this would require that total ammonia concentrations be held below about 0.20 mg/L.

4.2.5 NITRITE AND NITRATE

Ammonia and other nitrogenous materials in natural waters tend to be oxidized by aerobic bacteria, first to nitrite and then to nitrate. Therefore, all organic compounds containing nitrogen should be considered as potential nitrate sources, under aerobic conditions. Organic nitrogen compounds enter the environment from wild animal and fish excretions, dead animal tissue, human sewage, and livestock manure. Inorganic nitrates come primarily from manufactured fertilizers containing ammonium nitrate and potassium nitrate, and from nitrate-based explosives and rocket fuels.

In oxygenated waters, nitrite is rapidly oxidized to nitrate (Equation 4.3), so normally there is little nitrite present in surface waters. The reaction is reversible; under conditions of oxygen depletion, which often occur in groundwater, nitrate may be reduced to nitrite.

$$2NO_2^- + O_2 \rightleftarrows 2NO_3^- \tag{4.3}$$

Both nitrite and nitrate are important nutrients for plants, but both are toxic to fish (but not nearly as toxic as NH_3) and humans at sufficiently high concentrations. Nitrates and nitrites are very soluble, do not sorb readily to mineral and soil surfaces, and are very mobile in the environment, moving without significant loss when dissolved in surface waters and groundwater. Consequently, where soil nitrate levels are high, contamination of groundwater by nitrate leaching is a serious problem. Unlike ammonia, nitrites and nitrates do not evaporate and remain in water until they are consumed by plants and microorganisms.

RULES OF THUMB

1. In unpolluted, oxygenated surface waters, oxidized nitrogen is normally in the form of nitrate, with only trace amounts of nitrite.
2. Measurable nitrite concentrations are more common in groundwater because of low oxygen concentrations in the soil's subsurface.
3. Nitrate and nitrite both leach readily from soils to surface waters and groundwaters.
4. High concentrations (>1–2 mg/L) of nitrate or nitrite in surface waters or groundwater generally indicate agricultural contamination from fertilizers and manure seepage.
5. Greater than 10 mg/L of total nitrite and nitrate in drinking water is considered a human health hazard.

4.2.5.1 Health Issues

Drinking water standards for nitrate are strict because nitrates can be reduced to nitrites in human saliva and in the intestinal tracts of infants during the first 6 months of life. Nitrite oxidizes iron in blood hemoglobin from ferrous iron (Fe^{2+}) to ferric iron (Fe^{3+}). The resulting compound, methemoglobin, cannot carry oxygen and may cause a blood oxygen deficiency condition called methemoglobinemia. It is especially dangerous in infants (blue baby syndrome) because of their small total blood volume.

Example 2: Chemical Oxygen Demand Caused by Sodium Nitrite Disposal

A chemical company applied for a permit to dispose 250,000 gal of water containing 500 mg/L of nitrite into a municipal sewer system. The manager of the municipal wastewater treatment plant had to determine whether this waste might be detrimental to the operation of his plant. Calculate the increase in NBOD (see Chapter 3, Section 3.8.3) in the sewer system that would be caused by the chemical company's proposed wastewater release.

ANSWER

Under oxidizing conditions that exist in the treatment plant, nitrite is oxidized to nitrate as in Equation 4.3. The consumption of oxygen shown in Equation 4.3 makes the use of nitrite compounds to deoxygenate water useful as a rust-inhibiting additive in boilers, heat exchangers, and storage tanks. When nitrite is added to a wastewater stream, it becomes a source of the part of chemical oxygen demand called NBOD (Chapter 3, Section 3.8.3). More oxygen will be needed to maintain aerobic treatment steps at their optimum performance level. In addition, it will produce additional nitrate that may have to be denitrified before it can be discharged.

CALCULATION

In a wastewater stream of 250,000 gal containing 500 mg/L of nitrite, the net weight of nitrite is

$$500 \times 10^{-3} \text{ g/L} \times 250,000 \text{ gal} \times 3.79 \text{ L/gal} = 474,000 \text{ g of nitrite}$$

From Equation 4.3, stoichiometric consumption of oxygen is 1 mole of oxygen (32 g) for each 2 moles of nitrite (92 g) that is oxidized, resulting in a 32 g/92 g = 0.35 ratio of O_2 to NO_2^- by weight. Therefore, 474,000 g of nitrite will potentially consume

$$474,000 \text{ g } NO_2^- \times 0.35 = 165,000 \text{ g of dissolved oxygen} = \text{additional COD load}$$

Whether this additional NBOD load in the wastewater influent is of concern to the treatment plant operators will depend on the operating specifications of the treatment plant.

In addition, from Equation 4.3, stoichiometric production of nitrate is 2 moles (124 g) for each 2 moles (92 g) of nitrite oxidized, resulting in a 124 g/92 g = 1.35 ratio of NO_3^- to NO_2^- by weight. Therefore, oxidation of 474,000 g of nitrite will produce

$$474,000 \text{ g } NO_2^- \times 1.35 = 639,000 \text{ g (1409 lb) of nitrate}$$

Depending on the limit for nitrate in the treatment plant's discharge permit, additional capacity for denitrification might be required.

4.2.6 Methods for Removing Nitrogen from Wastewater

After the activated sludge treatment stage, municipal wastewater generally still contains some nitrogen in the forms of organic nitrogen and ammonia. Additional treatment may be required to remove nitrogen from the waste stream.

4.2.6.1 Air-Stripping Ammonia

Additional treatment by air stripping (see Example 1) can follow the activated sludge process. pH must be raised, possibly with lime (CaO), to about 10 or 11 to convert all ammoniacal nitrogen to the volatile NH_3 form. At higher temperatures, the removal efficiency will be increased. Lime (CaO) is often the least expensive way to raise the pH, but it generates $CaCO_3$ sludge. Sodium hydroxide (NaOH) is more expensive but does not generate sludge. Scaling, icing, and air pollution are some of the potential disadvantages of air stripping, whereas an advantage is that raising the pH with lime also precipitates phosphorus, if present, in the form of calcium phosphate compounds.

4.2.6.2 Nitrification–Denitrification

This is a two-step process:

1. In the nitrification step, ammonia and organic nitrogen are biologically oxidized by nitrifying bacteria (*Nitrosomonas* and *Nitrobacter*) to nitrate $\left(NO_3^-\right)$ under strongly aerobic conditions (ORP* = +100 to +350 mV). The oxidation of all nitrogen species to nitrate is achieved by extensive aeration of the sewage:

$$2NH_4^+ + 3O_2 \xrightarrow[\text{aerobic}]{\text{Nitrosomonas}} 4H^+ + 2NO_2^- + 2H_2O \qquad (4.4)$$

$$2NO_2^- + O_2 \xrightarrow[\text{aerobic}]{\text{Nitrosomonas}} 2NO_3^- \qquad (4.5)$$

2. In the denitrification step, the ORP must be lowered to more reducing conditions (ORP = +50 to −50 mV in the presence of denitrifying bacteria, to biologically reduce nitrate to nitrogen gas [N_2] under anaerobic conditions). The ORP should not go below −50 mV to prevent the formation of sulfides with accompanying foul odors (see Section 4.4.3). Nitrification–denitrification also requires a source of carbon nutrients for the bacteria. Water that is low in total organic carbon (shown as CH_2O in Equation 4.6) may require the addition of methanol or other carbon source.

$$4NO_3^- + 5CH_2O + 4H^+ \xrightarrow[\text{anaerobic}]{\text{denitrifying bacteria}} 2N_2(g) + 5CO_2(g) + 7H_2O \qquad (4.6)$$

---

* ORP= oxidation-reduction potential

4.2.6.3 Breakpoint Chlorination

Chlorination can be used to remove dissolved ammonia and ammonium nitrogen from wastewater by the chemical reactions shown in Equations 4.7a,b and 4.8. The chemical reaction of ammonia with dissolved chlorine results in denitrification by first converting ammonia to chloramine (NH_2Cl) (Equations 4.7a and 4.7b). With continued addition of $Cl_2(aq)$, nitrogen gas is formed (Equation 4.8). Any chloramine remaining serves as a weak disinfectant and is relatively nontoxic to aquatic life.

$$NH_3(aq) + Cl_2(aq) \rightarrow NH_2Cl(aq) + Cl^- + H^+ \qquad (4.7a)$$

$$NH_4^+(aq) + Cl_2(aq) \rightarrow NH_2Cl(aq) + Cl^- + 2H^+ \qquad (4.7b)$$

$$2NH_2Cl(aq) + Cl_2(aq) \rightarrow N_2(g) + 4H^+ + 4Cl^- \qquad (4.8)$$

In Equations 4.7a and 4.7b, the ratio by weight of chlorine to NH_3–N is 5:1 and ammonia is converted stoichiometrically to monochloramine (NH_2Cl) at a 1:1 molar ratio. Lesser amounts of dichloramine ($NHCl_2$) and trichloramine (NCl_3; also called nitrogen trichloride) may also be formed, depending on pH and small excesses of chlorine. Adding more chlorine leads to conversion of all the chloramines to nitrogen gas. The overall reactions for complete nitrification of ammonia by chlorine oxidation are

$$2NH_3(aq) + 3Cl_2(aq) \rightarrow N_2(g) + 6H^+ + 6Cl^- \qquad (4.9a)$$

$$2NH_4^+ + 3Cl_2(aq) \rightarrow N_2(g) + 8H^+ + 6Cl^- \qquad (4.9b)$$

Because water is present, a lesser amount of nitrate is also formed by Equations 4.9c and 4.9d:

$$NH_3(aq) + 4Cl_2(aq) + 3H_2O \rightarrow NO_3^- + 9H^+ + 8Cl^- \qquad (4.9c)$$

$$NH_4^+ + 4Cl_2(aq) + 3H_2O \rightarrow NO_3^- + 10H^+ + 8Cl^- \qquad (4.9d)$$

Reactions 4.9a through 4.9d go to theoretical completion at a molar ratio of 3:2 and a weight ratio of 7.6:1 of $Cl_2(aq)$ to either NH_3–N(aq) or NH_4^+–N(aq). This process is called breakpoint chlorination (see Figure 4.3). Reactions 4.9a through 4.9d are very fast and remove both ionized (NH_4^+) and unionized (NH_3) forms of ammonia.

- In Region A, easily oxidizable substances such as Fe^{2+}, H_2S, and organic matter are oxidized. Ammonia reacts to form chloramines. Organics react to form chloroorganic compounds.
- In Region B, adding more chlorine oxidizes chloramines to N_2O and N_2.
- At the breakpoint, virtually all chloramines and a large part of chloroorganics have been oxidized.
- In Region C, further addition of chlorine results in a free chlorine residual of $HOCl$ and OCl^-.

RULES OF THUMB

1. The rate of ammonia removal by breakpoint chlorination is most rapid at pH 8.3.
2. The rate decreases at higher and lower pH and at temperatures below 30°C.
3. Since Equations 4.9a through 4.9d produce H⁺ and lower the pH, additional alkalinity as lime might be needed if the concentration of total ammonia >15 mg/L. Add alkalinity as $CaCO_3$ in a weight ratio of about 11:1 of $CaCO_3$ to total ammonia–nitrogen (NH_3–N).
4. The chlorine breakpoint (Figure 4.3) occurs theoretically at a Cl_2:NH_3–N weight ratio of 7.6:1.
5. In actual practice, ratios of 10:1 to 15:1 may be needed if oxidizable substances other than NH_3 are present (such as Fe^{2+}, Mn^{+2}, S^{-2}, and organics).
6. Whenever wastewater containing organic matter is treated with chlorine, there is the potential of producing harmful chlorinated substances, particularly trihalomethanes (Chapter 3, Section 3.11.5.1). If this is unacceptable, as in drinking water sources, the waste stream must be pretreated to remove the organic matter.

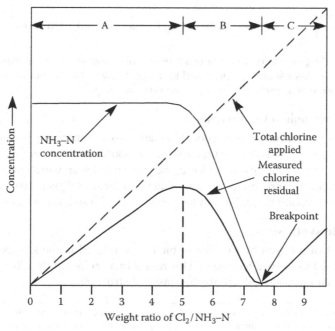

FIGURE 4.3   Breakpoint chlorination curves showing removal of ammonia from wastewater.

Example 3: Calculate the Chlorine Needed to Remove Ammonia

A waste treatment plant handles 1,500,000 L/day of sewage that contains an average of 50 mg/L of NH_3–N. How many grams of Cl_2(aq) must be present daily in the wastewater to remove all of the ammonia? Ignore Equations 4.9c and 4.9d, as they do not significantly affect the calculation.

ANSWER

By Equation 4.9a, three moles of chlorine gas are needed for every two moles of total ammonia–nitrogen.*

$$2NH_3(aq) + 3Cl_2(aq) \rightarrow N_2(g) + 6H^+ + 6Cl^-$$

Molecular weights are $Cl_2 = 71$ and $N = 14$.

$$3 \text{ moles of } Cl_2 = 3 \times 71 = 213 \text{ g}$$
$$2 \text{ moles of } N = 2 \times 14 = 28 \text{ g}$$

Thus, the stoichiometric weight ratio is $213/28 = 7.6$ g Cl_2 per gram of N (as ammonia). One mole of NH_3 contains 14 g of N and 3 g of H. Thus, 50 mg/L of NH_3 contains $14/17 \times 50$ mg/L = 41.2 mg/L of N. In 1,500,000 L there will be

$$1,500,000 \text{ L} \times 41.2 \text{ mg/L} = 61,800,000 \text{ mg N or } 61,800 \text{ g N/day}$$

The theoretical amount of chlorine required is

$$\frac{7.6 \text{ g } Cl_2}{1 \text{ g N}} = 61,800 \text{ g N} = 470 \text{ kg } Cl_2/\text{day or about 1036 lb/day}$$

Depending on the quantity of other oxidizable substances in the wastewater, the plant operator should be prepared to use up to twice this amount of chlorine, but note Rule of Thumb number 6 on the previous page.

4.2.6.4 Ammonium Ion Exchange

Ammonium ion exchange is a good alternative to air stripping because certain exchange resins, such as the natural zeolite clinoptilolite, are selective for ammonium ions. It might be necessary to lower the pH to 6 or lower, which converts 99.9% of the ammoniacal nitrogen to the ammonium ion form. NH_4^+ is exchanged for Na^+ or Ca^{2+} on the resin. The zeolite can be regenerated with sodium or calcium salts.

4.2.6.5 Biosynthesis

The removal of nitrogen by filtering of biomass produced in the sewage treatment system will reduce suspended solids. This results in a net loss of the nitrogen species that have been incorporated into the biomass cell structure.

---

* Since the calculation is based on the moles of nitrogen (N), which are the same as the total moles of $NH_3 + NH_4^+$ regardless of the relative proportions of the two species, the molar ratio of 3/2 for Cl_2/NH_3 applies also to total ammonia, $Cl_2/NH_{3,total}$.

4.3 PHOSPHORUS

4.3.1 BACKGROUND

Phosphorus is a common element in igneous and sedimentary rocks and in sediments, but it tends to be a minor element in natural waters because most inorganic phosphorous compounds have low solubility. Dissolved concentrations are generally in the range of 0.01–0.1 mg/L and seldom exceed 0.2 mg/L. The environmental behavior of phosphorus is largely governed by the low solubility of most of its inorganic compounds, its strong sorption to soil particles, and the fact that it is an essential nutrient for most life forms—animal, plant, and microbial.

Because of its low dissolved concentrations, phosphorus is usually the limiting nutrient for aquatic vegetation in natural waters. Wherever P is the limiting nutrient, any new source that increases the P concentration will increase growth of aquatic vegetation. In the absence of human activities, the dissolved phosphorous concentration in environmental waters is normally low enough to limit algal growth. However, because phosphorus is essential to metabolism, it is always present in animal wastes and sewage. Phosphorus in wastewater effluent and stormwater runoff is frequently the main cause of algal blooms and other precursors of eutrophication.

4.3.2 MAJOR USES FOR PHOSPHORUS

Phosphorous compounds are used for corrosion control in water supply and industrial cooling water systems. Certain organic phosphorous compounds are used in insecticides. Perhaps the main commercial uses of phosphorous compounds have been in fertilizers and in the production of synthetic detergents. Before the use of phosphate detergents, most municipal wastewater inorganic phosphorus was contributed from human wastes; between 2 and 5 g/day per person was released to the wastewater influent in urine, feces, and food wastes. As a consequence of detergent use, the concentration of phosphorus in treated municipal wastewaters increased from 3 to 4 mg/L in predetergent days, to 10–20 mg/L in the 1960s and 1970s, when phosphates were used freely in detergents.

However, the use of phosphorus in detergents is increasingly being banned. Detergent formulations in the United States formerly contained large amounts of polyphosphates as builders, to sequester metal ions and soil, and soften the water. The widespread use of detergents instead of soap made a major contribution to the available phosphorus in domestic wastewater. In the 1950s, laundry detergent contained almost 10% phosphorus; by the end of the 1960s, it was up to 20% or more. However, because of increasing problems with excessive algal blooms occurring in lakes and rivers across the United States, phosphorus was banned in laundry detergents starting in the 1970s. All 50 states of the U.S. now ban phosphorus in laundry detergents in excess of 0.5%. Individual states have begun to ban phosphorus in dishwasher detergent as well.

Restrictions imposed by states and the EPA limiting the phosphorus content of cleaning agents, together with improved secondary and tertiary treatment requirements for municipal wastewater treatment plants, have reduced the average phosphorus concentration in treated effluent to the current level of about 1 mg/L or less.

4.3.3 Phosphorous Cycle

In a manner similar to carbon and nitrogen, phosphorus in the environment is cycled between organic and inorganic forms. An important difference is that, unlike the carbon and nitrogen cycles, both of which include gaseous phases that provide reliable global redistribution of these two nutrients, phosphorus has no comparable global redistribution mechanism. Some nitrogen always returns to the atmosphere by ammonia volatilization and microbial denitrification; carbon compounds are ultimately decomposed by oxidation processes to gaseous carbon dioxide. For phosphorus, the closest approaches to global redistribution are by bird migration and international shipping of fertilizers.*

Also important are the differences in earthbound mobility of the three essential nutrients (C, N and P). All exist in anionic forms (CO_3^{2-}/HCO_3^-, NO_2^-/NO_3^-, and, $H_2PO_4^-$/HPO_4^{2-}/PO_4^{3-}) which are not readily retained in soils by common cation exchange reactions (Chapter 6, Section 6.10.2). However, while most forms of P are in particulate form over the environmental range of pH values, C and N have many soluble forms. C, as carbonate anions, forms many soluble and moderately soluble salts, as waters with moderate to high alkalinity demonstrate; N, as nitrate and nitrate anions, forms only soluble salts. Therefore, carbonates and nitrates/nitrites leach more readily than phosphates from soils to surface waters and groundwaters.

Phosphate anions are largely immobilized in the soil by the formation of insoluble compounds, chiefly iron, calcium, and aluminum phosphates, and by sorption to soil particles (see Figure 4.4). Because nitrogen and carbon compounds leach from soils more readily than phosphorous compounds, nitrogen and carbon are more generally available than phosphorus to water vegetation, a condition that contributes to phosphorous-limited algal growth in most surface waters. The critical level of inorganic phosphorus for forming algal blooms can be as low as 0.01–0.005 mg/L under summer growing conditions, but is more frequently around 0.05 mg/L.

Organic compounds containing phosphorus are found in all living matter. Orthophosphate (PO_4^{3-}) is the only form readily used as a nutrient by most plants and organisms. There are just two major steps in the phosphorous cycle:

1. Conversion of organic phosphorus to inorganic phosphorus
2. Conversion of inorganic phosphorus to organic phosphorus

Both steps are bacterially mediated. Conversion of insoluble forms of phosphorus, such as calcium phosphate, $Ca(HPO_4)_2$, into soluble forms, principally PO_4^{3-}, is also carried out by microorganisms. Organic phosphorus in the tissues of dead plants and animals and in animal waste products is converted bacterially to soluble phosphate, PO_4^{3-}. The PO_4^{3-} thus released to the environment is taken up again into plant and animal tissue.

---

* There are studies that estimate wet and dry deposition from the atmospheric of phosphorus contained in airborne particulates from forest fires, wind erosion, volcanic eruptions, and so on. Atmospheric deposition of phosphorus appears to be significant in certain geographic regions. (For a review of recent literature, see Technical Memorandum, *Detailed Assessment of Phosphorus Sources to Minnesota Watersheds—Atmospheric Deposition: 2007 Update*, from Barr Engineering Company to the Minnesota Pollution Control Agency, June 29, 2007. http://archive.leg.state.mn.us/docs/2010/mandated/100654.pdf.)

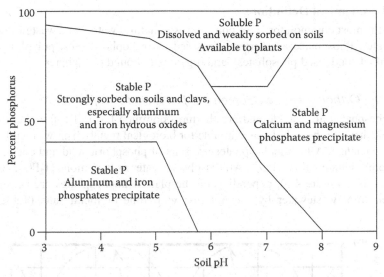

FIGURE 4.4 pH dependence of soluble (dissolved and labile) and stable phosphorus in soils.

RULES OF THUMB

1. In surface waters, phosphorous concentrations are influenced by sediments, which serve as a reservoir for sorbed and precipitated phosphorus. Sediments are an important part of the phosphorous cycle in streams. Bacteria-mediated exchange between dissolved and sediment-sorbed forms plays a role in making phosphorus available for algae and therefore contributes to eutrophication.

2. In streams, dissolved phosphorus from all sources, natural and anthropogenic, is generally present in low concentrations, around 0.1 mg/L or less.

3. The solubility of phosphates increases at low pH and decreases at high pH.

4. Particulate phosphorus (sorbed to sediments and precipitated as insoluble phosphorous compounds) is about 95% of the total phosphorus (TP) in most cases.

5. In carbonate soils, dissolved phosphorus can react with carbonate to form the mineral precipitate hydroxyapatite (calcium phosphate hydroxide), $Ca_{10}(PO_4)_6(OH)_2$.

4.3.4 Nomenclature for Different Forms of Phosphorus

There are many different P compounds of environmental interest; thus, in addition to their chemical names, the environmental literature uses several unique classifications based on their environmental behavior or analytical methods.

4.3.4.1   Chemical Definitions

By far, the most common forms of phosphorus in natural waters and wastewaters are phosphates. Phosphates are further classified as orthophosphates, polyphosphates (also called condensed phosphates), and organically bound phosphates.

4.3.4.1.1   Orthophosphates (Sometimes Just Called Phosphates)

Orthophosphates are soluble and are the most common form of P found dissolved in water: they are the only form available for biological metabolism without further chemical change. All orthophosphates are salts of phosphoric acid and contain one or more phosphate anions, PO_4^{3-}. All dissolve in water as the anions $H_2PO_4^-$, HPO_4^{2-}, and PO_4^{3-} (see Figure 4.5), depending on the pH. Orthophosphates are commonly removed from wastewater by precipitation with lime or metal salts (Table 4.1).

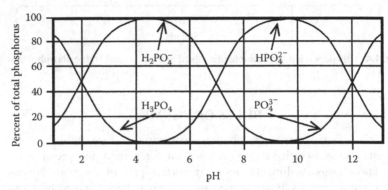

FIGURE 4.5   pH dependence of dissolved phosphate species (trihydrogen phosphate, H_3PO_4; dihydrogen phosphate, $H_2PO_4^-$; monohydrogen phosphate, HPO_4^{2-}; phosphate, PO_4^{3-}).

TABLE 4.1
Common Chemicals Used for Phosphorus Removal

Chemical	Formula	Typical Wt.% in Commercial Solutions
Aluminum sulfate (alum)	$Al_2(SO_4)_3 \cdot 14(H_2O)$	48%
Sodium aluminate	$Na_2Al_2O_4$	20%
Polyaluminum chloride (PAC)	$Al_nCl_{(3n-m)}(OH)_m$ e.g., $Al_{12}Cl_{12}(OH)_{24}$	51%
Ferric chloride	$FeCl_3$	37–47%
Ferrous sulfate or ferrous iron (pickle liquor)	Fe_2SO_4 or Fe^{2+}	Varies
Lime	CaO; $Ca(OH)_2$	Not a diluted solution; dry powder or concentrated liquid form as quicklime, CaO, or hydrated lime, Ca(OH)

To measure total phosphate, all species of phosphate are first chemically converted to orthophosphates before analysis. Examples of orthophosphates are

- Phosphoric acid, H_3PO_4: Mainly used for manufacture of fertilizer.
- Trisodium phosphate, Na_3PO_4 (also called trisodium orthophosphate or sodium phosphate): Once used extensively in detergents, but now widely banned or restricted for this use; it is still used as a water softener, cleaning agent, food additive (as an acidifier), stain remover, and degreaser.
- Disodium phosphate, Na_2HPO_4: A hygroscopic and water-soluble white powder used commercially as an anticaking additive in powdered materials, including food products.
- Monosodium phosphate (NaH_2PO_4): Used as a laxative and as a component of pH buffers.
- Diammonium phosphate ($(NH_4)_2HPO_4$): Used in fertilizers, as a fire retardant, as a yeast nutrient in winemaking and brewing mead, in purifying sugar, in soldering fluxes, and in dye solutions for wool.

4.3.4.1.2 Polyphosphates (Also Called Condensed Phosphates, Meaning Dehydrated)

Polyphosphates also are soluble, but they are not precipitated by lime or metal salts. They are resistant to chemical change, but can be hydrolyzed to orthophosphate. In the aquatic environment, hydrolysis slowly converts polyphosphates to orthophosphates. For analysis, they are converted to orthophosphates by boiling an acidified solution for 1–2 hours. For removal from a waste stream, biological treatment is required for conversion to orthophosphate. Examples of polyphosphates are

- Sodium hexametaphosphate, $Na_3(PO_4)_6$: Used for pH control; as a sequestrant in food additives, water softeners, and detergents; as a dispersing agent to break down clay and other soil types; and as a medicine in some homeopathic medical agents.
- Sodium tripolyphosphate, $Na_5P_3O_{10}$: Mainly used as an additive to detergents as a water softener, and used to retain moisture in food products and to improve the properties of a wide variety of products, such as paper, textiles, rubber goods, and flame retardants.
- Tetrasodium pyrophosphate, $Na_4P_2O_7$ (TSPP; also called sodium pyrophosphate or tetrasodium phosphate): Similar to other polyphosphates, TSPP finds wide use in consumer products as an additive, emulsifier, and dispersing agent in foods, toothpaste, detergents, and so on.

4.3.4.1.3 Organically Bound Phosphates (Also Called Sedimentary Phosphorus): Occurs in the Following Forms

- Phosphate minerals: Mainly hydroxyapatite, $Ca_5OH(PO_4)_3$.
- Nonoccluded phosphorus: Phosphate ions (usually orthophosphate) bound to the surface of SiO_2 or $CaCO_3$. Nonoccluded phosphorus is generally more soluble and more available than occluded phosphorus (below).

- Occluded phosphorus: Phosphate ions (usually orthophosphate) contained within the matrix structures of amorphous hydrated oxides of iron, aluminum, and amorphous aluminosilicates. Occluded phosphorus is generally less available than nonoccluded phosphorus.
- Organic phosphorus: Phosphorus incorporated with aquatic biomass, usually algal or bacterial.

4.3.4.2 Regulatory Definitions

For regulatory purposes, four different kinds of phosphorus, defined by their environmental behavior or analytical methods, are commonly used to describe P in surface runoff, streams, and lakes.

1. TP includes all chemical forms of P in a water sample, which should be the total of all filterable and particulate phosphorus forms described in items 2–4. It is probably the most often analyzed fraction of phosphorus because its value is used in many computer models to correlate phosphorus with eutrophic indicators in lakes and streams. Analytically, it is defined as the amount of phosphorus in a sample that is oxidized to orthophosphate by a method-specific oxidant.
2. Soluble P (SP; also called dissolved P) is defined as all the P-containing material in the sample that passes through a 0.45-μm filter, both organic and inorganic. It is sometimes subdivided into two subcategories:
   a. Soluble reactive phosphorus (SRP; also called dissolved inorganic phosphorus*) consists largely of inorganic orthophosphate (PO_4^{3-}). Orthophosphate is the phosphorus form that can be directly assimilated by algae, and the concentration of this fraction constitutes an index of the amount of phosphorus immediately available for algal growth. In water where phosphorus is the limiting nutrient, the concentration SRP should be very low to undetectable (<5 μg/L) because P is assimilated by algae as fast as it is supplied. If the SRP fraction increases, it can be inferred that phosphorus is not the limiting nutrient or is being supplied at rates faster than it can be utilized by the algae.
   b. Soluble unreactive phosphorus (SUP; also called soluble organic phosphorus) contains phosphorus forms that do not hydrolyze or dissolve under the conditions of the analytical test. SUP contains forms of organic phosphorus and inorganic polyphosphates that are not available to algae as a nutrient. SUP is measured as the difference between SP and SRP.

---

* The term "soluble reactive phosphorus" is preferred over "dissolved inorganic phosphorus" because this form of filtered phosphorus is not necessarily dissolved or inorganic. As used here,
  - *Soluble* does not include only dissolved P; it includes all phosphorus-containing material that passes through a 0.45 μm cellulose filter. This filter excludes most particulates, but some colloidal phosphorus may be present in the filtered fraction.
  - *Reactive* indicates that the phosphorus in the SRP fraction may not be only inorganic phosphorus, but could include any form of phosphorus, including some organic forms, that hydrolyzes and dissolves under the conditions of the analytical test.

3. Particulate phosphorus (also called sediment P) is all material containing P, inorganic and organic, and particulate and colloidal, that was captured on the filter. Typically, particulate forms will contain insoluble metal phosphates of iron, aluminum, calcium, magnesium (Figure 4.4), bacteria, algae, organic detritus, inorganic particulates such as clays, smaller zooplankton, and, occasionally, larger zooplankton, sediments, or large plant material.

   It is primarily P attached to sediment and insoluble organic matter. Particulate P is less bioavailable in aquatic systems in the short term, but becomes more bioavailable over time (see number 4, bioavailable phosphorus). Particulate phosphorus can be measured by filtering a known volume of water through a membrane filter and then digesting the filter, or obtained by subtracting total SP from TP.

4. Bioavailable phosphorus (also called algal available phosphorus) is the portion of total P that is available to algae as a nutrient. It typically is considered the most important form of P affecting water quality. Bioavailable P includes all of dissolved P and the part of particulate P from which plants or microbes can extract P for use as a nutrient.

4.3.5 MOBILITY IN THE ENVIRONMENT

4.3.5.1 Phosphorus in Soils and Sediments

Natural fertile soils typically contain 300 to 1000 ppm of total P. Soil systems are similar to water systems in that only a small portion of the total P is easily available to plants (see Figure 4.4). A small fraction of orthophosphate P is dissolved in soil moisture, and is the form immediately available to plants. When plants deplete dissolved orthophosphate in soils, dissolved P is replenished from the second major soil P pool, called labile P. Labile P is sorbed weakly to soil particles and organic matter. The third soil P pool, nonlabile or stable P, is sorbed strongly to soil particles in the form of iron and aluminum phosphates in acid soils, calcium phosphates in calcareous soils, and strongly to organic matter in all soils. Stable P is considered unavailable to plants and is released at a very slow rate to the labile and soluble P forms.

Phosphorus is an important plant nutrient and is often present in fertilizers to augment the natural concentration in soils. Phosphorus is also a constituent of animal wastes. Runoff from agricultural areas is a major contributor to TP in surface waters, where it occurs mainly in sediments because of the low solubility of its inorganic compounds and its tendency to sorb strongly to soil particles.

In natural soil and sediment systems, dissolved phosphorus is removed from solution by

- Precipitation
- Strong sorption to clay minerals and oxides of aluminum and iron
- Sorption to organic components of soil
- Assimilation by microbes and vegetation

In most soils, these removal mechanisms for dissolved phosphorus dominate over dissolving mechanisms, so that phosphorous compounds resist leaching. There is little movement of phosphorus with water drainage through most soils. P is mobilized in soils mainly by sorption to erosion sediments. Phosphorous transport into surface waters is controlled chiefly by preventing soil erosion and controlling sediment transport. In most soils, except for those that are nearly all sand, almost all the phosphorus applied to the surface in fertilizers or as plant residue is retained in the top 1–2 ft. The sorption capacity for phosphorus in good agricultural soils has been estimated to be in the range of 77 to over 900 lb/acre-ft of soil profile. Often, the total phosphorous removal capacity of a soil will exceed the planning life of a typical land application project. If the phosphorus-removing capacity of a soil becomes saturated, it usually can be restored in a few months, by adding calcium or iron amendments to replenish sorption sites and to precipitate dissolved phosphates. In addition, much sorbed and dissolved P can be removed by a few seasonal crop cycles.

Reducing (anaerobic) conditions, as in water-saturated soil, may increase phosphorous mobility because insoluble ferric iron, to which phosphorus is strongly sorbed, is reduced to soluble ferrous iron, thereby releasing sorbed phosphorus. In acid soils, aluminum and iron phosphates precipitate, whereas in basic soils, calcium phosphates precipitate. The immobilization of phosphorus is therefore dependent on soil properties, such as pH (Figure 4.4); aeration; texture; cation-exchange capacity; the amount of calcium, aluminum, and iron oxides present; and the uptake of phosphorus by plants.

Figure 4.4 shows some general relationships between soil pH and phosphorous reactions:

- In the acid pH range, dissolved phosphorus is predominantly $H_2PO_4^-$, and immobile phosphorus is bound with iron and aluminum compounds.
- In the basic pH range, dissolved phosphorus is predominantly HPO_4^{2-}, and immobile phosphorus is mainly in the form of calcium phosphate.
- Maximum availability of phosphorus for plant uptake (as well as leaching) occurs between pH 6 and 7.

4.3.5.2 pH Dependence of Dissolved Phosphate Anions

Phosphoric acid (H_3PO_4) dissociates stepwise in water according to Equation 4.10. Dissolved phosphate anions from other chemical sources (e.g., PO_4^{3-} from trisodium phosphate, Na_3PO_4) will behave in exactly the same way. Because each step of the dissociation releases H^+, the final equilibrium distribution of phosphate species is pH dependent. As pH becomes higher, the equilibrium of Equation 4.10 shifts increasingly to the right.

$$H_3PO_4 \xrightleftharpoons{H_2O} H_2PO_4^- + H^+ \xrightleftharpoons{H_2O} HPO_4^{2-} + 2H^+ \xrightleftharpoons{H_2O} PO_4^{3-} + 3H^+ \quad (4.10)$$

The pH dependence of the equilibrium concentrations of dissolved phosphate anions exhibits the pH dependence shown in Figure 4.5:

- Below pH 2, H_3PO_4 is the dominant species.
- Between pH 2 and 7, $H_2PO_4^-$ is the dominant species.
- Between pH 7 and 12, HPO_4^{2-} is the dominant species.
- Above pH 12, PO_4^{3-} is the dominant species.

4.3.5.3 Phosphorus on Agricultural Land

In unfertilized soils, soluble soil P, readily available to plants, is typically less than 1% of total soil P. Labile soil P (weakly bound to soil particles) is typically less than 5% of total soil P and is less tightly bonded than stable P (strongly bound to soil particles). Stable P is often more than 95% of total soil P. It includes tightly bonded P in secondary and primary minerals and in organic matter. In agriculture, fertilizers are used to increase the percentage of soluble P.

Most P fertilizers are composed of soluble P compounds, and some manure P is soluble. Application of fertilizer or manure P causes an initial dramatic increase in soluble P in the soil at the point of contact. Chemical equilibrium is rapidly reestablished as much of the added P enters the labile P pool. Over time, some of the P in the labile pool is converted into more stable organic and mineral forms. The immediate effect of P fertilization and manure P applications is to increase the capacity of the labile P pool to replenish solution P and total soil P. The net long-term effect depends on soil properties, P removal by crops, and P loss by other mechanisms.

4.3.5.4 Phosphorus in Surface Freshwaters

Phosphates in unpolluted surface waters come from decomposition of organic material and leaching of minerals containing phosphorus. If P concentrations are much higher than about 0.2 mg/L, one should suspect recent pollution from other sources such as fertilizer runoff, human and animal waste from failing septic systems, sewage treatment plants, livestock confinement areas, mass quantities of decomposing organic matter, industrial effluent, and detergent wastewater. Detergent wastewaters are still responsible for much of the phosphates polluting natural waters.

Phosphorus is often the limiting nutrient for the growth of aquatic vegetation. Excessive growth of vegetation in surface waters leads to depletion of oxygen, reduction of light transmission and water clarity, and production of algal toxins. These water quality changes can hurt fish populations, reduce water quality for recreation, and impart undesirable odors and tastes resulting in increased cost of treating water for domestic use. The progressive increase in nutrient concentration in water bodies that results in deterioration of water quality through overstimulation of aquatic vegetation is called eutrophication.

Eutrophication is a naturally occurring process that is often accelerated with intensification of agriculture or other practices that result in increased flow of nutrients to water bodies. Changes that take centuries in natural systems can take just decades with the high rates of P loss associated with some intensive agricultural systems. Accelerated eutrophication is one of the most obvious and persistent surface freshwater quality problems in the United States. Agriculture is one of the top two contributors to impairment of streams and among the top five contributors to impairment of lakes and reservoirs. In the long term, however, we need to be concerned about the relatively large amount of particulate P entering surface freshwaters.

Even when algal growth in lakes is temporarily limited by carbon or nitrogen instead of phosphorus, natural long-term mechanisms act to compensate for these deficiencies. Carbon deficiencies are corrected by CO_2 diffusion from the atmosphere, and nitrogen deficiencies are corrected by changes in biological growth mechanisms. Therefore, even if a sudden increase in phosphorus occurs, temporarily causing algal growth to be limited by carbon or nitrogen, eventually these deficiencies are corrected. Then, algal growth becomes proportional to the phosphorous concentration as the system once more becomes phosphorous limited.

4.3.6 REMOVAL METHODS FOR DISSOLVED PHOSPHATE

Current methods of phosphate removal are mainly based on chemical and biological precipitation procedures. Methods for removing phosphorus from waste streams and controlling phosphorus in natural waters are constantly evolving. A good summary of current technology may be found in the EPA's *Nutrient Control Design Manual* (EPA 2010).

4.3.6.1 Biological Phosphorus Removal

Biological phosphorus can be removed by operating an activated sludge process in an anaerobic–aerobic sequence. Certain bacteria respond to this sequence by accumulating large excesses of polyphosphate within their cells in volutin granules. During the anaerobic phase, phosphate is released. In the aerobic phase, the released phosphate and an additional increment is taken up and stored by the bacteria as polyphosphate, giving a net removal, coincident with organic removal and metabolism. Phosphate can be removed from the waste stream as sludge or through use of a second anaerobic step. During the second anaerobic step, the stored phosphate is released in dissolved form. Then, the bacterial cells can be separated and recycled, and the released soluble phosphate is removed by precipitation.

Biological P removal can sometimes attain effluent phosphorus concentrations less than 0.5 mg/L. However, because P that is strongly sorbed to suspended solids is unavailable to bacterial assimilation, TSS in the waste stream can retain large quantities of P, frequently around 5% by weight. Therefore, filtration or chemical precipitation steps are generally required in addition to biological P removal, to control the concentration of P bound to TSS in wastewater treatment effluent.

4.3.6.2 Phosphorus Removal by Chemical Precipitation

Chemical precipitation removes only orthophosphates, which typically are about 50–80% of the present. Most of the remaining phosphorus consists of polyphosphates, which do not react with the metal salts or lime used for chemical precipitation; polyphosphates require biological treatment. The colloidal and particulate phosphorus portion will generally be removed during solids separation processes. The soluble unreactive organic phosphorus fraction, mainly polyphosphates, is mostly hydrolyzed to orthophosphate during the precipitation process if it is biodegradable; if not biodegradable, it simply passes through the treatment process unchanged. Organically bound phosphorus usually makes up the smallest fraction of total influent phosphorus (<1 mg/L).

Municipal wastewater treatment plants in many areas are required to remove phosphorous in their treatment process. While the biological treatment process removes some phosphorus, in most cases precipitation as an insoluble metal phosphate is required to meet discharge regulations. This precipitation step is normally accomplished by the addition of a metallic salt such as ferric sulfate, ferric chloride, or aluminum sulfate in the primary or secondary clarifiers. The usual precipitants for removing phosphate are alum [$Al_2(SO_4)_3$], lime [$Ca(OH)_2$], ferric sulfate [$Fe_2(SO_4)_3$], and ferric chloride ($FeCl_3$). The choice of precipitant depends on the discharge requirements, wastewater pH, and chemical costs.

Pertinent reactions for the precipitation of phosphate with alum, ferric sulfate, ferric chloride, and lime are

$$\text{Alum: } Al_2(SO_4)_3 + 2HPO_4^{2-} \rightarrow 2AlPO_4(s) + 3SO_4^{2-} + 2H^+ \qquad (4.11)$$

$$\text{Ferric sulfate: } Fe_2(SO_4)_3 + 2HPO_4^{2-} \rightarrow 2FePO_4(s) + 3SO_4^{2-} + 2H^+ \qquad (4.12)$$

$$\text{Ferric chloride: } FeCl_3 + HPO_4^{2-} \rightarrow FePO_4(s) + 3Cl^- + H^+ \qquad (4.13)$$

$$\text{Lime: } 3Ca^{2+} + 2OH^- + 2HPO_4^{2-} \rightarrow Ca_3(PO_4)_2(s) + 2H_2O \qquad (4.14)$$

$$4Ca^{2+} + 2OH^- + 3HPO_4^{2-} \rightarrow Ca_4H(PO_4)_3(s) + 2H_2O \qquad (4.15)$$

$$5Ca^{2+} + 4OH^- + 3HPO_4^{2-} \rightarrow Ca_5(OH)(PO_4)_3(s) + 3H_2O \qquad (4.16)$$

Where effluent concentrations of phosphorus up to 1.0 mg/L are acceptable, the use of iron or aluminum salts in a wastewater secondary treatment system is often the process of choice. If very low levels of effluent phosphorus are required, precipitation at high pH by lime in a tertiary unit is necessary. The lowest levels of phosphorus are achieved by adding NaF with lime to form $Ca_5(PO_4)_3F$ (fluorapatite). The operating pH for phosphate removal with lime is usually above 11 because flocculation is best in this range.

If alkalinity is present, aluminum and iron ions are consumed in the formation of metal-hydroxide flocs. This may increase required dosages by up to a factor of 3. Calcium ions react with alkalinity to form calcium carbonate. Thus, the amount of precipitant needed for phosphate precipitation is controlled more by the alkalinity than the stoichiometry of the reaction. In the case of aluminum and iron precipitants, the reaction with alkalinity is not totally wasted because the hydroxide flocs assist in the settling and removal of metal-phosphate precipitates, along with other suspended and colloidal solids in the wastewater.

4.3.7 CASE STUDY 2: MODELING PHOSPHORUS BEHAVIOR IN DIFFERENT WATERSHEDS

Modeling the environmental behavior of phosphorus is largely a matter of quantifying all the input and loss mechanisms in different kinds of locations. Many watersheds are characterized by a mix of forest, agricultural, and urban land. The variability in land cover and P inputs create complex ecological effects and watershed management challenges that are commonly investigated with computer models (de la Crétaz and Barten 2007). Several different computer models are currently available, and it is important to custom fit the model to the watershed. It has been said that, "Though all computer models are wrong, some are useful" (de la Crétaz and Barten 2007) (see Chapter 1). This section reviews some aspects of P behavior that influence the design and boundary conditions of phosphorus models, to help make a modeling effort useful.

Under natural environmental conditions, with little agriculture or urban development, a watershed land surface is almost entirely pervious and vegetated. Phosphorus recycling tends to occur mostly within a relatively localized watershed ecosystem. If one of the three essential plant nutrients, carbon, nitrogen, and phosphorus, is in short supply, plant growth in the watershed exhausts that nutrient before the others and it becomes the growth-limiting nutrient. Carbon, as carbon dioxide, is always available to plants from the atmosphere, whereas N and P must be in dissolved forms to be taken up through plant roots. Therefore, plant growth is never limited by C, but always by the availability of a soluble form of either N or P. Other nutrients, such as sulfur, potassium, calcium, magnesium, and certain trace metals, are not considered here because they seldom are growth limiting for plants under natural conditions. Of the two potentially limiting nutrients, soluble P is more often the limiting nutrient than is soluble N, mostly because there usually are fewer sources of P, and P is generally the less mobile. For this reason, any increase in the ambient levels of P in a lake or stream usually accelerates plant growth and increases the potential for eutrophication. In a stable ecosystem, such as virgin forest or a noneutrophic lake, long-term changes in average net growth remain close to zero because time-averaged losses and inputs of P, as the growth-limiting nutrient, are approximately equal.

Under natural conditions in a stable ecosystem, such as a mature virgin forest:

1. Phosphorus is originally dissolved from P-bearing rocks by weathering processes.
2. Dissolved P, in the form of soluble orthophosphates, is taken up from soil by vegetation and stored in plant tissue.
3. Phosphorus is transferred to the tissue of animals that eat the plants.
4. Phosphorus is released back to the soil in the feces and urine of the animals, and when plants and animals die and decay.
5. Almost all the soluble phosphates not taken up by plants become immobilized in the soil by forming insoluble compounds (chiefly calcium, iron, and aluminum phosphates) and by sorption to soil particles and silicate clay minerals (Figures 4.4 and 4.6).
6. Surface water runoff and soil erosion are the most important processes that transport P. Runoff transports dissolved forms of P, whereas erosion transports sediment-sorbed P.
7. As long as the watershed surface remains pervious and sufficient vegetation is in place, storm runoff and soil erosion are minimized and little P is lost from the local watershed ecosystem.
8. The small amount of soluble P lost from the local ecosystem by soil erosion and surface runoff is replaced by P desorbed from soils and dissolved from minerals.

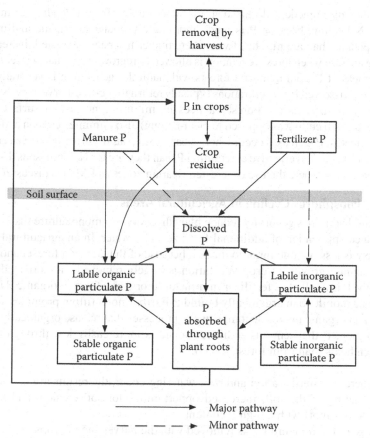

FIGURE 4.6 Movement of different forms of phosphorus on agricultural land.

When a natural watershed is disturbed by converting part of it to other land uses, the pattern of P recycling can be altered in ways that are detrimental to the watershed ecosystem and downstream water bodies. Urban and agricultural activities in particular are the major nonpoint sources of silt and runoff that transport P to streams and, ultimately, to lakes and reservoirs. Conversions of undisturbed vegetated lands, especially riparian areas, to either urban or agricultural uses are the changes that cause the greatest increases in P loading of streams and lakes. Nonpoint sources of nutrients are most commonly intermittent and are usually linked to seasonal agricultural activity or other irregular events such as construction or storm events.

4.3.7.1 Phosphorus Cycling in Virgin and Logged Forests

Forest soils bind P tightly and, in general, forests are sinks for P. For this reason, forests have an important function in protecting streams and downstream communities from P pollution. The large organic fraction of forest surface soil resists erosion and retains P very effectively. Phosphorus released by the decomposition of leaves and other forest litter is taken up and held in new forest vegetation growth. The small amount of P lost in stream flow is replaced by P from weathering and precipitation.

Forest-logging practices do not degrade water quality from P loading as much as it causes N loading, because P is much more tightly bound to organic and mineral soil compounds than are nitrates. Even where timber has been harvested, increases in P loading are rare when forest vegetation is allowed to regrow immediately after the cut. Thus, the loss of P from managed forests with natural regeneration is generally quite low. Stream-dissolved P concentrations typically return to pre-tree-harvest levels within about four years after cutting. Forest practices that minimize increases in surface runoff (partial and patch cutting as opposed to clear-cutting) also minimize erosion and export of P. The greatest P increases occur where clear-cutting has occurred. However, riparian areas in forests are less effective at retaining P than they are for N, for reasons discussed in Section 4.3.7.4, where the use of forested riparian zones as BMPs is discussed.

4.3.7.2 Phosphorus Cycling in Agricultural Areas

Agricultural activities generally reduce plant diversity to a monoculture that requires the repeated application of additional fertilizers and water. In an agricultural water-shed, there is less organic input to the soil because of the lack of a forest canopy and because annual growth of crop vegetation is largely removed. In cultivated fields, most of the P is applied as fertilizer in mineral (inorganic) form; inorganic P is more soluble and mobile than organically bound P. Furthermore, tilling promotes soil res-piration and organic matter decomposition, processes that release organically bound P. In addition, P that remains sediment-bound is more easily lost through erosion from open fields than from forested land.

- Where over-fertilization and over-watering occur, the results are excessive P loading of the soil, increased opportunities for soil erosion, and more surface runoff with a high P content.
- Most of the remaining P is removed with the harvesting of crops, requiring a repeat of the fertilization and watering cycles.
- The ideal situation is when P inputs from fertilization and natural retention match P removal from runoff and harvested crops, while careful irrigation practices and use of BMPs minimize water runoff and soil erosion.

4.3.7.3 Phosphorus Cycling in Urban Areas

Urban sources of P are mainly from fertilization of lawns and gardens and from dis-charges of wastewater treatment plants. For urban areas, P-modeling programs gener-ally assume that all surface water flow on impervious surfaces is transported unabated until it reaches a wastewater treatment plant, pervious soil, or some water body.

- Urban development in a forest watershed generally decreases the amount of vegetation and the proportion of pervious to impervious surface.
  - This increases the amount of soil erosion and surface runoff, resulting in increased amounts of P lost from the watershed and transported into streams and lakes.
    - Increased P losses can reduce the P content in the source ecosystem to the point where supplementary fertilization is needed and, at the same

time, increase the P content of downstream water bodies to the point where P becomes a pollutant, degrading water quality by promoting excessive algae and weed growth.

- Phosphorus control in urban runoff from grass and garden fertilization consists mainly of an efficient sewer system, vegetated buffer zones in riparian areas, and public education related to efficient fertilizer use.
- Discharge from wastewater treatment plants can generate localized high concentrations of P in receiving waters.
  - Municipal systems with secondary wastewater treatment typically remove 10–20% of the influent P and can reduce effluent concentrations to 3–5 mg/L. Tertiary treatment is capable of removing 99% of influent P.
  - The EPA-recommended stream water concentrations to avoid eutrophication are 0.1 mg-P/L for flowing waters and 0.05 mg-P/L for streams flowing into lakes.

4.3.7.4 Phosphorus Cycling in Watersheds with Mixed Forest, Agricultural, and Urban Land Uses

In one watershed modeling study (Radcliffe and Cabrera 2006), the land use distribution was estimated to be 3% urban, 28% agricultural, 62% forest and wetland, and 7% open water consisting mainly of a large lake into which the watershed drained. The model estimated that agriculture contributed 66% of the annual nonpoint source P load to the lake; urban land contributed 18% and forest land 16%. When the proportion of P leaving the watershed from different land uses was compared with the percentage of watershed area occupied by the land uses, it was found that

- The estimated P contribution from urban land was about six times the fraction of urban land area in the watershed.
- The estimated P contribution from agriculture was about twice its proportionate area.
- The estimated P contribution from forest land was about one-fourth its proportionate area.

The study concluded that better management of agricultural P should be a priority because agriculture contributes the greatest amount of the total P load to the lake. However, urban sources contribute more P per unit area than other land uses; therefore, nutrient removal at wastewater treatment plants should receive a high priority to reduce total P loading.

The behavior of P where land uses are mixed depends very strongly on topography, that is, on how the different land uses are geographically distributed in the watershed. In general, placing as much forest as possible within the surface flow paths from the urban and agricultural areas to all streams and lakes will minimize P loss to the water bodies.

However, the common BMP of establishing a vegetated buffer zone in riparian areas is not as effective for limiting P movement as it is for N movement. Runoff and erosion are the overland processes that transport P. Runoff transports dissolved P, whereas erosion transports sediment-sorbed P. Both forms are retained in forested buffer zones; dissolved P is retained by sorption to soils and organic matter and already-sorbed P is retained by sediment trapping. However, continued P loading in riparian zones can lead to a buildup of P in riparian buffer soils, potentially to the point where riparian forests eventually become a source rather than a sink for P.

The main reasons that riparian zones are less effective at retaining P than they are for N are

- Phosphorus binds preferentially to silts, clay, and the smaller sizes of organic matter, which are the first to erode during rainstorms and travel farther than larger particles. Hence, it is more likely that these fine particles will be transported through the buffer zone directly to streams.
- Smaller particles that are retained within the riparian zone have greater average surface area than coarser sizes and sorb more P per unit weight, compared to nonriparian watershed soil. This causes riparian buffer soils to become enriched in P relative to watershed source areas. As P loading increases in a riparian zone, so does the potential for P releases.
- Riparian zones are prone to flooding, which often creates anoxic conditions in the soils.
  - Anoxic conditions that accompany flooding can promote the release of dissolved, biologically available P from riparian sediments.
  - In contrast, anoxic conditions encourage the conversion of nitrates to nitrogen gas, returning nitrogen to the atmosphere rather than into a water body.

Many watersheds are characterized by a mix of forest, agricultural, and urban land. The variability in land cover and P inputs create complex ecological effects and watershed management challenges that are commonly investigated with computer models. Several different computer models are currently available, and it is important to custom fit the model to the watershed. It has been said that, "Though all computer models are wrong, some are useful." Even if the most useful model has been selected for a particular watershed, it should be expected that a period of testing and adjustment will be required for useful results.

4.4 NUTRIENT (N AND P) CONTAMINATION: EUTROPHICATION AND ALGAE PROBLEMS

The nutritional needs of aquatic plants are such that the quantity of nitrogen and phosphorus locally available in natural waters invariably controls the maximum potential growth of aquatic vegetation. Although phosphorus is more often the growth-limiting nutrient, depending on the specific water body characteristics, either nitrogen or phosphorus can be limiting. Water quality concerns often require concurrent controls on both nutrients. Thorough and up-to-date discussions of the roles of nitrogen and phosphorus as nutrients for aquatic plants may be found in Kadlec and Knight (2009).

Whenever the amount of nutrients in a stream or lake increases, the growth of aquatic plants and other organisms eventually also increases.* Excessive concentrations of N and P in a lake or stream degrade water quality by causing excessive plant growth, which leads to eutrophication.† The harmful effects of eutrophication are well documented. The progression from high-quality water to eutrophic water includes

- Accelerated growth of algae phytoplankton leading to harmful algal blooms
- Reduced water clarity from suspended solids
- Increased BOD from accumulating organic matter
- Diminished DO to levels harmful to aerobic aquatic organisms (hypoxia)
- Loss of submerged aquatic vegetation
- Increase of undecomposed organic matter in bottom sediments
- Anaerobic zones in bottom sediments and lower water column layers
- Bad water odors
- Potential fish kills
- Toxins from cyanobacteria (blue-green algae)
- Reduced water recreational value

Once eutrophication has started, it is hard to reverse, because bottom sediments become a reservoir for stable phosphorus, which can cycle between stable and labile forms, providing a continuing source of nutrient P long after the initial sources of excess P have been eliminated.

According to the EPA (EPA 2011),

Nitrogen and phosphorus pollution has the potential to become one of the costliest and the most challenging environmental problems we face. A few examples of this trend include the following:

1. 50% of U.S. streams have medium to high levels of nitrogen and phosphorus.
2. 78% of assessed coastal waters exhibit eutrophication.
3. Nitrate drinking water violations have doubled in eight years.
4. A 2010 USGS report on nutrients in ground and surface water reported that nitrates exceeded background concentrations in 64% of shallow monitoring wells in agriculture and urban areas, and exceeded the EPA's Maximum Contaminant Levels for nitrates in 7% or 2388 of sampled domestic wells. (USGS 2010).
5. Algal blooms are steadily on the rise; related toxins have potentially serious health and ecological effects.

---

* It is assumed here that other environmental characteristics, such as precipitation, flow rate, and suspended sediment load, remain relatively unchanged.
† Nitrogen in the form of ammonia is especially troublesome. In addition to stimulating eutrophication and being toxic to aquatic life as ammonia, it exerts a direct demand on dissolved oxygen when oxidized to nitrite and nitrate (see Sections 4.2.3 through 4.2.5); in the forms of nitrate and nitrate, it still serves as a plant nutrient promoting eutrophication.

To control eutrophication, the EPA makes the following recommendations:

- Total phosphates should not exceed 0.05 mg/L (as phosphorus) in a stream at a point where it enters a lake or reservoir.
- TP should not exceed 0.10 mg/L in streams that do not discharge directly into lakes or reservoirs.

4.4.1 ALGAL BLOOMS

Excess nitrogen and phosphorus cause rapid growth of algae. As algae grow, they consume DO and block sunlight from underwater plants, causing loss of important habitat for fish and other aquatic organisms. The underwater plants and excess algae eventually die, further decreasing DO as they decompose. Excessive algal growth because of nutrient contamination is called an algal bloom. In addition to eutrophication, algal blooms cause many other problems.

4.4.1.1 Health Impacts from Cyanobacteria (Blue-Green Algae)

A type of algae called cyanobacteria (blue-green algae), with characteristics of both plants and bacteria, was initially grouped with algae, but is now classified as bacteria based on cell structure and cell division. Cyanobacteria differ from other bacteria and resemble algae by containing photosynthetic pigments similar to those found in algae and plants. Although they are predominantly photosynthetic organisms, they are also capable of using organic compounds as a source of energy. Some cyanobacteria have a specialized structure called a heterocyst that can fix molecular nitrogen (Section 4.2.1). The ability to fix nitrogen gives these species a competitive advantage over other algae. Many cyanobacteria have gas vacuoles that allow them to remain in suspension and migrate to surface waters where there is plenty of light for photosynthesis. Cyanobacteria are ubiquitous in the aquatic environment; the same conditions in surface water that produce algal blooms can produce blooms of cyanobacteria.

Of the thousands of cyanobacteria species, 46 are known to produce potent toxins harmful to humans and animals, primarily hepatotoxins (which cause liver damage) or neurotoxins (which cause nerve damage) (EPA 2006). The most studied and among the most toxic and frequently encountered of the cyanotoxins is microcystin-LR, one of over 80 known toxic microcystin variants (congeners) produced by cyanobacteria. The World Health Organization (WHO 2004) has set a provisional guideline value of 1 μg/L for microcystin-LR in water used for recreation and domestic and agricultural water supplies.

4.4.1.2 Water Treatment for Microcystins from Cyanobacteria

Microcystin toxins mostly remain contained (>95%) within cyanobacteria cells. Any damage, such as that caused by some treatment options (e.g., chlorination and copper sulfate application), can cause cell rupture and produce cell leakage, resulting in an increase in the concentration of dissolved toxin in the waste stream. This behavior is critical for treatment, because dissolved toxin is not removed by conventional treatment technologies, while healthy cyanobacteria cells are successfully removed by several conventional treatments, such as coagulation and flocculation, which remove the cells intact and without damage. The resulting sludge containing toxic cyanobacteria must then be isolated from the treatment process since cells contained in sludge will break down rapidly and release dissolved toxin.

Other water treatment techniques (reverse osmosis, sedimentation, riverbank filtration, slow sand filtration, disinfection, granular and powdered activated carbon, ozonation, chlorination, and ultraviolet radiation) are effective to varying degrees, depending the site-specific nature of the cyanobacteria bloom, at removing the most common cyanobacteria and their toxins in drinking water. In general, a combination of different methods is required; when an appropriate combination of techniques is used, more than 95% of toxins are eliminated in finished water.

Nevertheless, controlling the source water to limit nutrient concentrations and make it unsuitable for algal growth remains the first choice for treatment of algal blooms.

RULES OF THUMB

1. The critical level of inorganic phosphorus for algae bloom formation can be as low as 0.01–0.005 mg/L under summer growing conditions, but more frequently is around 0.05 mg/L.
2. P concentrations
   a. Between 0.003 and 0.1 mg/L will generally stimulate plant growth.
   b. <0.03 mg/L are typically considered to be unpolluted.
   c. <0.1 mg/L should be maintained to avoid accelerated eutrophication.
   d. >0.1 mg/L create a potential for eutrophication.
3. Lakes are nitrogen-limited if the weight ratio of total nitrogen to total phosphorus (TN/TP) is <13, nutrient-balanced if 13 < TN/TP < 21, and phosphorous-limited if N/P > 21. Exact ranges depend on the particular algae species. Most lakes are phosphorous-limited; in other words, additional phosphorus is needed to sustain further algal growth.
4. Different N/P ratios and pH values favor the growth of different kinds of algae.
5. Low N/P ratios favor N-fixing cyanobacteria (blue-green algae).
6. High N/P ratios, often achieved by controlling phosphorous input by means of additional wastewater treatment, cause a shift from cyanobacteria to less objectionable algae species.
7. Lower pH (or increased CO_2) gives green algae a competitive advantage over cyanobacteria.
8. To control eutrophication, EPA recommends that total phosphates should not exceed 0.05 mg/L (as phosphorus) in a stream where it enters a lake or reservoir, or 0.10 mg/L in streams that do not discharge directly into lakes or reservoirs.

4.4.2 CASE STUDY 3: CAUSES AND LIMITATIONS OF ALGAL GROWTH IN STREAMS

(This case study is based on the report by Summers 2008.)

During 2007–2008, the West Virginia Department of Environmental Protection studied the Greenbrier River to determine which habitat factors promoted growth of rooted, filamentous algae. The study was initiated by reports that excessive algal growth was harming recreational uses of the river during the summer and degrading drinking water quality from suppliers using river water. Instream grab samples were analyzed for total and dissolved phosphorus, total nitrogen, alkalinity, hardness, and other parameters.

Both the chemical and physical conditions in the Greenbrier River—including hardness, alkalinity, temperature, clarity, and substrate—proved to be ideal for growth of filamentous algae. The water chemistry results revealed elevated levels of dissolved phosphorus and nitrogen in areas of excessive algae growth, with phosphorus being the limiting nutrient. Identifying the nutrient sources for specific algal blooms revealed that algae in the Greenbrier River resulted primarily from dissolved phosphorus contained in municipal sewage treatment plant effluent and dissolved nitrogen from several different sources, including agriculture, municipal discharges, and failing septic systems.

However, the study raised an important question: Why, if the algae bloom was driven primarily by sewage treatment discharges, were algal blooms absent on many other rivers in the state that appeared to have similar sewage treatment discharges and excess nutrient conditions? A query of a water quality database for West Virginia watersheds showed a significant difference in alkalinity and hardness values for streams with and without algal blooms. In brief, the findings were as follows:

- Streams with alkalinity less than 30 mg/L had no algae blooms. However, some streams with sufficient phosphorus and nitrogen and with alkalinity greater than 30 mg/L, also had no algal blooms.
- The hardness level in streams with adequate alkalinity (>30 mg/L), phosphorus, and nitrogen, but no algae, always tended to be higher than hardness in the Greenbrier River. For example, on another West Virginia river below a sewage treatment plant discharge, where nutrient levels were high but there was no algal bloom, hardness was 250–400 mg/L; on the Greenbrier River, where nutrient levels were similar but algal growth was severe, hardness never exceeded 100 mg/L.

Hardness is a measure of Ca^{+2} and Mg^{+2} concentrations (Chapter 3, Section 3.6), suggesting that calcium and magnesium were affecting algal growth in West Virginia rivers. The report that this case study summarizes (Summers 2008) contains a literature review of the role of calcium and magnesium in algal growth, finding that

- Low alkalinity (<25 mg/L) keeps phosphorus from being available as a nutrient. An alkalinity of greater than 50 mg/L is recommended for productive aquaculture ponds.

- Calcium and magnesium cations react with dissolved phosphorus to form stable (essentially insoluble) inorganic calcium and magnesium phosphate precipitates (Figure 4.4 and Sections 4.3.3 and 4.3.4.2) that are not available as nutrients for algal growth.

The study concluded that a minimum alkalinity is needed to make the phosphorous available for plant uptake, and that at higher hardness levels, common in many West Virginia streams, low-solubility salts of calcium and magnesium phosphate precipitate, making phosphorus much less available for algae development. For West Virginia streams, hardness levels greater 150 mg/L appear to inhibit growth of filamentous algae and some suppression of growth may begin when hardness exceeds 100 mg/L. Another benefit of high hardness levels is that they favor larger N/P ratios, discouraging formation of toxic cyanobacteria (blue-green algae) (see Rules of Thumb numbers 6 and 7 on page 165).

The first rule of algae control is to minimize nutrient releases at their sources when possible (phosphorus at sewage treatment plants in this case). This study calls attention to other environmental controls of algal growth and suggests the potentially useful control method of increasing water hardness where source controls may not be feasible or adequate.

4.4.3 Case Study 4: Conditions Favoring Nitrogen over Phosphorus as the Limiting Nutrient

(This case study is based on the report by Lewis et al. 2008.)

Much of Section 4.4 discusses situations where phosphorus is the limiting nutrient, although it is pointed out that nitrogen can be limiting under certain situations (Rules of Thumb numbers 3 and 4 on page 165). This case study illustrates such a situation.

In water bodies where the N/P weight ratio is low (less than about 25), nitrogen-fixing cyanobacteria (blue-green algae) have a competitive advantage over algae that cannot fix atmospheric or dissolved nitrogen (N_2). As long as the N/P weight ratio remains greater than about 21, P will be the limiting nutrient, even with a low N/P ratio, where cyanobacteria thrive. However, cyanobacteria require ample sunlight for fixing N and any condition that limits available light, such as high water turbidity, may also limit the amount of N available where cyanobacteria are important producers.

This study examined nutrient management options for a reservoir near Denver, Colorado, that sometimes had problems with algal blooms during the late summer months. The existing strategy of regulating point and nonpoint sources of anthropogenically generated P in the reservoir watershed was not providing sufficient control. It appeared that reducing P concentrations were not suppressing algae concentrations, as would be expected for cases where P was the limiting nutrient. The study found that the N/P ratio was generally low and that N usually was the limiting nutrient, for two main reasons:

1. The watershed was largely undeveloped so that only about 15% of the P in the reservoir originated from readily controllable point-source discharges. Most of the P entering the reservoir came from drainage from

P-rich alluvial soils in the watershed that could not easily be controlled. Although N sources had not been extensively studied, springtime N and P measurements indicated that the N/P ratio in the reservoir was seldom greater than 5.

2. Although N-fixing cyanobacteria were briefly observed in mid-summer, their growth appeared to be limited because of inadequate light and low DO, conditions associated with repeated deep mixing of the reservoir water column. Stable thermal stratification of the reservoir during the summer, which would have promoted reservoir turnover only twice a year during the onset of spring and winter, was hindered by the relatively shallow water in the reservoir and water releases from the reservoir outlet. The frequent deep mixing carried N-fixing cyanobacteria down into a deep zone with limited light and depleted DO, while carrying low-DO water to the surface; the net result was minimal production of N-fixing cyanobacteria.

Given the conditions in the reservoir that caused N to be the limiting nutrient, low N/P weight ratios, and insufficient fixation of N by N-fixing cyanobacteria, the full use of P for plant growth could not occur. This created a surplus of unused P in the lake; any reduction in the daily P input to the reservoir could not produce a corresponding reduction in algal growth until the surplus P was consumed. Based on phytoplankton nutrient requirements, the study concluded that the total P inventory of the reservoir water column would need to be reduced by about 50% before the P surplus was reduced enough that suppression of phytoplankton biomass by reducing P inflows could begin. No incremental water-quality benefits could be expected from P source controls until the onset of P limitation, at about 50% of current concentrations. This estimate did not take into account the additional cache of unused P weakly bound to the bottom sediments, which would be more slowly released to provide an additional source of surplus P, further reducing the effectiveness of watershed P controls.

The study concluded that continuing the strategy of trying to control watershed P might not be the most useful option for managing algae in the reservoir, and that switching to N controls might be more successful. However, controlling algal growth under N-limiting conditions can be more difficult than when P is the limiting nutrient. To quote the report:

> N management is likely to be much more expensive for point sources because it involves nitrification followed by denitrification, typically with organic carbon supplementation (methanol), whereas P control for point sources is achieved by precipitation (Metcalf and Eddy 2003). Also, watershed management of N may be difficult because of the high mobility of nitrate in groundwater. Even so, expensive N management may be preferable to cheaper but ineffective P management.

The authors of the report emphasize that nutrient management should not be based on control only of the limiting nutrient because environmental changes can occur that reverse the situation. For example, in this particular case, a change in reservoir management that produced a more stable thermal stratification of the reservoir could create conditions more favorable for N-fixing cyanobacteria and cause P to become the limiting nutrient. In general, management of both N and P sources is likely to be the most successful way to control algal growth and eutrophication in lakes and streams.

4.5 SULFIDE (S^{2-}) AND HYDROGEN SULFIDE (H_2S)

4.5.1 BACKGROUND

Hydrogen sulfide (written H_2S_{aq} or H_2S_{gas}, depending on whether it is in dissolved or gaseous form) is a moderately water-soluble, colorless gas with a distinctive disagreeable odor of rotten eggs. In water, it is formed in a two-step reaction when sulfide anion (S^{2-}) reacts reversibly with water to form hydrosulfide anion (HS^-) and dissolved H_2S_{aq} (Equations 4.17 and 4.18).

$$S^{2-} + H_2O \rightleftarrows HS^- + OH^- \tag{4.17}$$

$$HS^- + H_2O \rightleftarrows H_2S_{aq} + OH^- \tag{4.18}$$

In water exposed to air, some dissolved H_2S_{aq} will partition into the air as H_2S_{gas}. Equation 4.19 is the overall reversible reaction of sulfide with water, written with only the sulfide species.

$$S^{2-} \rightleftarrows HS^- \rightleftarrows H_2S_{aq} \rightleftarrows H_2S_{gas} \tag{4.19}$$

S^{2-} and HS^- are dissolved, nonvolatile anions with no odor. H_2S_{aq} is dissolved gas (solubility = 4000 mg/L at 20°C) that can give water a musty or swamp-like taste at concentrations less than 1 mg/L. The odor of water with as little as 0.5 mg/L of hydrogen sulfide is detectable as sewer-like by most people. At 1–2 ppm, dissolved hydrogen sulfide gives water an intense rotten egg odor and makes the water corrosive to metals and cement. In general, any detection of hydrogen sulfide by taste or the offensive rotten eggs odor will make water unacceptable for consumption, even though it is still well below a dangerous toxicity level. Analytically, the three dissolved sulfur species, S^{2-}, HS^-, and H_2S_{aq}, are collectively called total sulfide, or simply sulfide.

H_2S_{gas} released from water solutions is a moderately soluble gas with the characteristic odor of rotten eggs; since it is slightly heavier than air (density$_{H_2S}$ / density$_{air}$ = 1.2), H_2S_{gas} tends to accumulate at the bottom of poorly ventilated spaces. Because inhaled H_2S_{gas} is highly toxic (comparable to hydrogen cyanide gas), corrosive to metals and cement (at concentrations as low as 1 mg/L), and flammable (flammable range in air: 4.3–46%; burns in oxygen to form the irritating gas sulfur dioxide, SO_2), it is the species of greatest environmental concern. Although the dissolved form has no EPA drinking water MCL, because ingestion usually does not pose a health risk,* acute toxicity from inhalation of the gaseous form, H_2S_{gas}, is

---

* Due to its repulsive odor and taste at below toxic concentrations, the EPA and the World Health Organization have deemed it unlikely that a person could consume a harmful dose of hydrogen sulfide from drinking water, even accidentally.

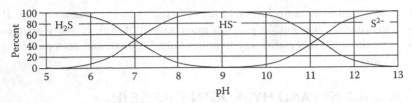

FIGURE 4.7 Distribution of hydrogen sulfide species in water as a function of pH; T = 20°C. H_2S_{aq} is the most toxic and volatile form; S^{2-} and HS^- are nonvolatile and much less toxic. $H_2S_{aq} > 2.0$ mg/L constitutes a long-term hazard to fish.

high; eye irritation can begin at concentrations of 10–20 ppmv, and serious eye and lung damage around 50–100 ppmv (see Section 4.5.7).

At equilibrium, the distribution among dissolved sulfide species (S^{2-}, HS^-, and H_2S_{aq}) depends mainly on the pH and somewhat on the temperature of the water (see Figures 4.7 and 4.8). The proportion of dissolved hydrogen sulfide gas increases with decreasing pH; wherever water pH is less than about 7.5, the potential for H_2S_{gas} formation is increased.

- At <pH 5, at least 99% of dissolved sulfide is in the form of H_2S_{aq}, the unionized form.
- At pH 7, dissolved sulfide is 50% HS^- and 50% H_2S_{aq}.
- At pH 9, about 99% is in the form of HS^-.
- S^{2-} becomes measurable only above pH 10.

4.5.2 SOURCES

There are two main natural sources of H_2S in the aquatic environment: the reduction of mineral sulfates and sulfites to sulfide (Equation 4.20), and the anaerobic decomposition of organic matter containing sulfur (Equation 4.21). Both mechanisms require reducing, or anaerobic, conditions, and both are strongly accelerated by the presence of sulfur-reducing bacteria (SRB).

Hydrogen sulfide enters the atmosphere as a gas from both natural and human sources. Natural sources include volcanoes, hot sulfur springs, marsh gases, anaerobic biodegradation of sulfates, and sulfur-containing organic matter. Human sources include petroleum refineries, electric power plants burning coal or fuel oil containing sulfur, oil and gas wells, pulp and paper mills, sewage treatment plants, large confined animal feeding operations, Portland cement kilns, municipal waste landfills, and coke ovens. Hydrogen sulfide is also produced indirectly through acid mine drainage (see Chapter 5, Section 5.9.1), in which the dissolution of metal sulfides yields sulfate ion, which then can be reduced to hydrogen sulfide by SRB (Equation 4.21).

Natural sources account for about 90% of the total global emissions to the atmosphere, and human activities for the remaining 10%. The concentration of hydrogen sulfide in the air in unpolluted areas is low, typically between 0.03 and 0.1 μg/m³, although concentrations above 100 μg/m³ have been reported near industrial plants. Atmospheric H_2S reacts with atmospheric moisture and oxygen to form acidic sulfur compounds that fall to earth in acid rain.

The most common oxidized form of sulfur found in environmental waters is sulfate ion, SO_4^{2-}, which is only formed under aerobic conditions. In sulfide ions (HS^- and S^{2-}) and dissolved hydrogen sulfide (H_2S_{aq}), sulfur is in a reduced form; these chemical species are only found in anaerobic waters (reducing conditions, devoid of oxygen). In aerobic surface waters, hydrosulfide anion, HS^-, is oxidized to sulfate ion by DO (Equation 4.20); in anaerobic water, HS^- reacts to form H_2S_{aq} by Equation 4.18.

$$HS^- + 2O_2(aq) \rightleftarrows SO_4^{2-} + H^+ \qquad (4.20)$$

Groundwaters often contain sulfide, where it arises from water contact with soluble sulfide minerals and from anaerobic bioreduction of dissolved sulfates. In addition, reducing conditions are common in groundwater where DO has been consumed by biodegradation of organic matter and is not readily replaced because groundwater is relatively isolated from contact with the atmosphere. Certain bacteria (called sulfate-reducing bacteria, abbreviated SRB), which are commonly found in most soils, can greatly accelerate the formation of S^{2-} under anaerobic conditions by bioreduction of sulfates (Equation 4.21). In anaerobic surface waters where sulfate is present along with organic matter and/or SRB, sulfate is reduced to sulfide anion (S^{2-}) (Equation 4.21), which then produces dissolved H_2S_{aq} by Equations 4.17 and 4.18. Odorous water is often an indicator of this process. Another indicator is the formation of blackened soils, sludge, and sediments, which often accompanies the reaction of dissolved hydrogen sulfide with dissolved iron to form precipitated black ferrous sulfide (FeS), along with other metal sulfides.

$$SO_4^{2-} + \text{organic matter} + SRB \rightarrow S^{2-} + H_2O + CO_2 + SRB \text{ growth} \qquad (4.21)$$

In anaerobic surface waters, sulfide from Equation 4.21 is a common product of wetlands and eutrophic lakes and ponds. Aerobic surface waters are oxygenated by oxygen diffusion from the atmosphere and by oxygen released from photosynthesizing aquatic plants; oxygen diffusion is enhanced in fast-flowing, turbulent, and cascading water. In standing water, where oxygen diffusion from the atmosphere is slow and where dead organic matter (leaf and grass litter, insect and animal waste, etc.) can accumulate, biodegradation of the organic matter can consume DO faster than it is replenished by diffusion and create anaerobic conditions (see Chapter 3, Section 3.8).

RULES OF THUMB

In water, S^{2-} reacts by:

$$S^{2-} + 2H_2O \rightleftarrows OH^- + HS^- + H_2O \rightleftarrows H_2S(g) + 2OH^- \qquad (4.22)$$

1. Raising the pH shifts the equilibrium to the left, converting the malodorous gas H_2S into odorless and nonvolatile HS^- and S^{2-}.
2. Lowering the pH shifts the equilibrium to the right, converting odorless and nonvolatile HS^- and S^{2-} into H_2S gas.
3. Lowering the temperature shifts the equilibrium to the right (more H_2S) at any pH.

4. Well water, groundwater, or stagnant surface water that smells of H_2S (rotten eggs) is usually a sign of SRB.
5. Water conditions promoting the formation of H_2S are

> Sulfate, > 60 mg/L
>
> Oxidation–reduction potential, < –60 mV
>
> pH < 6

4.5.3 FORMATION OF H_2S IN WELLS, DETENTION PONDS, WETLANDS, AND SEWERS

Water conditions promoting the formation of H_2S are sulfate >60 mg/L (or the presence of sulfur-containing organic matter such as protein), oxidation–reduction potential <200 mV, pH <6–7, and the presence of SRB. These conditions frequently occur in standing or slowly moving water, such as in detention ponds, wetlands, and sewers, where organic litter can accumulate and where the water or soil contains sulfate. In water wells, which are normally protected from organic litter, the odor of H_2S sometimes occurs because of the presence of sulfate in the water and SRB at the air–water interface in the well.

Since surface waters seldom contain more than 8–12 mg/L of DO (more often less), decay of organic matter can quickly reduce DO to anaerobic levels (<1 mg/L). Such waters often develop a bottom layer of black sediments containing iron and other metal sulfides along with organic matter in various stages of decay. In still water, oxygen diffusion into this sediment layer is slow and anaerobic conditions can be maintained with minimal water cover (<10 inch deep). If all water is removed and the soil allowed to dry, diffusion of oxygen into the sediment quickly oxidizes the sulfides to sulfate and the H_2S disappears.

The blackening of soils, wastewater, sludge, and sediments in locations with standing water, along with the odor of rotten eggs, is an indication that sulfide is present. The black material results from a reaction of H_2S_{aq} and S^{2-} with dissolved iron and other metals to form precipitated FeS, along with other metal sulfides.

RULES OF THUMB

1. Well water smelling of H_2S is usually a sign of SRB. Look for a water redox potential below –200 mV and a sulfate, SO_4^{2-}, concentration in groundwater >60 mg/L.
2. A typical concentration of H_2S_{aq} in unpolluted surface water is <0.25 mg/L.
3. H_2S_{aq} >2.0 mg/L constitutes a chronic hazard to aquatic life.
4. In aerated water, H_2S_{aq} is bio-oxidized to sulfates and elemental sulfur.
5. Unionized H_2S_{aq} is volatile and air-strippable. The ionized forms, HS^- and S^{2-}, are nonvolatile.

4.5.4 Hydrogen Sulfide Odor

The human nose is very sensitive to the odor of H_2S_{gas} at low air concentrations. Since a portion of H_2S dissolved in water will always partition into the atmosphere, the odor of hydrogen sulfide gas is usually the first warning that the water is contaminated with sulfides. The solubility of H_2S in water is about 4 g/L. Although individual sensitivities vary greatly, the taste and odor threshold for H_2S_{aq} has been estimated to be a water concentration between 0.05 and 0.1 mg/L. At concentrations less than 1 mg/L, dissolved H_2S_{aq} gives water a musty odor; at 1–2 mg/L it smells like rotten eggs. However, at air concentrations greater than 100–150 ppm, right around the threshold for eye and lung damage (see Section 4.5.7), H_2S_{gas} deadens the olfactory senses and becomes odorless to humans, greatly increasing the danger of accidental exposure.

4.5.5 Removing Dissolved Sulfides and H_2S Gas from Water

Most methods for removal of sulfide contaminants from water use some form of oxidation, although filter/sorption media like activated carbon, catalytic carbon, or manganese greensand are sometimes used when total sulfide concentrations are small (<1 mg/L for activated carbon; <6 mg/L for greensand). Since a significant fraction of dissolved sulfide is in anionic form at pH > 7 (see Figure 4.8), anion exchange systems are sometimes useful for treatment if the pH is high enough.

Oxidation methods may utilize aeration (e.g., Equations 4.23 and 4.24) or chemical treatment (chlorine, ozone, permanganate, hydrogen peroxide, etc.; e.g., Equations 4.25 through 4.27), which convert sulfides to dissolved sulfate and solid

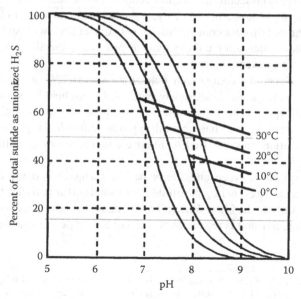

FIGURE 4.8 Fraction of hydrogen sulfide in unionized form (H_2S_{aq}) as a function of temperature and pH.

elemental sulfur.* Elemental sulfur, written S^0, precipitates as a solid and can be removed by filtering.

$$2\,S^{2-} + 2\,O_2(aq) \rightarrow SO_4^{2-} + S^0(s) \tag{4.23}$$

$$H_2S_{aq} + \tfrac{1}{2}O_2(aq) \rightarrow S^0(s) + H_2O \tag{4.24}$$

$$H_2S_{aq} + Cl^- \rightarrow 2\,HCl + S^0(s) \tag{4.25}$$

$$H_2S_{aq} + H_2O_2(aq) \rightarrow 2\,H_2O + S^0(s) \tag{4.26}$$

$$S^{2-} + 4\,H_2O_2(aq) \rightarrow SO_4^{2-} + 4\,H_2O \tag{4.27}$$

RULES OF THUMB

1. Chemical oxidation is recommended for treating concentrations of sulfide >6 mg/L, to form compounds that do not cause foul taste or odors in drinking water.
2. The most commonly used chemical oxidant is chlorine.
    a. Most effective when water pH is 6.0–8.0.
    b. Often applied as household bleach or sodium hypochlorite.
    c. For pH < 7.5, start with 2–3 mg/L chlorine for every 1.0 mg/L sulfide. Recommended contact time is at least 20 minutes. Increase concentration and/or contact time if necessary.
    d. For pH 7.5–8, start with 5–10 mg/L chlorine for every 1.0 mg/L sulfide. Recommended contact time is at least 20 minutes. Increase concentration and/or contact time if necessary.
    e. For pH > 8, use ozone or hydrogen peroxide instead of chlorine solution; discuss concentrations and contact time with supplier.
    f. Filter water after chlorination to remove suspended elemental sulfur or excess chlorine.
3. Liquid chlorine solutions can deteriorate by 20–25% per month in storage. Always use fresh, newly purchased household bleach or 12% sodium hypochlorite. Due to the short shelf-life of these products, it is not a good idea to purchase more than you are immediately going to use.
    a. Filtration may be used to remove concentrations of sulfide <6 mg/L.
    b. Manganese greensand filters have a manganese oxide coating that oxidizes hydrogen sulfide gas to solid sulfur particles that are retained in the filter. When all of the manganese oxide is consumed, the greensand is regenerated with potassium permanganate.

---

* The most common form of solid elemental sulfur is orthorhombic sulfur, in which layers of cyclic S_8 rings are stacked together. However, sulfur has many cyclic allotropes, up to at least S_{20}. The molecular unit of elemental sulfur is usually written as S_8 or S^0; the latter will be used here so as not to imply any particular structure of elemental sulfur.

    c. Activated carbon filters (granular or carbon block) use carbon particles that have been processed to greatly increase their porosity and surface area available for sorption. They can be used to remove trace amounts (<0.3 mg/L) of hydrogen sulfide by sorbing the gas to the carbon surface. Regeneration of the activated carbon is not usually practical and periodic replacement is necessary.

    d. Catalytic carbon filters are made by impregnating the surfaces of activated carbon with reactive chemicals (generally proprietary recipes containing organic or inorganic oxidants). Catalytic carbon filters combine the adsorptive properties of conventional activated carbon with the ability to catalyze, in the presence of DO, the chemical oxidation of sorbed contaminants.

4.5.6 CASE STUDY 5: TREATING ODOROUS WATER WELLS INFECTED WITH SRB BY SHOCK CHLORINATION

Water wells and distribution systems may become infected by SRB, a condition resulting in bacterial reduction of dissolved sulfate in the water to hydrogen sulfide gas, and made evident by the rotten egg odor of the water. If SRB are present in a well or distribution system, it is generally necessary to shock chlorinate the entire system, including the aquifer zone around the outside of the well bore.

Disinfecting a well by shock chlorination involves

1. Filling the well and associated plumbing with a 150–250-ppm solution of chlorine in water.
2. Surging the well to force chlorine solution into the aquifer zone around the well.
3. Allowing a contact time of 12–24 hours.
4. Pumping the well to flush out the oxidizing solution until the chlorine residual is gone.

This usually will eliminate the odor problem, at least temporarily. The odor often returns, maybe within a few days, several weeks, or even the following year, because disinfection of a well or distribution system is seldom complete. Some SRB are generally protected from contact with the chlorine solution in well joints and crevices and outside the well in the surrounding aquifer. These may survive the chemical treatment and regenerate, requiring that the procedure be repeated. If the odor still recurs after 2 or 3 repeated shock-chlorination procedures, either a program of periodic shock chlorination must be maintained or some method of external treatment can be used, such as aeration or chemical chlorination in an external storage tank. Since SRB cannot survive in aerobic conditions, aeration in an open or vented holding tank will help to transfer dissolved H_2S to the atmosphere and maintain aerobic conditions, thereby stopping further production. This works best at pH < 7, where dissolved sulfide is mainly present as air-strippable dissolved H_2S gas (see Figure 3.10).

4.5.7 TOXICITY AND REGULATION OF H_2S

Human health risks from environmental hydrogen sulfide are principally via inhalation, where the gas is rapidly absorbed into the bloodstream through the lungs. Inhaled H_2S is very toxic, comparable to inhaling hydrogen cyanide gas. Although toxicity data from oral ingestion are lacking, it is unlikely that anyone could consume a harmful dose of hydrogen sulfide in drinking water because a concentration high enough to be a drinking water health hazard also makes the water extremely unpalatable. For these reasons, the EPA does not regulate hydrogen sulfide in drinking water. The odor of water with as little as 0.5 ppm of hydrogen sulfide concentration is detectable by most people. Concentrations less than 1 ppm can give water a musty or swampy taste and odor. At 1–2 ppm, hydrogen sulfide gives water an offensive rotten egg odor and makes the water corrosive to metals and concrete. In general, any detection of hydrogen sulfide by taste or the offensive rotten egg odor will make water unacceptable.

> **RULES OF THUMB**
>
> 1. 0.00047 ppmv: Recognition threshold, the concentration at which 50% of humans can detect the characteristic odor of hydrogen sulfide, normally described as resembling a rotten egg.
> 2. <10 ppmv: Recommended exposure limit = 8 hours/day. Long-term, low-level exposure may result in fatigue, loss of appetite, headaches, irritability, poor memory, and dizziness.
> 3. 10–20 ppmv: Borderline concentration for eye irritation.
> 4. 50–100 ppmv: Likelihood of serious eye and lung damage.
> 5. 100–150 ppmv: Olfactory nerve becomes paralyzed, sense of smell disappears, and awareness of danger may be lost.
> 6. 320–530 ppmv: Pulmonary edema begins with the possibility of death.
> 7. 530–1000 ppmv: Affects central nervous system to cause cessation of breathing.
> 8. 800 ppmv: Lethal concentration for 50% of humans after 5-minute exposure (LC50).
> 9. >1000 ppmv: Immediate collapse with loss of breathing.

4.5.7.1 TYPICAL STATE STREAM STANDARDS FOR UNDISSOCIATED H_2S

See Appendix A, Section A.2.26.

4.5.7.2 Health Effects

See Appendix A, Section A.2.26.

4.6   ODORS OF BIOLOGICAL ORIGIN IN WATER (MOSTLY FROM HYDROGEN SULFIDE AND AMMONIA PRODUCTS)

The most common inorganic gases found in water are carbon dioxide (CO_2), methane (CH_4), hydrogen (H_2), hydrogen sulfide (H_2S), ammonia (NH_3), carbon disulfide (CS_2), sulfur dioxide (SO_2), oxygen (O_2), and nitrogen (N_2). In addition, many different organic gases in water arise from biodegradation of organic matter. Of these dissolved inorganic and organic gases, those with an offensive odor generally contain N and/or S combined with H, C, and/or O, such as H_2S, NH_3, CS_2, and SO_2.

Hydrogen sulfide (H_2S), from the anaerobic reduction of sulfate (SO_4^{2-}) by bacteria (Equations 4.21 and 4.22 followed by 4.17 and 4.18), is usually the most prevalent odor in anaerobic natural waters and sewage. Sulfate enters natural waters from sulfate minerals and aerobic decomposition of organic material; it is formed in domestic wastewater by the aerobic biodegradation of sulfur-containing proteins. In addition to H_2S, other disagreeable odorous compounds may be formed by anaerobic decomposition of various organic materials. The particular compounds that are formed depend on the types of bacteria and organic materials present. Table 4.2 contains information about the most common odor-producing bacteria, and Table 4.3 lists a number of common odiferous inorganic and organic compounds produced by these bacteria, with their odor characteristics and odor threshold concentrations. Sewage carrying industrial wastes may contain other volatile organic chemicals that can contribute additional odors.

4.6.1   CHEMICAL CONTROL OF ODORS

Depending on the odor-causing compound, chemical control of odors may be accomplished by a combination of pH control; eliminating the causes of reducing conditions; introducing chemical oxidation and aeration; sorption to activated charcoal; air stripping of volatile species; and chemical conversion (often microbially mediated, as in nitrification).

4.6.1.1   pH Control

4.6.1.1.1   Hydrogen Sulfide

To treat odors from H_2S, raise the pH (by adding NaOH or lime) to shift the equilibrium of Equation 3.33 to the left. This converts gaseous H_2S to the nonodorous ionic forms HS^- and S^{2-}. However, the pH must be maintained above 9 for complete odor removal. Normally, odor control by removing the sulfur compounds is more practical.

Lowering the pH (by adding acid) shifts the equilibrium of Equation 3.36 to the right, converting the ionic forms to gaseous H_2S. At low pH, the gas can be removed from the water by air stripping (see the following example). This, of course, does not destroy the H_2S; it moves it from the water to the air. Figure 3.11 gives the fraction of hydrogen sulfide that is in the volatile form of H_2S at different pH values and temperatures. Note that H_2S behaves the opposite of NH_3 (Figure 3.8). Stripping efficiency is increased with decreasing pH and lower temperatures.

TABLE 4.2
Common Odor-Producing Bacteria

Bacteria Type	Representative Bacteria	Nutrients Consumed	Respiration	Favorable Growth Conditions	Odor-Producing Products	Odors Produced	Typical Habitat	Visual Indications
Sulfate-reducing (SRB) and sulfur-reducing	Sulfate-reducing: *Desulfovibrio, Desulfobacter, Syntrofobacter*; Sulfur-reducing: *Desulfuromonas*	SO_4^-; short-chain organic acids, cellulose, sugars	Anaerobic	Cold, anaerobic water with neutral to acidic pH (except for specialized thermophilic sulfate reducers in hot springs and ocean vents)	H_2S, methyl mercaptan, dimethyl sulfide [DMS, $(CH_3)_2S$], dimethyldisulfide [DMDS, $(CH_3S)_2$]	Septic, rotten egg, decayed vegetables	Stagnant water in anaerobic ponds, wastewater treatment, and septic systems	Light-colored foam on surface of water, later turning dark brown or black
Actinobacteria	*Mycobacteria Nocardia Antinomyces Frankia Streptomyces*	High-weight organics like cellulose, chitin[a], proteins, waxes, paraffins, rubber	Aerobic	Warm, damp soil and sewage	2-methylisoborneol[b] (MIB), geosmin[c], mucidone[d]	Earthy, musty, soil after rain	On surface of decomposing sewage and animal wastes; also in many soils	Blue-gray to light green powdery or filamentous layer on decomposing sewage
Myxobacteria (slime bacteria)	*Myxococcus Angiococcus Stigmatella Sorangium*	High-weight organics, including chitin and, sometimes, cellulose	Aerobic	Organic-rich wastes and soils with neutral to alkaline pH and moderate to warm temperatures	2-methylisoborneol[b] (MIB); geosmin[c]	Earthy, musty	Soils and animal wastes	Sometimes form macroscopic, fruiting bodies that look like slimy droplets whose color depends on the species

| Cyanobacteria (blue-green algae) | Microcystis Oscillatoria Anacystis Anabaena | H_2O, CO_2, N_2; orthophosphates; sometimes H_2S, S_2O_3, H_2 | Aerobic | Warm, stagnant to slow-moving water with high levels of nutrients and dissolved oxygen | NH_3, DMS, DMDS, MIB, methyl mercaptan, n-butyl mercaptan, geosmin | Always: pungent ammonia. Moderate quantities: earthy, musty. Large quantities: septic, rotten vegetables | Warm ponds, lakes, wetlands, and seawater | Dark green to yellowish-brown scum on the surface of water |

[a] Long-chain polymer of a n-acetylglucosamine, found in cell walls of fungi and shells of arthropods and insects

[b] exo-1,2,7,7-Tetramethyl-bicyclo[2.2.1]heptan-2-ol, characterized by earthy, musty odor

[c] 4,8a-dimethyldecalin-4a-ol, characterized by earthy, musty odor

[d] 3-isobutyl-6-ethyl-alpha-pyrone, characterized by musty odor and taste

TABLE 4.3

Odor Characteristics and Threshold Concentrations in Water of Some Common Organic Compounds

Substance	Formula	Odor Threshold Concentration in Water (mg/L)	Odor Characteristics
Acetajdehyde	CH_3CHO	0.004	Pungent, fruity
Allyl mercaptan	$CH_2=CHCH_2SH$	0.00005	Very disagreeable; strong garlic, coffee
Ammonia	NH_3	0.037	Sharp, pungent
Amyl mercaptan	$CH_3(CH_2)_3CH_2SH$	0.0003	Unpleasant, putrid
Benzyl mercaptan	$C_6H_5CH_2SH$	0.0.00019	Unpleasant, strong
Butylamine	$C_2H_5(CH_2)_2NH_2$	—	Sour, ammoniacal
Cadaverine	$NH_2(CH_2)_5NH_2$	—	Putrid, decaying flesh
Chlorine	Cl_2	0.010	Pungent, irritating, suffocating
Chlorophenol	ClC_6H_4OH	0.00018	Medicinal, phenolic
Crotyl mercaptan	$CH_3CH=CHCH_2SH$	0.000029	Skunk-like
Dibutylamine	$(C_4H_9)_2NH$	0.016	Fishy
Diethyl sulfide	$(CH_3CH_2)_2S$	0.000025	Nauseating, ethereal
Diisopropylamine	$(C_3H_7)_2NH$	0.0035	Fishy
Dimethylamine	$(CH_3)_2NH$	0.047	Putrid, fishy
Dimethyl sulfide	$(CH_3)_2S$	0.0011	Decayed vegetables
Diphenyl sulfide	$(C_6H_5)_2S$	0.000048	Unpleasant
Ethylamine	$C_2H_5NH_2$	0.83	Ammoniacal
Ethyl mercaptan	C_2H_5SH	0.00019	Decayed cabbage
Hydrogen sulfide	H_2S	0.00047	Rotten eggs
Indole	C_8H_6NH	—	<20 ppm: floral >20 ppm: mothballs, fecal
Methylamine	CH_3NH_2	0.021	Putrid, fishy
Methyl mercaptan	CH_2SH	0.0011	Decayed cabbage
Propyl mercaptan	$CH_3(CH_2)_2SH$	0.000075	Unpleasant
Putrescene	$NH_2(CH_2)_4NH_2$	—	Putrid, nauseating
Pyridine	C_6H_5N	0.0037	Disagreeable, irritating
Skatole	C_9H_8NH	0.0012	Fecal, nauseating
Sulfur dioxide	SO_2	0.009	Pungent, irritating
Tert-butyl mercaptan	$(CH_3)_3CSH$	0.00008	Skunk, unpleasant
Thiocresol	$CH_3C_6H_4SH$	0.001	Rancid, skunk-like
Thiophenol	C_6H_5SH	0.000062	Putrid, nauseating
Triethylamine	$(C_2H_5)_3N$	0.08	Ammoniacal, fishy

4.6.1.1.2 *Ammonia*

pH control of odors from ammonia is opposite to that for hydrogen sulfide. To remove odor caused by ammonia with air stripping, the pH must be raised (see Example 1 and Section 4.2.3).

4.6.1.2 Oxidation

The addition of oxidizing agents such as Cl_2, NaOCl, H_2O_2, O_2, $KMnO_4$, or ClO_2 will oxidize hydrogen sulfide to odorless sulfate ion, SO_4^{2-}, and ammonia to odorless nitrogen compounds, including elemental nitrogen, N_2.

RULES OF THUMB

1. The usual chlorine dose for odor control is 10–50 mg/L.
2. 8.9 mg of chlorine is required to oxidize 1 mg of hydrogen sulfide, H_2S.

4.6.1.3 Eliminate Reducing Conditions Caused by Decomposing Organic Matter

This often means mechanically cleaning out the organic slime and sludge in a well, sewer, drain, or wetland. Mechanical cleaning will aid the use of oxidizing agents. Eliminating anaerobic conditions can also be accomplished sometimes by aerating the water. Increasing the DO level will shift conditions from reducing to oxidizing as the oxygen diffuses into the reducing zone of the water. Mixing currents in the water help this process. If the reducing zone is thick and the water stagnant and motionless, aeration control might be very slow. Drying out wet soil that has an H_2S odor also allows oxygen to diffuse into the organic matter, changing the decomposition processes from anaerobic to aerobic.

4.6.1.4 Sorption to Activated Charcoal

Sorption to powdered or granular activated charcoal is a reliable last resort for removing bad tastes and odors. Powdered charcoal can be added as a slurry directly to a waste stream. Gases can be passed through a canister filled with granulated activated charcoal.

EXERCISES

1. A wastewater treatment plant removed ammonia with a nitrification–denitrification process that also reduced alkalinity. For each gram of NH_3-N removed, the process also removed 7.14 g of alkalinity-$CaCO_3$. If the plant was designed to remove 25 mg/L of NH_3-N and the total throughput was 250,000 gal/day, how many pounds of caustic soda (sodium hydroxide, NaOH) must be added each day to restore the alkalinity to its original value before ammonia removal?

2. A wastewater flow contains 30 g/L total ammonia nitrogen and has a 10 g/L discharge limit. An air-stripping tower is to be used. Its temperature ranges from 20°C to 30°C and the pH is normally about 9. At what pH must the stripper be operated?

REFERENCES

de la Crétaz, A. L., and Barten, P. K. 2007. *Land Use Effects on Streamflow and Water Quality in the Northeastern United States*. Boca Raton, FL: CRC Press, Taylor & Francis Group, p. 344.

EPA. 2000. *Nutrient Criteria Technical Guidance Manual: Rivers and Streams*. Washington, DC: United States Environmental Protection Agency, Office of Water, Office of Science and Technology. EPA-822-B-00-002, July 2000.

EPA. 2006. *Regulatory Determinations Support Document for Selected Contaminants from the Second Drinking Water Contaminant Candidate List (CCL 2)*. EPA Report 815-D-06-007. Washington, DC: Office of Ground Water and Drinking Water, U.S. Environmental Protection Agency, Chapter 15, pp. 14–15.

EPA. 2010. *Nutrient Control Design Manual*. Washington, DC: United States Environmental Protection Agency, Office of Water, Office of Science and Technology. EPA/600/R-10/100, August 2010.

EPA. 2011. Memorandum, 3/16/2011, *Working in Partnership with States to Address Phosphorus and Nitrogen Pollution through Use of a Framework for State Nutrient Reductions*, From: Nancy K. Stoner, Acting Assistant Administrator To: Regional Administrators, Regions 1–10. http://water.epa.gov/scitech/swguidance/standards/criteria/nutrients/upload/memo_nitrogen_framework.pdf.

Kadlec, R. H., and Knight, R. L. 2009. *Treatment Wetlands*, 2nd ed. Boca Raton, FL: CRC Press, Taylor & Francis Group, p. 1016.

Lewis, W. M., Saunders III, J. F., and McCutchan Jr., J. H. 2008. "Application of a Nutrient-Saturation Concept to the Control of Algal Growth in Lakes." *Lake and Reservoir Management* 24: 41–46. http://cires.colorado.edu/limnology/pubs/pdfs/Pub178.pdf.

Radcliffe, D. E., and Cabrera, M. L. (eds.) 2006. *Modeling Phosphorus in the Environment*. Boca Raton, FL: CRC Press, Taylor & Francis Group, p. 432.

Summers, J. 2008. *Assessment of Filamentous Algae in the Greenbrier River and Other West Virginia Streams*, WVDEP-DWWM. December 17, 2008, www.dep.wv.gov/WWE/watershed/wqmonitoring/Documents/Greenbrier/Assessment_Filamentous_Algae_Greenbrie_%20River.pdf.USGS. 2007. *Nutrients in the Nation's Streams and Groundwater: National Findings and Implications*, Fact Sheet 2010-3078, U.S. Geological Survey, 2010. http://pubs.usgs.gov/fs/2010/3078.

USGS. 2010. *The Quality of Our Nation's Waters—Nutrients in the Nation's Streams and Groundwater, 1992–2004*. U.S. Geological Survey Circular 1350, p. 174.

WHO. 2004. *Guidelines for Drinking-Water Quality*, 3rd ed. *Volume 1: Recommendations*. Geneva: WHO.

In the U.S., the most commonly occurring metals at Superfund sites are lead, chromium, arsenic, zinc, cadmium, copper, and mercury.

Evanko and Dzombak, *Remediation of Metals-Contaminated Soils and Groundwater*, 1997

5 Behavior of Metal Species in the Natural Environment

5.1 METALS IN WATER

A casual glance at the periodic table shows that most of the elements (about three-fourths) are metals or metalloids.* It often happens in environmental literature that little or no distinction is made between metals and metalloids, especially for the metalloids arsenic, selenium,[†] and antimony, which often are classified with metals in tables of standards, environmental discharge limits, chemical properties, and so on. This book follows this common practice.

---

* Metalloids are those elements in periodic table groups 3A through 6A that have electrical and chemical properties intermediate between those of metals and nonmetals. They are B, Si, Ge, As, Se (see also footnote †), Sb, Te, and Po. For regulatory purposes, it is sometimes useful to group metals and metalloids together, as when they share the same analytical method (e.g., ion-coupled plasma spectroscopy).

[†] Selenium (Se) is most commonly listed as a nonmetal, similar in chemistry to sulfur. However, some crystalline forms resemble metals closely enough that it is sometimes called a metalloid. *Standard Methods* (Eaton et al. 1995) includes selenium in a table of metals and metalloids that are analyzed by electrothermal atomic absorption spectrometry and many analytical laboratories include selenium in their lists of metals analyses.

To discuss their chemical behavior, the elemental metals may be divided into three general classes:

1. Alkali metals: Li, Na, K, Rb, Cs, and Fr (periodic table group 1A).
2. Alkaline metals: Be, Mg, Ca, Sr, Ba, and Ra (periodic table group 2A).
3. Metals not in the alkali or alkaline groups. These include the transition metals (all the periodic table group B metals) and the metals and metalloids that appear in groups 3A through 6A.* Some metals are also classified in ways not based primarily on periodic table groups, the so-called trace or heavy metals.†

Metals in environmental waters arise from both natural and anthropogenic sources. In many cases, anthropogenic inputs of metals exceed natural inputs. Living organisms require some metals as essential nutrients, including calcium, sodium, potassium, magnesium, iron, zinc, chromium, cobalt, copper, nickel, manganese, molybdenum, and selenium. Excessive levels or certain oxidation states of some essential metals, however, are detrimental to living organisms. In addition to non-nutrient metals generally recognized as toxic, such as antimony, arsenic, beryllium, cadmium, lead, and mercury, health-based water quality standards will also include the nutrient metals chromium, copper, nickel, selenium, and zinc, all of which can be toxic at too-high levels or in certain oxidation states.‡

When metal atoms combine chemically with other metal atoms, the result is a metal solid, either a pure elemental metal solid or an alloy of different metals. When metal atoms combine with nonmetal atoms, the products are nonmetallic compounds that range from ionic salts like sodium chloride to volatile, liquid organometallic compounds like dimethylmercury.

Metals and metal-containing compounds in natural waters may be in dissolved or solid (particulate) forms,§ depending on water quality parameters of pH, oxidation–reduction potential (ORP), and the presence of dissolved anion species, such as sulfide or carbonate, that can form compounds with metal ions. See Chapter 3 to learn more about these water quality parameters. This chapter discusses the environmental

---

* Periodic table groups 3A through 6A contain both metals and nonmetals.
† The term "heavy metals" is often encountered in texts and reports, usually meaning metals with atomic numbers equal to or greater than that of Cu (atomic no. 29), especially metals exhibiting toxicity. However, the term "heavy metals" has no precise definition and its use is inconsistent (Duffus 2002). Another class designation often used is trace metals, generally used for those metals found in the Earth's crust with average concentrations less than 1%. Nearly all the metals are included in this class; the exceptions (with average crustal concentrations greater than 1%) being Na, K, Ca, Mg, Fe, and Al.
‡ Paracelsus (1493–1541), a Renaissance physician sometimes called the father of toxicology, observed that, "All things are poison and nothing is without poison, only the dose permits something not to be poisonous," which is often paraphrased as, "The dose makes the poison." Paracelsus did not know about oxidation states.
§ In this chapter, particulate forms include colloids, which are small aggregates of molecules that are small and light enough to remain suspended and dispersed in water, typically less than 2 micrometers (μm) in diameter. Colloidal suspensions often are clear-like solutions and appear to be homogeneous. They bridge the properties of dissolved species and particulates large enough to settle out in water. While truly dissolved species cannot be filtered out of solution, colloids cannot pass through filters with a pore size smaller than the colloid particles.

behavior of metals in their different forms and how different water parameters influence this behavior.

Examples of dissolved forms* are

- Metal cations: Ca^{2+}, Fe^{2+}, K^+, Al^{3+}, Ag^+, and so on
- Metal complexes†: $Zn(OH)_4^{2+}$, $Au(CN)_2^-$, $Ca(P_2O_7)^{2-}$, PuEDTA, and so on
- Organometallics: $Hg(CH_3)_2$, $B(C_2H_5)_3$, $Al(C_2H_5)_3$, and so on

Examples of solid forms are

- Weathered and eroded mineral sediments (clays, oxides, hydroxides, sulfides, carbonates, silicates, and so on)
- Precipitated solid oxides, hydroxides, sulfides, carbonates, silicates, and so on
- Colloidal aggregates of metal species, small enough to remain suspended in still water but large enough to be filterable
- Cations and complexes sorbed to mineral sediments and organic matter

Metal species undergo continuous changes between dissolved and solid forms, and the rates of sorption, desorption, precipitation, and dissolution processes depend on pH, ORP, other chemical species present, and the chemical composition of bottom and suspended sediments. Sorption to sediments and precipitation of dissolved metal species remove metals from the water column and store them in the sediments, where they are less biologically available. Desorption and dissolution return metals to the water column, where they become biologically available again and where water flow may carry them to new locations, where sorption and precipitation may recur. Metals may be desorbed or dissolved from sediments when the water undergoes changes in salinity, ORP, or pH. Chapter 6 covers additional aspects of sorption and desorption processes.

In the water environment, nonradioactive metals are of greatest environmental and health concern when in dissolved forms, where they are more mobile and more biologically available than metals in particulate forms. However, ingestion and inhalation of particulates containing metals can also be a serious health hazard.

---

* It is shown in this chapter that metal cations (metal anion complexes also) dissolved in water always attract a hydration shell of water molecules because of electrostatic attractions. Although a metal cation dissolved in water is often written as an elemental ion, for example, Al^{3+}, it actually is a cluster consisting of the metal ion enclosed within progressively larger surrounding shells of water molecules (see Section 5.3.4 and Figure 5.3). To indicate that a dissolved metal carries a hydration shell with it, the suffix "aq" (for *aqueous*) may be appended to it, for example, Al^{3+}(aq).

† A complex is a dissolved chemical species formed by the association of a cation with one or more anions or neutral species (such as water) that contain nonbonding electron pairs. The cation has room for one or more electron pairs in its valence shell and the anions or neutral species it connects with (called ligands) have nonbonding electron pairs in their outer shells that can fill the cation electron shell vacancies. Remember that chemical bonds consist of electron pairs shared by the bonded atoms. A complex differs from a covalent compound in that the ligand brings both electrons of the bonding pair to the bond, while covalent compounds are formed when the connected atoms contribute one electron each. Cation–ligand bonds are called coordinate bonds, and the complexes formed may be called coordination compounds.

Radioactive metals may be hazardous because of their ionizing emissions as well as their chemical toxicity, and are potentially harmful in both dissolved and particulate forms, even at a distance without radioactive metal species entering the body.

5.2 MOBILITY OF METALS IN WATER/SOIL ENVIRONMENTS

The mobility of metals in water/soil environments is largely governed by whether they are in dissolved or solid forms (see the Rules of Thumb on the next page). Other site-specific environmental conditions, such as surface water and groundwater flow rate, rainfall and wind severity, soil types, and slope of terrain, are generally also relevant.

The important metal species within the solid and dissolved forms are as follows:

- Solid elemental metal precipitates: These may be particles of colloid size (<2 µm diameter) or larger.* Colloids remain suspended in water and are mobilized by water movement. Larger particles may settle out and require stronger flows to move them as sediments.
- Solid metal compounds: These also may be colloid size or larger and are formed by weathering of minerals and by reactions of dissolved metal cations with water and other dissolved species, such as carbonate, hydroxide, and sulfide.
- Dissolved metal cations with their associated hydration shell.
- Dissolved metal compounds: These include ionic carbonate and hydroxide complexes.
- Metal species sorbed to solid soil and sediment surfaces: These include species sorbed from the dissolved state as well as settleable and colloidal solids. Sorption processes may be reversible to some degree, resulting in a retardation of dissolved metal movement relative to water flow, or irreversible, resulting in immobilization of metal species, except for mobility by erosion mechanisms.

RULES OF THUMB

1. Dissolved forms of metals move with surface water and groundwater flows.
2. Metals in particulate form can be transported as sediments by wind or suspended in moving water.
3. Both dissolved and particulate forms of metals may sorb to organic soil solids, where they may be immobilized or carried along with eroding soils.
4. Particulate metals pose the greatest environmental risks when they encounter environmental conditions that increase their solubility.

---

* The operational distinction between solid and dissolved forms is based on a filtering process. When a water sample is passed through a 0.45-micron filter, chemical species that pass through are generally considered to be dissolved; species retained on the filter are considered to be solids. Colloids are a special class of matter, often formed as fine precipitates, with properties on the border of dissolved and solid states (see Chapter 6, Section 6.8).

5.3   GENERAL BEHAVIOR OF DISSOLVED METALS IN WATER

Exactly what are dissolved metals anyway, that is, those written with (aq) attached, such as Zn(aq)? They certainly are not tiny pure metal atoms that have somehow been extracted from solid metals or minerals and become dispersed throughout a water sample, even though much common usage would not contradict this simplistic concept. Consider the phrase "metal solubilities" as found in many handbooks, or the use of Environmental Protection Agency (EPA) and state standards for "metals" in drinking and environmental waters, or limits on "metal concentrations" in industrial discharge permits; none of these convenient simplifications suggest that the real nature of dissolved metals always involves other chemical species. In fact, it can be misleading for nonspecialists to think in terms of the solubility of pure metals. To illustrate this, consider the following two common rules of solubility:

1. pH solubility rule: Metals become more soluble and tend to dissolve as water pH is lowered (more acidic); as water pH is raised (more basic), metals become less soluble and tend to precipitate.
2. ORP solubility rule: Metals are more soluble and tend to dissolve under reducing conditions (anaerobic, anoxic, and negative ORP), while under oxidizing conditions (aerobic, oxic, and positive ORP) they are less soluble and tend to precipitate.

We will see that these rules are often not true. Furthermore, the solubility rules do not call attention to the fact that the expression "solubility of metals" seldom refers to just pure elemental metals; it most often refers to the solubility of different metal compounds that are formed in a water sample. For example, metals are sometimes less soluble at high pH because metal cations (e.g., Cu^{2+}) often (but not always) react in high-pH water to form low-solubility metal hydroxides and oxides, for example, $Cu(OH)_2$. At low pH, where hydroxide concentrations are smaller, they may remain as soluble hydrated cations.

5.3.1   THE EXAMPLE OF IRON

To illustrate some flaws in the two solubility rules above, consider the common metal iron (Fe). Pure metallic iron is rarely found in nature because iron reacts readily with many other elements. In oxidizing aqueous environments (positive ORP), like shallow stream sediments, solid iron is often found as ferric iron hydroxide, $Fe(OH)_3$. In reducing aqueous conditions (negative ORP), like anoxic lake sediments, solid iron is more likely to be found as ferrous iron carbonate ($FeCO_3$). Iron, as $FeCO_3$, is more soluble at acidic pH 6 than at basic pH 11, and appears to follow the "more soluble at lower pH" rule (at least between pH 6 and 11). On the other hand, iron in the form of $Fe(OH)_3$ is more soluble at pH 11 than at pH 6, in apparent contradiction to this rule. In further disagreement with the pH solubility rule, Figures 5.1 and 5.2 show that the solubility of both of these iron compounds (and many other metal compounds) is a minimum at a particular pH, different for each

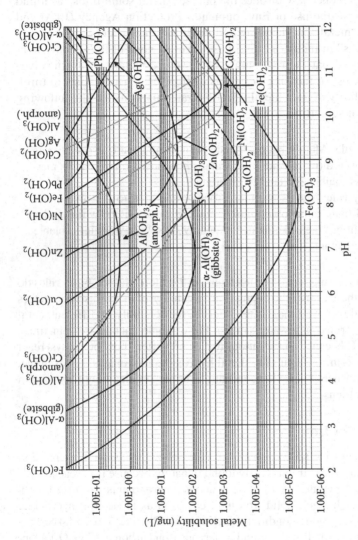

FIGURE 5.1 Maximum concentration of dissolved metal in equilibrium with solid metal hydroxide. As an example, the $Fe(OH)_3$ curve represents the total concentrations of Fe present in all dissolved iron species (Fe^{3+}, $Fe(OH)^{2+}$, $Fe(OH)_2^+$, $Fe(OH)_3^0$, $Fe(OH)_4^-$, and so on) in equilibrium with solid $Fe(OH)_3$ as a function of pH. Curves are identified as the uncharged solid metal hydroxide. Selected curves are shaded for clarity.

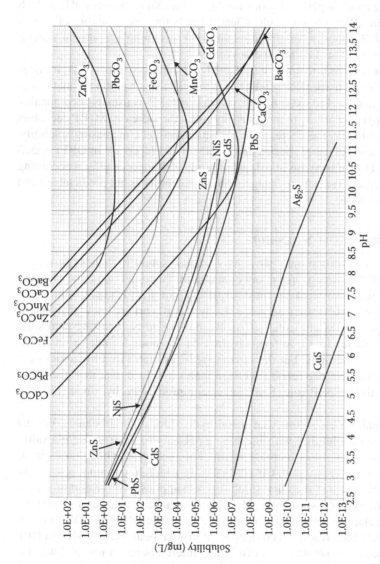

FIGURE 5.2 Maximum concentration of dissolved metal in equilibrium with solid metal carbonates and sulfides. As an example, the $CdCO_3$ curve represents the total concentrations of Cd present in all dissolved cadmium species (Cd^{2+}, $CdCO_3^0$, $Cd(CO_3)_2^{2-}$, and so on) in equilibrium with solid $CdCO_3$ as a function of pH. Curves are identified by the solid metal carbonate or sulfide. Selected curves are shaded for clarity.

(about pH 8 for $Fe(OH)_3$ and pH 11 for $FeCO_3$), and increases at both higher and lower pH values.

Figures 5.1 and 5.2 also show that, below about pH 9.5, ferrous iron as $FeCO_3$, which is formed under reducing conditions, is more soluble than is ferric iron as $Fe(OH)_3$, formed under oxidizing conditions, in agreement with the ORP solubility rule. On the other hand, at pH 12, ferrous $FeCO_3$ is less soluble than $Fe(OH)_3$, which disagrees with the ORP solubility rule. Clearly, predicting the solubility of iron at various pH or ORP values, as might be required for designing a treatment system, requires knowledge of the chemical species of iron.

Another factor that tends to upset the pH and ORP solubility rules is that, in water environments, changes in the ORP often change the pH also. Oxidation of metal species by exposure to dissolved oxygen (DO; increase in ORP) is usually accompanied by a decrease in pH. According to the solubility rules, the increase in ORP decreases metal solubility, whereas the decrease in pH acts oppositely, increasing solubility. Conversely, reduction of metal species due to lack of DO (decrease in ORP) is usually accompanied by an increase in pH, again producing opposite effects according to the solubility rules. The actual change in metal solubility is not predictable from the solubility rules alone.

5.3.2 Predicting Metal Solubilities

With such obvious limitations, why have metal solubility rules persisted? Figures 5.1 and 5.2 show that, for many common metal compounds, the pH solubility rule often works for acidic solutions (pH < 7); in this pH range, solubilities generally increase as pH values decrease. Early metal workers knew that metals and metal compounds would usually dissolve more readily in strong acid solutions. Also, we will see that iron forms hydroxide compounds under both reducing and oxidizing conditions, but the ferrous (reduced state) iron hydroxides are much more soluble than the ferric (oxidized state) hydroxides at any given pH, in agreement with the ORP rule. Some other common metal hydroxides behave similarly.

Figures 5.1 and 5.2 also show that metal hydroxides and carbonates, but not sulfides, decrease in solubility with increasing pH only to a particular pH value, above which their solubilities increase again. This behavior, which varies in its details for each metal compound, cannot be easily generalized into a few simple rules. We will also see that the ORP solubility rule applies to some common metals, but not to all. In other words, the pH and ORP solubility rules may sometimes work to predict relative solubilities, but only within restricted pH and ORP ranges, which are different for each metal compound. It just happens that the pH and ORP ranges where the rules may work are often encountered in environmental waters.

However, environmental and, especially, waste treatment conditions may expose metal species to a wider range of pH and ORP conditions than are commonly encountered. Then, the simple pH and ORP solubility rules cannot be relied on. Precipitation of dissolved metals is the most common technique for treating metal-contaminated wastewaters and remediating metal-contaminated sites. The simplistic solubility rules will not always lead to the best precipitation treatments for removing dissolved metals from water, nor can they predict when there will be exceptions to these rules. To successfully treat metal-contaminated water, environmental professionals must understand metal solubilities on a more detailed level.

5.3.3   FOUR IMPORTANT THINGS TO KNOW ABOUT METAL BEHAVIOR IN WATER

1. *The solubilities of different metal compounds may be influenced differently by changes in pH and ORP.* An important principle of chemistry is that, at the molecular level, the reactivity of every chemical, whether atom, molecule, or ion, depends mostly on just two factors—its outer-shell electron structure and its chemical environment, especially the nature of other chemical species with which it can react. Changing the ORP changes the electron structure of the metal compound by adding or removing electrons in the outer shell; changing the pH changes the chemical composition of the water environment by altering the H^+/OH^- ratio.

   a. The outer-shell electron structure of a chemical species, that is, the number and distribution of the electrons in its outermost electron orbitals, is influenced mainly in water solutions by the ORP. Because each different metal compound has a unique electron structure, each will have a unique response to changes to ORP. For different metals with very different electron structures, the influence of changes in ORP on metal solubilities will also differ greatly.

   b. A change in pH always changes the chemical environment of water solutions because it changes the actual chemical composition of the water itself. Water molecules always dissociate to some degree, depending on the temperature, into hydrogen ions (H^+) and hydroxide ions (OH^-) (see Chapter 3). When the pH of water decreases (more acidic), the hydrogen ion concentration increases and the hydroxide ion concentration decreases. When the pH increases (more basic), the opposite occurs: The hydrogen ion concentration decreases and the hydroxide ion concentration increases. Many (but not all) metal species react with hydroxide ions to form compounds of low solubility, so that raising the pH increases the probability of those reactions, but only to a certain point, as we will see below.

2. *The solubility of a metal depends on the other atoms to which it is bonded.* Most accurately, metal solubilities should always be stated in terms of the water solubility of specific metal compounds, for example, metal hydroxides, metal carbonates, metal sulfides, and metal phosphates. Metals become dissolved when soluble compounds containing metal atoms come in contact with water. The dissolved metal species are, for the most part, ionic, that is, electrically charged, and may contain various combinations of atoms obtained from the original solid compound and from water.

   The dissolution reaction is driven primarily by electrostatic attractive forces between ionic metal species and polar water molecules (see Section 2.8). (Polar molecules are neutral overall, but contain both positive and negative charged regions.) Although dissolved metals may partly consist of some electrically neutral species, driven into solution by thermal energy and polar attractions, these are always the least soluble of the dissolved species, and in lowest abundance. (See Pankow [1991] and Stumm and Morgan [1996] for more detailed descriptions of the dissolution process.)

3. *Solid metal compounds always dissolve by reacting with water (hydrolysis) or with other dissolved substances in water (e.g., acids); changes in pH and ORP can either inhibit or enhance these reactions.* Metals in solid form, whether pure or combined with other elements as compounds, are always electrically neutral and can only dissolve when they chemically react with water* or with substances already dissolved in water, such as acids. Once dissolved in water, metal-containing species can further react with other substances dissolved in water, such as hydrogen, hydroxide, carbonate, or sulfide ions, to form a variety of products, each having its own solubility characteristics. As described in point 1b above, even initially pure water always contains different molecular and ionic species. These are mainly water molecules (H_2O), hydrogen ions (H^+) that react immediately with water molecules to form hydronium ion (H_3O^+), and hydroxide ions (OH^-), all of which can react with dissolved metal species to form metal hydroxides and oxides with solubilities that depend on pH and ORP. Depending on the concentrations of all the substances present, any given change in the pH and/or ORP may either increase or decrease the concentration of dissolved metal species.

4. *Dissolved metal species differ from precipitated solids by having a surrounding shell of water molecules and a repulsive electric charge that hinder close contact.* Dissolved metal cation species resist coalescing into solid precipitates because their hydration shells hinder close approaches and their positive charges create a repulsive electrostatic force between neighboring hydrated cations. If their positive charges are neutralized by bonding with anions, as described below, they no longer attract hydration spheres or have electrostatic repulsions and can coalesce to form lower-solubility solids. Further reaction with anions can result in negatively charged anionic metal complexes that behave similarly to positively charged metal cation species, with hydration shells and repulsive forces that result in increased solubility.

5.3.4 HYDROLYSIS REACTIONS

No matter what other substances are present in a water solution, the most abundant compounds in contact with dissolved metals are generally water species: water molecules, hydroxide anions, and hydrogen cations. Reactions of metal cations with water species are called hydrolysis. Hydrolysis reactions are the usual criteria for assessing whether a metal is soluble or insoluble under certain ORP and pH conditions. When a metal such as iron is said to be less soluble under oxidizing conditions and

---

* Hydrolysis is considered a chemical reaction.

more soluble under reducing conditions, what usually is meant is that the compounds formed by reaction of the metal cation with water under oxidizing conditions are less soluble. Under reducing conditions, iron does not hydrolyze significantly in water and can remain as a dissolved cation (with a hydration shell). The same is true for pH conditions; metals tend to be less soluble at high pH because their cations often react with hydroxide anions in high-pH water to form low-solubility hydroxides and oxides. At low pH, where hydrogen cation concentrations are high and hydroxide anion concentrations are low (see Chapter 3), dissolved metals may remain as soluble hydrated cations.

The simplest form of a dissolved metal is a pure metal cation (positively charged ion), such as Fe^{3+} or Zn^{2+}. However, pure metal cations cannot persist as such in water solutions. Any charged species in solution will always interact with other charged or polar species by electrical forces. Because water molecules are polar, each positive metal cation electrostatically attracts the negative end of nearby water molecules (the oxygen end; see Chapter 2), accumulating a cloud-like hydration shell of water molecules around it (as described by Equation 5.1 and illustrated in Figure 5.3).

$$M^{+n} \xrightarrow{\ H_2O\ } M(H_2O)_x^{+n} \qquad\qquad (5.1)$$

where M is a metal cation, n is the number of positive charges on the cation, and x is the maximum number of water molecules in the innermost hydration shell.

Typically, one to six water molecules form the innermost spherical layer of water molecules directly adjacent to the metal cation, depending on the size and charge of the central metal cation. Successive hydration spheres surround the innermost sphere, each extending farther out, attracted to the cation less and less strongly.

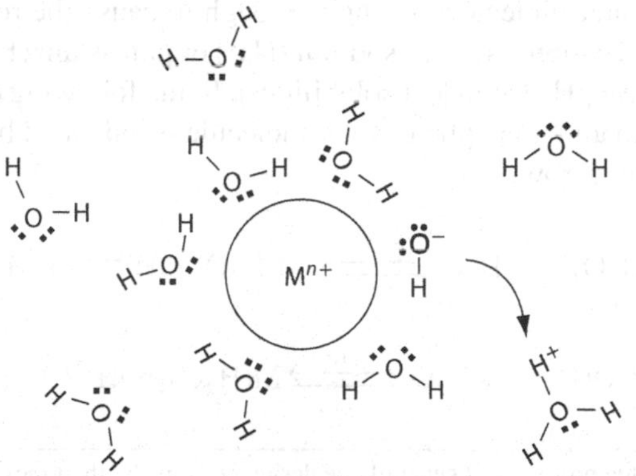

FIGURE 5.3 Water molecules form a hydration shell around a dissolved metal cation. Molecules in the hydration shell can lose a proton to bulk water molecules, as indicated by the arrow, leaving a hydroxide group bonded to the metal and forming a positively charged hydronium ion, H_3O^+, in the bulk water. By transferring an H^+ to a water molecule the hydrated metal behaves as an acid (see Chapter 3). Eventually, after several such proton transfers, the metal may precipitate as an uncharged metal hydroxide of low solubility.

Water molecules in the innermost sphere are most strongly attracted to metal cations having the smallest radii and greatest positive charge.

Metal cations with their cloud of water molecules are said to be hydrated. Using iron as an example, the chemical notation for a hydrated ferric iron cation may be written as $Fe(H_2O)_x^{3+}$, where x represents the number of water molecules in the innermost hydration layer. A simpler way to indicate a dissolved cation is to append the suffix "aq" (for aqueous) to it, as in $Fe^{3+}(aq)$. In this text and many others, whenever a dissolved cation is written without any indication of its hydration shell, that is, Fe^{3+}, the more accurate designation is to be assumed, that is, $Al^{3+}(aq)$.

Depending on the strength of the electrostatic attraction between the metal cation and the water molecules, the water molecules closest to the metal cation (in the innermost hydration shell) may bond as a ligand, forming a metal–water complex (see footnote †️ on page 183). The strength of the electrostatic attraction depends on the magnitude of the cation charge, the cation radius, and, to a lesser extent, the electronegativity of the metal. The strongest bonds are formed with cations having the smallest radius, greatest positive charge, and electronegativity greater than 1.8 (Wulfsburg 1987).

5.3.5 Hydrated Metals as Acids

Hydrated metal ions can behave as acids by releasing protons (H^+) from their water ligands to the surrounding free H_2O molecules, forming acidic hydrated protons,* H_3O^+, $H_5O_2^+$, and so on (see Figure 5.3 and Chapter 3). The stronger the attraction between the metal cation and the oxygen on the water ligands, the weaker is the attraction between the ligand water oxygens and the hydrogens to which they are bonded. This results in protons being released more readily from the water ligands to the surrounding water molecules and causing the hydrated metal cation to behave like a stronger acid. As the proton-release process continues stepwise, eventually a neutral metal hydroxide is formed.†️ Only metal cations with a charge of +2 or greater can attract polar water molecules strongly enough to cause the release of hydrogen ions; singly charged cations, such as sodium (Na^+) or potassium (K^+), will not do this (and they do not have pH-dependent solubilities). In the following equations, the loss of a hydrogen ion to an inner sphere water molecule is indicated by the $-H^+$ symbol over the equilibrium arrows.

$$M(H_2O)_6^{+n} + H_2O \underset{}{\overset{-H^+}{\rightleftharpoons}} M(H_2O)_5OH^{(+n-1)} + H_3O^+ \tag{5.2}$$

$$M(H_2O)_5OH^{+(n-1)} + H_2O \underset{}{\overset{-H^+}{\rightleftharpoons}} M(H_2O)_4(OH)_2^{+(n-2)} + H_3O^+ \tag{5.3}$$

---

* For simplicity, hydrated protons will generally be designated only by their most common form, H_3O^+. Because solvent water molecules are normally not included when balancing a chemical reaction, the stoichiometry of acid–base reactions remains unchanged, regardless of how many water molecules are shown attached to an acidic proton, that is, whether the hydrated proton is designated by H^+, H_3O^+, $H_5O_2^+$, and so on.

†️ It is shown in Sections 5.6 and 5.7 that, as pH increases, the hydroxide groups attached to a metal cation can exceed the number needed to achieve neutrality, creating metal hydroxide anions whose solubility then begins to increase as additional hydroxide groups become attached.

For example, with Fe^{3+}, it takes three proton-transfer steps to form neutral ferric hydroxide:

$$Fe(H_2O)_6^{3+} + H_2O \xrightarrow{-H^+} Fe(H_2O)_5OH^{2+} + H_3O^+ \qquad (5.4)$$

$$Fe(H_2O)_5OH^{2+} + H_2O \xrightarrow{-H^+} Fe(H_2O)_4(OH)_2^+ + H_3O^+ \qquad (5.5)$$

$$Fe(H_2O)_4(OH)_2^+ + H_2O \xrightarrow{-H^+} Fe(H_2O)_3(OH)_3 + H_3O^+ \rightleftarrows Fe(OH)_3(s) \atop + H_3O^+ + 3H_2O \qquad (5.6)$$

In Equation 5.6, the loss of a third H^+ makes a neutral ferric hydroxide species that no longer attracts a hydration sphere around it, and the three remaining inner shell water molecules are released to the bulk water. Adding Equations 5.4 through 5.6 gives the overall reaction:

$$Fe(H_2O)_3^{3+} + 3H_2O \rightleftarrows Fe(OH)_3(s) + 3H_3O^+ \qquad (5.7)$$

With each step, the hydrated metal is progressively deprotonated, forming polyhydroxides and becoming less and less soluble. At the same time, the solution becomes increasingly acidic due to the formation of more H_3O^+. Eventually, the metal may precipitate as a low-solubility hydroxide. The degree of acidity induced by metal hydration is greatest for cations having the greatest electronegativity, which are those of high charge and small size. All metal cations with a charge of +3 or more are moderately strong acids. This process is one source of acidic water draining from mines.[*]

5.3.6 Influence of pH on the Solubility of Metals

Equations 5.2 through 5.7 are all reversible, with H_3O^+ on the right side. This means that the equilibria of these reactions shift to the left if the solution concentration of H_3O^+ is increased (by adding more acid) and to the right if the concentration of H_3O^+ is decreased (by adding a base). Thus, the formation of metal hydroxides by hydration of metal cations is sensitive to the solution pH. Considering the overall reaction, we see that lowering the pH (increasing the concentration of H_3O^+) shifts the equilibrium of to the left, tending to dissolve any solid metal hydroxide that has precipitated. Raising the pH (increasing the concentration of OH^-) consumes H_3O^+ (by $OH^- + H_3O^+ \rightarrow 2H_2O$) and shifts the equilibrium of Equation 5.7 to the right, precipitating more insoluble metal hydroxide. Thus, one may say that the metal becomes more soluble at lower pH and less soluble at higher pH, even though what actually occurs is that the hydrated metal species form increasingly more of the electrically neutral and less soluble $Fe(OH)_3$ at higher pH values.

---

[*] Metal hydrolysis is usually a relatively minor source of acidity compared to oxidation of metal pyrites (see Section 5.9.1).

An equivalent explanation of why metal hydroxide formation is pH dependent is to observe that when the pH is raised, shifting the equilibria of Equations 5.2 through 5.7 to the right, the concentration of hydroxide anions adjacent to hydrated metal cations is increased. This increases the negative charges that are the source of the attractive forces pulling H^+ ions from water molecules in the inner hydration sphere of the metal cation into solution, and facilitates the stepwise conversion of inner-layer water molecules into chemically bound hydroxide groups. Each newly bound hydroxide ion (carrying a single negative charge) reduces the positive charge on the metal species by one unit, as in Equations 5.2 and 5.3. As long as the metal hydroxide species are charged, they electrostatically attract water molecules and repel one another, thereby remaining dissolved. However, when enough negative hydroxide ions have become chemically bound to hydrated metal cations to neutralize them, as in Equation 5.6, the electrostatic repulsion between the metal species disappears and their attraction to polar water molecules is greatly reduced.

When this occurs, any remaining inner-layer water molecules are no longer strongly attracted to the metal cation and are released to the surrounding water. In addition, the electrically neutral metal species no longer repel one another and can aggregate and precipitate as a solid. For example, with hydrated ferric iron (Fe^{3+}), a sequence of three H^+ releases from the inner hydration sphere forms, first $Fe(OH)^{2+}$, then $Fe(OH)_2^+$, and finally neutral $Fe(OH)_3$, which then precipitates as solid $Fe(OH)_3$ (although a small amount of neutral $Fe(OH)_3$ will remain in solution).

However, if the pH is raised too high, precipitated metal hydroxides can redissolve (see Figure 5.2). At high pH values, a neutral metal hydroxide may form a complex with OH^- anions to become negatively charged. This increases the metal anion solubility because, in the same manner as metal cations, metal anions are electrostatically attracted to water molecules (this time the attraction is to positively charged regions of the water dipoles) and electrostatically repel other metal anions. For example, precipitated neutral $Fe(OH)_3$ can react with OH^- anions as follows:

$$Fe(OH)_3 + OH^- \rightleftharpoons Fe(OH)_4^-$$ (5.8)

$$Fe(OH)_4^- + OH^- \rightleftharpoons Fe(OH)_4^{2-}$$ (5.9)

As more hydroxide groups are added, forming more highly charged metal polyhydroxide anions, the metal anions become more soluble because higher ionic charges result in stronger attractions to polar water molecules. As shown in Figure 5.2, the value of pH, where solubility is a minimum and begins to increase again with increasing pH, varies from metal to metal.

5.3.7 Amphoteric Behavior

Another aspect of metal solubility, which is seen in Figures 5.1 and 5.2, is that, if the pH is raised beyond the point of forming a neutral metal hydroxide or carbonate, the solubility begins to increase again and precipitated metal hydroxides and carbonates can redissolve. Substances that have a minimum solubility at a particular pH, becoming more soluble at both higher and lower pH values, are called ampho-teric. Most metal hydroxides and carbonates are amphoteric. Note, in Figure 5.2, that metal sulfides do not exhibit amphoteric behavior.

When pH is raised in water solutions containing precipitated metal hydroxides, by the addition of sodium, calcium, or magnesium hydroxide salts, the solutions contain increasingly larger amounts of free hydroxide anions (OH^-). When the pH is high enough, free hydroxide anions in solution can react (bond) with neutral metal hydroxides, causing a neutral metal hydroxide to become a negatively charged metal hydroxide, as in Equations 5.8 through 5.9, resulting in increased solubility.

In Figure 5.1, using the ferric hydroxide curve as an example, the vertical axis labeled "Metal Solubility" represents the sum of the milligrams/liter of iron atoms contained in all of the dissolved $Fe(OH)_3$ species (Fe^{3+}, $Fe(OH)^{2+}$, $Fe(OH)_2^+$, $Fe(OH)_3^0(aq)$, $Fe(OH)_4^-$, $Fe(OH)_5^{2-}$, and so on) in equilibrium with solid $Fe(OH)_3(s)$. As shown in Figure 5.1, the value of pH where solubility begins to increase with increasing pH varies for different metal hydroxides.

In Section 5.4.2 the amphoteric behavior of metal carbonates is discussed. Increasing the pH by adding carbonate anions instead of hydroxide anions produces metal carbonates that react similarly to metal hydroxides, becoming less soluble as the positive charge on the metal carbonate complex decreases. The metal carbonates reach a solubility minimum when a neutral metal carbonate is formed, and begin to redissolve because of increasing solubility when negatively charged metal carbonate anion complexes are formed.

In Figure 5.2, using the cadmium hydroxide curve as an example, the vertical axis labeled "Metal Solubility" represents the sum of the milligrams/liter of cad-mium atoms contained in all of the dissolved cadmium carbonate species (Cd^{2+}, $Cd(CO_3)^0(aq)$, $Cd(CO_3)_2^{2-}$, $Cd(CO_3)_3^{4-}$, and so on) in equilibrium with solid $CaCO_3(s)$.

RULES OF THUMB

1. Only polyvalent cations (e.g., Fe^{3+}, Zn^{2+}, Mn^{2+}, and Cr^{3+}) have large enough charges to attract water molecules strongly enough to act as acids, by causing the release of H^+ from water molecules in the hydra-tion sphere. Monovalent cations, such as Na^+, do not act as acids at all.
2. The interactions of metal cations with water (Equations 5.2 through 5.7) cause the solubility of metal species in water to be dependent on pH and ORP.
   a. Low pH (high H_3O^+ concentration and high acidity) increases metal solubility by shifting the equilibria of Equations 5.2 through 5.7 to the left, decreasing the formation of less-soluble metal polyhydroxides.
   b. High pH (low H_3O^+ concentration and low acidity) decreases metal solubility by shifting the equilibria of Equations 5.2 through 5.7 to the right, increasing the formation of less-soluble metal polyhydroxides.

    c. Low ORPs (reducing conditions, low-to-zero DO levels, where electron donors are more common than electron acceptors) increase the solubility of many metals (see Section 5.1) by promoting lower oxidation numbers for metal cations (lower positive charge, e.g., Fe^{2+} rather than Fe^{3+}). For cations with lower positive charge, the equilibria of Equations 5.2 through 5.7 are maintained more strongly to the left, resulting in less formation of low-solubility polyhydroxides.

    d. High ORPs (oxidizing conditions, high DO levels, where electron acceptors are more common than electron donors) decrease the solubility of many metals by promoting higher oxidation numbers for metal cations (higher positive charge, e.g., Fe^{3+} rather than Fe^{2+}). For cations with higher positive charge, the equilibria of Equations 5.2 through 5.7 are maintained more strongly to the right, resulting in greater formation of low-solubility polyhydroxides.

3. The presence of dissolved species such as sulfide or carbonate, which form low-solubility compounds with metal cations, can largely negate the above generalizations by competing with hydroxide formation.

Example 1: Effect of Dissolved Metal on Alkalinity

A sample of groundwater contains a high concentration of dissolved iron, about 20 mg/L. At the laboratory, alkalinity is measured to be 150 mg/L as $CaCO_3$. Is this laboratory measurement of alkalinity likely to accurately represent the groundwater alkalinity? Alkalinity is discussed in Chapter 3, Section 3.4.3.

ANSWER

Soluble inorganic iron is in the ferrous form, Fe^{2+}. Because of its small positive charge and relatively large diameter, loss of protons from the hydration sphere is not a significant process for hydrated ferrous iron Fe^{2+} (denoted in Equation 5.10 by $Fe(H_2O)_6^{2+}$). However, when a groundwater sample is exposed to air, oxygen (an electron acceptor) dissolving from the atmosphere can oxidize Fe^{2+} to the ferric form, Fe^{3+}. This process is often enhanced by aerobic iron bacteria (see reaction step 2 in Figure 5.4). Depending on the pH, hydrated Fe^{3+} can lose protons from its hydration sphere to any bases present, including water molecules (to form H_3O^+) and hydroxide ions (to form H_2O). As hydrated Fe^{3+} loses protons, it forms ferric hydroxide species, and, at the same time, the increase in solution H_3O^+ and decrease in solution OH^- make the solution more acidic. The acidic behavior of hydrated Fe^{3+} occurs to a greater extent at higher pH, where proton loss from the inner hydration sphere is enhanced. Equation 5.10 represents the overall oxidation reaction that converts dissolved ferrous iron to precipitated ferric hydroxide:

$$4Fe(H_2O)_6^{2+} + O_2 \rightleftarrows 4Fe(OH)_3(s) + 14H_2O + 8H^+ \tag{5.10}$$

$Fe(OH)_3$ is a yellow to red-brown precipitate often seen on rocks and sediments in surface waters with high iron concentrations.

The molar concentration of H^+ formed by Equation 5.8 can be up to two times the Fe^{2+} molar concentration, depending on the final pH. Each H^+ released will neutralize a molecule of base, consuming some alkalinity, by reactions such as

$$H^+ + OH^- \rightleftarrows H_2O \tag{5.11}$$

$$H^+ + HCO_3^- \rightleftarrows H_2CO_3 \tag{5.12}$$

$$H^+ + CO_3^{2-} \rightleftarrows HCO_3^- \tag{5.13}$$

We will assume a worst-case scenario with respect to affecting the alkalinity, where the pH is high enough that the equilibrium of Equation 5.10 goes essentially to completion to the right side as additional oxygen dissolves from the atmosphere. The atomic weights of hydrogen and iron are 1 and 56 g/mol, respectively. If the equilibrium of Equation 5.10 is completely to the right, 1 mole (56 g) of Fe^{2+} will produce 2 moles of H^+ (2 g). At the time of sampling, the concentration of dissolved Fe^{2+} (as $Fe(H_2O)_6^{2+}$) was about 20 mg/L and all is eventually oxidized to Fe^{3+} (as $Fe(OH)_3$). The molar concentration of iron is

$$\frac{0.020 \text{ g/L}}{56 \text{ g/mol}} = 0.00036 \text{ mol/L or } 0.36 \text{ mmol/L}$$

By Equation 5.10, the moles of H^+ produced are two times the moles of iron:

Moles of H^+ = 2×0.36 mmol/L = 0.72 mmol/L

Grams of H^+ = 0.72 mmol/L $\times$ 1 mg/mmol = 0.72 mg/L

We must now determine what effect this quantity of H^+ will have on the alkalinity. Alkalinity is measured in terms of a comparable quantity of $CaCO_3$. The molecular weight of $CaCO_3$ is 100 g/mol, and it dissolves to form the doubly charged ions Ca^{2+} and CO_3^{2-}. Alkalinity is a property of the CO_3^{2-} anion, which consumes acidity by accepting two H^+ cations:

$$2H^+ + CO_3^{2-} \rightleftarrows H_2CO_3 \tag{5.14*}$$

Therefore, 0.72 mmol/L of H^+ will react with $0.72/2 = 0.36$ mmol/L of CO_3^{2-}, and 0.36 mmol/L of $CaCO_3$ is required as a source of the CO_3^{2-}. From the definition of alkalinity, the change in alkalinity is equal to the change in concentration of $CaCO_3$, in milligrams per liter.

$$0.36 \text{ mmol/L of } CaCO_3 = 0.36 \text{ mmol/L} \times \frac{100 \text{ mg}}{1 \text{ mmol}}$$

$$= 36 \text{ mg/L} = \text{change in alkalinity}$$

---

* Equation 5.14 can be obtained by adding Equations 5.12 and 5.13.

The groundwater alkalinity at the time of sampling is equal to the laboratory-measured alkalinity plus the alkalinity lost by hydrolysis of ferrous iron (Equation 5.10).

The maximum possible value for the original alkalinity of the groundwater before exposure to air was

$$150 \text{ mg/L} + 36 \text{ mg/L} = 186 \text{ mg/L as CaCO}_3$$

The laboratory alkalinity measurement was lower than the actual groundwater alkalinity, with a maximum error of about 21%.

In the above example, the pH was assumed high enough to maintain the equilibrium of Equation 5.10 entirely to the right. Under these conditions, essentially all the H^+ added to the solution will react by Equations 5.12 and 5.13 to form H_2CO_3. Thus, there would be no significant net change in aqueous H^+ (as H_3O^+) and little corresponding change in pH. In practice, the changes in concentrations brought about by Equations 5.10 through 5.14 will never cause the equilibria of the reactions to go completely to the right. Thus, the H^+ added will never react completely with the alkalinity and there will always be at least a small net increase in the H_3O^+ concentration, accompanied by a corresponding small decrease in pH.

This example illustrates the pH buffering effect of alkalinity. The addition of H^+ to the solution by hydration of metal ions, as in Equation 5.10, will not change the pH greatly as long as some alkalinity remains, because the added H^+ is taken up by carbonate species in the water. This is also true for the addition of H^+ from other sources, such as mineral and organic acids.

5.4 ADJUSTING pH TO REMOVE METALS FROM WATER BY PRECIPITATION

5.4.1 HYDROXIDE PRECIPITATION

Hydroxide precipitation is commonly accomplished in water treatment by raising the pH with sodium hydroxide (NaOH, caustic soda), calcium hydroxide ($Ca(OH)_2$, hydrated lime), or magnesium hydroxide ($Mg(OH)_2$). As seen in Sections 5.3.5 and 5.3.6, raising the pH facilitates the release of H^+ from hydrated metal cations and leads to the stepwise conversion of inner-layer water molecules into chemically bound hydroxide groups. Each newly bound hydroxide ion (carrying a single negative charge) reduces the positive charge on the metal species by one unit, as in Equations 5.2 through 5.7. As long as the metal hydroxide species are charged, they electrostatically attract water molecules and repel one another, thereby remaining dissolved. However, when enough negative hydroxide ions have become chemically bound to hydrated metal cations to neutralize them, as in Equation 5.6, the electrostatic repulsion between the metal species disappears and their attraction to polar water molecules is greatly reduced.

When metal species are not electrically charged, they can aggregate and precipitate as a solid. For example, with hydrated ferric iron (Fe^{3+}), a sequence of three H^+ releases from the inner hydration sphere forms, first $Fe(OH)^{2+}$, then $Fe(OH)_2{}^+$, and finally neutral $Fe(OH)_3$, which then precipitates as solid $Fe(OH)_3$ (although a small amount of neutral $Fe(OH)_3$ will remain in solution).

To summarize, as the pH increases,

- The release of hydrogen ions from hydrated cations becomes easier.
- The average number of hydroxide ions bonded to Fe^{3+} increases and the net positive charge on the metal species decreases.
- The quantity of solid $Fe(OH)_3$ increases and the concentration of dissolved charged ferric hydroxide species decreases.

The result is that the solubility of iron hydroxide decreases.

If the water pH is made to decrease, the process reverses. As pH decreases,

- Hydronium ions (H_3O^+) in solution release H^+ back to hydroxide groups (OH^-) on the metal species, converting metal-bound hydroxide groups back to neutral water molecules.
- The average number of hydroxide ions bonded to Fe^{3+} decreases and the net positive charge on the metal species increases.
- The quantity of solid $Fe(OH)_3$ decreases and the concentration of dissolved charged ferric hydroxide species increases.

The result is that the solubility of iron hydroxide increases.

Thus, one may say that the metal becomes more soluble at lower pH and less soluble at higher pH, even though what actually occurs is that the hydrated metal species form increasingly more of the electrically neutral and less-soluble $Fe(OH)_3$ at higher pH values.

5.4.2 CARBONATE PRECIPITATION

Carbonate precipitation is commonly accomplished in water treatment by adding calcium carbonate ($CaCO_3$, limestone) or sodium carbonate (Na_2CO_3, soda ash). Calcium carbonate is useful only below about pH 7 because of the low solubility of limestone at higher pH. Sodium carbonate is used when it necessary to treat at higher pH values than is possible with limestone. The chemical reaction for metal-carbonate precipitation is shown by Equation 5.23 in Case Study 3 in Section 5.9.3.

Figure 5.2 shows that metal carbonates, such as metal hydroxides, are amphoteric and have a solubility minimum at a particular pH, increasing in solubility at both higher and lower pH values. In a manner similar to hydroxide precipitation, when dissolved carbonate compounds are present, dissolved bicarbonate (HCO_3^-) and carbonate (CO_3^{2-}) anions attach progressively to metal cations as pH increases, reducing the positive charge on the cationic species and eventually forming a neutral metal carbonate that can precipitate. Carbonate precipitation yields lower solubilities than hydroxide precipitation for several metals, notably, lead, cadmium, and ferrous iron. Also, the pH range where the lowest solubilities are achieved tends to be broader than with hydroxide precipitation, allowing more tolerant pH control during treatment. The broader solubility curves also improve the likelihood that, when regulatory limits are not lower than about 0.01 mg/L, they can be met without raising the pH above 9, possibly avoiding the need for pH adjustment after treatment.

5.4.3 SULFIDE PRECIPITATION

Sulfide precipitation is commonly accomplished in water treatment by adding calcium sulfide (CaS), sodium sulfide (Na_2S), sodium hydrosulfide (NaHS), or ferrous sulfide (FeS). Figure 5.2 shows that metal sulfide solubilities behave very differently from those of metal hydroxides and carbonates. Metal sulfides do not exhibit amphoteric behavior like metal hydroxides or carbonates, and do not have a pH where solubility is a minimum. Instead, they decrease steadily in solubility as pH increases. In addition, over most of the pH range, particularly below about pH 10, metal sulfides have much lower solubilities than metal hydroxides or carbonates. The reasons for these differences are related to how the sulfide anion (S^{2-}) differs from hydroxide and carbonate in its electrostatic attractions to dissolved metal species. However, the precipitation process is similar; dissolved sulfide anions (HS^- and S^{2-}) attach progressively to metal cations as pH increases, eventually forming a neutral metal sulfide that can precipitate. The chemical reaction for metal-sulfide precipitation is shown by Equation 5.22 in Case Study 3 in Section 5.9.3.

An important consideration when precipitating metals by sulfide precipitation is that the pH must be high enough (greater than about pH 8) to prevent a reaction of sulfide anions with water, which forms toxic hydrogen sulfide gas (H_2S) with its characteristic rotten egg smell. If a sulfide precipitation process requires pH values much lower than about 8, it could be necessary to somehow contain the hydrogen sulfide gas that will be produced, in order to protect workers. Also, sulfide reagents require hazardous material handling, and discharge of sulfide-enriched water may not be acceptable in many situations. It has been said that sulfide precipitation is the gold standard for treatment of metal-contaminated water, if you can tolerate the smell.

5.5 YOUR RESULTS MAY VARY

At this point, it is necessary to present a cautionary "your results may vary" statement. The curves in Figures 5.1 and 5.2 are based on theories of metal solubility, using data from experimental measurements of solubility under different conditions. First, theories are always incomplete representations of reality, and necessarily employ simplifications, some known and others unknown; second, experimental measurements are always subject to error; third, actual water samples will seldom match exactly all the conditions assumed by the theoretical model. For example, during water treatment by hydroxide precipitation, carbonate or bicarbonate anions from hydrolysis of carbon dioxide dissolved from the atmosphere are always present, and metal carbonates are generally less soluble than their corresponding hydroxides. The precipitation of unaccounted-for metal carbonates can result in lower residual metal concentrations than predicted. On the other hand, the presence of metal-complexing species, such as cyanide, acetate, and amino acids, arising from secretions and decay products of water organisms, can result in greater than predicted metal solubilities.

The presence of unaccounted-for species is a common problem and can strongly affect measured solubilities. Metal solubilities reported in the literature will often differ by several orders of magnitude for the same metal under supposedly the same conditions. The curves in Figures 5.1 and 5.2 should be treated as qualitatively correct, but do not be surprised if another literature source provides somewhat different solubility versus pH values for the same metal compound. The curves can be used to decide on the first steps in a precipitation process, but the procedure always will have to be optimized by bench tests and even some on-site trial-and-error tests.

RULES OF THUMB

Advantages and Disadvantages of Different Metal Precipitation Treatment Methods

Hydroxide Precipitation

Advantages

- Fast reaction rates; shorter detention times.
- Easily automated pH control.
- Relatively inexpensive reagents.
- Low enough solubilities for many metal regulatory limits.
- Iron and manganese hydroxide precipitates can scavenge other metal ions (especially, Ag, Al, As, Co, Cu, Mo, Ni, Pb, V, and Zn) by adsorption and coprecipitation, achieving lower concentrations by reducing their apparent solubilities.

Disadvantages

- Produces large amount of sludge.
- Solubility of some metals may not be low enough (e.g., Pb, Ag, and Zn), but see last bullet under "Advantages".
- Minimum solubility occurs at different pH values for different metals, increasing at higher and lower pH (amphoteric behavior).
- Some metals have a narrow pH range for minimum solubility, possibly requiring closer pH control.
- Treatment of mixed-metal-contaminated water may require batch processing.

(Continued)

Advantages and Disadvantages of Different Metal Precipitation Treatment Methods (*Continued*)

Carbonate Precipitation

- Easily automated pH control.
- Relatively inexpensive reagents.
- Many metals have a broad pH range for minimum solubility, allowing less-stringent pH control.
- Lower solubilities than hydroxide precipitation for many metals.
- Low solubility of some metals at lower pH allows treatment at lower pH.

- Longer reaction times than hydroxide precipitation require longer detention times.
- Minimum solubility occurs at different pH values for different metals and increases at higher and lower pH values (amphoteric behavior).
- Treatment of mixed-metal-contaminated water may require batch processing.

Sulfide Precipitation

- Fast reaction rates; shorter detention times.
- Lower solubilities than hydroxide or carbonate precipitation for most metals.
- Not amphoteric; metal solubilities decrease progressively as pH increases. Simpler treatment of mixed-metal-contaminated water.
- Low solubilities at lower pH allow treatment at lower pH, if hydrogen sulfide releases OK.

- Potential for unacceptable hydrogen sulfide gas releases when treatment pH is less than about 8.
- Potential for unacceptable residual sulfide in treatment effluent.
- Reagents require hazardous materials handling.
- Higher capital and operating costs than hydroxide and carbonate precipitation.

5.6 HOW ORP AFFECTS THE SOLUBILITY OF METALS

Similar to pH, ORP also can influence metal solubilities. Changing the ORP changes the number of electrons in the outer orbitals of metal atoms, affecting how metal atoms react with other substances and frequently influencing metal solubility. (See Chapter 3, Section 3.5 for further discussion of ORP.)

To use the example of iron again, in deoxygenated water where the ORP has fallen to negative values, dissolved iron is present in the ferrous state. Because of its smaller positive charge, the ferrous iron cation (Fe^{2+}) attracts polar water molecules less strongly than does the ferric iron cation (Fe^{3+}) and, therefore, has a smaller and less strongly bound hydration sphere. Therefore, hydrated Fe^{2+} does not form low-solubility hydroxides as readily as hydrated Fe^{3+} and the solubility curve in Figure 5.1 for $Fe(OH)_2$ is positioned mostly higher (more soluble) and displaced to higher pH values than that for $Fe(OH)_3$.

In addition, the ORP may influence other substances present in the water, such as sulfate, in a manner more significant than its direct effect on metal species. For example, unlike iron, the electron structure of some other metals is not particularly sensitive to ORP changes within common environmental conditions. However, if sulfate is present, a negative ORP (anaerobic conditions) can reduce sulfate to sulfide, which reacts with many metals to form metal sulfides of very low solubility.

As an example, lead is one metal that is not very sensitive to ORP changes. In pure water exposed to the atmosphere, the solubility of lead species* does not change greatly with the ORP. However, if water that contains sulfate becomes anaerobic, the low ORP reduces sulfate to sulfide and dissolved lead species can precipitate as insoluble lead sulfide. In the absence of sulfate or sulfide, dissolved lead species remain relatively soluble at negative ORPs.

5.6.1 ORP AND pH ARE INTERACTIVE

ORP and pH interact, a fact not considered in the pH and ORP solubility rules (Section 5.3). In water environments, changes in ORP often affect pH also. Reversible reactions, like those initiated by changes in pH or ORP, generally yield products that act to oppose the forward reaction and restore an equilibrium state. Thus, oxidation of metal species by exposure to DO and increases in ORP (which the ORP rule predicts will generally reduce metal solubility) is usually accompanied by a release of H^+ that acts to decrease the pH (which the pH rule predicts will generally increase metal solubility). Conversely, reduction of metal species due to lack of DO (decrease in ORP) is usually accompanied by an increase in pH because of the formation of OH^-. In reactions where anaerobic (reducing) conditions would increase metal solubility, the formation of additional OH^- acts oppositely to limit the reaction.

5.7 ORP-SENSITIVE AND ORP-INSENSITIVE METALS

For the reasons above, it is useful to classify metals as ORP sensitive or ORP insensitive, according to how strongly their solubility is influenced by changes in ORP, within a range normally achievable under environmental or water treatment conditions.

5.7.1 COMMON ORP-SENSITIVE METALS: CHROMIUM, COPPER, IRON, MANGANESE, MERCURY, MOLYBDENUM, THALLIUM, URANIUM, AND VANADIUM

ORP-sensitive metals, that is, chromium (Cr), copper (Cu), iron (Fe), manganese (Mn), mercury (Hg), molybdenum (Mo), thallium (Tl), uranium (U), and vanadium (V), are those that can undergo changes in oxidation state (outer-orbital electron structure) under common environmental or water treatment conditions, often resulting in solubility changes. These metals change solubility when ORP conditions change sufficiently. All except molybdenum, vanadium, and uranium tend to be less soluble under oxidizing conditions than under reducing conditions (as long as sulfide is not present). Molybdenum, vanadium, and uranium tend to be more soluble under oxidizing

---

* Lead dissolved in water exposed to the atmosphere, like other dissolved metals, can always potentially react with hydroxide and carbonate anions to form various hydroxide and carbonate species. Hydroxide ions are always available from dissociation of water molecules (Chapter 3, Equation 3.1) and carbonate ions are present from dissolved carbonate and bicarbonate ions formed when carbon dioxide dissolved from the atmosphere reacts with water (Chapter 3, Section 3.4).

conditions because they can form soluble oxy-anions, similarly to the ORP-sensitive metalloids discussed below. These general principles are qualified by the phrase "tend to be" because the solubility of metals depends on more than one variable. Coincident changes in other variables, such as temperature, pH, and other dissolved species present, may, in combination, influence metal solubility more than ORP.

As an example, in pure water at negative ORP (reducing conditions) and pH less than about 7, iron is present as the soluble ferrous cation Fe^{2+}. Above pH 7, ferrous hydroxide, $Fe(OH)_2$, is formed by hydrolysis in a manner analogous to the formation at positive ORPs of ferric hydroxide, $Fe(OH)_3$, and the solubility of $Fe(OH)_2$ is pH dependent, like that of $Fe(OH)_3$. At its minimum solubility (about pH 10.7), $Fe(OH)_2$ is about 150 times more soluble than $Fe(OH)_3$ at its minimum (about pH 8). Thus, the solubility of iron hydroxides depends on both pH and ORP.

Examination of Figures 5.1 and 5.2 shows the following:

- At any given pH, reduced ferrous hydroxide is more soluble than oxidized ferric hydroxide.
- Above about pH 10.6, the solubility difference between the reduced and oxidized forms is rather small (around a factor of 2–4).
- Between pH 7 and pH 9, the solubility difference is much larger, by a factor between 10^5 and 10^7.
- Below pH 5 and above pH 11, the oxidized form is more soluble than the reduced form is at pH 10.6.

The other ORP-sensitive metals behave similarly. Under reducing conditions, they are generally present as soluble hydrated cations at lower pH values and as moderately soluble neutral hydroxides that go through a solubility minimum at some higher pH.

Under oxidizing conditions and pH greater than about 5.5, ORP-sensitive metals react in water to form low-solubility hydroxides and carbonates, often resulting in a pH-dependent upper limit to dissolved metal concentrations in environmental surface waters. For example, under oxidizing conditions, surface water concentrations of dissolved iron cannot exceed the solubility of $Fe(OH)_3$, as indicated in Figure 5.1. Thus, stream concentrations of dissolved iron are limited by precipitation of insoluble $Fe(OH)_3$, dissolved manganese by precipitation of insoluble MnO_2, dissolved chromium by precipitation of insoluble $Cr(OH)_3$, and dissolved copper by precipitation (in the presence of carbonate) of insoluble copper hydroxy carbonates ($Cu_2(OH)_2(CO_3)$, malachite, and $Cu_3(OH)_2(CO_3)_3$, azurite).

5.7.2 COMMON ORP-INSENSITIVE METALS: ALUMINUM, BARIUM, CADMIUM, LEAD, NICKEL, AND ZINC

The metals aluminum (Al), barium (Ba), cadmium (Cd), lead (Pb), nickel (Ni), and zinc (Zn), do not change their oxidation state within the ORP conditions common in the environment or water treatment systems. Although the ORP does not affect their solubilities significantly, it may affect other anions, like sulfide, with which the metals can react. Also, because of weaker electrostatic attractions to water molecules, these metals do not react strongly with water to form low-solubility hydroxides.

Under oxidizing conditions and the absence of anions with which they can react, these metals tend to remain as dissolved cations or moderately soluble neutral hydroxides. In the presence of other reactants in water, they can form carbonates, phosphates, sulfides, and the like, in addition to oxides/hydroxides, whose solubilities depend more on pH than on ORP.

Although reducing conditions have little direct effect on these metals as cations, under reducing conditions, with sufficient sulfide present, all can form sulfides of low solubility.

5.7.3 COMMON ORP-SENSITIVE METALLOIDS: ARSENIC AND SELENIUM

Arsenic (As) and selenium (Se) tend to behave oppositely to most of the ORP-sensitive metals. They are more soluble under oxidizing conditions than under reducing conditions.

Under oxidizing conditions, they form the oxy-anions arsenate (AsO_4^{3-}) and selenite (SeO_3^{2-}), which react with Fe, Mn, and Pb cations to form moderately soluble compounds. Under reducing conditions, selenium forms insoluble elemental selenium and iron selenide ($FeSe_2$); with sulfide present, arsenic forms sulfides of very low solubility.

5.8 METAL WATER QUALITY STANDARDS

The possibility that environmental changes in pH or ORP might change the solubility of metal species in surface and groundwaters has to be considered when water quality standards for metals are determined. A stream that meets standards for aquatic life at one location might undergo a change in pH or ORP a little farther downstream that causes some metal parameter to increase to a hazardous concentration. Depending on the designated uses of the water body in question, standards may be written to anticipate potential environmental changes in metal solubilities that could be environmentally hazardous.

A metal water quality standard may be written for the dissolved, potentially dissolved, or total recoverable (also called *total*; see bullet below) forms. When stream standards are written for dissolved metals, it is assumed that metals in solid forms (adsorbed or precipitated) will not contribute significantly to local environmental risks. When standards are written for metals as potentially dissolved, or total recoverable/total, it is assumed that environmental protection requires that sorbed or precipitated metals also be considered as potential risks, because downstream pH and ORP might change in a manner that could dissolve some metal solids.

- *Dissolved*: The sample is filtered on site immediately after or, preferably, during collection through a 0.40- or 0.45-μm filter, and the filtrate is then acidified to pH 2 for preservation before analysis. Acidification after filtering prevents precipitation of any dissolved metal in the sample filtrate before analysis. This procedure omits from the analysis metals that were filtered out as solid precipitates or sorbed on suspended sediments.
  - In general, the dissolved method more accurately measures (compared to total or total recoverable analyses) the ionic form of metals that is

the most toxic to aquatic life, while excluding less toxic solid forms. Also, because they are generally considered more mobile and biologically available, concentrations of dissolved metals are useful for fate and transport and risk analyses studies.

- It is essential that the sample be filtered immediately after or during collection. A true measure of dissolved metals in the source water cannot be obtained after transporting an unfiltered sample to a laboratory. An unfiltered sample collected to determine dissolved metals in the source water cannot be acidified for preservation because the lower pH could cause some precipitated metals in the original sample to dissolve. It cannot be transported without acidification because potential gain or loss of dissolved CO_2 or O_2 can result in changes in pH or ORP that could dissolve or precipitate metals en route, resulting in a measurement not representative of the source water.
- *Potentially dissolved*: Sample is acidified to pH 2, held for 8–96 hours, and then filtered through a 0.45-μm filter and analyzed. This procedure is intended to simulate the possibility that metals bound in suspended sediments might be transported into more acidic environmental conditions and partially dissolve. It measures the metals dissolved at the time of sampling as well as a portion of the metals initially bound to suspended sediments and released during the holding period at low pH.
  - Although adoption of the potentially dissolved method for effluent monitoring may overstate the availability of ionic metals in an effluent, it is also possible that the dissolved method would potentially understate the availability of ionic metals once an effluent has mixed with receiving waters. As a conservative measure, potentially dissolved metal limits are often used in effluent monitoring and National Pollution Discharge Elimination System (NPDES) permits when these are based on dissolved metals standards for surface waters and when it may be expected that the pH of the receiving water may be or become lower than the pH of the discharge water.
- *Total recoverable and Total*: Initially, the EPA listed two different analytical methods to measure total recoverable (Trec) and total metals.
  - *Total recoverable*: In the original analytical method, the sample was acidified to pH 2 and analyzed without filtering. This procedure was intended to measure all metals that were dissolved and weakly bound to suspended sediments. The method was later modified to include metals more strongly associated with sediments by starting with a volume reduction step, followed by refluxing (acid digestion) in hot, dilute mineral acids for 30 minutes,* then diluting with reagent water to the original volume, and filtering before analysis.

---

* Unfiltered samples with turbidity <1 NTU do not require digestion. Very low turbidity means that the suspended sediment concentration is very small and that essentially all metals in the sample are dissolved species, which do not need to be digested for analysis.

- *Total*: This analysis differs from the total recoverable analysis by the severity of the acid refluxing step. The unfiltered sample is first concentrated by volume reduction and then refluxed repeatedly with dilute mineral acids (with continued addition of acid) until all metals are dissolved. The sample then is brought back to its original volume with reagent water and filtered. This procedure is intended to measure all metals contained in the sample, including metals in all solid forms (associated with organic suspended matter, sediments, and precipitates) as well as metals dissolved in the water. In addition to water quality analyses, total metal analysis is used for soils, solid wastes, mining leachate, mineral characterization, toxic metal content of foods, and other products, and in risk analysis when tracking exposure pathways from a contaminant source to different life forms.
- The EPA has found that there was no statistical difference between the Total and Total Recoverable methods in results obtained from equivalent samples. Therefore, the EPA recommended "For effluent guidelines, for permitting under NPDES, and for other purposes in EPA's water programs, the terms *total metal* and *total recoverable metal* may be used interchangeably to reflect that it is the hard mineral acid digestion procedure that is used" (EPA 1998).

Total recoverable metals analysis is an older procedure that is still used for both surface water standards and discharge limits where the water pH may reasonably be expected to become lower at some other location. The EPA now recommends that dissolved metals analyses be used for water quality standards because it is believed that dissolved metal concentrations represent the toxicity of a metal in the water column more closely and more reproducibly than total recoverable metal concentrations.* Therefore, total recoverable metal standards are being replaced in many states by site-specific dissolved metal standards for surface waters and by potentially dissolved metal limits in wastewater discharges.

The EPA has published a guide for converting measured dissolved metal concentrations into values for total recoverable metal concentrations, and vice versa, to assist in setting discharge limits based on either analytical method (EPA 1996). In the toxicity tests used to derive metal toxicity criteria, some fraction of the metal was dissolved and some fraction was bound to particulate matter. The conversion factors are based on what the metal toxicity criteria would be if the toxicity studies for a given metal had been based only on measurements of dissolved concentrations. Since the dissolved fraction more closely approximates the biologically available fraction than does the particulate fraction, converting total recoverable values into dissolved has the net effect of lowering the water quality criteria concentrations, that is, conservatively making standards more stringent.

---

* The primary mechanism for toxicity to organisms that live in water is by adsorption to or uptake across the gills, a process that requires metal to be in a dissolved form. Although particulate metal may exhibit some toxicity, it appears to be substantially less toxic than dissolved metal.

RULES OF THUMB

For a particular metal in a single-water sample:

1. A total metal or total recoverable metal analysis should always yield a larger concentration than a dissolved metal analysis, because dissolved metal is a subset of total metal.
2. A potentially dissolved analysis should always yield a concentration in the range between the dissolved and either total or total recoverable analyses.
3. Subtract the dissolved concentration of the metal from its total or total recoverable concentration to estimate the concentration contained in the solid materials.
4. The metal concentration associated with solid materials is the maximum concentration by which a potentially dissolved, total, or total recoverable analysis can exceed the dissolved analysis.
5. Allow about 5–10% error in reported laboratory results.

5.9 APPLICATIONS

Several useful applications of the principles governing metal solubilities are presented in the case studies below. Another important application, the treatment of urban stormwater runoff to remove dissolved metals, is described in Chapter 11, Section 11.1.

5.9.1 CASE STUDY 1: ACID ROCK DRAINAGE

The main cause of acid rock drainage (ARD)* is oxidation of iron pyrite. Iron pyrite, FeS_2, is the most widespread of all sulfide minerals and is found in many ore bodies. The somewhat complicated, but very efficient, process by which ARD develops is first described below in detail and then summarized to help make the overall process more clear.

During mining operations, particularly coal mining, iron pyrite in the ore becomes exposed to air and water, resulting in the formation of sulfuric acid and ferrous ion:

$$FeS_2(s) + \frac{7}{2}O_2 + H_2O \rightleftarrows Fe^{+2} + 2SO_4^{-2} + 2H^+ \tag{5.15}$$

The rate of reaction in Equation 5.15 is mainly limited by the availability of oxygen, since there is usually sufficient moisture in mine wastes and mine workings. Reaction 5.15 releases acidity (H^+), sulfate, and soluble ferrous iron into the water, along with other sulfide-bound metals and metalloids (As, Cd, Cu, Pb, Sb, Se, Zn, U, etc.).

---

* Also called by the less general term "acid mine drainage." Strictly speaking, acid mine drainage refers to water exiting from underground mine workings, while acid rock drainage includes water that drains through tailings piles and other waste rock depositories.

Next, dissolved ferrous iron (Fe^{+2}) is oxidized slowly by DO to ferric iron (Fe^{+3}), consuming some acidity:

$$Fe^{+2} + \frac{1}{4}O_2 + H^+ \rightleftarrows Fe^{+3} + \frac{1}{2}H_2O \qquad (5.16)$$

Above pH 4 and in the absence of iron-oxidizing bacteria, Equation 5.16 is the slowest (rate-limiting) step in the reaction sequence. However, below pH 4 and in the presence of iron-oxidizing bacteria, the rate of Equation 5.16 is greatly accelerated by bio-oxidation, a million-fold or more. The ferric iron formed in Equation 5.16 can further oxidize pyrite, as in Equation 5.17, where ferric iron is reduced back to Fe^{+2}, releasing much more acidity.

$$FeS_2(s) + 14Fe^{+3} + 18H_2O \rightleftarrows 15Fe^{+2} + 2SO_4^{-2} + 16H^+ \qquad (5.17)$$

In Equation 5.17, eight times more acidity is generated when ferric iron oxidizes pyrite than when DO serves as the oxidant in Equation 5.15 (16 equivalents compared to 2 equivalents, per mole of FeS_2). In the pH range from 2 to 7, pyrite oxidation by Fe^{+3} (Equation 5.17) is kinetically favored over abiotic oxidation by oxygen (Equation 5.16). In addition, Equation 5.17 returns soluble Fe^{+2} to the reaction cycle where it again consumes acidity via Equation 5.16 and leads to a cyclic repeat of Equation 5.17.

Overall, four equivalents of acid are formed for each mole of FeS_2 oxidized in the cyclic reaction sequence of Equations 5.16 and 5.17. If bacterially mediated oxidation is occurring in Equation 5.16, the reaction cycle can be accelerated over a million-fold.

Ferric iron also hydrolyzes (reacts with water), releasing more acid to the water and forming insoluble ferric hydroxide, which can coat streambeds with the yellow-orange deposits known as yellow boy:

$$Fe^{+3} + 3H_2O \rightleftarrows Fe(OH)_3(s) + 3H^+ \qquad (5.18)$$

Precipitated $Fe(OH)_3$ serves as a reservoir for dissolved Fe^{+3}. If the generation of Fe^{+3} by Equation 5.17 is stopped because of lack of oxygen, then Fe^{+3} is supplied by dissolution of solid $Fe(OH)_3$ and is available to react via Equation 5.17.

5.9.1.1 Summary of Acid Formation in ARD

The steps of acid formation in ARD are summarized below and illustrated in Figure 5.4.

Step 1: Iron pyrite, dissolved or solid, is oxidized by DO (Equation 5.15), lowering the pH and producing Fe^{+2} and SO_4^{-2}.

Step 2: Fe^{+2} formed in step 1 is oxidized slowly by DO to Fe^{+3} (Equation 5.16). This is the rate-limiting step in the reaction sequence in the absence of

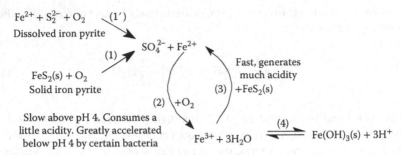

FIGURE 5.4   Reaction scheme for generation of acid rock drainage by pyrite oxidation.

iron-oxidizing bacteria. The abiotic rate decreases with lower pH. However, iron-oxidizing bacteria can greatly accelerate this step when the pH falls below 4.

Step 3: Fe^{+3} from step 2 is reduced rapidly back to Fe^{+2} by pyrite (Equation 5.17) generating much acidity. Ferrous iron, Fe^{+2}, generated in step 3 reenters the reaction cycle via step 2.

Step 4: A portion of ferric iron, Fe^{3+}, reacts with water (Equation 5.18) to form ferric hydroxide precipitate, $Fe(OH)_3(s)$, releasing more acidity. When the Fe^{3+} concentration diminishes, some $Fe(OH)_3$ precipitate can dissolve, serving as a reservoir for replenishing Fe^{3+} and maintaining the acid producing cycle.

As pH is lowered, step 1 becomes less important and the abiotic rate of step 2 decreases. However, step 2 can be greatly accelerated by certain iron-oxidizing bacteria (*Metallogenium, Ferrobacillus, Thiobacillus*, and *Leptospirillum*). These bacteria derive energy from the oxidation of Fe^{+2} to Fe^{+3}. Below pH 4, iron-oxidizing bacteria can catalyze step 2, speeding up the overall reaction rate by a factor as large as one million and lowering the pH to 2 or less. Furthermore, these bacteria can tolerate high concentrations of dissolved metals (e.g., 40,000 mg/L Zn and Fe; 15,000 mg/L Cu) before experiencing toxic effects. They thrive in ARD waters as long as a minimal amount of oxygen is present. Once bacterial acceleration of iron pyrite oxidation occurs, it is hard to reverse.

RULES OF THUMB

1. Oxidation of iron pyrite is the most acidic of all common weathering reactions. The production of ARD can be a rapid, self-propagating, cyclic process that is accelerated by low pH and the presence of iron-oxidizing bacteria. The process will continue as long as oxygen, pyrite, and water are present.
2. In locations where ARD encounters sufficient calcite minerals, the production of acidity may be offset by the neutralization reaction (see Chapter 3 and Section 5.9.1.3):

$$CaCO_3 + 2\,H^+ \rightleftarrows Ca^{2+} + CO_2(g) + H_2O \qquad (5.19)$$

which also consumes sulfate produced by Equation 5.17 by precipitating solid calcium sulfate, Equation 5.20.

$$SO_4^{2-} + Ca^{2+} \rightleftarrows CaSO_4(s) \qquad (5.20)$$

3. An indicator of calcite neutralization of ARD is that such waters generally contain 1500–2200 mg/L sulfate.

5.9.1.2   Noniron Metal Sulfides Do Not Generate Acidity

The abiotic oxidation by dissolved O_2 of noniron metal sulfides does not generate significant amounts of acidity. The metals are released as dissolved cations, but acidity is not produced, for example:

$$CuS(s) + 2O_2 \rightleftarrows Cu^{+2}(aq) + SO_4^{-2}(aq)$$

$$ZnS(s) + 2O_2 \rightleftarrows Zn^{+2}(aq) + SO_4^{-2}(aq)$$

$$PbS(s) + 2O_2 \rightleftarrows Pb^{+2}(aq) + SO_4^{-2}(aq)$$

Two possible reasons for the lack of acid formation when noniron sulfides are oxidized are

1. The oxidation state of sulfur is different in iron pyrite than in other sulfides, occurring as S_2^{-2} in iron pyrite and as S^{-2} in other sulfides. The respective oxidation reactions of these two sulfur forms indicate that acid is produced only with S_2^{-2}, because H_2O is required for the complete reaction to form acid:

   (Iron pyrite) $S_2^{-2} + 7/2O_2 + H_2O \rightleftarrows 2\,SO_4^{-2} + 2\,H^+$

   (Other sulfides) $S^{-2} + 2O_2 \rightleftarrows SO_4^{-2}$

2. Cu^{+2}, Zn^{+2}, Pb^{+2}, and so on, do not hydrolyze as extensively as Fe^{+3} does, so noniron sulfides do not react significantly by reactions equivalent to Equation 5.17:

$$Fe^{+3} + 3H_2O \rightleftarrows Fe(OH)_3(s) + 3H^+$$

Ferric iron, Fe^{+3}, can oxidize other metal sulfides, such as ZnS (sphalerite), CuS (covellite), PbS (galena), and chalcopyrite ($CuFeS_2$), in a similar fashion to its oxidation of FeS_2, releasing metal cations into the water, but not generating acidity.

5.9.1.3   Acid-Neutralizing Potential of Soil

Acid-neutralizing potential (ANP)* is a measure of how effectively the alkalinity (neutralization potential) in a soil or rock sample can neutralize the acid-producing

---

* Also called acid–base potential (ABP).

potential resulting from the presence of pyrite in the sample. The ANP is equal to the equivalents of calcium carbonate ($CaCO_3$, calcite) that are present in the sample in excess of the equivalents needed to neutralize the amount of acid that could potentially be produced from oxidation of pyritic sulfur also present in the sample.

5.9.1.4   Determining the Acid-Neutralizing Potential

The presence of iron pyrite, oxygen, and water will not necessarily result in acid formation if sufficient alkalinity to buffer the water is also present. The likelihood that a soil or rock sample will generate acidity when exposed to air and water may be estimated with a semiempirical definition of the ANP (Equation 5.21). This equation might be used, for example, to evaluate a soil cap used to cover and impound waste materials. If precipitation passing through the soil cap were made more acidic because of iron pyrite in the soil, it might mobilize metals or other contaminants in the wastes and create a new hazard. Equation 5.21 assumes that the acid-neutralizing and acid-forming materials are uniformly mixed.

$$ANP = (alkalinity) - (31.25)\ (wt\%\ \ pyritic\ \ sulfur) \qquad (5.21)$$

where the ANP units are tons of acid-neutralizing substances (as $CaCO_3$-equivalents) per 1000 tons of solid material.

Theoretically, if the ANP is positive, leachate from the sample should to be basic. In practice, the value of ANP ensuring nonacidic leachate varies from site to site. The difficulty of obtaining uniform mixing of the amending materials is often one reason for this. A general guideline is to apply a safety factor and require that the ANP be at least +3. On the other hand, any rock or earth material with a negative ANP is likely to have acidic leachate. If the ANP is −5 or more negative, the earth material has an acid-neutralizing deficiency of at least 5.0 tons $CaCO_3$/1000 tons material, and may be considered a potentially hazardous waste.

Figure 5.5 illustrates this behavior. Note that, while some calcite remains and the ANP is positive, pore water pH and sulfate concentration are not strongly affected by sulfide oxidation. When all the calcite is consumed and the ANP becomes negative, acid-neutralization (buffering) ceases and pH falls below 3, and sulfate concentration rises sharply.

RULES OF THUMB

1. If the ANP is positive, leachate from the sample is likely to be basic. Theoretically, rock waste with a positive ANP (greater acid-neutralizing potential than acid-forming potential) can weather forever without releasing acidic water.
2. If the ANP is negative, leachate is likely to become acidic eventually.
3. If the ANP is −5, or still more negative, the earth material is likely to be acidic enough to be defined as a potentially hazardous waste.

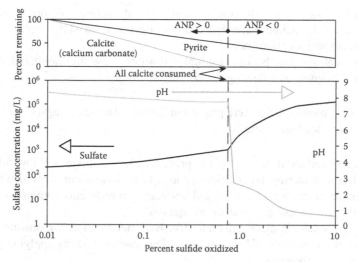

FIGURE 5.5 Generalized figure based on a model simulation showing how the ANP affects pH and sulfate concentration in soil pore water, as pyrite and calcite are consumed during acid rock drainage. The characteristic 10% sulfide rock contains both pyrite and calcite. In the top graph, the pyrite line illustrates the percent of acid-generating potential remaining, and the calcite line represents the percent of ANP remaining, as the percent of rock sulfide oxidized increases. The bottom graph shows the corresponding changes in soil pore water pH and sulfate concentration. (Adapted from EPA, *Management and Treatment of Water from Hard-Rock Mines*, Engineering Issue, 2005.)

5.9.2 CASE STUDY 2: IN SITU STABILIZATION OF METAL CONTAMINANTS IN SOILS

The responsible party of a former mine site in a remote location was required to remediate an area containing buried mine tailings so that it would cease contaminating a nearby stream. After considering several remediation alternatives, including excavation and landfilling, soil washing, phytoremediation, and isolation, they decided to proceed with in situ stabilization, or chemical immobilization, to prevent the metals from moving off-site.

The plan was to add chemical amendments to the soil that would react with contaminant metals and form low-solubility metal species that would be less mobile in groundwater or surface water. Because the soils to be treated were in contact with both groundwater and surface water, the amendments had to work in both reducing and oxidizing conditions. The kinds of amendments required depend on the kinds of metal contaminants to be treated.

Metal contaminants in soils occur mainly as

- Metals precipitated as low-solubility hydroxides, oxides, phosphates, carbonates, and sulfides.
- Crystallized soluble metal salts such as sulfates and chlorides; these dissolve readily into water at any pH.

- Metal cations sorbed to organic and mineral surfaces in soil; these are mobilized by acidic water because hydrogen cations (H^+) displace them from surface sorption sites.
- Metal cations dissolved in residual soil water; these are mobilized whenever they encounter flowing water.

All of these metal species were present in the buried mine tailings, so that amendments were needed that would

- Increase soil alkalinity and pH to prevent metal compounds of low solubility from contacting water acidic enough to dissolve them. This also stabilizes metal cations sorbed to soil surfaces. Amendments commonly used for this are limestone, lime, or phosphates.
- Sequester soluble metal salts and dissolved metal cations by causing them to form compounds of low solubility. This can be done by applying phosphates or carbonates.
- Immobilize dissolved metals by sorption to cation exchange surfaces.

Also, because even insoluble forms of metals can be mobilized by erosion in surface water runoff, it is necessary to reduce soil erosion by applying seeds, nutrients, and soil stabilizers to help increase vegetation and prevent soil erosion. Surface grading to reduce steep slopes might also be needed.

Since the tailings covered a large surface area, it was also necessary to search for the cheapest forms of amendments that would still be effective. Many common waste materials have been studied for use as soil amendments that would alter heavy metals to less-soluble and less-bioavailable forms. These include

- Organic amendments high in sorption capacity (municipal biosolids, composts, and manures)
- Alkaline materials (limestone, cement kiln dust, fly ash, phosphoric acid, and rock phosphate)
- Commercial fertilizers

Where it is applicable, in situ stabilization of heavy metals is a preferred technique for remediation of contaminated sites. It can be accomplished with any combination of the following methods:

- Chemically amend the contaminated soils to raise the soil pH, so metals do not dissolve in acid runoff.
- Chemically amend the contaminated soils with reactants that form insoluble compounds with dissolved metals, so they precipitate in place.
- Increase vegetation cover and, if necessary, grade the surface to reduce soil erosion.

All of these stabilization methods were applied to the site:

- The tailings were covered with a 10-foot soil cap of clean fill dirt mixed with sufficient limestone ($CaCO_3$) to raise the pH and produce an ANP of +5.
- A low-permeability clay barrier was installed upgradient of the tailings to deflect acidic groundwater around and beneath the tailings, so that leaching of metals into groundwater was reduced and a larger fraction of water passing through the tailings was from surface precipitation.
- In addition to raising the pH, the limestone provided carbonate that reacted with many of the metals to precipitate low-solubility metal carbonates (Section 5.4.2).
- The surface of the soil cap was planted to increase vegetative cover.

The soil stabilization methods have satisfactorily reduced the dissolved metal concentrations in downgradient groundwater and the continued effectiveness of the treatment is being followed with monitoring wells.

5.9.3   CASE STUDY 3: TREATMENT OF METAL-CONTAMINATED GROUNDWATER WITH A SULFATE-REDUCING PERMEABLE REACTIVE BARRIER

A sulfate-reducing permeable reactive barrier (SRPRB) is another type of passive treatment system used to treat acidic groundwater containing dissolved metal contaminants, such as from acid mine drainage. A passive (or near-passive) treatment system is one that removes dissolved metals with minimal or no operator control, use of external power sources, or addition of chemical reagents. Passive treatment is often the most cost-effective technology for sites where long-term (decades or longer) treatment is required and where continuous oversight becomes prohibitively costly. For an overview of other passive treatment technologies, for acid mine drainage, see EPA (2005), Costello (2003), and Younger (2000).

An SRPRB is a near-passive subsurface treatment system used to raise groundwater pH and reduce dissolved metal and sulfate concentrations. It is based on directing groundwater through a permeable reactive barrier containing limestone and sulfate-reducing bacteria (SRB) in an organic nutrient medium. Permeable reactive barriers designed to enhance bacterial sulfate reduction and metal sulfide precipitation can be used to mitigate acid mine drainage and the associated release of dissolved metals into environmental waters.

The use of bacterially mediated sulfate reduction in permeable reactive barriers is an alternative technique for the remediation of acid mine drainage. Permeable reactive barriers are installed in the path of migrating mine drainage water by excavating vertically from the ground surface down through the contaminated depth of the aquifer and replacing the removed material with a permeable reactive medium. The fill medium is designed to raise the groundwater pH, maintain anaerobic conditions, and support SRB growth. These bacteria continuously reduce sulfate in the groundwater to sulfide, which reacts with dissolved metals to precipitate them as low-solubility metal sulfides. The medium provides dissolved C, N, and P nutrients that are necessary for growth and reproduction of bacteria. The metal-contaminated groundwater entering the barrier must provide a sufficiently high concentration of the sulfate (SO_4^{2-}) that is required for reduction to sulfide (S^{2-}).

Under anaerobic conditions, SRB catalyze an oxidation–reduction reaction in which organic carbon is oxidized and groundwater sulfate is reduced to sulfide. The reduction of SO_4 produces H_2S and the oxidation of carbon produces CO_2, which reacts with water to form HCO_3^-, resulting in an increase in both alkalinity and pH. These reaction steps are presented in the generalized Equations 5.21 through 5.23.

Bioreaction:

SRB + sulfate + organic carbon → sulfide + bicarbonate (alkalinity) + SRB growth

$$SRB + 2CH_2O(aq) + SO_4^{2-} \rightarrow HS^- + H^+ + 2HCO_3^-(aq) + SRB \text{ growth} \qquad (5.22)*$$

Chemical reactions:

Dissolved metal + sulfide → precipitated metal sulfide + acidity
Dissolved metal + bicarbonate → precipitated metal carbonate + acidity

$$(Metal)^{2+} + HS^- \rightarrow (Metal)S(s) + H^+ \qquad (5.23)$$

$$(Metal)^{2+} + HCO_3^- \rightarrow (Metal)CO_3(s) + H^+ \qquad (5.24)$$

The oxidation of organic matter also releases ammonia and dissolved phosphate, which is utilized by the bacteria or released into the environment. An increase in H_2S concentrations coupled with the low solubility of metal sulfides results in the removal of dissolved metals as metal sulfides as precipitates within the SRPRB.

The conceptual design of an example SRPRB treatment system is shown in Figure 5.6. Many details of the construction are site specific, but the basic concepts are generally applicable. It consists of a permeable subsurface reactive wall with two anaerobic treatment zones through which groundwater passes sequentially. To help maintain anaerobic conditions within the treatment zones, a 1-foot-thick layer of high-porosity concrete sand is placed over the top of the SRPRB to serve as a groundwater overflow conduit and this is covered with an impermeable geomembrane liner below a surface layer of soil, to slow oxygen diffusion from the atmosphere.

- *Zone 1*: The first zone is an SRPRB bioreactor consisting of organic matter that serves as a slow-release source of bacterial nutrients (straw and hardwood sawdust) mixed with 0.5- to 3-inch cobble to sustain porosity, along

---

* At near-neutral pH, the major dissolved carbonate species is bicarbonate (HCO_3^-) (see Chapter 3, Section 3.3.2, Figure 3.2).

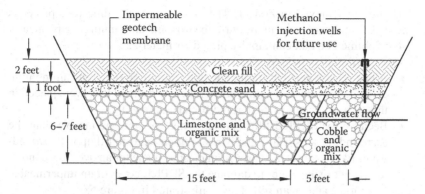

FIGURE 5.6   Sulfate-reducing permeable barrier wall for removing dissolved metals from groundwater by precipitation. The figure is a cross-section through the barrier parallel to the flow of groundwater. The vertical and perpendicular (to the page) dimensions are chosen to intercept all the contaminated plume portion of the groundwater. The thickness of the barrier parallel to the flow of groundwater is chosen to provide long-enough residence times in each zone for completion of the biological and chemical reactions occurring within them.

with a small amount of cow manure added as a source of bacteria. The barrier fill material provides readily bioavailable carbon and nitrogen for quick starting and long-term maintaining the bacterial growth.

- Within this anaerobic zone, a population of bacteria develops that consumes organic carbon and metabolically reduces sulfate to sulfide, a process that also generates alkalinity and raises the pH. The sulfide reacts with dissolved metals to precipitate them as low-solubility metal sulfides within the first zone. Because the deposition of metal sulfides (along with trapping sediment transported by the groundwater) will tend, over time, to reduce the permeability of Zone 1, it is necessary to include sufficient cobble to maintain adequate water flow.
- The effectiveness of this anaerobic zone depends on maintaining porosity while providing adequate contact time for metal sulfide precipitation. The barrier is designed so that water will remain within Zone 1 at least 10 hours for precipitation of most metal sulfides. The metals not readily precipitated in Zone 1 are ferrous iron, manganous manganese, and aluminum. These generally require longer retention times and a pH greater than is typically achieved in Zone 1, because of the acidic groundwater flow entering the barrier. Better conditions for precipitating ferrous iron, manganous manganese, and aluminum (longer retention time and pH > 6) are provided in Zone 2. By delaying precipitation of most of the iron, manganese, and aluminum until Zone 2, clogging of Zone 1 pore spaces is reduced.
- *Zone 2*: The second zone is filled with the same mixture as Zone 1, except that cobbles are replaced with crushed limestone 2 to 6 inches in size. In addition, retention time in Zone 2 is about three times longer than that in Zone 1. The limestone adds further alkalinity to the water, resulting in an additional pH increase, and precipitates iron, manganese, and aluminum

that are not removed in Zone 1. The large size of the limestone pieces is needed to maintain porosity because the iron and aluminum precipitate as a high-volume floc that can rapidly plug finer materials.

- *Some additional considerations:*
  - A fast-release nutrient starter solution containing sugar and ammonium chloride was sprayed on the barrier fill materials as they were placed in the excavation to stimulate rapid bacterial growth.
  - Because the SRPRB fill material is always more permeable than the surrounding subsurface, the barrier region tends to fill up with ground-water like a bathtub and provision must be made for an overflow conduit to prevent rising water to damage the SRPRB cover of an impermeable membrane and clean soil. This is illustrated in Figure 5.6.
  - A system of monitoring wells is required for measuring water quality in upgradient and downgradient groundwater to assess the performance of the SRPRB. Injection wells within the SRPRB permit additional liquid nutrient, such as methanol, to be added to the fill material when needed.

An SRPRB constructed as described above has operated successfully, up to the time of this publication (at least 6 years), to treat ARD at a remediated mining site, lowering groundwater metal concentrations to within state environmental regulations.

5.9.4 Case Study 4: Identifying Metal Loss and Gain Mechanisms in a Stream

(This case study is modeled after parts of a report by Balistrieri et al 1995.)

Background: Measurements of pH and metal concentrations in a stream receiving ARD indicated that the stream was acidic and had elevated levels of chloride and metals just below the confluence with the mine drainage. The metal and chloride concentrations generally decreased again farther downstream, but with small increases at some locations.

The downstream changes in metal concentrations could be due to

- Evaporation of stream water
- Dilution by higher-quality water entering the stream
- Chemical reactions such as sorption or precipitation
- Groundwater enriched in dissolved metals entering the stream

Problem: How can one evaluate which mechanisms—dilution, evaporation, new water inputs, or chemical reactions—cause changes in stream metal concentrations?

Solution: Collect samples from several selected sites on the stream, including one site above the confluence with the mine drainage, and analyze them for the metals of interest, and also a conservative constituent that is comparatively unreactive. In this case, we will use Cu and Zn as the metals of interest and chloride anion (Cl^-) as the conservative constituent. A conservative constituent, like chloride, does not adsorb or precipitate significantly; if we assume no new sources, then its concentration will change only because of dilution and water evaporation.

Take two chloride solutions with different starting concentrations of chloride and mix them in various ratios. When the chloride concentration of any of the mixtures is plotted against the mixing ratio (1:1, 2:1, 3:1, and so on), the concentration of chloride in each mixture should lie on the straight line that connects the highest and lowest chloride concentrations. This is true for any conservative constituent. In this example, flow downstream of the confluence of the stream and the mine drainage serves to mix their chloride concentrations and each location downstream represents a particular mixing ratio, where the chloride can only change by evaporation of stream water or by new water inputs (containing no chloride).

Figure 5.7 is a line diagram of a stream receiving mine drainage. Sampling sites 1 through 10 are indicated schematically. It is known that the mine discharge (sampling point 2) carries higher concentrations of dissolved copper, zinc, and chloride than are in the stream upstream at sampling point 1.

- Site 1 is in the stream above the confluence with the mine drainage.
- Site 2 is in the mine drainage just before it enters the stream.
- Site 3 is in the stream just below the confluence.
- Sites 4–10 are sequentially positioned farther downstream from the confluence.

For each sampling site, plot the dissolved metal concentration versus the chloride concentration and connect the points for sites 1 and 2, the sites having the lowest and highest chloride and metal concentrations, with a straight line, as in Figures 5.8 and 5.9. Chloride and metal concentrations are lowest at site 1, above the confluence of the stream and the mine drainage, and highest at site 2, in the mine drainage just before it flows into the stream. The straight line between these site points represents a wide range of possible dilution factors arising from the mine drainage mixing into the stream flow and from tributary drainages into the stream. This line represents the mixing curve for the conservative analyte chloride.

If a metal concentration at a site downstream of the confluence lies on the chloride mixing line (within 25% of the predicted value), the metal concentration at that site results from the same causes as the site concentration of chloride, that is, dilution or evaporation. If the point lies below the line, loss mechanisms other than dilution have contributed to the metal concentration decrease at that site. If the point lies above the line, enrichment mechanisms other than evaporation have caused a metal

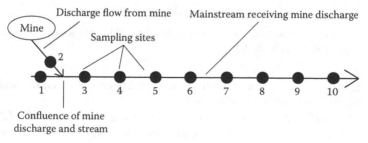

FIGURE 5.7   Line diagram of a stream with mine discharge and sampling locations.

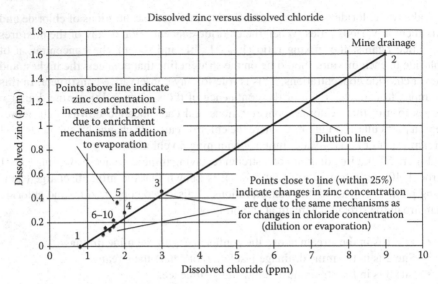

FIGURE 5.8　Plot of dissolved zinc versus dissolved chloride ratios. Numbered points indicate sampling sites in Figure 5.7.

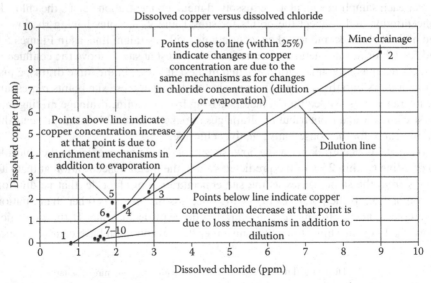

FIGURE 5.9　Plot of dissolved copper versus dissolved chloride ratios. Numbered points indicate sampling sites in Figure 5.7.

concentration increase. This will be true even if additional dilution or evaporation occurs downstream from the confluence.

For Zn, Figure 5.8 shows that all points except for site 5 lie within 25% of the line-predicted value. At site 5, Zn concentrations are about 80% higher than predicted for a conservative substance, which is probably caused by inputs of Zn from bank erosion and shallow groundwater inflows carrying dissolved zinc between site 5 and the confluence of the receiving stream and the mine drainage.

For Cu, Figure 5.9, sites 3 and 4 below the confluence show a decrease that is mainly due to dilution. As with Zn, Cu concentrations at site 5 are about 80% higher than what was predicted from simple mixing, again suggesting a source of metals between sites 4 and 5. The Cu increase is still apparent at site 6, although additional dilution has occurred. At sites 7–10, Cu concentrations are about 50–73% lower than predicted by simple mixing. This suggests that chemical removal processes such as precipitation and adsorption to sediments are effective in this region of the stream.

Sediment measurements, not indicated in Figures 5.7 through 5.9, showed that the stream sediments are enriched significantly in Cu from the mine drainage confluence all the way downstream to site 10. On the other hand, the sediments are not much enriched in Zn just below the confluence, but Zn sediment enrichment increases farther downstream below site 5. The small amount of Zn in sediments just below the confluence and enrichment of Zn in sediments farther downstream suggest that solid phase transport as stream sediment may be an important transport mechanism for Zn.

EXERCISES

1. Is a solution of ferric sulfate, $Fe_2(SO)_3$, in water expected to be acidic, basic, or neutral? Explain your answer.
2. Figure 5.2 shows that the solubilities in water of metal hydroxides decrease as pH increases up to a point. The solubility plots go through a minimum and then increase if pH continues to increase.
   a. Explain why this occurs.
   b. Would you expect a plot of metal sulfides to show similar behavior? Explain your answer.
3. A groundwater sample contains dissolved Fe^{2+}. What difficulties might be encountered in using this sample for determining the in situ pH of the groundwater?
4. In Figure 5.2, the least-soluble hydroxides, below the pH where the minimum solubility is reached, are $Fe(OH)_3$, $Al(OH)$, and $Cr(OH)_3$. These are all trivalent, whereas all the divalent and univalent hydroxides are more soluble. Discuss why this behavior is expected. The small discrepancies with $Cr(OH)_3$ and $Cu(OH)_2$ may be due to differences in cation size or to experimental uncertainty.
5. An industrial discharge permit has a limits for chromium, copper, and cadmium of 0.01 mg/L each. Assuming that the untreated discharge will always exceed these limits, what difficulties are encountered if you try to comply by using only pH adjustment? Use Figure 5.2 and assume that the only complexing anion present is OH−. How might you meet the limit for all three metals?

REFERENCES

Balistrieri, L. S., Gough, L. P., Severson, R. C., and Archuleta, A. 1995. "The Biogeochemistry of Wetlands in the San Luis Valley, Colorado: The Effects of Acid Drainage from Natural and Mine Sources." Paper presented at the Summitville Forum Proceedings, Colorado Geological Survey, Special Publication 38, edited by H. H. Posey, J. A. Pendleton, and D. Van Zyl, pp. 219–26.

Costello, C. October 2003. *Acid Mine Drainage: Innovative Treatment Technologies.* U.S. Environmental Protection Agency Technology Innovation Office. http://clu-in.org/s .focus/c/pub/i/1054/.

Duffus, J. H. 2002. "Heavy Metals a Meaningless Term?" *Pure Appl. Chem.* 74 (5): 793–807.

Eaton, A. D., Clesceri, L. S., and Greenberg, A. E. (eds.) 1995. *Standard Methods for the Examination of Water and Wastewater*, 19th ed. Washington, DC: American Public Health Association, American Water Works Association, and Water Environment Federation.

EPA. June 1996. *The Metals Translator: Guidance for Calculating a Total Recoverable Permit Limit from a Dissolved Criterion*, United States Environmental Protection Agency, Office of Water (4305), EPA 823-B-96-007. www.dep.wv.gov/WWE/permit/individual/ Documents/365_dissmetals.pdf.

EPA. 1998. *Memo from William Telliard Titled: Total vs. Total Recoverable Metals*, August 19, 1998. www.envexp.com/download/techsupport/articles/Total%20vs%20 Recoverable.pdf.

EPA. 2005. *Management and Treatment of Water from Hard-Rock Mines.* Engineering Issue.

Evanko, C. R., and Dzombak, D. A. 1997. *Remediation of Metals-Contaminated Soils and Groundwater*, Technology Evaluation Report, TE-97-01, Carnegie Mellon University, Department of Civil and Environmental Engineering, Pittsburgh, PA, October 1997.

Pankow, J. F. 1991. *Aquatic Chemistry Concepts.* Boca Raton, FL: Lewis Publishers.

Stumm, W., and Morgan, J. J. 1996. *Aquatic Chemistry*, 3rd ed. New York: John Wiley & Sons.

Wulfsburg, G. 1987. *Principles of Descriptive Inorganic Chemistry.* Belmont, CA: Brooks/ Cole–Wadsworth.

Younger, P. 2000. "The Adoption and Adaptation of Passive Treatment Technologies for Mine Waters in the United Kingdom." *Mine Water Environ.* 19: 84–97.

Modern large-scale farming has been called a method for converting petroleum into food.

Chapter 4, Section 4.2.1

6 Soil, Groundwater, and Subsurface Contamination

6.1 NATURE OF SOILS

Soil can be either an obstacle to contaminant movement or a contaminant transporter, whereas water is always a potential conveyor of contaminants. The stationary soil matrix slows the passage of groundwater and provides solid surfaces to which contaminants can sorb, delaying or stopping their movement. On the other hand, soil can also move—carried by wind, water flow, and construction equipment. Movement of soil, like movement of water, transports the contaminants it carries. Predicting and controlling pollutant behavior in the environment require understanding how soil, water, and contaminants interact, which is the subject of this chapter.

6.1.1 SOIL FORMATION

Soil is the weathered and fragmented outer layer of the earth's solid surface, initially formed from the original rocks and then amended by growth and decay of plants and organisms. The initial step of conversion of rock to soil is destructive "weathering." Weathering is the disintegration and decomposition of rocks by natural physical, chemical, and biological processes.

223

6.1.1.1 Physical Weathering

Physical weathering causes fragmentation of rocks, which results in an increase in the exposed surface area and, thereby, the potential for further, more rapid, weathering. Common causes of physical weathering are the following:

- Expansion and contraction caused by environmental heating and cooling.
- Stress forces caused by mineral crystal growth and the expansion and contraction of water when it freezes and melts in cracks and pores.
- Penetration of tree and plant roots.
- Scouring and grinding by abrasive particles carried by wind, water, and moving ice.
- Unloading forces that arise when rock-confining pressures are lessened by geologic uplift, erosion, or changes in fluid pressures. Unloading can cause cracks at thousands of feet below the surface.

6.1.1.2 Chemical and Biological Weathering

Chemical and biological weathering of rocks causes changes in their mineral composition. Common causes of chemical weathering are the following:

- Hydrolysis and hydration reactions (water reacting with minerals)
- Oxidation (usually by oxygen in the atmosphere and in water) and reduction (usually by microbes) reactions
- Dissolution and dissociation of minerals
- Precipitation reactions within crystals and at their boundaries (e.g., the formation of solid oxides, hydroxides, carbonates, and sulfides) that cause stresses because of volume changes
- Loss of mineral components by leaching and volatilization
- Chemical exchange processes, such as cation exchange

Physical, chemical, and biological weathering processes often produce loose materials that can be deposited elsewhere after being transported by wind (aeolian deposits), running water (alluvial deposits), or glaciers (glacial deposits).

6.1.1.3 Secondary Mineral Formation

The next steps, after the rocks have been fractured and broken down, are the formation of secondary minerals (e.g., clays, mineral precipitates, etc.) and changes caused by plants and microorganisms. Secondary minerals are formed within the soil by chemical reactions of the primary (original) minerals. The reactions forming secondary minerals are always in the direction of greater chemical stability under local environmental conditions. These reactions are facilitated by the presence of water, which dissolves and mobilizes different components of the original rocks, allowing them to react to form new compounds.

6.1.1.4 Roles of Plants and Soil Organisms

Plants and soil organisms play many complex roles. Roots form extensive networks permeating soil, which can exert pressure that compresses aggregates in one location and separates them in another. Water uptake by roots causes differential dehydration in soil, initiating soil shrinkage and opening of many small cracks.

The plant root zone in the soil is called the rhizosphere, which is the soil region where plants, microbes, and other soil organisms interact. Soil organisms include thousands of species of bacteria, fungi, actinomycetes, worms, slugs, insects, mites, and so on. The number of organisms in the rhizosphere can be 100 times larger than that in non-rhizosphere soil zones. Root secretions and dead roots promote microbial activity that produces humic cements. Root secretions contain various sugars and aliphatic, aromatic, and amino acids, as well as mucigel, a gelatinous substance that lubricates root penetration. These substances and dead root material are the nutrients for rhizosphere microorganisms. The root structure itself provides a surface area for microbial colonization.

6.2 SOIL PROFILES

A vertical profile through soil, Figure 6.1, tells much about how the soil was formed. It usually consists of a succession of more or less distinct layers, or strata. The layers can form from aeolian or alluvial deposition of material, or from in situ weathering processes.

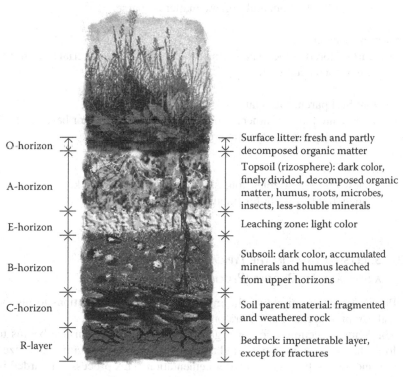

O-horizon — Surface litter: fresh and partly decomposed organic matter

A-horizon — Topsoil (rizosphere): dark color, finely divided, decomposed organic matter, humus, roots, microbes, insects, less-soluble minerals

E-horizon — Leaching zone: light color

B-horizon — Subsoil: dark color, accumulated minerals and humus leached from upper horizons

C-horizon — Soil parent material: fragmented and weathered rock

R-layer — Bedrock: impenetrable layer, except for fractures

FIGURE 6.1 Generalized soil profile showing the horizon sequence.

6.2.1 SOIL HORIZONS

When soil layers develop in situ by the weathering processes described previously, they form a sequence called horizons. The horizons are designated by the U.S. Department of Agriculture by the capital letters O, A, E, B, C, and R, in order of the farthest distance from the surface (see Figure 6.1).

O-horizon: Organic
- The top horizon; starts at the soil surface.
- Formed from surface litter.
- Dominated by fresh or partly decomposed organic matter.

A-horizon: Topsoil
- The zone of greatest biological activity (rhizosphere).
- Contains an accumulation of finely divided, decomposed, organic matter, which imparts a dark color.
- Clays, carbonates, and most metal cations are leached out of the A-horizon by downward percolating water; less-soluble minerals (such as quartz) of sand or silt size become concentrated in the A-horizon.

E-horizon: Leaching zone (sometimes called the A-2 horizon)
- Light-colored region below the rhizosphere where clays and metal cations are leached out and organic matter is sparse.

B-horizon: Subsoil
- Dark-colored zone where downward migrating materials from the A-horizon accumulate.

C-horizon: Soil parent material
- Fragmented and weathered rock, either from bedrock or base material that has deposited from water or wind.

R-layer: Bedrock
- Below all the horizons; consists of consolidated bedrock.
- Impenetrable, except for fractures.

6.2.2 SUCCESSIVE STEPS IN THE TYPICAL DEVELOPMENT OF A SOIL AND ITS PROFILE (PEDOGENESIS)

1. Physical disintegration (weathering) of exposed rock formations forms the soil parent material, the C-horizon.
2. The gradual accumulation of organic residues near the surface begins to form the A-horizon, which might acquire a granular structure, stabilized to some degree by organic matter cementation. This process is retarded in desert regions where organic growth and decay are slow.

3. Continued chemical weathering (oxidation, hydrolysis, etc.), dissolution, and precipitation begin to form clays.
4. Clays, soluble salts, chelated metals, and so on, migrate downward through the A-horizon, carried by permeating water, to accumulate in the B-horizon.
5. The C-horizon, now below the O-, A-, and B-horizons, continues to undergo physical and chemical weathering, slowly transforming into B- and A-horizons, deepening the entire horizon structure.
6. A quasi-stable condition is approached in which the opposing processes of soil formation and soil erosion are more or less balanced.

6.3   ORGANIC MATTER IN SOIL

Soil organic matter influences the weathering of minerals, provides food for soil organisms, and provides sites to which ions are attracted for ion exchange. Only two types of organisms can synthesize organic matter from nonorganic materials: certain bacteria called autotrophs and chlorophyll-containing plants. Organic matter is developed in soil by the metabolism of wastes and decay products of plants and soil organisms. For example, the metabolism of soil fungi produces excellent complexing agents such as oxalate ion and chelating organic acids like citric acid. These promote the dissolution of minerals and increase nutrient availability. Some soil bacteria release the strong organic chelating agent 2-ketogluconic acid, which reacts with insoluble metal phosphates to solubilize the metal ions and release soluble phosphate, a plant nutrient.

Another example of the role played by soil organic matter is the oxalate ion metabolism of certain soil organisms. In calcium soils, oxalate forms calcium oxalate, $Ca(C_2O_4)$, which then reacts with precipitated metals (particularly Fe or Al) to complex and mobilize them. The reaction with precipitated aluminum is

$$3H^+ + Al(OH)_3(s) + 2Ca(C_2O_4) \rightarrow Al(C_2O_4)_2^-(aq) + 2Ca^{2+}(aq) + 3H_2O \quad (6.1)$$

Because hydrogen ions are consumed, this reaction reduces acidity and raises the pH of acidic soil. It also weathers minerals by dissolving some metals and provides Ca^{2+} as a plant and biota nutrient. Similar processes with silicate minerals release K^+ and other nutrient cations.

The amount of organic matter in soil has a strong influence on soil properties and on the behavior of soil contaminants. For example, plants compete with soil for water. In sandy soils, the pore space is large and the particle surface area is small. Water is not strongly adsorbed to sands and is easily available to plants. However, in sandy soils the water drains off quickly. On the other hand, water binds strongly to the organic matter in soil, therefore soils with high organic content hold more water, but the water is less available to plants.

RULE OF THUMB

The amount of organic matter is typically less than 5% in most soils, but is critical for the plant productivity. Peat soils can have 95% organic matter. Mineral soils can have less than 1% organic matter.

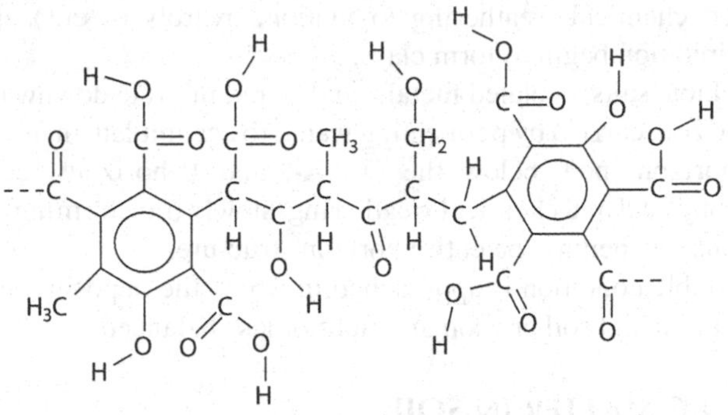

FIGURE 6.2 Characteristic structural portion of an unionized humic or fulvic acid.

6.3.1 HUMIC SUBSTANCES

The most important organic substance in soil is humus, a collection of variously sized polymeric molecules consisting of soluble fractions (humic and fulvic acids)* and an insoluble fraction (humin). Humus is the near-final residue of plant biodegradation and consists largely of protein and lignin. Humus is the remains of the degradation of the more easily degradable components of plant biomass, leaving only the parts most resistant to further degradation. Humic materials are not well defined chemically and have variable composition. Percent by weight for the most abundant elements are C: 45–55%, O: 30–45%, H: 3–6%, N: 1–5%, and S: 0–1%. The exact chemical structure depends on the source plant materials and the history of biodegradation.

Humic and fulvic acids are soluble (under normal environmental conditions) organic acid macromolecules containing many –COOH and –OH functional groups that partially ionize in water, acting as weak dibasic and tribasic acids, releasing H+ ions and providing negative charge centers on the macromolecule to which cations are strongly attracted (see Figure 6.2). Humic materials are the most important class of natural soil complexing agents and are found where vegetation has decayed.

6.3.2 SOME PROPERTIES OF HUMIC MATERIALS

6.3.2.1 Binding to Dissolved Species

Humic materials are effective in removing metals from water by sorption to negative charge sites, mainly at the structural oxygen atoms. Polyvalent metal cations are sorbed especially strongly. The cation-exchange capacity of humic materials can

---

* The difference between a humic and a fulvic acid is an operational one: they can be separated from one another by lowering the solution pH to 1 with hydrochloric acid. This precipitates the humic acids and leaves fulvic acids in solution. Although average molecular weights of humic compounds vary widely (between ~20,000–3,000) depending on the origin of the soil, fulvic acids have lower molecular weights and are somewhat more acidic than humic acids (Shinozuka et al. 2004).

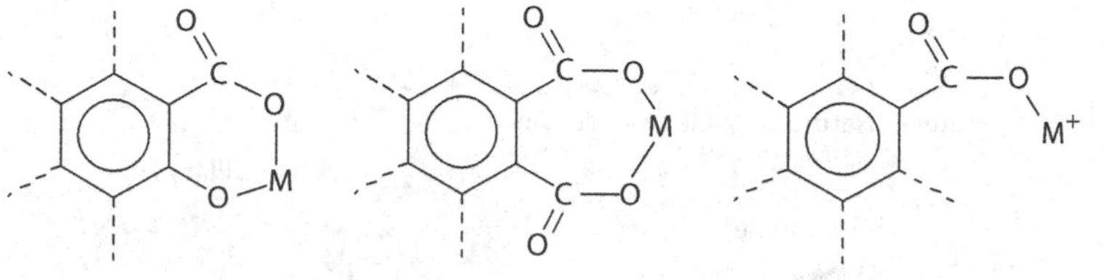

Chelation between
carboxyl and hydroxyl
groups

Chelation between
two carboxyl groups

Complexation with
carboxyl group

FIGURE 6.3   Several ways that metal ions (M^{2+}) can bind to humic and fulvic acids.

be as high as 500 meq/100 mL. Humic materials may sorb metals like uranium in concentrations 10,000 times greater than adjacent water. Humic materials also bind organic pollutants, especially less-soluble compounds like DDT and atrazine. Much of the utility of wetlands for water treatment arises from their high concentrations of humic materials. Figure 6.3 shows several ways by which metal cations bind to humic and fulvic acids.

6.3.2.2   Light Absorption

Humic materials absorb sunlight in the blue region (transmitting yellow) and can transfer absorbed solar energy to sorbed molecules, initiating reactions. This energy transfer process can be effective in degrading pesticides and other organic compounds.

6.4   SOIL ZONES

In discussions of groundwater movement, the soil subsurface is commonly divided into three zones, based on their air and water content (see Figure 6.4). In the region from the ground surface down to an aquifer water table, soils contain mostly air in the pore spaces, with some adsorbed and capillary-held water, which is called the water-unsaturated zone or vadose* zone. In the region from the top of the water table to bedrock, soils contain mostly water in the pore spaces, which is called the saturated zone (or phreatic zone).

Between the vadose and saturated zones, there is a transition region called the capillary zone, where water is drawn upward from the water table by capillary forces. The thickness of the capillary zone depends on the soil texture—the smaller the pore size, the greater the capillary rise. In fine gravel (2–5 mm grain size), the capillary zone will be of the order of 2.5 cm thick. In fine silt (0.02–0.05 mm grain size), the capillary zone can be 200 cm or greater.

The saturated zone lies above the solid bedrock, which is impermeable except for fractures and cracks. The region of the subsurface overlying the bedrock is generally an unconsolidated, porous, granular mineral material.

---

* From the Latin *vadosus*, meaning shallow.

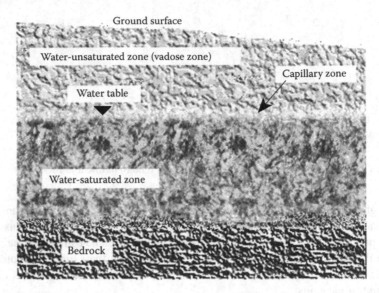

FIGURE 6.4   Soil zones and groundwater in the subsurface region.

6.4.1 AIR IN SOIL

Air in soil has a different composition from atmospheric air, mainly because of bio-degradation of organic matter by soil organisms, plant respiration* in the root zone, and the slow rate of exchange between air in the atmosphere and air in subsurface pore spaces. Biodegradation occurs in many small steps, but the net overall reaction is shown in Equation 6.2, where organic matter in soil is represented by the approximate generic unit formula $\{CH_2O\}$. An actual molecule of soil organic matter would have a formula that is approximately some whole number multiple of the $\{CH_2O\}$ unit.

$$\{CH_2O\} + O_2 \rightarrow CO_2 + H_2O \tag{6.2}$$

Equation 6.2 shows that, for each $\{CH_2O\}$ unit contained within a larger organic molecule, one CO_2 molecule and one H_2O molecule are produced by biodegrada-tion. Oxygen from soil pore space air is consumed and CO_2 is released by microbial metabolism and plant root zone respiration.

Much of the soil air is semi-trapped in pores and cannot readily equilibrate with the atmosphere. As a result, the O_2 content in soil pore space air is decreased from its atmospheric value of 21% to about 15% and CO_2 content is increased from its atmospheric value of about 0.03% to about 3%. This, in turn, increases the dissolved CO_2 concentration in groundwater, making it more acidic. Acidic groundwater con-tributes to the weathering of soils, especially calcium carbonate ($CaCO_3$) minerals.

---

* Plant photosynthesis removes carbon dioxide from the atmosphere (or water for aquatic plants) and uses it to build carbon-based plant structural matter, releasing oxygen. Plant respiration (like animal respiration) takes oxygen from the atmosphere (or water for aquatic plants) to generate the energy needed, releasing carbon dioxide and water.

When soil becomes water-saturated, as in the saturated zone, many changes occur:

1. Oxygen becomes used up by respiration of microorganisms.
2. Anaerobic processes lower the oxidation potential of water so that reducing conditions (electron gain) prevail, whereas oxidizing conditions (electron loss) dominate in the unsaturated zone.
3. Certain metals, particularly iron and manganese, become mobilized by chemical reduction reactions (Equations 6.3 through 6.5) that change them from insoluble to soluble forms (see Chapter 4):

$$Fe(OH)_3(s) + 3H^+ + e^- \rightleftarrows Fe^{+2}(aq) + 3H_2O \qquad (6.3)$$

$$Fe_2O_3(s) + 6H^+ + 2e^- \rightleftarrows 2Fe^{+2}(aq) + 3H_2O \qquad (6.4)$$

$$MnO_2(s) + 5H^+ + 2e^- \rightleftarrows Mn^{+2}(aq) + 2H_2O \qquad (6.5)$$

Groundwater, moving under gravity and pressure gradients, can transport dissolved Fe^{2+} and Mn^{2+} into zones where oxidizing conditions prevail, for example, by surfacing to a spring, stream, or lake. There, Equations 6.3 through 6.5 are reversed and the metals redeposit as solid precipitates, mainly $Fe(OH)_3$ and MnO_2. Precipitation of $Fe(OH)_3$ often causes "red water" and red or yellow deposits on rocks and soil. MnO_2 deposits are black. These deposits can clog underdrains in fields and water treatment filters.

6.5 CONTAMINANTS BECOME DISTRIBUTED IN WATER, SOIL, AND AIR

In the environment, contaminants always contact air, water, and soil, which comprise the three physical phases of gas, liquid, and solid. No matter where it originates, a contaminant moves across the interfaces between water, soil, and air phases to become distributed, to different degrees, into every phase it contacts. Partitioning of a pollutant from one phase into other phases serves to deplete the concentration in the original phase and increase it in the other phases. The movement of contaminants through soil is a process of continuous redistribution among the different phases it encounters. It is a process controlled by gravity, capillarity, sorption to surfaces, colloid movement, and volatility.

6.5.1 VOLATILIZATION

The main partitioning process from liquids and solids to air is volatilization, which moves a contaminant across the liquid–air or solid–air interface into the atmosphere or into air in soil pore spaces. Volatilization is an important partitioning mechanism for compounds with high vapor pressures. For liquid mixtures such as gasoline, the most volatile components are lost first, causing the composition and properties of the remaining liquid mixture to change over time. For example, the most volatile components of gasoline are generally the smallest molecules in the mixture. The remaining larger molecules have stronger London attractive forces. Hence, as gasoline weathers and loses the smaller molecules by volatilization, its vapor pressure decreases and its viscosity and density increase.

Dissolved ionic molecules are never volatile because of strong attractions, both electrostatic and London, to polar water molecules. Consequently, there is negligible partitioning of ionic pollutants (e.g., dissolved NH_4^+, HS^-, NH_3^-, metal cations, etc.) from water to air.

6.5.2 SORPTION

The main partitioning process from liquids and air to solids is sorption, which moves a contaminant across the liquid–solid or air–solid interface to organic or mineral solid surfaces. For nonionic molecules (e.g., benzene, DDT, dioxin, perchloroethylene, pyrene, etc.), sorption from the water phase to solid surfaces is generally inversely proportional to their solubility, being greatest for compounds of low solubility. Soluble ionic molecules can be an exception to this principle in the presence of surfaces with surface charges, like humic substances (Section 6.3.1) and clays. The binding of dissolved ionic molecules to charged surface sites is the basis of ion exchange and some membrane techniques (Chapter 3, Section 3.11.6.7) for removing dissolved metals and other ionic contaminants from water. When a molecule becomes sorbed to a surface, its electron shell structure is changed. Therefore, a contaminant sorbed to a surface will undergo chemical and biological transformations at different rates and by different pathways than if it were dissolved. This is the basis for many devices used to catalyze chemical reactions, such as automobile catalytic converters.

6.6 PARTITION COEFFICIENTS

The tendency for a nonionic pollutant to move from one phase to another is often quantified by the use of a partition coefficient, also called a distribution coefficient. Partition coefficients are chemical specific, and they can be measured directly or, in some cases, estimated from other properties of the chemical. The simplest form of a partition coefficient is the ratio of the pollutant concentration in phase 1 to its concentration in phase 2:

$$K_{1,2} = \frac{\text{concentration in phase 1}}{\text{concentration in phase 2}} = \frac{C_1}{C_2} \tag{6.6}$$

This expression assumes that a linear relation exists between the concentrations of a substance in different phases, and is often satisfactory for low to moderate concentrations. The phase of the denominator, C_2, is referred to as the reference phase. In environmental work, water is commonly used as the reference phase to maintain some consistency in published values.

Using water as the reference phase, a linear relation yields Equations 6.7 through 6.9 for partitioning between water and air, water and free product (original physical form of the pollutant, such as a layer of oil floating on a river or above the groundwater table; sometimes called the bulk pollutant), and water and soil. In these equations, the pollutant may be a pure substance, like benzene, or one component from a mixture, like benzene from free product liquid gasoline. Each value of K depends on properties of the particular pollutant and the temperature. K_d also depends on the type of soil.

Air–water partition coefficient:

$$C_a = K_H C_w \tag{6.7}$$

K_H is the air–water partition coefficient (also known as Henry's law constant). C_a and C_w are the pollutant concentrations in air and water, respectively.

Free product–water partition coefficient:

$$C_p = K_p C_w \tag{6.8}$$

K_p is the free product–water partition coefficient.
C_p and C_w are the pollutant concentrations in the free product and water, respectively.

Soil–water partition coefficient:

$$C_s = K_d C_w \tag{6.9}$$

K_d is the soil–water partition coefficient.
C_s and C_w are the pollutant concentrations sorbed on soil and dissolved in water, respectively.

6.6.1 AIR–WATER PARTITION COEFFICIENT (HENRY'S LAW)

Henry's law, $C_a = K_H C_w$, describes how a substance distributes itself at equilibrium between water and air;* a larger Henry's law constant indicates a greater tendency to volatilize from water. The units of Henry's law constant, $K_H = C_a/C_w$, depend on the units that are used to express concentrations in air and water. For the case of oxygen gas, O_2, at 20°C:

- When air and water concentrations both have the same units,

$$K_H(O_2, 20°C) = 26 \text{ (unitless)} \tag{6.10}$$

---

* Henry's law constants are tabulated in many references, such as *Handbook of Chemistry and Physics*, CRC Press; Lide (1991 and later); Lyman et al. (1990); Mackay and Shiu (1981); and Sander (1999). There are also computer programs that calculate Henry's law constant from other chemical properties.

- For water concentration in mol/L or mol/m³, and air in atmospheres,

$$K_H(O_2, 20°C) = 635\,L \cdot atm/mol = 0.635\,m^3 \cdot atm/mol \qquad (6.11)$$

- For water concentration in mg/L and air in atmospheres,

$$K_H(O_2, 20°C) = 0.0198\,L \cdot atm/mg \qquad (6.12)$$

Example 1: Estimating Some Relative Physical Properties

Suppose you need to compare the relative vapor pressures, water solubilities, and Henry's law constants of the compounds tabulated below, but are only able to find melting point data. Estimate the relative values of these parameters based on their melting points and structures.

- Vapor pressure (P_v) is a measure of the tendency for molecules to partition from a pure substance into the atmosphere.
- Solubility (S_w) is a measure of the tendency of molecules to partition from a pure substance into water.
- Henry's law constant (K_H) is a measure of a compound's tendency to partition between water and air. It may be considered the vapor pressure of a substance when it is dissolved in water.

Compound	Structure
Phenol: Melting temperature = 41.0°C	OH (structure)
1,2,3,5-Tetrachlorobenzene: Melting temperature = 54.5°C	(structure)
1,2,3,5-Tetrachlorobenzene: Melting temperature = 140°C	(structure)

DEFINITIONS

The *vapor pressure* of a substance varies inversely with intermolecular attraction. Substances with high vapor pressure have weak intermolecular attractive forces.

Therefore, vapor pressure will tend to vary inversely with melting point, because a high melting point indicates strong intermolecular attractive forces.*

The *solubility* of a substance varies with polarity and the ability to form hydrogen bonds to water molecules. The more polar the molecules of a substance and the more hydrogen bonding to water, the more soluble it will be, because of stronger attraction to water molecules. It also varies with molecular weight. Higher molecular weight tends to decrease a compound's solubility because London attractive forces are stronger, attracting the compound molecules to one another more strongly.

The *Henry's law constant* of a substance depends on two different properties. It varies inversely with solubility and directly with vapor pressure. In general, solubility will dominate in determining K_H.

ANSWER

Vapor pressure: (Lowest to highest vapor pressure will be, to a first approximation, in the order of highest to lowest melting point.)

> 1, 2, 4, 5-tetrachlorobenzene < 1, 2, 3, 5-tetrachlorobenzene < phenol

Solubility: (Lowest to highest solubility will be in the order of lowest to highest polarity, highest to lowest molecular weight, and fewest to most hydrogen bonds to water.)

Least soluble is 1,2,4,5-tetrachlorobenzene (nonpolar because of symmetry; high molecular weight, no hydrogen bonds).

More soluble is 1,2,3,5-tetrachlorobenzene (less symmetrical, therefore more polar; same molecular weight as 1,2,4,5-tetrachlorobenzene, no hydrogen bonds).

Most soluble is phenol (most polar; the only compound that can form hydrogen bonds to water; lowest molecular weight of all).

Henry's law constant: Lowest to highest K_H should vary according to highest to lowest solubility and lowest to highest vapor pressure. Phenol, with the highest polarity, greatest solubility, and lowest melting point, will have the lowest K_H. Because the solubilities of 1,2,3,5-tetrachlorobenzene and 1,2,4,5-tetrachlorobenzene are so similar, their K_{HS} are expected to be similar also. However, their vapor pressures differ by a factor of more than 10. The much higher vapor pressure of 1,2,3,5-tetrachlorobenzene might give it a slightly higher K_H than 1,2,4,5-tetrachlorobenzene.

> Phenol < 1, 2, 4, 5-tetrachlorobenzene ≤ 1, 2, 3, 5-tetrachlorobenzene

The measured values in Table 6.1 confirm these relative vapor pressure, solubility, and K_H estimates.

---

* The correspondence between melting point and vapor pressure is only approximate, because the melting point may also depend on the crystal lattice energy, a function of the molecular geometry of the solid. However, it can serve as a first approximation where more accurate data are not available.

TABLE 6.1
Measured Values for Melting Point, Vapor Pressure, Solubility, and Henry's Law Constant

Compound	MW (g)	T_m (°C)	P_v (atm)	S_w (mg/L)	K_H (atm m³/mol)
Phenol	94.1	41.0	2.6×10^{-4}	82,000	4.0×10^{-7}
1,2,3,5-Tetrachlorobenzene	215.9	54.5	1.4×10^{-5}	3.5	5.8×10^{-3}
1,2,4,5-Tetrachlorobenzene	215.9	140.0	5.3×10^{-6}	2.2	2.6×10^{-3}

Note: T_m = melting point; P_v = vapor pressure; S_w = solubility in water; K_H = Henry's law constant.

Example 2

Rank the four compounds below in the order of increasing Henry's law constant (tendency to partition from water into air).

I	II	III	IV
1-chloro-3-methylbenzene	3-methylphenol (m-cresol)	3-methylaniline	3-methylanisole

ANSWER

All the compounds are similar in molecular weight and their structures differ only in the top functional group. The molecule having the group with the weakest attractive force to water will most readily partition from water into air. Therefore, we want to rank them by their relative attractions to water, that is, solubilities.

  I. Has no oxygen, nitrogen, or fluorine for H-bonding to water and is the least polar. It will volatilize the most readily.
  IV. Has an oxygen, but no hydrogen is attached to it for H-bonding to an oxygen on a water molecule, although a hydrogen on a water molecule can H-bond to the oxygen on IV. It will be next in volatility from water to air.
  II. & III. Both can H-bond. II can form one H-bond, while III can form two H-bonds. In addition, hydrogens on water molecules can H-bond to the

oxygen on II and the nitrogen on III. Therefore, II is third in volatility and III is the least volatile in water.

Order of Henry's law constants: I > IV > II > III

Their measured Henry's law constants (dimensionless) are the following:

$$I, K_H = 1.38; IV, K_H = 1.17; II, K_H = 0.0034; III, K_H = 0.0015$$

A larger Henry's law constant means greater tendency to volatilize from water (see Section 6.6.1).

Example 3

If soil pore water is measured to contain 3.2 mg/L of oxygen at 20°C, what is the concentration, in mg/L and in atmospheres, of oxygen in the air of the soil pore space?

ANSWER

$$\text{For } O_2 \text{ at } 20°C, K_H = 26 \text{ (Equation 5.10)} = \frac{C_a}{3.2\,\text{mg/L}};$$

$$C_a = (26)(3.2\,\text{mg/L}) = 83.2\,\text{mg/L}$$

With different units,

$$K_H = 0.0198\,\text{L} \cdot \text{atm/mg (Equation 5.12)} = \frac{C_a}{3.2\,\text{mg/L}}$$

$$C_a = (0.0198\,\text{L} \cdot \text{atm/mg})(3.2\,\text{mg/L}) = 0.063\,\text{atm}$$

Since the normal atmospheric partial pressure of oxygen at sea level is about 0.2 atm, this result, showing that the soil oxygen concentration is lower than atmospheric concentration, indicates the presence of microbial activity in the soil that has consumed oxygen.

Example 4: BOD and Henry's Law

A certain sewage treatment plant located on a river typically removes 100,000 lb $(4.54 \times 10^7 \text{ g})$ of biodegradable organic waste each day. If there were a plant upset and it became necessary to release one day's waste into the receiving river, how many liters of river water could potentially be contaminated to the extent of totally depleting the water of all oxygen?

ANSWER

An approximate chemical equation suitable for biodegradation of organic matter is Equation 6.2 from Section 6.4.1:

$$\{CH_2O\} + O_2 \rightarrow CO_2 + H_2O$$

Assume that the river water is initially saturated with oxygen from the air at 20°C and that, after the spill, no additional oxygen dissolves from the atmosphere, a worst-case scenario.

NECESSARY DATA

Atm. pressure at the treatment plant = 0.82 atm

Vapor pressure of water at 20°C = 0.023 atm

Percentage of O_2 in dry air = 21%

From Equation 6.11, $K_h(O_2) = 635 \cdot$ atm/mol

CALCULATION

Organic matter is biodegraded, consuming oxygen according to Equation 6.2, which shows that one mole of O_2 is consumed for each mole of CH_2O (assumed to be the major structural unit of organic matter) biodegraded. The molecular weight of CH_2O is 30 g/mol. Therefore,

$$\text{Moles of } CH_2O \text{ in sewage} = \frac{4.54 \times 10^7 \text{ g}}{30 \text{ g/mol}} = 1.5 \times 10^6 \text{ mol}$$

$$= \text{moles of } O_2 \text{ consumed}$$

$$\text{Atmospheric pressure} = P_{total} = 0.82 \text{ atm} = P_{O_2} + P_{N_2} + P_{H_2O}$$

$$P_{dry\ air} = \text{atmospheric pressure} - \text{partial}$$

$$\text{pressure of water vapor}$$

$$= P_{total} - P_{H_2O}$$

$$\text{Therefore, } P_{O_2} = (0.21)(P_{total} - P_{H_2O})$$

$$= (0.21)(0.82 \text{ atm} - 0.023 \text{ atm})$$

$$= 0.17 \text{ atm}$$

Use Henry's law to find the concentration of dissolved O_2 in the river.

$$K_H(O_2) = 635 L \cdot \text{atm/mol} = \frac{C_a}{C_w} = \frac{0.17 \text{ atm}}{C_w}$$

$$C_w = [O_2(aq)] = \frac{0.17 \text{ atm}}{635 \text{ atm} \cdot \text{L/mol}} = 2.7 \times 10^{-4} \text{ mol/L}$$

or $[O_2(aq)] = (2.7 \times 10^{-4} \text{ mol/L}) \times (32 \text{ g / mol}) = 8.6 \text{ mg/L}$

In saturated water at 20°C and 0.82 atm total pressure, $[O_2, aq] = 2.7 \times 10^{-4}$ mol/L

$$\text{Liters of river water depleted of } O_2 = \frac{1.5 \times 10^6 \text{ mol } O_2 \text{ consumed}}{2.7 \times 10^{-4} \text{ mol } O_2/L \text{ in river}} = 5.6 \times 10^9 L$$

This is a worst-case scenario. Less oxygen than calculated would be lost from the river because of continuous replenishing of oxygen to the river by partitioning from the atmosphere. The rate at which this occurs would depend on several factors, such as the surface-to-volume ratio of the river, water turbulence and cascades, wind velocity, and temperature.

Note that both vapor pressure (proportional to C_a) and solubility (proportional to C_w) of a pure solid or liquid generally increase with temperature, but vapor pressure always increases faster. Therefore, the value of K_H increases with temperature, indicating that, for a gas partitioning between air and water, the atmospheric portion increases and the dissolved portion decreases when the temperature rises. This is consistent with the observation that the water solubility of gases decreases with increasing temperature.

RULE OF THUMB

If a tabulated value for K_H cannot be found, it may be estimated roughly by dividing the vapor pressure of a compound by its aqueous solubility. For some compounds, tabulated values of vapor pressure and solubility may be easier to find than K_H values.

$$K_H = \frac{C_a \,(\text{partial pressure in atmospheres})}{C_w \,(\text{mol/L})}$$

$$\approx \frac{\text{vapor pressure (atm)}}{\text{aqueous solubility (mol/L)}} \tag{6.13}$$

In this case, the units of K_H are $\text{atm} \cdot \text{L mol}^{-1}$

Example 5

Estimate Henry's law constants for chlorobenzene and bromomethane using vapor pressure and aqueous solubility.

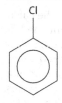

Chlorobenzene: P_v (25°C) – 1.6 × 10^{-7} atm; C_w (25°C) – 4.5 × 10^{-3} mol/L
K_H (25°C) ≈ P_v/C_w = 1.6 × 10^{-2} atm/4.5 × 10^{-3} mol/L = 3.6 L·atm/mol
This happens to exactly match the experimental value. To put K_H into a dimensionless form, divide by RT (R = universal gas constant, T = temperature in degrees Kelvin), equivalent to multiplying by 0.0414 mol/atm:

$$K_H' = \frac{3.6\,\text{L} \cdot \text{atm/mol}}{(0.0821\,\text{L} \cdot \text{atm/mol} \cdot \text{K})(298\,\text{K})} = 0.15$$

Bromomethane: P_v (liq, 25°C) = 1.8 atm; C_w (1 atm, 25°C) = 0.16 mol/L.

At 25°C, bromomethane has a vapor pressure >1 atm, so it is a gas. Since the solubility is given at 1 atm partial pressure, all gases should be used at a vapor pressure of 1 atm.
K_H (25°C) ≈ P_v/C_w = 1 atm/0.16 mol/L = 6.3 L·atm/mol

$$K_H' = \frac{6.3\,\text{L} \cdot \text{atm/mol}}{(0.0821\,\text{L} \cdot \text{atm} / \text{mol} \cdot \text{K})(298\,\text{K})} = 0.26$$

Example 6

Estimate relative Henry's law constants for the compounds in Example 1 using vapor pressure and solubility data from Table 6.1. Compare your answer with the measured K_H values in Table 6.1.

ANSWER

From Equation 6.13, the Henry's law constant varies directly with vapor pressure and inversely with aqueous solubility.

$$K_H \approx \frac{\text{vapor pressure (atm)}}{\text{aqueous solubility (mol/L)}}$$

Use vapor pressure and solubility data from Table 6.1 to estimate approximate Henry's law values.

$$K_H(\text{phenol}) \approx \frac{2.6 \times 10^{-4}\ \text{atm}}{8.2 \times 10^4\ \text{mg/L}} \times \frac{94.1\,\text{g/mol}}{10^{-3}\,\text{g/mg}} = 3.0 \times 10^{-4}\ \text{L} \cdot \text{atm/mol}$$

$$K_H(1,2,3,5\text{-tetrachlorobenzene}) \approx \frac{1.4 \times 10^{-5}\ \text{atm}}{3.5\,\text{mg/L}} \times \frac{215.9\,\text{g/mol}}{10^{-3}\,\text{g/mg}} = 0.86\,\text{L} \cdot \text{atm/mol}$$

$$K_H(1,2,4,5\text{-tetrachlorobenzene}) \approx \frac{5.3 \times 10^{-6}\ \text{atm}}{2.2\,\text{mg/L}} \times \frac{215.9\,\text{g/mol}}{10^{-3}\,\text{g/mg}} = 0.52\,\text{L} \cdot \text{atm/mol}$$

$K_H(1,2,3,5\text{-tetrachlorobenzene}) > K_H(1,2,4,5\text{-tetrachlorobenzene}) \gg K_H(\text{phenol})$

Table values*

$$K_H(\text{phenol}) = 4.0 \times 10^{-4}\,\text{L} \cdot \text{atm/mol}$$
$$K_H(1,2,3,5\text{-tetrachlorobenzene}) = 5.8\,\text{L} \cdot \text{atm/mol}$$
$$K_H(1,2,4,5\text{-tetrachlorobenzene}) = 2.6\,\text{L} \cdot \text{atm/mol}$$

The calculated value for phenol is very close to the measured value. The calculated values for the chlorobenzenes are in the correct order but are about one-fifth too small. It cannot be determined why the calculated values for 1,2,3,5-tetrachlorobenzene and 1,2,4,5-tetrachlorobenzene differ from the measured values without examining the original experimental data. It might be due to experimental error based on the difficulty of accurately measuring such low vapor pressures. This could also explain why the Henry's law constant of the polar 1,2,3,5-tetrachlorobenzene appears to be higher than that of the nonpolar 1,2,4,5-tetrachlorobenzene. The absolute difference in the constants is small and could be accounted for by the limits of error in the vapor pressure measurements, where an uncertainty factor of 3 or 4 is not unusual in tabulated values of small vapor pressures.

It might seem counterintuitive that phenol, the compound with the highest vapor pressure in the pure state, has the lowest Henry's law constant. This example shows the importance of the environment immediately surrounding the molecule.

---

* Reference: *Canadian Water Quality Guidelines*, Task Force on Water Quality Guidelines of the Canadian Council of Resource and Environment Ministers, March 1987.

Phenol is much more strongly attracted to water molecules than to other phenol molecules. Consequently, it enters the vapor state more readily from pure liquid phenol than from a water solution.

6.6.2 SOIL–WATER PARTITION COEFFICIENT AND THE FREUNDLICH ISOTHERM

Partitioning from water into soil surfaces can be limited by the available soil surface area. The rate of transfer will slow as the soil surface becomes saturated. For this reason, the partitioning of a compound between water and soil may deviate from linearity. This is particularly true for the partitioning of organic compounds. To account for nonlinearity, the corresponding partition coefficient is often written in a modified form called the Freundlich isotherm. The modification consists in introducing an empirically determined exponent to the C_w term.

$$C_s = K_d C_w^n \tag{6.14}$$

where
C_s is the concentration of sorbed organic compound in solid phase (mg/kg)
C_w is the concentration of dissolved organic compound in water phase (mg/L)
$K_d = C_s/C_w^n$ is the partition coefficient for sorption
n is the empirically determined exponential factor

When $C_w \ll C_s$ (the common case for organic compounds of low water solubility), then n is close to unity for most organics. However, n is typically temperature dependent.

Equation 6.14 can be written as an equation for a straight line by taking the logarithm of both sides, as in

$$\log C_s = \log K_d + n \log C_w \tag{6.15}$$

It is important not to extrapolate the Freundlich isotherm too far beyond the range of experimental data.

Example 7: Use of the Freundlich Isotherm

A power company planned to discharge their power plant cooling water into a small lake. Before purchasing the lake, they tested the water and found the pesticide 2,4-D (2,4-dichlorophenoxyacetic acid) at 0.8 ppt (0.8 parts per trillion, or 0.8×10^{-9} g/L), just a little below the permitted limit of 1 ppt. The company calculated that their operation would raise the water temperature in the mixing zone near their discharge from 5°C to about 25°C. Should they anticipate a problem?

ANSWER

2,4-D has low solubility and is denser than water. It will sink in the lake and become sorbed on the bottom sediments. The potential problem is whether the expected increase in temperature will cause the 2,4-D limit to be exceeded because of

TABLE 6.2
Values for Freundlich Isotherm Parameters of 2,4-D

Temperature (°C)	n	K_d	$\log K_d$
5	0.76	6.53	0.815
25	0.83	5.20	0.716

additional 2,4-D partitioning into the water from the bottom sediments. It is a case for the Freundlich isotherm, because the empirical constant n is a function of temperature and will cause a change in K_d when the temperature changes. A Web search found a study (Means and Wijayratne 1982) that measured Freundlich isotherm values for 2,4-D at 5°C and 25°C (Table 6.2). Use Equation 6.14:

$$C_s = K_d C_w^n$$

At 5°C: $C_s = (6.53)(0.8 \times 10^{-9} \text{ g/L})^{0.76} = (6.53)(1.22 \times 10^{-7}) = 797 \times 10^{-9} \text{ g/kg}$

The calculation indicates that there are 797 ppt of 2,4-D sorbed on the sediments at 5°C. Note that the concentration of 2,4-D sorbed to sediments is about 1000 times larger than the dissolved concentration. Even if a temperature rise to 25°C causes a large percentage increase in the dissolved portion, the percentage loss from the sediment fraction will be 1000 times smaller. Assume that the sediment concentration at 25°C is essentially the same as that at 5°C. This allows an approximate calculation of C_w at 25°C.

At 25°C: $797 \times 10^{-9} \text{ g/kg} = (5.20)(C_w)^{0.83}$

$$C_w(25°C) = \left(\frac{797 \times 10^{-9}}{5.20} \right)^{\frac{1}{0.83}} = 6.2 \times 10^{-9} \text{ g/L} = 6.2 \text{ ppt}$$

The expected temperature rise will cause 2,4-D to desorb from the bottom sediments and raise the water concentration well over the permitted limit of 1 ppt.

6.6.3   DETERMINING K_D EXPERIMENTALLY WITH THE FREUNDLICH ISOTHERM

The Freundlich equation, Equation 6.14, can be used to determine K_d experimentally, as follows:

1. Prepare samples having several different concentrations of dissolved contaminant in equilibrium with soil from the site of interest.
2. Measure the contaminant concentrations in water, C_w, in each sample.
3. Measure the corresponding contaminant concentrations sorbed to soil, C_s.
4. Plot $\log C_s$ versus $\log C_w$ to get a straight line with slope $= n$ and intercept $= \log K_d$.

Example 8

Prepared water samples containing different concentrations of benzene were equilibrated with clean soil from a site under study. Equilibrium concentrations of dissolved and sorbed benzene are shown in Table 6.3. Find K_d for benzene in this soil.

TABLE 6.3
Benzene Partitioning Data for Soil and
Water in Equilibrium

Dissolved Benzene C_w (mg/L)	Sorbed Benzene C_s (mg/kg)
6.59	2.2
10.00	3.1
33.28	8.1
34.57	9.6
68.31	15.00
88.89	26.00
183.74	44.00
340.54	89.00
452.30	119.00
674.79	130.00
819.56	188.00
955.95	247.00

ANSWER

1. Determine the base-10 logarithms of all the concentration values.

Logarithms of C_w and C_s

log C_w	log C_s
0.819	0.342
1.000	0.491
1.522	0.908
1.539	0.982
1.834	1.176
1.949	1.415
2.264	1.643
2.532	1.949
2.655	2.076
2.829	2.114
2.914	2.274
2.980	2.393

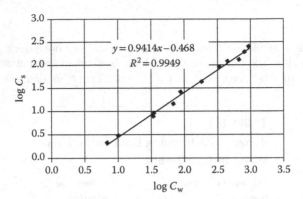

$$y = 0.9414x - 0.468$$
$$R^2 = 0.9949$$

FIGURE 6.5   Freundlich isotherm for benzene partitioning between water and soil.

2. Plot $\log C_w$ versus $\log C_s$ and fit a straight line through the points. The formula of the line is $\log C_s = \log K_d + n \log C_w$. Therefore, the slope of the line
is equal to n and the y-axis intercept is equal to $\log K_d$. The resulting plot is
shown in Figure 6.5.
   The equation of the least-squares fitted line is $\log C_s = 0.941 \log C_w - 0.468$. Therefore,

$$n = 0.941$$
$$\log K_d = -0.468$$
$$K_d = 0.340\,L/kg$$

6.6.4   ROLE OF SOIL ORGANIC MATTER

Values for K_d are extremely both site-specific and chemical-specific, because the
extent of sorption depends on several physical and chemical properties of both the
soil and the sorbed chemical. For dissolved neutral organic molecules, such as fuel
hydrocarbons, sorption to soils is controlled mostly by sorption to the organic portion
of the soil. Therefore, the value of the soil–water partition coefficient, K_d, for neutral
organic molecules depends on the amount of organic matter in the soil. Expressing
the soil–water partition coefficient in terms of soil organic carbon (K_{oc}), rather than
total soil mass (K_d), can eliminate a large part, but not all, of the site-specific variability in K_d for organic pollutants.

   The amount of organic matter in soil is usually expressed as either the weight
fraction of organic carbon, f_{oc}, or the weight fraction of organic matter, f_{om}. The
amount of organic matter in typical mineral soils is generally between about 1% and
10%, but is typically less than 5%. In wetlands and peat-soils, it can approach 100%.
Since soil organic matter is approximately 58% carbon, f_{oc} typically ranges between
0.006 and 0.06. Some characteristic values of f_{oc} for a range of different soil types
are given in Table 6.4.

   There is a critical lower value for the fraction of organic carbon in soil, f_{oc}^*
(dependent on the dissolved organic compound), below which sorption to inorganic
matter becomes dominant. Typical values for f_{oc}^* are between about 10^{-3} and 10^{-4}
(0.1–0.01%).

TABLE 6.4
Typical Values of Fraction of Organic Carbon in Different Soils

Type of Soil	Typical f_{oc} (wt. fraction)
Coarse soil	0.04
Silty loam	0.05
Silty clayey loam	0.03
Clayey silty loam	0.005
Clayey loam	0.004
Sand	0.0005
Glaciofluvial	0.0001

Note: These typical values were collected from many sources. They should only be used as crude estimates when site-specific measurements cannot be obtained. The range of measured values for a single soil type can span a factor of 100.

If we define $K_{oc} = \dfrac{C_{oc}}{C_w}$, where C_{oc} is the concentration of contaminant sorbed to organic carbon and C_w is the concentration of contaminant dissolved in water, then the soil–water partition coefficient becomes

$$K_d = K_{oc} f_{oc} \qquad (6.16)$$

This relation is useful as long as f_{oc} is greater than about 0.001.

RULES OF THUMB

1. It is often found that typical soil organic matter is about 58% carbon. Using this value to derive a relation between the fraction of organic carbon and the fraction of organic matter gives $f_{oc} \approx 0.58 f_{om}$.
2. Soil organic carbon content can vary by a factor of 100 for similar soils. Although there have been tabulations of f_{oc} values according to soil type (e.g., Maidment 1993), it is much better to use values measured at the site for use in Equation 6.16.

Example 9

Find K_{oc} for benzene in the soil of Example 8. The soil contains 1.1% organic matter.

ANSWER

From Example 8, $K_d = 0.34$ L/kg. The soil organic matter was measured to be 1.1%, or $f_{om} = 0.011$. This may be converted to the fraction of organic carbon (f_{oc}) by the approximate rule of thumb

$$f_{oc} = 0.58\, f_{om}$$

Therefore, $f_{oc} = 0.58 \times 0.011 = 0.0064$. Since $K_d = K_{oc}f_{oc}$, we have

K_{oc}(benzene) = 0.340/0.0064 = 53 L/kg.

6.6.5 Octanol–Water Partition Coefficient, K_{OW}

For laboratory experiments, the liquid compound octanol, an eight-carbon organic alcohol, $CH_3(CH_2)_7OH$, has been accepted as a good surrogate for the organic carbon fraction of soils. The partitioning behavior of organic compounds between octanol and water is similar to that between the organic carbon fraction of soil and water. The basic steps for measuring the octanol–water partition coefficient, K_{ow}, are as follows:

1. Combine octanol and water in a bottle. Octanol forms a separate phase floating on top of the water.
2. Add the organic contaminant (e.g., carbon tetrachloride, CCl_4), shake the mixture, and let the phases separate.
3. Measure the contaminant concentrations in the octanol phase and in the water phase.

Then

$$K_{ow} = \frac{C_{octanol}}{C_{water}} \tag{6.17}$$

An empirical equation that relates K_{ow} and organic carbon to K_d is

$$K_d = f_{oc}bK_{ow}^a \tag{6.18a}$$

or

$$\log K_d = a \log K_{ow} + \log f_{oc} + \log b \tag{6.18b}$$

where the empirically determined constants a and b depend on the organic compound.

Equation 6.19, derived from Equations 6.16 and 6.18a, is a relation that allows calculation of K_{oc} in terms of K_{ow}.

$$K_{oc} = bK_{ow}^a \tag{6.19}$$

The usefulness of these relations is evident in the rules of thumb for K_{ow}. Knowing the value of K_{ow} for a compound allows one to qualitatively estimate many of its environmental properties.

RULES OF THUMB FOR K_{ow}

High $K_{ow} \geq 1000$	**Low $K_{ow} \leq 500$**
The higher the K_{ow} is for a compound	The lower the K_{ow} is for a compound
• The higher is the sorption to soil	• The lower is the sorption to soil
• The higher is the bioaccumulation	• The lower is the bioaccumulation
• The lower is the biodegradation rate	• The higher is the biodegradation rate
• The lower is the water solubility	• The higher is the solubility
• The lower is the mobility	• The higher is the mobility

Example 10

Consider the pesticide DDT in terms of the rules of thumb for K_{ow}. For DDT, $K_{ow} = 3.4 \times 10^6$, which is well within the high value range for K_{ow}. From the rules of thumb, one can predict that DDT is less soluble in water (its measured solubility is 25 µg/L), is slowly biodegraded, is persistent in the environment (low mobility and slow biodegradation), is strongly adsorbed to soil, and is strongly bioaccumulated.

On the other hand, phenol has $K_{ow} = 30.2$, a low value. Phenol is highly water-soluble (its measured solubility is 8.3×10^4 mg/L), is rapidly biodegraded, is not persistent in the environment, is weakly sorbed to soil, is highly mobile, and is weakly bioaccumulated.

6.6.6 ESTIMATING K_D USING MEASURED SOLUBILITY OR K_{OW}

Literature values for K_d measurements vary considerably because they are very site specific. Considerable effort has been expended in finding more consistent approaches to soil sorption. The EPA has published a comprehensive evaluation of the soil–water partition coefficient, K_d (USEPA 1999).

The following empirical observations have led to methods for estimating K_d from more easily measured parameters:

- Water solubility is inversely related to K_d; the lower the solubility, the greater the K_d.
- Since molecular polarity correlates with solubility, molecules whose structure indicates low polarity (hence, low solubility) may be expected to have a high K_d.
- For compounds of low solubility, such as fuel hydrocarbons, sorption is controlled primarily by interactions with the organic portion of the solid sorbent.
- The surface area of the solid is important. The larger the surface area, the larger the K_d.
- A simple way to estimate the tendency for an organic compound to partition between water and organic solids in the soil is to measure K_{ow}, the partition coefficient for the organic compound between water and octanol. The larger the K_{ow}, the larger the K_d.

Because K_d is site specific, it is generally preferable to calculate it from the more easily obtained quantities K_{ow}, K_{oc}, f_{oc}, and solubility. K_{ow} and solubility (S) are easily measured in a laboratory, and K_{oc} may be normalized to f_{oc}, percentage of organic carbon in soil, which is an easily measurable site parameter. There are linear relationships between log K_{oc}, log K_{ow}, and S (Lyman et al. 1990) that vary according to the class of compounds tested (e.g., chlorinated compounds, aromatic hydrocarbons, ionizable organic acids, pesticides, etc.). These relations can be used to calculate K_{oc} from K_{ow} or S in the absence of measured K_{oc} data. The EPA has reviewed the soil–water partitioning literature and selected or calculated the most reliable values in their judgment. In these calculations, the EPA used the following relations.

For nonionizable, semi-volatile organic compounds (Group 1 in Table 6.5),

$$\log K_{oc} = 0.983 \log K_{ow} + 0.00028 \qquad (6.20)$$

For nonionizable, volatile organic compounds (Group 2 in Table 6.5),

$$\log K_{oc} = 0.7919 \log K_{ow} + 0.0784 \qquad (6.21)$$

In addition, the EPA suggests the use of the following equation for estimating K_{oc} from solubility (S) or bioconcentration factors (BCF):

$$\log K_{oc} = 0.681 \log BCF + 1.963 \qquad (6.22)$$

Values for K_{oc}, K_{ow}, S, and BCF for many environmentally important chemicals have been collected in look-up tables for the use in the EPA's soil screening guidance procedures (USEPA 1996). Table 6.5 is adapted from these EPA tables.

Example 11

BENZENE SPILL

A benzene leak soaked into a patch of soil. To determine how much benzene was in the soil, several soil samples were taken in a grid pattern across the area of maximum contamination. From analysis of these samples, the average benzene concentration in the soil was 2422 mg/kg (ppm). The soil contained 2.6% organic matter. From Table 6.5, log Kow(benzene) = 2.13. If rainwater percolates down through the contaminated soil, what concentration of benzene might initially leach from the soil and be found dissolved in the water? Assume that the water and soil are in equilibrium with respect to benzene.

CALCULATION

The general approach to this type of problem is

1. Find a tabulated value of K_{oc} for the chemical of concern, or calculate it from tabulated values for K_{ow} or S.
2. Obtain a measurement of f_{oc} or f_{om} in soil at the site or estimate it from the soil type.
3. Calculate K_d from the above quantities.

TABLE 6.5
Chemical Properties Used for Calculating Partition Coefficients and Retardation Factors

CAS No.	Compound	Chemical Group[a]	S, Solubility of Pure Compound (mg/L)	HLC, Henry's Law Constant (atm·m³/L)	H, Henry's Law Constant (unitless)	log K_{OW}	log K_{OC}	K_{OC}, Organic Carbon Partition Coefficient (L/kg)	T_{bp}, Normal Boiling Point (°C)	T_{mp}, Normal Melting Point[b] (°C)
83-32-9	Acenaphthene	1	4.24E+00	1.55E-04	6.36E-03	3.92	3.85	7.08E+03	288	93
120-12-7	Anthracene	1	4.34E-02	6.50E-05	2.67E-03	4.55	4.47	2.95E+04	324	215
71-43-2	Benzene	2	1.75E+03	5.55E-03	2.28E-01	2.13	1.77	5.89E+01	178	5.5
56-55-3	Benzo(a)anthracene	1	9.40E-03	3.35E-06	1.37E-04	5.70	5.60	3.98E+05	376	84
205-99-2	Benzo(b)fluoranthene	1	1.50E-03	1.11E-04	4.55E-03	6.20	6.09	1.23E+06	380	168
207-08-9	Benzo(k)fluoranthene	1	8.00E-04	8.29E-07	3.40E-05	6.20	6.09	1.23E+06	401	217
50-32-8	Benzo(a)pyrene	1	1.62E-03	1.13E-06	4.63E-05	6.11	6.01	1.02E+06	380	176
117-81-7	Bis(2-ethylhexyl) phthalate	1	3.40E-01	1.02E-07	4.81E-06	7.30	7.18	1.51E+07	347	-55
56-23-5	Carbon tetrachloride	2	7.93E+02	3.04E-02	1.25E+00	2.73	2.24	1.74E+02	177	-23
57-74-9	Chlordane	2	5.60E-02	4.86E-05	1.99E-03	6.32	5.08	1.20E+05	329	106
108-90-7	Chlorobenzene	2	4.72E+02	3.70E-03	1.52E-01	2.86	2.34	2.19E+02	207	-45
67-66-3	Chloroform (trichloromethane)	2	7.92E+03	3.67E-03	1.50E-01	1.92	1.60	3.98E+01	168	-64
218-01-9	Chrysene	1	1.60E-03	9.46E-05	3.88E-03	5.70	5.60	3.98E+05	379	258
50-29-3	DDT	1	2.50E-02	8.10E-06	3.32E-04	6.53	6.42	2.63E+06	278	109
53-70-3	Dibenzo(a,h)anthracene	1	2.49E-03	1.47E-08	6.03E-07	6.69	6.58	3.80E+06	395	269
84-74-2	Di-n-butylphthalate	1	1.12E+01	9.38E-10	3.85E-08	4.61	4.53	3.39E+04	323	-35
95-50-1	1,2-Dichlorobenzene	2	1.56E+02	1.90E-03	7.79E-02	3.43	2.79	6.17E+02	234	-17

(Continued)

TABLE 6.5 (*Continued*)
Chemical Properties Used for Calculating Partition Coefficients and Retardation Factors

CAS No.	Compound	Chemical Group[a]	S, Solubility of Pure Compound (mg/L)	HLC, Henry's Law Constant (atm·m³/L)	H, Henry's Law Constant (unitless)	log K_{OW}	log K_{OC}	K_{OC} Organic Carbon Partition Coefficient (L/kg)	T_{bp}, Normal Boiling Point (°C)	T_{mp}, Normal Melting Point[b] (°C)
106-46-7	1,4-Dichlorobenzene	2	7.38E+01	2.43E-03	9.96E-02	3.42	2.79	6.17E+02	231	53
75-35-4	1,1-Dichloroethlyene	2	2.25E+03	2.61E-02	1.07E+00	2.13	1.77	5.89E+01	152	-123
156-59-2	cis-1,2-Dichloroethlyene	2	3.50E+03	4.08E-03	1.67E-01	1.86	1.55	3.55E+01	168	-80
156-60-5	trans-1,2-Dichloroethlyene (DCE)	2	6.30E+03	9.38E-03	3.85E-01	2.07	1.72	5.25E+01	161	-50
78-87-5	1,2-Dichloropropane	2	2.80E-03	2.80E-03	1.15E-01	1.97	1.64	4.37E+01	187	-70
60-57-1	Dieldrin	2	1.95E-01	1.51E-05	6.19E-04	5.37	4.33	2.14E+04	323	176
121-14-2	2,4-Dinitrotoluene	1	2.70E+02	9.26E-08	3.80E-06	2.01	1.98	9.55E+01	310	71
115-29-7	Endosulfan	2	5.10E-01	1.12E-05	4.59E-04	4.10	3.33	2.14E+03	357	106
72-20-8	Endrin	2	2.50E-01	7.52E-06	3.08E-04	5.06	4.09	1.23E+04	381	200
100-41-4	Ethylbenzene	2	1.69E+02	7.88E-03	3.23E-01	3.14	2.56	3.63E+02	209	-95
206-44-0	Fluoranthene	1	2.06E-01	1.61E-05	6.60E-04	5.12	5.03	1.07E+05	347	108
86-73-7	Fluorene	1	1.98E+00	6.36E-05	2.61E-03	4.21	4.14	1.38E+04	299	115
76-44-8	Heptachlor	1	1.80E-01	1.09E-03	4.47E-02	6.26	6.15	1.41E+06	318	96
58-89-9	γ-HCH (Lindane)	2	6.80E+00	1.40E-05	5.74E-04	3.73	3.03	1.07E+03	314	113
193-39-5	Indeno(1,2,3-cd)pyrene	1	2.20E-05	1.60E-06	6.56E-05	6.65	6.54	3.47E+06	432	162
74-83-9	Methyl bromide	2	1.52E+04	6.24E-03	2.56E-01	1.19	1.02	1.05E+01	136	-94
75-09-2	Methylene chloride	2	1.30E+04	2.19E-03	8.98E-02	1.25	1.07	1.17E+01	156	-96

CAS	Compound	Group[a]								[b]
91-20-3	Naphthalene	1	3.10E+01	4.83E-04	1.98E-02	3.36	3.30	2.00E+03	255	80
108-95-2	Phenol	1	8.28E+04	3.97E-07	1.63E-05	1.48	1.46	2.88E+01	235	41
79-34-5	1,1,2,2-Tetrachloroethane (PCA)	2	2.97E+03	3.45E-04	1.41E-02	2.39	1.97	9.33E+01	215	-44
127-18-4	Tetrachloroethylene (PCE, PERC)	2	2.00E+02	1.84E-02	7.54E-01	2.67	2.19	1.55E+02	201	-22
108-88-3	Toluene	2	5.26E+02	6.64E-03	2.72E-01	2.75	2.26	1.82E+02	195	-95
8001-35-2	Toxaphene	1	7.40E-01	6.00E-06	2.46E-04	5.50	5.41	2.57E+05	347	65 to 90
120-82-1	1,2,4-Trichlorobenzene	2	3.00E+02	1.42E-03	5.82E-02	4.01	3.25	1.78E+03	252	17
71-55-6	1,1,1-Trichloroethane (1,1,1-TCA)	2	1.33E+03	1.72E-02	7.05E-01	2.48	2.04	1.10E+02	175	-30
79-00-5	1,1,2-Trichloroethane (1,1,2-TCA)	2	4.42E+03	9.13E-04	3.74E-02	2.05	1.70	5.01E+01	197	-37
79-01-6	Trichloroethylene (TCE)	2	1.10E+03	1.03E-02	4.22E-01	2.71	2.22	1.66E+02	183	-85
67-66-3	Trichloromethane (chloroform)	2	7.92E+03	3.67E-03	1.50E-01	1.92	1.60	3.98E+01	168	-64
75-01-4	Vinyl chloride	2	2.76E+03	2.70E-02	1.11E+00	1.50	1.27	1.86E+01	126	-154
108-38-3	m-Xylene	2	1.61E+02	7.34E-03	3.01E-01	3.20	2.61	4.07E+02	211	-48
95-47-6	o-Xylene	2	1.78E+02	5.19E-03	2.13E-01	3.13	2.56	3.63E+02	214	-25
106-42-3	p-Xylene	2	1.85E+02	7.66E-03	3.14E-01	3.17	2.59	3.89E+02	211	13

[a] Group 1: Semivolatile nonionizing organic compounds. Fitted to log K_{oc} = 0.983, log K_{ow} + 0.00028. Group 2: Volatile organic compounds, chlorobenzenes, and certain chlorinated pesticides. Fitted to log K_{oc} = 0.7919, log K_{ow} + 0.0784.

[b] Compounds solid at soil temperature are defined as those with a melting point >20°C. Compounds liquid at soil temperature are defined as those with a melting point <20°C.

Source: Adapted from USEPA, Soil Screening Guidance: Technical Background Document. Washington, DC, Office of Emergency and Remedial Response, EPA/540/R95/128, 1996.

Benzene is a volatile, nonionizable organic compound (refer to Group 2 in Table 6.5). Use $K_d = \dfrac{C_s}{C_w} = K_{oc}f_{oc}$ and Equation 5.21:

$$\log K_{oc} = 0.7919\log K_{ow} + 0.0784$$

$$\log K_{oc} = 0.7919(2.13) + 1.7651$$
$$K_{oc} = 58.2$$

Since $f_{oc} \approx 0.58f_{om} = 0.58(0.026) = 0.015$, we have

$$K_d = \frac{C_s}{C_w} = K_{oc}f_{oc} = 58.2(0.015) = 0.88\,L/kg$$

The average benzene concentration in the soil was 2422 mg/kg (ppm). Suppose 1 L of pure water were added to a soil sample weighing 1 kg. After the benzene has partitioned between the soil and water, we will have $C_s + C_w = 2422$ ppm. Therefore,

$$K_d = \frac{C_s}{C_w} = \frac{2422 - C_w}{C_w} = 0.88$$
$$0.88C_w = 2422 - C_w$$
$$0.88C_w + C_w = 2422 \text{ mg/L}$$
$$1.88C_w = 2422 \text{ mg/L}$$
$$C_w = \frac{2422 \text{ mg/L}}{1.88} = 1288 \text{ mg/L} \approx 1300 \text{ mg/L to 2 significant figures.}$$

Note that a substantial portion of the sorbed benzene partitions into the water, indicating that benzene has sufficient solubility to be mobile in the environment. This is predictable from the Rules of Thumb for K_{ow}, in which K_{ow}(benzene) = 2.13 (from Table 6.5) is classified as a low value, giving benzene correspondingly high solubility and mobility.

6.7   MOBILITY OF CONTAMINANTS IN THE SUBSURFACE

The sorption* of a contaminant from water to a solid is a reversible reaction. Just as the contaminant has some probability of sorbing from water to a surface that it contacts, a sorbed contaminant also has a probability of desorbing from the surface back into the water. For strongly sorbed contaminants, the probability of sorption is much greater than the probability of desorption, but both processes continually take place, although with different rate constants. The rate constants of sorption (k_{sorb}) and desorption (k_{desorb}) depend on the strength of the bonds holding the sorbed compound to the surface. The rates of sorption and desorption depend on their respective rate constants and on the concentrations of dissolved and sorbed contaminants.

---

* Sorption: The act of becoming sorbed; to sorb: to take up and hold, as by sorption, a process which includes both adsorption (held on the surface) and absorption (drawn into the interior).

$$\text{Rate of sorption} = k_{sorb}C_w \qquad (6.23)$$

$$\text{Rate of desorption} = k_{desorb}C_s \qquad (6.24)$$

where
  C_w is the contaminant concentration in water
  C_s is the contaminant concentration sorbed to soil
  k_{sorb} and k_{desorb} are the rate constants, which depend on the binding strength of sorption

For strongly sorbed contaminants, $k_{sorb} \gg k_{desorb}$. For weakly sorbed contaminants, $k_{sorb} \ll k_{desorb}$.

Consider the case of a dissolved contaminant sorbing onto initially clean soil particles. Until equilibrium is reached, C_w is continuously decreasing and C_s is continuously increasing as dissolved contaminant molecules move from the water to soil surfaces. Therefore, according to Equations 6.23 and 6.24, the rate of sorption continuously decreases and the rate of desorption (which initially started at zero) continuously increases. Eventually, the two rates must reach the condition of equilibrium, where both rates are equal.

The partition coefficient K_d quantifies the equilibrium condition of sorption, where sorption and desorption occur at the same rate. When both rates are equal

$$k_{sorb}\,C_w = k_{desorb}\,C_s$$

and

$$\frac{k_{sorb}}{k_{desorb}} = \frac{C_s}{C_w} = K_d \qquad (6.25)$$

Equation 6.25 shows that if $k_{sorb} \gg k_{desorb}$, which is the case of strong sorption, then at equilibrium $C_s \gg C_w$, K_d has a large value, and more of the contaminant is sorbed than is dissolved. If $k_{sorb} \ll k_{desorb}$, the reverse is true. Thus, strongly sorbed contaminants accumulate to higher concentrations on the soil than do more weakly sorbed contaminants.

In terms of contaminant mobility, this means that strongly sorbed contaminants remain sorbed to soil surfaces longer before desorbing, on average, than do weakly sorbed contaminants and, consequently, move more slowly through the subsurface than does the groundwater in which they are dissolved. The movement through the subsurface of dissolved contaminants in groundwater is analogous to the movement of analytes through a chromatograph column. Because each analyte binds to the column wall (called the stationary phase) with a unique binding strength, different analytes move through the column at different velocities and eventually become separated in space along the column.

Contaminants dissolved in groundwater are similarly retarded in their down-gradient movement relative to the flow of groundwater. Soil serves as the chromatographic stationary phase. The extent of retardation is related to their value of K_d. The front of a moving groundwater contaminant plume will contain the fastest-moving contaminants with the lowest values of K_d (having the weakest sorption strengths) and, moving back upgradient through the plume, one will encounter progressively slower moving contaminants with progressively larger values of K_d.

RULES OF THUMB

The mobility of dissolved organic compounds in groundwater depends on K_d.

- If $K_d = 0$, the organic compound does not sorb to the soil it passes through and moves at the groundwater velocity.
- If $K_d > 0$, movement of the organic compound is retarded. Its velocity relative to that of groundwater is inversely related to the value of K_d; the larger the K_d, the slower the velocity of the contaminant relative to the velocity of the groundwater (see Section 6.7.1).

Example 12

Consider the case where groundwater contaminated with benzene and toluene is moving through subsurface soils containing 1.6% organic carbon. Compare benzene and toluene qualitatively with respect to their mobility in the subsurface. Use data from Table 6.5.

ANSWER

From Table 6.5, K_{oc}(benzene) = 58.9 L/kg and K_{oc}(toluene) = 182 L/kg. Use $K_d = K_{oc} f_{oc}$, K_d(benzene) = 58.9 × 0.016 = 0.942 and K_d(toluene) = 182 × 0.016 = 2.91.

K_d(toluene) is approximately three times larger than K_d(benzene), indicating that it sorbs more strongly to the soil and will move significantly slower than benzene through the subsurface.

6.7.1 RETARDATION FACTOR

A retardation factor for contaminant movement can be calculated using K_d (Fetter 1993; Freeze and Cherry 1979). The retardation factor, R, for a contaminant is defined as

$$R = \frac{\text{average linear velocity of groundwater}}{\text{average linear velocity of contaminant}} \tag{6.26}$$

A retardation factor of 10 means that the contaminant moves at one-tenth of the average velocity of the groundwater. The average linear velocity of the contaminant is defined as its velocity measured at the point where its concentration is equal to 1/2 of its concentration at the contaminant source.

TABLE 6.6
Calculated Retardation Factors and Mobility Classifications

R	Examples	K_d	Mobility Classification
<3	Methylene chloride, MTBE, 1,2-DCA	<0.03	Highly mobile
3–9	Benzene, 1,1-DCA, chloroform	0.03–1.2	Mobile
9–30	Ethylbenzene, toluene, xylenes	1.2–4.3	Intermediate
30–100	Styrene, pyrene, lindane	4.3–15	Less mobile
>100	Naphthalene, dioxin, heptachlor	>15	Immobile

Note: Assumed values are $f_{oc} = 0.01$, $\rho = 2.0$ g/cm^3, $h = 0.3$.

Assuming a linear partition coefficient, $K_d = \frac{C_{soil}}{C_w}$, the retardation factor becomes (Fetter 1993)

$$R = 1 + \frac{\rho K_d}{h} \tag{6.27}$$

or, for a Freundlich partition coefficient, $K_d = \frac{C_{soil}}{C_w^n}$

$$R = 1 + \left(\frac{K_d}{n}\right)\left(\frac{\rho}{h}\right)\left(C_w^{1/(n-1)}\right) \tag{6.28}$$

where
ρ = soil bulk density (g/cm^3), typically 1.5–1.9 g/cm^3
h = effective soil porosity,* typically 0.35–0.55
n = empirical Freundlich exponential factor from Equation 6.14

Some retardation factors and qualitative mobility classifications determined from data in Table 6.5 are presented in Table 6.6. The table was developed using Equations 6.16 and 6.27 with typical values for soil f_{oc}, effective porosity, and bulk density.

6.7.2 EFFECT OF BIODEGRADATION ON EFFECTIVE RETARDATION FACTOR

Equations 6.27 and 6.28 are based only on sorption to organic carbon and do not consider the influence on dissolved contaminant mobility of processes like biodegradation, ion exchange, precipitation, and chemical changes. As shown in Figure 6.6, biodegradation causes additional retardation to plume movement.

---

* See Section 6.7.4.

RULES OF THUMB

1. Dissolved contaminants generally move more slowly than groundwater because of sorption, ion exchange, precipitation, and biodegradation, but the retardation factor calculated from Equation 6.28 or 6.29 considers sorption only.
2. The movement rate of biodegradable compounds is overestimated by the sorption retardation factor, especially over long periods of time.
3. Sorption to soil not only slows the growth of a contaminant plume, but also slows the rate of cleanup by pump-and-treat methods and increases the water volume that must be extracted to achieve acceptable residual concentrations.

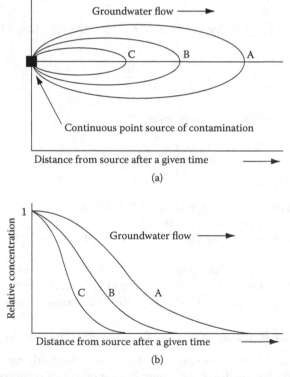

FIGURE 6.6 Positions of a contaminant plume front as affected by different flow conditions: (a) Horizontal cross section through plumes from a continuous point source. (b) Decrease in concentration with distance from the source. Curve A represents a nonretarded plume, curve B represents a plume retarded by sorption, and curve C represents a plume retarded by sorption and biodegradation.

Example 13: Calculation of a Sorption Retardation Factor

Calculate the retardation factor, R, for 1,2-dichloroethane (1,2-DCA) in a soil with 2.7% total organic carbon (TOC), dry bulk density of 1.7 g/cm³, and effective soil porosity of 40%.

ANSWER

$$R = 1 + \frac{rK_d}{h}, \quad \text{and} \quad K_d = K_{oc} f_{oc}$$

where
$K_{oc} = 17.4$ (from Table 6.5)
r = dry bulk soil density = 1.7 g/cm³
h = effective porosity of soil = 0.40

$$f_{oc} = 2.7\% = \frac{2.7}{100} = 0.027$$
$$K_d = 17.4 \times 0.027 = 0.470$$
$$R = 1 + \frac{1.7 \times 0.470}{0.40} = 3.0$$

Divide the groundwater velocity by 3.0 to get the velocity with which 1,2-DCA moves through this particular soil.

Example 14

Calculate the retardation factor for tetrachloroethene (PCE*) in a silty clayey loam soil, using data from Table 6.5. Soil density is 2.5 g/cm³ and effective porosity is 31%. Percentage of organic carbon was not reported.

ANSWER

For tetrachloroethene, $K_{oc} = 155$.
Since no measured value for f_{oc} is available, a value from Table 6.4 can be used. For a silty clayey loam soil, $f_{oc} \approx 0.03$.

$$K_d = 155 \times 0.03 = 4.7 \text{ and } R = 1 + \frac{2.5 \times 4.7}{0.31} = 38$$

Tetrachloroethene has a K_{oc} nearly 10 times that of 1,2-DCA in Example 13, which means that it is much less water-soluble. Thus, it is less mobile and has a higher retardation factor.

---

* Other common synonyms for tetrachloroethene are tetrachloroethylene, perchloroethylene, ethylene tetrachloride, carbon bichloride, carbon dichloride, and PERC.

6.7.3   A MODEL FOR SORPTION AND RETARDATION: WHY PUMP-AND-TREAT MAY NOT WORK

Consider the result of Example 13, where the retardation factor for 1,2-DCA was found to be 3.0 in a particular soil. Since the movement of the contaminant is slowed by a factor of 3 relative to the groundwater velocity, it might appear that flushing three pore volumes of water through the impacted soil would completely desorb 1,2-DCA from that part of the subsurface. In other words, the retardation factor might erroneously be interpreted as the number of groundwater pore volumes that must be flushed through the contaminated zone to desorb a contaminant from the impacted soil.

In fact, most less-soluble organic contaminants can never be completely flushed from the soil, because there normally is some fraction of the contaminant that becomes almost irreversibly bound to the organic matter in soil. The part of the contaminant that behaves according to its K_d value governs the movement of the front of the contaminant plume, but there is usually a portion that binds to the soil more strongly and moves more slowly. An aging process causes the fraction of contaminant that can be desorbed by flushing to decrease with time. This part cannot be removed by water flushing in any reasonable time. It can remain in the soil as a slowly diminishing source of groundwater contamination for tens, or even hundreds, of years. This is why pump-and-treat remediation methods are seldom successful at removing the last part of contamination from soil. When desorption of a contaminant becomes very slow and pump-and-treat methods become ineffective, other approaches such as bioremediation may be needed to achieve required cleanup levels.

A two-step conceptual model for sorption has been proposed (Alexander 1995) that is consistent with time-dependent irreversible sorption and other observations, such as a decrease in biodegradation rates with time.

6.7.3.1   A Two-Step Conceptual Model for Sorption

Step 1: A compound with large K_d is initially adsorbed rapidly from water to the external surfaces of soil particles. Measurements of K_d are normally based on the sorption behavior of the initially sorbed contaminant. K_d represents the concentration ratio between sorbed and dissolved contaminant if the elapsed time between sorption and measurement is not too long. The dissolved fraction is available to microorganisms for biodegradation, and to organisms such as fish and humans that might be susceptible to its toxic effects.

Step 2: As time passes, a portion of the surface-sorbed compound begins to diffuse into micropores in the solid surface, moving away from the surface into the particle interior. Here the compound becomes sequestered within the soil in locations that are remote from the surface and where it is less available for desorption.

Thus, aging appears to be associated with continuous diffusion into more remote sites on the solid particle, where the molecules are retained and rendered less accessible to biological, chemical, and physical changes. After 1–10 years, depending on site-specific soil and pollutant conditions, a large fraction of the sorbed pollutant will not desorb from the soil.

Another limitation of the retardation factor is that the equations unrealistically assume that the soil matrix is homogeneous, which is rarely the case. Nevertheless, the results are still useful for rough estimates and for estimating relative mobilities of dissolved contaminants.

6.7.4 SOIL PROPERTIES

The subsurface environment contains inorganic minerals, organic humic materials, air, and water. Also found are plant roots, microorganisms, and burrowing animals, not to mention building foundations, utility service lines, and other man-made structures. All of these disturbances of the natural soil texture can affect the movement of contaminants through the subsurface.

The physical properties of soil that have the greatest effect on the movement of water and contaminants are true porosity, effective porosity, particle size range, and hydraulic conductivity. Representative values for these properties are given in Tables 6.7 through 6.9.

TABLE 6.7
Representative Values of Effective Porosity for Some Soil Types

Soil Type	Effective Porosity (%)
Well-sorted sand or gravel	25–40
Sand and gravel, mixed	20–35
Medium sand	15–30
Glacial sediments	5–20
Silt	1–20
Clay	1–2

TABLE 6.8
Soil Particle Size Range for Some Soils

Soil Type	Particle Size Range (mm)
Clay	<0.002
Silt	0.002–0.04
Very fine sand	0.04–0.10
Fine sand	0.10–0.20
Medium sand	0.20–0.40
Coarse sand	0.40–0.90
Very coarse sand	0.9–2.0
Fine gravel	2.0–10.0
Medium gravel	10.0–20.0
Coarse gravel	20–40
Very coarse gravel	40–80

TABLE 6.9

Representative Values of Water Hydraulic Conductivity for Some Soils

Soil Type	Hydraulic Conductivity for Water: Typical Range (cm/s)
Clay	10^{-6} to 10^{-9}
Silt	10^{-3} to 10^{-7}
Fine sand	10^{-2} to 10^{-5}
Medium sand	10^{-1} to 10^{-4}
Coarse sand	1 to 10^{-4}
Gravel	10^2 to 10^{-1}

- *True porosity* is the ratio of the volume of empty space (pore volume) to total volume of soil. It can be expressed as a simple ratio or as a percentage. For a soil sample in which 1/4 of the total volume is empty (or void) space, the porosity may be expressed as 0.25 or 25%.
- *Effective porosity* is the ratio of the volume of effective void space to the total volume of material, generally expressed as a percentage. Effective porosity accounts for the fact that pore spaces that are not connected, or are too small for the fluid to overcome capillarity, do not contribute to fluid movement. The difference between true porosity and effective porosity is most noticeable in clay, where true porosity is very high, 34–60% (it can hold a lot of water). Effective porosity, however, is very low, 1–2%, making clay relatively impermeable.
- *Soil particle size range* determines how soil particles pack together and, thereby, the average soil pore size. Pore size strongly affects the available soil surface area as well as the capillary attraction between the soil and liquids. The smaller the pore size, the stronger is the capillary attraction and the greater is the total soil surface area within a given volume of soil. The larger the surface area, the larger is the volume of liquid immobilized by sorption to soil surfaces. Capillary attractions are an additional force that acts to retard the movement of liquids through soil and can immobilize a fraction of the liquid. Where pore size is small enough, the distance between adjacent soil particles can be small enough that capillary attraction can extend across significant fractions of the pore volume. Thus, silt retards the movement of liquids more than coarser soils like sand and gravel, and will immobilize a larger quantity of liquid than will an equal volume of coarser soil. Clay also is effective at retarding the movement of liquids but, because of its low effective porosity, cannot immobilize large volumes of liquid by sorption and capillarity.
- *Hydraulic conductivity* indicates the ability of subsurface material to transmit a particular fluid. For example, the hydraulic conductivity of a given soil to transmit water is greater than it would be to transmit a more viscous fluid such as diesel fuel. The hydraulic conductivity of a very porous soil, like loose sand, is greater than for a less porous soil, such as fine silt.

6.8   PARTICULATE TRANSPORT IN GROUNDWATER: COLLOIDS

Contaminants move in groundwater systems as dissolved species in the water, as flowing free-phase liquids, or combined with moving particulates. Particulates that can move through soils with groundwater must be small enough to move through the soil pore spaces. Such particulates are generally less than 2.0 μm in diameter and are called colloids. Colloids are a special class of matter with properties that lie between those of the dissolved state and the solid or immiscible liquid states.

Colloids have a high surface-area-to-mass ratio due to their very small size. Groundwater concentrations of colloidal materials can be as high as 75 mg/L, corresponding to as many as 10^{12} particles/L. This represents a large surface area available for transporting sorbed contaminants.

There are many sources of colloidal material in groundwater. Colloids are formed in soil when fragments of soil, mineral, or contaminant particles become detached from their parent solid because of weathering, or by physical abrasion caused by rock movement or drilling and other construction activities. Then they may be carried to the groundwater when water from irrigation or precipitation percolates downward through the soil. Colloids are often introduced directly into groundwater from landfills. Colloids can also form within groundwater as fine precipitates when dissolved minerals in groundwater undergo pH or redox potential changes, and they can form as emulsions of small droplets from immiscible liquids, such as free-phase hydrocarbons, which exhibit many characteristics of particulate colloids, including the ability to sorb dissolved contaminants to the droplet–water interface.

6.8.1   COLLOID PARTICLE SIZE AND SURFACE AREA

When the particle size is reduced to 1–2 μm or smaller, surface forces arising from the surface charge or London force attractions begin to exert a significant influence on the particle behavior. Consider the effect of reducing the particle size of a given mass of solid. A cube of any material that is 10 mm on a side has a smooth surface area of 6.0×10^{-4} m^2. Cut it in half in each of the three directions perpendicular to its faces to get eight cubes, which are each 5 mm on a side. The total surface area is now 12.0×10^{-4} m^2. Continue subdividing until the cubes are 1 μm on an edge. The total surface area is now 6.0 m^2, an increase of 10,000 times over the original cube.

Montmorillinite, a clay mineral, in the dispersed state may break down into small plate-like particles only one unit cell in thickness, about 10^{-9} m. Its specific surface area is about 800 m^2/g. A monolayer of 10 g of this material would cover a football field.

6.8.2   PARTICLE TRANSPORT PROPERTIES

Contaminants of low solubility can move with groundwater dispersed as colloids or attached by sorption or occlusion with colloids, resulting in an unexpected mobility of the low-solubility material. When contaminants are attached to colloids, their transport behavior is determined by the properties of the colloid, and not the sorbed contaminant. In a low-velocity flow of groundwater, particles larger than 2 μm tend to settle by gravity. Particles smaller than 0.1 μm tend to sorb readily to larger soil particles, becoming retarded or immobilized. Thus, particles in the range 0.1–2.0 μm are the most mobile in groundwater.

6.8.3 ELECTRICAL CHARGES ON COLLOIDS AND SOIL SURFACES

The explanation for how water chemistry affects colloid behavior lies in the behavior of charged particles in water solutions. Soil surfaces in an aquifer generally have a net negative charge due to the dominance of silicates in the minerals, which have exposed electronegative oxygen atoms. Colloidal particles also are usually charged, with a sign dependent on the nature of the colloid and the water pH.

- Metal oxide colloids tend to be positively charged.
- Sulfur and the so-called noble metals* tend to be negatively charged.
- Organic macromolecules, such as humic materials, proteins, or resins used in water treatment for flocculation, acquire a charge that depends on the pH.
  - At low pH, hydrogen ions bind to the molecules and make the colloid positive.
  - At high pH, the binding of hydroxyl ions makes the charge negative. For such particles, there always is a particular pH, called the point-of-zero charge (PZC), where the number of surface-bound hydrogen and hydroxyl ions is equal, resulting in no net charge.

6.8.3.1 Electrical Double Layer

In natural water, the charge on colloids and on soil surfaces attracts dissolved ions into a configuration known as an electric double layer. Consider the double layer that forms adjacent to a negatively charged mineral surface. The first layer forms when dissolved cations (positive ions) are attracted to the oppositely charged mineral surfaces. Then anions (negative ions) are attracted to the positive ion region of the first layer, to form a second, more diffuse, oppositely charged layer surrounding the first. The net result is an inner layer of cations surrounded by an outer layer of anions. Positively charged surfaces acquire an inner layer of anions and an outer layer of cations. The inner layer effectively neutralizes the surface charge and the outer layer effectively neutralizes the inner layer. Subsequent layers can form in principle, but they are too diffused to have any effect. The electric double layer strongly influences the processes of adsorption and coagulation.

6.8.3.2 Sorption and Coagulation

If two colloid particles come close enough together, London attractive forces will pull them together into a larger particle and start coagulation. A similar phenomenon happens if a colloidal particle comes close enough to a soil surface. London attractive forces will pull them together and the colloid becomes sorbed to the soil surface.

---

* Metals that are resistant to oxidation and corrosion under common environmental conditions are sometimes referred to as noble metals. In addition to gold, silver, and platinum, they may include copper, mercury, aluminum, palladium, rhodium, iridium, tantalum, and osmium.

Two colloidal particles of the same material have the same sign of charge in the outer part of their double layer and therefore repel one another. For two colloidal particles to coagulate, they must collide with enough energy to force past their repulsive double layers and approach close enough for London attractive forces to be effective. The same is true for adsorption to soil particle surfaces.

Brownian motion provides the collision energy for overcoming the electrical repulsion of the double layer. In high-energy collisions, particle momentum may overcome charge repulsion and allow particles to approach close enough to enter the zone of London attraction, where adsorption and coagulation can occur.

- At high ionic strength (high dissolved ion concentrations; high total dissolved solids [TDS]), the double layer is thin because the ion atmosphere is dense and there is more charge per unit volume. The higher ion charge density can neutralize the net particle charge with a thinner layer.
- At low ionic strength, a thicker ionic charge layer is required to neutralize the charge on colloidal particles.

Thus, colloidal particles can approach charged surfaces and other colloids more closely at high rather than low TDS concentrations; high TDS concentrations allow London attractions, which become stronger at shorter distances, to be more effective. There are several familiar illustrations of this principle:

- When river water carrying colloidal clay reaches the ocean, the salt water (high TDS) induces coagulation. This is a major cause of silting in estuaries.
- Applying a styptic pencil stops bleeding from small cuts. The styptic pencil contains aluminum salts, often alum (aluminum sulfate). Dissolving high concentrations of aluminum salt into the wound raises the TDS concentration in the external blood and initiates coagulation of colloidal proteins in blood.
- Colloidal material can be "salted out" by dissolving salts into the solution.
- The high TDS of typical landfill leachate provides optimal conditions for immobilizing entrained colloids by sorption and coagulation.

In summary, colloid sorption behavior is influenced by

- Forces acting on colloidal particles
  - Electrostatic attraction and repulsion
  - London attractions
  - Brownian motion
- Properties of the groundwater
  - Ionic strength (related to TDS and conductivity)
  - Ionic composition
  - Flow velocity
- Properties of the colloids
  - Size
  - Chemical nature
  - Concentration

- Properties of the soil matrix
  - Geologic composition
  - Particle size distribution
  - Soil surface area

RULES OF THUMB

1. Usually the most important factor governing colloid behavior is groundwater chemistry and the least important is flow velocity.
2. Coagulation and sorption of colloids are more efficient under high TDS conditions.
3. Generally, colloids are more mobile when the TDS concentration is low and less mobile when the TDS concentration is high.

6.9   CASE STUDY: CLEARING MUDDY PONDS

Ponds and small lakes sometimes become muddy (or turbid).* Some common causes are

- Stormwater runoff from fields with disturbed soil or sparse vegetation, such as construction sites or fields with active agricultural equipment.
- Shoreline erosion caused by strong winds and wave action, and by shoreline activity of people and animals.
- Animal activity in the water, cattle or wildlife coming to drink or cool off.
- Heavy use by waterfowl digging in shallow water and eating shoreline vegetation.
- High populations of certain fish that spawn and feed in the bottom sediments.

In addition to being unsightly, muddy water blocks sunlight penetration, limiting plant growth and reducing the amount of aquatic life supported by the pond's food chain; it even can clog the gills of fish, suffocating them. A pond muddied by a singular event like a particularly strong storm or a temporary construction activity will often clear after several days if left undisturbed; occasionally however, a muddy pond will fail to clear. When this happens, the pond water must be treated to restore its clarity.

In this case study, a period of severe rainstorm activity caused flooding of a construction site and washed a large quantity of sediment into a pond with about 3 acres of surface area. The average depth of the pond was about 8 ft. A quart jar of turbid water collected from the pond was allowed to stand undisturbed for 5 days. After this time, some sediment had settled to the bottom of the jar, but the water was still quite murky.

---

* This case study deals with water turbidity caused by soil particles and not turbidity caused by algae or other organic suspensions. Water colored by suspended soils is normally a chocolate brown or red-brown color, while water colored by algae is green.

Suspended sediment that has not settled out after this period is most likely to be colloidal clay particles carrying a negative surface charge that keeps them suspended, as described in Section 6.8. A common treatment for turbid ponds is to induce coagulation by increasing the TDS, especially with cations having high positive charges (+2 and +3). In addition to thinning the double layer, positively charged multivalent cations form bridges between the negatively charged clay particles, greatly assisting the coagulation process and speeding water clarification.

In principle, any salt may be added to facilitate coagulation. For environmental safety, however, only compounds not harmful to the ecosystem in the required concentrations can be used. The most commonly used chemicals for clearing muddy ponds are alum (aluminum sulfate, $Al_2(SO_4)_3$) and gypsum (calcium sulfate, $CaSO_4$). Alum is more effective than gypsum (requires smaller concentrations) because the aluminum cation (Al^{3+}) has a greater positive charge than the calcium cation (Ca^{2+}) and forms more extensive bridges between negatively charged clay particles. However, aluminum cations act as stronger acids than calcium cations (see Section 5.3.5) and may lower the pond pH to an unacceptable value. For this reason, buffered alum, with added hydrated lime (calcium hydroxide, $Ca(OH)_2$), should be used for application to ponds with low alkalinity (less than 100 mg/L), or hydrated lime can be added before adding alum (about 20 lb of hydrated lime per acre-foot).

The pond in this case study had an alkalinity of 320 mg/L as $CaCO_3$, high enough that buffered alum was not necessary. The dimensions of the ponds resulted in a volume of about 24 acre-feet. Rule of Thumb number 2 below indicates that 1200 lb of alum would be sufficient. A conservative application of 600 lb was used first, which cleared the pond to a measured underwater visibility of 2 ft after 2 days. Further treatment was deemed unnecessary.

RULES OF THUMB

1. For healthy aquatic life, underwater visibility should be at least 1–1.5 ft most of the time.

2. Where turbidity is caused by suspended clay or silt, about 50 lb of alum per acre-foot of water will generally clear most turbid ponds in about 1 week. Alum should be dissolved in water and sprayed over the pond surface on a calm day. For minimum risk to aquatic life, it is prudent to first add 1/3 to 1/2 of the calculated dose and wait 2 days to see if more alum is needed to increase underwater visibility to 1–1.5 ft.

3. If alkalinity > 50 mg/L and pH > 7, regular alum may be used. Otherwise, use buffered alum or first add about 20 lb of hydrated lime (calcium hydroxide) per acre-foot of water. However, large amounts of alum can lower the alkalinity and pH rapidly. If more than 300–400 lb of alum are required, the alkalinity and pH should be monitored frequently during addition.

4. Gypsum can also be used to clear ponds. It will not acidify the water as much as alum, but it requires a dose about 10 times heavier than alum. About 500 lb of gypsum will have the same effect as 50 lb of alum.
5. To calculate the volume of a pond in acre-feet, estimate its surface area in square feet, multiply by its average depth in feet, and divide this volume by 43,560 ft^3/acre-ft.

6.9.1　Pilot Jar Tests

Sometimes, instead of the trial-and-error approach described in the case study, it is better to use a more scientific approach to determine the correct amount of alum or gypsum (or other coagulant) for clearing a turbid pond. For example, if the pond has a pH beyond the range of 6–8, or the sediment is not common clay or silt but instead contains a large fraction of coal mine dust or, perhaps, arises from animal feedlot runoff, the doses that work for clay and silt might not be appropriate. In such cases, a better approach is to perform a set of jar tests where you test the effects of a series of progressively more concentrated doses of coagulant on samples of the water to be treated. In this way you can conveniently measure the effects of different coagulants at different doses, to find the most effective and economical way to treat the pond.

When working in the field, accurate scales and volumetric glassware are not always available for making up known concentrations of coagulant solutions. Fortunately, precise measurements are seldom critical since modest overdoses of coagulant are generally not harmful and often serve to speed the clearing process. Therefore, the recipes for the pilot tests given below use measurements based on tablespoons and 1-gallon plastic or glass jars (1-gallon plastic water bottles are fine). If alum is to be used, measure the alkalinity and pH of the pond.

6.9.1.1　Jar Test Procedure with Alum Coagulant

1. Collect six 1-gallon jars of turbid pond water and one 1-gallon jar of bottled drinking water.
2. Number the pond water bottles from 1 to 6.
3. Using a standard measuring spoon, mix one tablespoon of alum into the 1-gallon jar of drinking water, stirring until the alum forms a slurry.
4. Add one tablespoon of the alum slurry to pond water jar #1, two tablespoons to jar #2, three tablespoons to jar #3, and so on, for all six jars.
5. Allow the treated jars to stand undisturbed for at least 12 hours.
6. Identify the turbid water jar that required the least number of tablespoons of alum slurry to clear it.
7. Use the table on the next page to determine the proper dose of alum for treating the pond. For example, if the pH > 7, alkalinity > 100 mg/L, and jars #3, #4, #5, and #6 all cleared up after standing for 12 hours, jar #3 required the least amount of alum for clearing. Add 13 lb/acre-ft of hydrated lime to the pond, followed by 90 lb/acre-ft of alum.

Number of Tablespoons of Alum Slurry Added to Clear Sample in 12 hours	Amount of Alum to Be Added to the Pond (lb/acre-ft)	Amount of Hydrated Lime to Be Added before Adding Alum If pH > 7 and Alkalinity > 100 (mg/L)	Amount of Hydrated Lime to Be Added before Adding Alum If pH < 7 and Alkalinity < 100 (mg/L)
1	30	0	13
2	60	0	26
3	90	13	39
4	120	17	52
5	150	21	65
6	180	25	78

6.9.1.2 Jar Test Procedure with Gypsum Coagulant

1. Collect 12 1-gallon jars of turbid pond water and one 1-gallon jar of bottled drinking water.
2. Number the pond water bottles from 1 to 12.
3. Using a standard measuring spoon, mix two tablespoons of gypsum into the 1-gallon jar of drinking water, stirring until the gypsum forms a slurry.
4. Add one tablespoon of the gypsum slurry to jar #1, two tablespoons to jar #2, three tablespoons to jar #3, and so on, for all 12 jars.
5. Allow the treated jars to stand undisturbed for at least 12 hours.
6. Identify the turbid water jar that required the least number of tablespoons of gypsum slurry to clear it.
7. Use the table below to determine the proper dose of gypsum for treating the pond. For example, if all jars #8 and above cleared up after standing for 12 hours, jar 8 required the least amount of gypsum for clearing. Therefore, add 640 lb/acre-ft of gypsum to the pond.

Number of Tablespoons of Gypsum Slurry Added to Clear Sample in 12 Hours	Amount of Gypsum to Be Added to the Pond (lb/acre-ft)
1	80
2	160
3	240
4	320
5	400
6	480
7	560
8	640
9	720
10	800
11	880
12	960

6.10 ION EXCHANGE

Ion exchange is the reversible interchange of ions between a solid and a liquid. Hydrated ions on a solid are exchanged, equivalent for equivalent, for hydrated ions in solution. Cation exchange involves the interchange of positive ions. Anion exchange involves the interchange of negative ions.

In the natural environment, solid particles generally carry a surface charge, either positive or negative, which is true for both organic and inorganic solids. As water passes through soils, dissolved ions can leave the water to become attached to oppositely charged sites on soil surfaces. This displaces ions of the same charge sign previously attached to the surface, so that they become dissolved and mobile in the water. In general, ions of higher charge and smaller hydrated diameter will displace ions of lower charge and larger hydrated diameter. Larger hydrated diameter correlates with smaller ionic diameter and smaller ionic charge. Thus, smaller ions of the same charge have the largest hydrated diameters, as do ions of approximately the same ionic diameter but with smaller charge. Such ions (small ionic diameter and small charge) coordinate with more water molecules in their hydration sphere, resulting in a larger hydrated diameter. This is why nonhydrated sodium cation, Na^+, which is smaller than nonhydrated K^+, and has a smaller ionic charge than both Mg^{2+} and Ca^{2+}, causes greater swelling and loss of permeability of clayey soils than K^+, Mg^{2+}, or Ca^{2+} (see discussion of Section 6.11.5).

RULES OF THUMB

1. Dissolved ions with higher binding strength tend to displace surface-bound ions of lower binding strength.
2. Binding strength increases with larger nonhydrated ionic diameter (smaller hydrated diameter) of the ions.
3. Binding strength increases with the charge on the ions.
4. There is also a concentration effect. Continual high concentrations of any ion eventually displace most of the other ions having the same charge sign.

The order of cation binding strengths to a negatively charged surface is (strongest) $Cr^{3+} > Al^{3+} >> Ba^{2+} > Sr^{2+} > Ca^{2+} > Mg^{2+} >> Cs^+ > NH_4^+ > K^+ > Na^+ > H_3O^+ > Li^+$ (weakest).

For example, Cr^{3+} will displace Al^{3+} on a surface; Ca^{2+} will displace K^+; H_3O^+ will displace Li^+. There is no permanent change in the structure of the solid that serves as the ion-exchange material.

6.10.1 Why Do Solids in Nature Carry a Surface Charge?

Solid particle surfaces can acquire an electric charge in four ways. All four surface charge mechanisms can exist at the same time on mineral surfaces, and the latter two can exist at the same time on nonmineral (organic) surfaces also.

1. *Lattice imperfections*: During the crystal growth of silica minerals, an Al^{3+} cation may enter a lattice location intended for Si^{4+}, or a Mg^{2+} may substitute for Al^{3+} (all are isoelectronic third period cations and, thus, are of similar sizes), resulting in a net negative charge on the crystal.
2. *Differential solubilities*: Ions at different locations on the surface of slightly soluble salt crystals may have different tendencies to dissolve into water, resulting in either a negative or positive charge imbalance.
3. *pH-dependent chemical reactions at the particle surface*: Many solid surfaces (oxides, hydroxides, organics) contain ionizable functional groups, such as $-OH$, $-COOH$, or $-SH$. At high pH, these groups lose an H^+ (by $H^+ + OH^- \rightarrow H_2O$), becoming charged as $-O^-$, $-COO^-$, or $-S^-$. At low pH, these groups gain an H^+, becoming $-OH_2^+$, $-COOH_2^+$, or $-SH_2^+$.
4. *Adsorption of hydrophobic (low-solubility) or surfactant ions*: This can result in either positive or negative surfaces, and is not pH dependent.

RULES OF THUMB

For permanent surface charge:
1. Surface charge caused by lattice imperfections is permanent and is not pH dependent.
2. Permanent surface charge occurs on clays and most minerals.
3. Permanent surface charge on minerals and clays is generally negative.

For pH-dependent surface charge:
1. At high pH, a negatively charged surface prevails.
2. At low pH, a positively charged surface prevails.
3. At some intermediate pH, the pH-dependent surface charge is zero. This pH is called the point of zero charge (PZC).

6.10.2 Cation- and Anion-Exchange Capacity

Cations are attracted to the negative sites on a solid surface (see Section 6.10.1). Cation-exchange capacity (CEC) is defined as the total number of negatively charged sites in a material at which reversible cation adsorption and desorption can occur. Operationally, it is measured by determining the total concentration (usually in meq/100 g of dry soil) of all exchangeable cations sorbed. Thus, CEC is a measure of the reversible adsorptive capacity of a material for cations. At equilibrium, the total sorbed cation charge equals the total negative charge on the solid, resulting in overall neutrality. The portion of CEC not affected by pH changes is due to adsorption to permanently charged sites. The portion of CEC that increases with pH is caused by pH-dependent charged sites. Below about pH 5, H^+ ions are strongly bound to oxygen atoms at crystal edges, making these sites unavailable for cation adsorption. As pH increases above 5, H^+ ions are increasingly released into solution, making new sites available for cation adsorption.

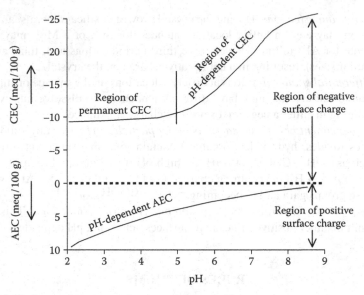

FIGURE 6.7 Ion-exchange capacity dependence on pH.

RULES OF THUMB

- pH-dependent CEC does not change much as pH increases up to about pH 5 (Figure 6.7).
- Above pH 5, CEC increases rapidly with pH.
- AEC increases as pH decreases. Gibbsite, kaolinite, goethite, and allophane clays exhibit small AECs. As pH rises above 3, AEC begins to decrease.

Anion-exchange capacity (AEC) arises mainly from protonation of hydroxyl groups on the surface of minerals and organic particles. It is mostly pH dependent.

$$\text{Surface-OH} + H^+ \rightarrow \text{surface-OH}_2^+$$

6.10.3 EXCHANGEABLE BASES: PERCENT BASE SATURATION

The primary exchangeable bases (exchangeable metal cations) are Na^+, Ca^{2+}, Mg^{2+}, and K^+. They usually occupy the majority of CEC sites in natural environments. The remaining CEC sites are occupied mainly by exchangeable H^+. The surface concentration of H^+ is pH dependent. Soil with exchangeable H^+ behaves like an acid, releasing H^+ into solution as a function of pH.

Percent base saturation is defined as the percentage of primary exchangeable base cations relative to the total sorbed cation concentration, at, or near, pH 7 ($CEC_{pH} = 7$):

$$\text{Percent base saturation} = (\text{primary exchangeable bases } CEC_{pH=7}) \times 100$$

If there is no pH-dependent surface charge (i.e., only permanent surface charge), $CEC_{pH} = 7$ will equal total exchangeable cations (including H^+) at any pH.

Example 15

Suppose a clay is measured to have the following cations (in meq/100 g):

$$Ca^{2+} = 16.2, Mg^{2+} = 4.4, K^+ = 0.1, Na^+ = 1.6, \text{ and } H^+ = 10.2$$

What is its percent base saturation?

ANSWER

At the pH of the CEC measurement, the total $CEC = 16.2 + 4.4 + 0.1 + 1.6 + 10.2 = 32.5$ meq/100 g. The 22.3 meq/100 g of base exchange capacity due to Ca^{2+}, Mg^{2+}, K^+, and Na^+ represents 68.6% of the total CEC. The 10.2 meq/100 g of exchangeable H^+ is 31.4% of the total CEC. Assume that there is no pH-dependent surface charge. Then

$$CEC_{pH=7} = 32.5 \text{ meq/100 g, and percent base saturation} = 68.6\%$$

Percent base saturation is related to the soil pH as follows:

- The higher the percent base saturation, the higher will be the pH (more sites have been vacated by H^+ and occupied by metal cations).
- The lower the percent base saturation, the lower the pH (more sites are occupied by H^+ and unavailable to metal cations).

Leaching of soils reduces base saturation but does not change CEC. Therefore, soil leaching tends to increase soil acidity.

RULES OF THUMB

1. Soil pH is correlated with the percentage of base saturation.
2. The higher the base saturation, the higher the soil pH. Nearly 90–100% base saturation indicates a soil pH around 7 or higher.
3. Low base saturation (<90%) indicates acidic soils.
4. A 5% decrease in base saturation causes a decrease in pH of about 0.1 unit.

Example 16

Estimate the pH of the soil of Example 15.

ANSWER

The percent base saturation was measured to be 68.6%. At pH 7, the percent base saturation is about 100% (Rule of Thumb 2). Therefore, 68.6% represents a decrease of

$$100 - 68.6 = 31.4\% \text{ from pH } 7$$

A 5% decrease in base saturation causes a decrease in pH of about 0.1 (Rule of Thumb 4). Therefore, $31.4/5 \approx 6$ units of 5% decrease in base saturation, or 6×0.1 units decrease in pH, that is, 0.6 pH units decrease. Soil pH is around $7 - 0.6 = 6.4$. If a soil with a low base saturation is made into a slurry with water, the water pH will be acidic, around 4–5 or lower. Because a low base saturation means that many sites are occupied by H^+, acidic water results from exchangeable H^+ being released from the soil into the solution. Similarly, passing acidic water through soil causes H^+ to exchange with soil-bound metal cations. The most weakly bound metals are displaced first and, if the pH is low enough, more strongly bound metals are displaced. At first, K^+ and Na^+ are removed, and then Ca^{2+} and Mg^{2+} go into solution. Trivalent ions, such as Al^{3+}, are the last to leave the soil and go into solution.

RULE OF THUMB

The presence of an elevated concentration of Al^{3+} (>0.1 mg/L) is a characteristic of acidic water (pH < 5).

6.10.4 CEC IN CLAYS

Clays sorb cations from the solution until the fixed total charge is reached that represents their CEC. The total charge of sorbed cations equals the total negative surface charge on the clay.

A clay particle may be regarded as a system composed of two parts:

1. A relatively large, insoluble negatively charged particle.
2. A loosely held swarm of sorbed exchangeable cations.

CEC depends on the clay crystalline layer structure. Clays have internal surfaces as well as external, because of their layered structure. In some clays, the layers are held together more strongly than in others. Water can penetrate into weakly bound layers and force them apart, causing the clay to swell.

Different clays can differ markedly in their layer structures. Montmorillonite is a strongly swelling clay. Kaolinite does not swell significantly. Clays that swell in water have higher CECs than nonswelling clays. Swelling separates the layers and increases the surface area available for ion exchange. Surface charge on the interlayer surfaces attracts additional cations. Between the layers, sorbed cations with large hydrated diameters (such as Na^+) exert separating forces that can greatly increase clay expansion. This decreases clay permeability to water (see Section 6.11.5).

RULE OF THUMB

A CEC for clay > 10–15 meq/100 g indicates some degree of layer expansion by water swelling.

6.10.5   CEC IN ORGANIC MATTER

Just 1% of humic organic material in a mineral soil contributes a CEC of around 2 meq/100 g of soil, which is about 4 times the CEC of an equal weight of clay.*

RULE OF THUMB

A rough estimate of the CEC of a soil can be made as follows:

1. Estimate or determine the percentages by weight of silicate clay and organic matter.
2. Multiply clay percentage by 0.5 to get the clay contribution.
3. Multiply the organic percentage by 2 to get the humus contribution.

Add the results to get the total CEC in meq/100 g.

Example 17

Suppose a soil contained 2% organic matter and 16% clay. Estimate its CEC.

ANSWER

The soil CEC may be estimated to be around $(2 \times 2) + (16 \times 0.5) = 12$ meq/100 g.

6.10.6   RATES OF CATION EXCHANGE

The rates of exchange depend on the clay type. Kaolinite does not swell in water and there is no inner-layer access. Exchange reactions in kaolinites are almost instantaneous because they occur only on the outer surface. Illites swell slightly, in which exchange reactions can take several hours because a small part of the exchange occurs between inner crystal unit layers, to which cation diffusion is slow. Montmorillonite expands considerably and most of the exposed surfaces are on the inner layers. Montmorillonites take still longer to reach ion-exchange equilibrium.

---

* Clay is much denser than humic matter. An equal weight of clay represents many fewer milliequivalents of CEC.

6.11 AGRICULTURAL WATER QUALITY

Most water quality–related problems in irrigated agriculture fall into four general types:

1. Salinity of irrigation water
2. Toxicity to plants due to specific ions in irrigation water
3. Nutrient imbalance
4. Low water infiltration rate through soil

Thus, an evaluation of irrigation water quality is commonly based on measuring the total salt content, specific ions that are potentially toxic, determinants of water infiltration rate (chiefly TDS and sodium adsorption ratio [SAR]), and observing crops for signs of nutrient imbalance. Measuring the following parameters will normally allow a basic evaluation of agricultural water quality:

pH	Calcium	Chloride
Alkalinity	Magnesium	Sulfate
TDS	Boron	Bicarbonate
SAR	Copper	Nitrate + nitrite
Sodium	Selenium	Total phosphorus

6.11.1 SALINITY*

In many cases, water salinity, measured in the field as electrical conductivity (EC) and in the laboratory as TDS, is the most important indicator of agricultural water quality. High salinity, due to high concentrations of dissolved salts (TDS), may reduce water availability to plants to the extent that crop yield is affected, because high TDS lowers the maximum osmotic pressure that plants can exert across their root membranes for absorbing water. This effect decreases the rate at which water can be moved from the soil into the plants. When plant roots encounter water with a too-high concentration of TDS, the osmotic pressure balance across the root membrane can even be reversed, driving water from within the plant out into the soil and desiccating the plant. The amount of water transpired through crops in a field is directly related to crop yield.

The high salt concentrations associated with high salinity also increase the risk of specific ion toxicity (Section 6.11.2). Note, however, that high salinity has a good side—it is beneficial with regard to higher water infiltration rates through soil (Table 6.10).

Conversely, low salinity of irrigation water has both good and bad sides—it allows higher osmotic pressure across the root membranes, resulting in easier water absorption, but a too-low salinity can reduce water availability to plants by decreasing the water infiltration rate (Table 6.10). Maintaining a balanced salinity is generally achieved by proper irrigation practices.

---

* The definition of salinity and its relation to TDS are found in Chapter 3, Section 3.9.3.

TABLE 6.10
Suggested Maximum Parameter Levels in Groundwaters Used for Crop Irrigation

General Problem	Problem Parameters	Units	Degree of Restriction		
			None	Moderate	Severe
Salinity	Total dissolved solids (TDS)	mg/L	475	475–2200	>2200
	Electrical conductivity (EC)	μS/cm	700	700–3000	>3000
Water infiltration	Adj. sodium adsorption ratio (SAR-adj.).[a] Dominant clay in soil (crystal lattice)[b] montmorillonite/smectites (2:1)	—	<6	6–9[c]	>9
	Illite/vermiculite (2:1)		<8	8–16[c]	>16
	Kaolinite/sesquioxides (1:1)		<16	16–24[c]	>24
Specific ion toxicity	Salinity (EC)	μS/cm	>500	500–200	<200
	Sodium (Na)[d][e]	mg/L	<70	70–200	>200
		SARadj	<3	3–9	>9
	Chloride (Cl)[d][e]	mg/L	<140	140–350	>350
	Boron (B)	mg/L	<0.75	0.75–2.0	>2.0
	Trace elements[f]	*Units*	*Suggested Maximum Value*		
	Aluminum (Al)	mg/L		5	
	Arsenic (As)	mg/L		0.1	
	Beryllium (Be)	mg/L		0.1	
	Cadmium (Cd)	mg/L		0.01	
	Cobalt (Co)	mg/L		0.05	
	Chromium (Cr)	mg/L		0.10	
	Copper (Cu)	mg/L		0.20	
	Fluoride (F)	mg/L		1.0	

(Continued)

TABLE 6.10 (*Continued*)
Suggested Maximum Parameter Levels in Groundwaters Used for Crop Irrigation

General Problem	Problem Parameters	Units	
	Iron (Fe)	mg/L	5.0
	Lithium (Li)	mg/L	2.5
	Manganese (Mn)	mg/L	0.20
	Molybdenum (Mo)	mg/L	0.01
	Nickel (Ni)	mg/L	0.20
	Lead (Pb)	mg/L	5.0
	Selenium (Se)	mg/L	0.02
	Vanadium (V)	mg/L	0.10
	Zinc (Zn)	mg/L	2.0

[a] SAR-adj. = adjusted SAR, see Section 6.11.5.2.

[b] Limits are for the dominant type of clay mineral in the soil, since structural stability varies between the various clay types. Problems are less likely to develop if water salinity is high and more likely to develop if water salinity is low.

[c] Use the lower value if EC < 400 µS/cm; use a middle value if EC = 400–1600 µS/cm; use the upper value if EC > 1600 µS/cm.

[d] Most tree crops and woody ornamentals are sensitive to sodium and chloride (use limits shown). Most annual crops are not sensitive.

[e] With sprinkler irrigation on sensitive crops, use about 75% of the given limits for sodium and chloride. Sodium or chloride in excess of 3 meq L^{-1} under certain conditions has resulted in excessive leaf absorption and crop damage.

[f] Toxicity and the suggested maximum value depend strongly on the crop. Normally not monitored unless a problem is expected. Several trace elements are essential nutrients in low concentrations.

Source: Data from FAO, "Water Quality for Agriculture," FAO Irrigation and Drainage Paper No. 29, Rev. 1, Food and Agriculture Organization of the United Nations, 1986, and Colorado water quality standards for agricultural uses.

6.11.2 Specific Ion Toxicity

Certain ions can accumulate in sensitive crops to concentrations high enough to cause plant damage and reduce yields. Specific ion toxicity arises mainly from chloride, sodium, and boron. Other trace elements that may also be toxic to plants are usually present in groundwater in such low concentrations that they seldom are a problem. In general, concentrations of concern for specific ion toxicity are lower for sprinkler irrigation than for surface irrigation because toxic ions can be absorbed directly into the plant through leaves wetted by the sprinkler water. Direct leaf absorption speeds the rate of accumulation of toxic ions in plants.

- *Chloride:* Although chloride is essential to plants in very low amounts,* it can cause toxicity to sensitive crops at higher concentrations (Table 6.10). Chloride is the most common source of specific ion toxicity in agriculture. Chloride mobility is not retarded significantly by soils nor fully blocked by root tissues. Therefore, it moves readily with soil–water to the root zone, where it is absorbed by plants and carried by transpiration to the leaves, where it accumulates. Symptoms such as leaf burn or drying of leaf tissue normally occur initially at the extreme tip of the leaves when the chloride concentration exceeds the plant's tolerance level.
- *Sodium:* Sodium is similar to chloride in its mobility through soils and passage across root membranes. However, toxicity symptoms from sodium differ from those of chloride in that leaf burn, scorch, and dead tissue occur initially along the outside edges of leaves, instead of the extreme leaf tip as with chloride. Symptoms appear first on older leaves, starting at the outer edges and, as the severity increases, moving progressively inward between the veins toward the leaf center. Because of the common factor of high sodium, plant deterioration that is due to poor infiltration of water to the root zone (usually caused by a too-high SAR, discussed in Section 6.11.5.1) is sometimes mistaken for sodium toxicity. However, with sodium-sensitive crops (mostly tree crops and woody ornamentals), sodium toxicity begins at a lower SAR value than would be expected to cause a permeability problem.
- *Boron:* Boron, like chloride, is essential in low amounts, but boron toxicity begins at lower concentrations than for chloride (Table 6.10). Toxicity can occur on sensitive crops at concentrations less than 1.0 ppm. Surface water rarely contains enough boron to be toxic but well water or springs may contain toxic amounts because of contact with boron-containing minerals, especially near geothermal areas and earthquake faults.

6.11.3 Nutrient Imbalance

The main cause of nutrient imbalance in irrigation water is excessive nitrogen. Although nitrogen is an essential plant nutrient that is required in relatively large

---

* Chloride is an essential plant nutrient, but in such small quantities that it frequently is considered nonessential.

amounts, excessive concentrations can stimulate too high a growth rate, resulting in weak supporting stalks, delayed plant maturity, and poor overall crop quality.

Nutrient nitrogen exists in water mainly as nitrate anion (NO_3^-), although ammonium cation (NH_4^+) and nitrite anion (NO_2^-) may also be present (see Chapter 4, Section 4.2). The concentration of NO_3–N in most surface and groundwater is usually less than 5 mg/L, but in some cases of groundwater it may exceed 50 mg/L NO_3–N.

Most crops are relatively unaffected by a nitrogen concentration less than 30 mg/L-N. However, too-high nitrogen levels may promote vegetation growth but reduce fruit production. Also, sensitive crops like sugar beets, grapes, apricot and citrus trees, many grains, and avocados may experience deterioration of quality at nitrogen concentrations above 5 mg/L-N. Below 5 mg/L-N, nitrogen is generally beneficial, even for nitrogen-sensitive crops, but can still stimulate nuisance growth of algae and aquatic plants in streams, lakes, canals, and drainage ditches (see Chapter 4, Section 4.4.3).

6.11.4 IRRIGATION WATER QUALITY GUIDELINES

Table 6.10 lists potential problem parameters for determining irrigation water quality. Most of the parameters listed as trace elements need to be monitored only for certain sensitive crops.

6.11.5 LOW INFILTRATION RATE (HIGH RELATIVE SODIUM AND LOW SALINITY)

Soil texture is an important factor in determining soil permeability and water infiltration rates. A low water infiltration rate may reduce water flow through soil so that the root zone does not receive sufficient water for optimal plant growth.* Soil permeability is strongly dependent on the presence of dissolved salts in the soil water, especially those of sodium, calcium, and magnesium; the composition of soil water is, in turn, strongly dependent on the composition of water used for irrigation. The electrostatic fields around cations and anions exert stabilizing forces that help to maintain soil aggregation and support a soil structure that is sufficiently porous for adequate water infiltration. The infiltration rate generally increases with increasing water salinity and decreases with either decreasing salinity or increasing sodium content relative to calcium and magnesium. Irrigation water quality is commonly described in terms of its salinity hazard (using TDS or EC) and its sodium hazard (using an empirical quantity called the *SAR*† described in Section 6.11.5.1). These two factors, salinity and SAR, must be considered together to properly evaluate the net effect on water infiltration rate.

---

* An infiltration rate of 3 mm/h is considered low, while a rate of 12 mm/h is relatively high. This section is concerned with how irrigation water quality can affect infiltration rates; note however, that infiltration rates can be affected by factors other than water quality, including soil texture and type of clay minerals.
† An empirical quantity called the SAR, based on relative sodium, calcium, and magnesium concentrations, has been developed to estimate the effect of water quality on soil infiltration rates (see Section 6.11.5.1). An improved version of the SAR, called SAR-adj., is discussed in Section 6.11.5.2.

- High TDS water (>5 dS/cm) helps to maintain soil aggregation and higher infiltration rates. Note, however, from Section 6.11.1, that high TDS also lowers osmotic pressure across root membranes.
- Low TDS water (<0.5 dS/m) is "corrosive" in that it leaches soluble sorbed and precipitated salts from the soil, causing soil dispersion into smaller particles that can fill smaller pore spaces and reduce infiltration rates. Very low TDS water (<0.2 dS/m) usually results in low infiltration rates, regardless of the SAR (see Section 6.11.1).
- Excessive sodium in irrigation water also promotes soil dispersion and structural breakdown if the sodium content is too high relative to other dissolved cations, especially calcium and magnesium. This can occur even when total TDS is in a range that normally would not cause excessive soil dispersion if sodium levels were lower, because of the sodium cation's large radius of hydration (see Section 6.11.5.1). Because sodium, calcium, and magnesium are by far the most abundant cations in most natural waters, the effects of other cations can often be neglected for their effect on soil dispersion. The SAR, developed to estimate the effect of water quality on soil infiltration rates, is based on relative sodium, calcium, and magnesium concentrations.

6.11.5.1 Sodium Adsorption Ratio

The SAR indicates the amount of sodium present in soil, relative to calcium and magnesium. If the sodium fraction is too large, soil permeability of soils containing certain clays may be low and the movement of water through the soil is restricted. SAR is important for plant growth because its magnitude may indicate the availability of soil pore water to plant roots.

Most clays are made of crystalline sheets of silicon oxide and metal oxide, arranged in stacked layers. These are classified as either type 1:1 or type 2:1, depending on their layer structure. Clays with 2:1 structure (e.g., smectite, illite, montmorillonite, etc.) carry negative surface charges and attract dissolved cations (and also polar water molecules) that are sorbed into the space between the layers, causing the clays to swell and increase in volume. Clays with 1:1 structure (e.g., kaolinite, serpentine, etc.) are not charged and do not swell significantly when wet. When soil with high 2:1 clay content (>30% by weight) is irrigated with high SAR/low EC water, the decrease in water permeability can be severe; 1:1 clay soils irrigated with similar water do not lose permeability.

A sodium ion has a large radius of hydration, which means that its hydration sphere of bound water molecules is larger than for most other cations. If the sodium concentration is large compared to other dissolved metals, the degree of swelling is correspondingly greater. Swelling reduces the average soil pore size and causes a decrease in water permeability. If the sodium ion is replaced by ion exchange with cations having a smaller radius of hydration, soil dispersion is reduced and soil permeability increased. When soil with high 2:1 clay content (>30% by weight) is irrigated with high SAR/low EC water, the decrease in water permeability can be severe; 1:1 clay soils irrigated with similar water do not lose permeability.

In most natural waters, Ca^{2+}, Mg^{2+}, and Na^+ are by far the most abundant cations, so other cations can often be neglected for their effect on soil dispersion. Also, Ca^{2+} and Mg^{2+}, being doubly charged, are more tightly sorbed to clay surfaces than sodium and, so, are preferentially sorbed. They also have smaller radii of hydration and cause less soil dispersion. The relative amounts of sodium, calcium, and magnesium sorbed to soil are proportional to the amounts dissolved in irrigation or groundwater. By measuring the concentrations of sodium, calcium, and magnesium in water used for irrigation, the risk of decreased soil permeability can be evaluated by calculating the SAR using the empirical Equation 6.29.

$$SAR = \frac{[Na^+]}{\sqrt{\dfrac{[Ca^{2+}]+[Mg^{2+}]}{2}}} \qquad (6.29)$$

In Equation 6.29, $[Na^+]$, $[Ca^{2+}]$, and $[Mg^{2+}]$ represent water concentrations expressed in meq/L (see Section 1.3.4).

6.11.5.2 Adjusted Sodium Adsorption Ratio

For most irrigation waters, the standard SAR formula, Equation 6.29, is suitable for determining a potential sodium hazard. However, Equation 6.29 does not take into account changes in SAR caused by changes in water quality that may occur during or after irrigation, under conditions of high alkalinity and/or high bicarbonate concentrations. While the solubility of sodium remains high over a very wide range of environmental conditions, the solubilities of calcium and magnesium do not. Changes in water quality during or after irrigation can cause calcium and magnesium to precipitate while the dissolved sodium concentration remains unchanged. This will increase the magnitude of the SAR and, hence, the risk of diminished infiltration.

Causes of changes in water quality during or after irrigation include:

- Increased concentrations of dissolved calcium and magnesium salts due to absorption of "pure" water by plants between irrigation intervals.
- Decreased concentrations of dissolved calcium and magnesium salts due to dilution by applied water.
- Increased concentrations in dissolved calcium and magnesium salts due to dissolution of soil minerals. Dissolution is promoted by dilution and by carbon dioxide dissolved in the soil–water
- Decreased concentrations in dissolved calcium and magnesium salts due to precipitation of calcium and magnesium salts. Precipitation will occur when the concentrations of calcium, magnesium, and carbonate are sufficient to exceed the solubility products (see Appendix B) of their corresponding salts.

While sodium concentrations are little affected by changes in water quality, concentrations of magnesium and especially calcium can change significantly, resulting in different SAR measurements before and after the application of irrigation water.

As stated earlier, water conditions of high alkalinity and/or high bicarbonate concentrations promote precipitation of calcium salts, which lowers calcium and magnesium concentrations and increases the SAR.

RULE OF THUMB

An adjusted SAR (SAR-adj.) is recommended instead of the standard SAR, for irrigation water with high concentrations of alkalinity (pH > 8.5) and bicarbonate (HCO_3^- > 200 mg/L).

6.11.5.3   Calculation of Adjusted Sodium Adsorption Ratios

The standard SAR (Equation 6.29) is suitable for typical irrigation water, but it does not take into account the effects of water salinity or of changes in SAR that may occur during or after irrigation because of changes in the solubility of calcium. Irrigating with water having high alkalinity and/or high bicarbonate concentrations can cause calcium to precipitate as calcium carbonate while, at the same time, the dissolved sodium concentration remains unchanged.

This increases the magnitude of the SAR and, hence, the risk of diminished infiltration. A more accurate sodium hazard calculation has been developed that accounts for the effects of salinity and calcium precipitation (Ayers and Westcot 1994; Suarez 1981). This more accurate method generates a value known as the SAR-adj.*

Although the simpler standard SAR (Equation 6.29) is useful for most irrigation water and is simpler to calculate, the SAR-adj. is preferred, because it includes the effects of salinity and potential calcium precipitation. The SAR-adj. differs from the standard SAR only by using a modified value of calcium, labeled Ca_x. It is calculated by

$$\text{SAR-adj.} = \frac{[Na]}{\sqrt{\dfrac{[Ca_x] + [Mg]}{2}}} \qquad (6.30)$$

In Equation 6.30, all concentrations are in meq/L and Ca_x is a modified calcium value that adjusts the measured calcium concentration in irrigation water to what the effective calcium concentration would be in soil water because of chemical changes that occur after irrigation.

Ca_x includes the effects on the sodium hazard introduced by salinity, dissolved carbon dioxide, and the precipitation or dissolution of calcium (based on the bicarbonate/calcium ratio), and assumes that no magnesium is precipitated.

---

* Different efforts to improve the SAR calculation have produced some confusion concerning how to identify the different results. In technical reports and laboratory data, the terms most often used for expressing the sodium hazard are *SAR and SAR-adj.*
  • SAR normally means the standard SAR calculated with Equation 6.29.
  • SAR-adj. is the term used for the calculation method described in this chapter. It not only accounts for the potential of calcium to precipitate based on alkalinity, but also includes the effects of salinity. It is currently the most accurate way to determine the sodium hazard of irrigation water.

Its calculation is complicated, making use of equations given in Suarez (1981) along with the extended Debye–Huckel equation to account for salinity, but accurate approximations are available. Two convenient methods are illustrated in Examples 19–21, one that uses a look-up table for Ca_x, Table 6.11, and another that calculates Ca_x with Equation 6.31 or 6.32.*

$$Ca_x = 10^{\left\{0.283 - \left(0.667 \times \log_{10} \frac{[HCO_3]}{[Ca]} + 0.022 \times EC\right)\right\}}$$ (6.31)

Equation 6.31 may also be written:

$$\log_{10} Ca_x = \left\{0.283 - \left(0.667 \times \log_{10} \frac{[HCO_3]}{[Ca]} + 0.022 \times EC\right)\right\}$$ (6.32)

In Equations 6.31 and 6.32, $[HCO_3]$ and $[Ca]$ are in meq/L and EC is in dS/m (1 dS/m = 1000 µS/cm). The rationale for Equation 6.32 may be found in Lesch and Suarez (2009) and in Stowell and Gelernter (n.d.).

RULE OF THUMB

For most irrigation water, the standard SAR calculated with Equation 6.29 is satisfactory; SAR and SAR-adj. are normally within 10% of one another, which generally makes little difference in the resulting sodium hazard. It is only under extreme conditions of high bicarbonate and low EC concentrations that the difference becomes significant (see Equations 6.11 through 6.13).

6.11.5.4   What SAR Values Are Acceptable?

Potential soil infiltration problems of irrigation water cannot be adequately assessed based on SAR alone. This is because the swelling potential of low TDS water is greater than high TDS water with the same SAR or SAR-adj. values. Therefore, a more accurate evaluation of the sodium infiltration/permeability hazard requires using the EC together with SAR and SAR-adj.

Acceptable SAR or SAR-adj. values[†] for irrigation water are dependent on the particular water and soil characteristics and on the EC of irrigation water. As noted

---

* A convenient spreadsheet programmed to calculate SAR-adj. has been placed on the Internet (Stowell and Gelernter, n.d.): http://www.paceturf.org/index.php/journal/index_new/C50/.
† In the following discussions, the limiting values for SAR are the same in magnitude whether calculated for the standard SAR or for SAR-adj.; SAR-adj. just estimates a more appropriate numerical value for calcium when loss of calcium by precipitation is likely. Therefore, Figure 6.8 and Table 6.12 are suitable for either SAR or SAR-adj.

in Section 6.11.5, high-salinity water tends to stabilize soil structure and maintain permeability. Thus, high EC limits the deleterious effects of sodium (Figure 6.8). It should be noted that water with EC greater than about 3000 µS/cm is of poor quality for irrigation regardless of the SAR value (Section 6.11.1).

RULES OF THUMB

1. High SAR/low EC values in irrigation water are likely to significantly reduce the water permeability in soils containing >30% by weight of 2:1 clays (e.g., smectite, illite, montmorillonite, etc.).
2. Water having an SAR value less than 3 will not diminish the soil permeability when the EC is greater than about 200 µS/cm (0.2 dS/m).
3. Water having SAR values between 3 and 6 and EC between 500 and 2000 µS/cm may require care with irrigation methods. Soil permeability may be diminished.
4. Water having SAR values greater than 10 and EC less than 500 µS/cm will require considerable care with irrigation methods. Soil permeability may be too low to allow sufficient water to reach plant roots.

Some uses of standard SAR and SAR-adj. are illustrated by the following examples.

Example 18: Determination of Standard Sodium Adsorption Ratio

Determine the standard SAR, using Equation 6.29, for two different water supplies, A and B. Measured concentrations were

Supply A	Ca^{2+} = 4.6 mg/L	Mg^{2+} = 1.1 mg/L	Na^+ = 144 mg/L	TDS = 255 mg/L
Supply B	Ca^{2+} = 108 mg/L	Mg^{2+} = 33 mg/L	Na^+ = 63 mg/l.	TDS = 1050 mg/L

ANSWER

1. Determine the Ca^{2+}, Mg^{2+}, and Na^+ concentrations in meq/L (see discussion of equivalent weight in Chapter 1).

	Equivalent Weight (g/eq or mg/meq)	Concentration			
		Water Supply A		Water Supply B	
Constituent		(mg/L)	(meq/L)	(mg/L)	(meq/L)
Ca^{2+}	20.0	4.6	0.23	108	5.40
Mg^{2+}	12.15	1.1	0.090	33	2.72
Na^+	23.0	255	11.1	63	2.74

TABLE 6.11

Look-Up Table for the Modified Calcium Concentration, Ca$_x$ to Be Used in Calculations of Sar-Adj., Given as a Function of EC and Bicarbonate/Calcium Ratio (HCO$_3$/Ca), Both Measured in the Irrigation Water Before Application to Crops

EC of applied water (dS/m) HCO$_3$/Ca(meq/L)	Ca$_x$ (meq/L)											
	0.1	0.2	0.3	0.5	0.7	1.0	1.5	2.0	3.0	4.0	6.0	8.0
.05	13.20	13.61	13.92	14.40	14.79	15.26	15.91	16.43	17.28	17.97	19.07	19.94
.10	8.31	8.57	8.77	9.07	9.31	9.62	10.02	10.35	10.89	11.32	12.01	12.56
.15	6.34	6.54	6.69	6.92	7.11	7.34	7.65	7.90	8.31	8.64	9.17	9.58
.20	5.24	5.40	5.52	5.71	5.87	6.06	6.31	6.52	6.86	7.13	7.57	7.91
.25	4.51	4.65	4.76	4.92	5.06	5.22	5.44	5.62	5.91	6.15	6.52	6.82
.30	4.00	4.12	4.21	4.36	4.48	4.62	4.82	4.98	5.24	5.44	5.77	6.04
.35	3.61	3.72	3.80	3.94	4.04	4.17	4.35	4.49	4.72	4.91	5.21	5.45
.40	3.30	3.40	3.48	3.60	3.70	3.82	3.98	4.11	4.32	4.49	4.77	4.98
.45	3.05	3.14	3.22	3.33	3.42	3.53	3.68	3.80	4.00	4.15	4.41	4.61
.50	2.84	2.93	3.00	3.10	3.19	3.29	3.43	3.54	3.72	3.87	4.11	4.30
.75	2.17	2.24	2.29	2.37	2.43	2.51	2.62	2.70	2.84	2.95	3.14	3.28
1.00	1.79	1.85	1.89	1.96	2.01	2.09	2.16	2.23	2.35	2.44	2.59	2.71
1.25	1.54	1.59	1.63	1.68	1.73	1.78	1.86	1.92	2.02	2.10	2.23	2.33
1.50	1.37	1.41	1.44	1.49	1.53	1.58	1.65	1.70	1.79	1.86	1.97	2.07
1.75	1.23	1.27	1.30	1.35	1.38	1.43	1.49	1.54	1.62	1.68	1.78	1.86

2.00	1.13	1.16	1.19	1.22	1.26	1.31	1.36	1.40	1.48	1.54	1.63	1.70
2.25	1.04	1.08	1.10	1.14	1.17	1.21	1.26	1.30	1.37	1.42	1.51	1.58
2.50	0.97	1.00	1.02	1.06	1.09	1.12	1.17	1.21	1.27	1.32	1.40	1.47
3.00	0.85	0.89	0.91	0.94	0.96	1.00	1.04	1.07	1.13	1.17	1.24	1.30
3.50	0.78	0.80	0.82	0.85	0.87	0.90	0.94	0.97	1.02	1.06	1.12	1.17
4.00	0.71	0.73	0.75	0.78	0.80	0.82	0.86	0.88	0.93	0.97	1.03	1.07
4.50	0.66	0.68	0.69	0.72	0.74	0.76	0.79	0.82	0.86	0.90	0.95	0.99
5.00	0.61	0.63	0.65	0.67	0.69	0.71	0.74	0.76	0.80	0.83	0.88	0.93
7.00	0.49	0.50	0.52	0.53	0.55	0.57	0.59	0.61	0.64	0.67	0.71	0.74
10.00	0.39	0.40	0.41	0.42	0.43	0.45	0.47	0.48	0.51	0.53	0.56	0.58
20.00	0.24	0.25	0.26	0.26	0.27	0.28	0.29	0.30	0.32	0.33	0.35	0.37
30.00	0.18	0.19	0.20	0.20	0.21	0.21	0.22	0.23	0.24	0.25	0.27	0.28

Source: Adapted from Suarez, D. L., *Soil Sci. Soc. Am. J.*, 45, 469, 1981.

Notes: Ca_x = effective concentration of calcium in soil water following irrigation. Ca = measured concentration of calcium in irrigation water before irrigation. Units of Ca_x, HCO_3, and Ca are meq/L and EC is dS/m.

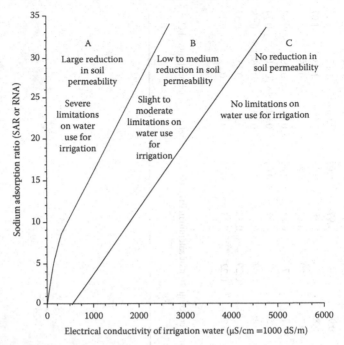

FIGURE 6.8 Effects of SAR or SAR-adj. and EC on soil permeability. Zone A exhibits large reductions in soil permeability to water and will severely restrict the amount of water available to plant roots. Zone B has low to moderate reductions in soil permeability and may require care in irrigation practices and choice of crops. Zone C water has little or no reduction in soil permeability. Note, however, that water with EC greater than 3000 μS is of poor quality for agriculture because of osmotic stress, regardless of soil permeability.

2. Compute the SAR of the two waters.
   a. For Supply A:

$$SAR(A) = \frac{[Na^+]}{\sqrt{\dfrac{Ca^{2+} + Mg^{2+}}{2}}} = \frac{11.1}{\sqrt{\dfrac{0.23 + 0.090}{2}}} = 27.8$$

   Water Supply A has a very high SAR value (27.8), and the water would be generally unacceptable for irrigation. Note that the TDS value for water Supply A is moderate (255 mg/L) and that Supply A water is generally excellent for domestic use.
   b. For Supply B:

$$SAR(B) = \frac{2.74}{\sqrt{\dfrac{5.40 + 2.72}{2}}} = 1.36$$

Water Supply B has a low SAR value (1.36) and a TDS value (1050 mg/L, equivalent to about 1750 μS/cm, see Chapter 3, Section 3.12.3) that is beneficial for agriculture and would be an excellent source of irrigation water. The TDS value is high for potable water and water supply B is an inferior source of domestic water compared with Supply A. Supply B will require more treatment to produce acceptable drinking water.

Example 19: Calculation of SAR-adj. for High-Calcium Bicarbonate Water by the Look-Up Table Method

Laboratory analysis of a sample of high-calcium bicarbonate water, from a source being considered for irrigation, returned the following data:

Parameter	Concentration (mg/L)	Concentration (meq/L)
Na	820	35.7
Ca	553	27.6
Mg	149	12.2
HCO$_3$	641	10.5
EC	2.15 dS/m	2.15 dS/m

1. Calculate the ratio $\dfrac{HCO_3}{Ca}$ from the water analysis for use in Table 6.11:

$$\frac{HCO_3}{Ca} = \frac{10.5}{27.6} = 0.380$$

2. Note the value of EC in the water analysis: EC = 2.15 dS/m.
3. Use these results in Table 6.11 to find Ca$_x$. Use the closest values available; interpolation is usually not necessary:
   HCO$_3$/Ca = 0.380; use the 0.40 row in the table.
   EC = 2.15 dS/m; use the 2.2 column.
   Reading across the HCO$_3$/Ca = 0.40 row to the EC = 2.0 column yields the value Ca$_x$ = 4.11.
4. Insert the value of Ca$_x$ from step 3 and the values for Mg and Na from the water analysis into Equation 6.30.

$$adj\ RNa = \frac{[Na]}{\sqrt{\dfrac{[Ca_x] + [Mg]}{2}}} = \frac{35.7}{\sqrt{\dfrac{[4.11] + [12.2]}{2}}} = 12.5$$

The standard SAR, calculated from Equation 6.29 with [Ca] = 27.6, equals 8.42, significantly smaller than SAR-adj. = 12.5. Using SAR = 8.00 and EC = 2.15 dS/m, Table 6.12 estimates no restrictions on the use of the irrigation water. However, the more accurate SAR-adj. = 12.5 and EC = 2.15 dS/m estimates slight to moderate restrictions on the use of the irrigation water.

TABLE 6.12
Effect of Sodium Hazard and Electrical Conductivity on Water Infiltration Rate

	Degree of Restriction on Use of Water for Irrigation at Different EC Ranges		
SAR or SAR-adj.	None	Slight to Moderate	Severe
0–3	>0.7 dS/m (700 µS/cm)	0.7–0.2 dS/m (700–200 µS/cm)	<0.2 dS/m (200 µS/cm)
3–6	>1.2 dS/m (1200 µS/cm)	1.2–0.3 dS/m (1200–300 µS/cm)	<0.3 dS/m (300 µS/cm)
6–12	>1.9 dS/m (1900 µS/cm)	1.9–0.5 dS/m (1900–500 µS/cm)	<0.5 dS/m (500 µS/cm)
12–20	>2.9 dS/m (2900 µS/cm)	2.9–1.3 dS/m (2900–1300 µS/cm)	<1.3 dS/m (1300 µS/cm)
20–40	>5.0 dS/m (5000 µS/cm)	5.0–2.9 dS/m (5000–2900 µS/cm)	<2.9 dS/m (2900 µS/cm)

Example 20: Calculation of SAR-adj. for High-Calcium Bicarbonate Water by Equation 6.33

Using the same irrigation water analysis as in Example 19, calculate SAR-adj. using Equation 6.33.

$$\log_{10} Ca_x = \left\{ 0.283 - \left(0.667 \times \log_{10} \frac{[HCO_3]}{[Ca]} + 0.022 \times EC \right) \right\} \qquad (6.33)$$

$$\log_{10} Ca_x = \left\{ 0.283 - \left(0.667 \times \log_{10} \frac{[10.5]}{[27.6]} + 0.022 \times 2.15 \right) \right\} = 0.516$$

$$Ca_x = 10^{0.516} = 3.28$$

$$adj\ RNa = \frac{[Na]}{\sqrt{\dfrac{[Ca_x] + [Mg]}{2}}} = \frac{35.7}{\sqrt{\dfrac{[3.28] + [12.2]}{2}}} = 12.8$$

The small difference in SAR-adj. values determined in Examples 19 and 20 arises from interpolation approximations when using Table 6.11.

Example 21: Comparison of SAR-adj. and SAR for a Typical Irrigation Water

Laboratory analysis of a sample of typical irrigation water returned the following data:

Parameter	Concentration (mg/L)	Concentration (meq/L)
Na	53	2.30
Ca	277	13.8
Mg	64	5.26
HCO$_3$	85	1.39
EC	2.34 dS/m	2.34 dS/m

Calculate the standard SAR and compare with SAR-adj. calculated with Equation 6.33.

$$SAR = \frac{[Na]}{\sqrt{\dfrac{[Ca]+[Mg]}{2}}} = \frac{[2.30]}{\sqrt{\dfrac{[13.8]+[5.26]}{2}}} = 0.745$$

$$\log_{10} Ca_x = \left\{ 0.283 - \left(0.667 \times \log_{10} \frac{[HCO_3]}{[Ca]} + 0.022 \times EC \right) \right\}$$

$$\log_{10} Ca_x = \left\{ 0.283 - \left(0.667 \times \log_{10} \frac{[1.39]}{[13.8]} + 0.022 \times 2.34 \right) \right\} = 0.896$$

$$Ca_x = 10^{0.896} = 7.88$$

$$SAR\text{-adj.} = \frac{[Na]}{\sqrt{\dfrac{[Ca_x] + [Mg]}{2}}} = \frac{2.30}{\sqrt{\dfrac{7.88 + 5.26}{2}}} = 0.897$$

For this low-calcium bicarbonate irrigation water, the standard SAR and SAR-adj. values are quite close and both indicate no sodium hazard and no restriction for irrigation use.

6.11.6   Case Study: Changes over Time in SAR of Coal-Bed Methane–Produced Water

(This case study is based on the report by Sessoms et al. 2002.)

Produced water from coal-bed methane (CBM) production in the Powder River Basin in Wyoming often contains high concentrations of dissolved sodium bicarbonate, and the surface water and soils of the region contain high concentrations of dissolved calcium (Rice et al. 2000). In this laboratory study, experiments were performed that simulated exposing typical CBM-produced water to the atmosphere and discharging it into receiving surface waters characteristic of the Powder River Basin. The experiments demonstrated that if CBM-produced water containing high concentrations of sodium bicarbonate is discharged into regional surface waters or applied to regional soils as irrigation water, soluble sodium bicarbonate in the water can initiate the precipitation of calcium carbonate, resulting in

- Significant increases in SAR values
- Small decreases in EC values
- Increases in pH

The research findings indicated that, after CBM-produced water is brought to the surface, SAR values tend to increase with time. The processes of evaporation, concentration, and precipitation within a stream channel, impoundment, or soil are likely to make final SAR values significantly different from those in the CBM-produced water shortly after it has been discharged from the well. Therefore, SAR should be assessed at the irrigated location after CBM-produced water has been used for irrigation, and not necessarily at the location of initial discharge or exposure to the atmosphere. For NPDES permit monitoring, the SAR-adj. should be used to account for calcium carbonate precipitation and a specific point of compliance selected to allow for in-stream changes in water chemistry.

Exercises

1. What are soil horizons?
2. The pores of soil contain a significant amount of air. How does the composition of soil air differ from that of atmospheric air and why is it so?
3. Many soils have ion-exchange properties. What does this mean?
4. A chemical spill has contaminated a lake with carbon tetrachloride (CCl_4). For carbon tetrachloride, Table 6.5 shows that log K_{ow} = 2.73 (thus, K_{ow} = 5.4×10^2). The maximum concentration of carbon tetrachloride that a certain species of fish can accumulate with a low probability of health risks is reported to be 10 ppm. Assuming that the fatty tissue of these fish behaves similarly to octanol with respect to carbon tetrachloride partitioning, at what value should the water quality standard for carbon tetrachloride be set in this lake to protect this fish?
5. A mountain stream is classified for Cold Water Aquatic Life use. The average atmospheric pressure at the stream's elevation of 6600 feet is about 0.8 atm. A very low BOD and turbulence in the shallow stream assures that oxygen in the stream water is in equilibrium with oxygen in the atmosphere. What is the DO (concentration of dissolved oxygen in mg/L) of the stream? Use Henry's law. The level of approximation allows you to neglect the vapor pressure of water and to assume a temperature of 20°C. Compare your answer with Table 3.5 in Chapter 3.

REFERENCES

Alexander, M. 1995. "How Toxic Are Toxic Chemicals in Soil?" *Environmental Science & Technology* 29 (11): 2713–7.

Ayers, R. S., and D. W. Westcot. 1994. *Water Quality for Agriculture*. FAO Handbook 29. http://www.fao.org/docrep/003/t0234e/t0234e00.html (valid 6/9/2010).

Canadian Council of Resource and Environmental Ministers (CCREM). 1987. *Canadian Water Quality Guidelines*. Ottawa, ON: Inland Waters Directorate, Environmental Canada.

FAO. 1986. "Water Quality for Agriculture." FAO Irrigation and Drainage Paper No. 29, Rev. 1, Food and Agriculture Organization of the United Nations.

Fetter, C. W. 1993. *Contaminant Hydrogeology*. New York: Macmillan.

Freeze, R. A., and J. A. Cherry. 1979. *Groundwater*. Upper Saddle River NJ: Prentice-Hall, Inc.

Lesch, S. M., and D. L. Suarez. 2009. "Technical Note: A Short Note on Calculating the Adjusted SAR Index." *Transactions of the American Society of Agricultural and Biological Engineers* 52 (2): 493–6, ISSN 0001-2351.

Lide, David R., ed. (1990 or later). *Handbook of Chemistry and Physics*. 71st ed. Boca Raton, FL: CRC Press.

Lyman, W. J., W. F. Reehl, and D. H. Rosenblatt. 1990. *Handbook of Chemical Property Estimation Methods*. 2nd printing. Washington, DC: American Chemical Society.

Mackay, D., and W. Y. Shiu. 1981. "A Critical Review of Henry's Law Constants for Chemicals of Environmental Interest." *Journal of Physical and Chemical Reference Data* 10 (4): 1175–99.

Maidment, D. R., ed. 1993. *Handbook of Hydrology*. New York: McGraw-Hill, Inc.

Means, J. C., and R. Wijayratne. 1982. "Role of Natural Colloids in the Transport of Hydrophobic Pollutants." *Science* 215: 968–70.

Rice, C. A., M. S. Ellis, and J. H. Bullock. 2000. "Water Co-Produced with Coalbed Methane in the Powder River Basin, Wyoming: Preliminary Compositional Data." Open-file Report 00-372. Denver, CO: U.S. Department of the Interior, U.S. Geological Survey.

Rook, J. 1974. "Formation of Haloforms during Chlorination of Natural Waters." *Journal of the Society for Water Treatment Examination* 23: 234.

Sessoms, H. N., J. W. Bauder, K. Keith, and K. E. Pearson. 2002. *Chemical Changes in Coalbed Methane Product Water Over Time, 2002*. Bozeman, MT: Department of Land Resources and Environmental Sciences, Montana State University. http://waterquality.montana.edu/docs/methane/cbmwater.pdf.

Shinozuka,T., M. Shibata, and T. Yamaguchi. 2004. "Molecular Weight Characterization of Humic Substances by MALDI-TOF-MS." *Journal of Mass Spectrometry Society of Japan* 52 (1): 29–32.

Stowell, L., and W. Gelernter. n.d. "Tools for Estimating Sodium Hazard Based on Irrigation Water Quality Reports." Report sponsored by PACE Turf LLC. http://www.paceturf.org/index.php/journal/index_new/C50/.

Suarez, D. L. 1981. "Relation between pHc and Sodium Adsorption Ratio (SAR) and an Alternative Method of Estimating SAR of Soil or Drainage Waters." *Soil Science Society of American Journal* 45: 469–75.

USEPA. 1996. *Soil Screening Guidance: Technical Background Document.* Washington, DC: Office of Emergency and Remedial Response. EPA/540/R95/128.

USEPA. 1999. *Understanding Variation in Partition Coefficient, K_d, Values,* Vol. 1 (EPA 402-R-99-004A) and Vol. 2 (EPA 402-R-99-004B).

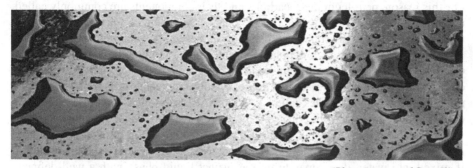

(Photo courtesy of Gary Witt)

In contact with water, nonpolar liquids do not dissolve appreciably, but remain as separate, immiscible liquid phases called NAPLs.

Chapter 7, Section 7.1

7 General Properties of Nonaqueous Phase Liquids and the Behavior of Light Nonaqueous Phase Liquids in the Subsurface

7.1 TYPES AND PROPERTIES OF NONAQUEOUS PHASE LIQUIDS

According to the U.S. Environmental Protection Agency (USEPA 2011), as of March 2011, about 498,295 releases from underground storage tanks (USTs) have been reported nationwide. Cleanups have been initiated at 471,756 of these sites, and 407,680 sites have been cleaned up. The backlog of sites still to be cleaned up is 90,615. The majority of these USTs contained petroleum hydrocarbons, which are a mixture consisting mainly of compounds having low solubilities in water. When added to water, a liquid with low water solubility remains mostly isolated from the water in a separate phase, with a visible boundary between the two liquids.

For this reason, petroleum hydrocarbons and their components are commonly called nonaqueous phase liquids (NAPLs). This chapter and Chapter 8 are an introduction to understanding the environmental behavior of these important pollutants.

NAPLs are nonpolar or low-polarity liquids that are minimally soluble in water. Gasoline and diesel fuels, oils, chlorinated solvents, and pesticides are examples. In contact with water, nonpolar liquids do not dissolve appreciably, but remain as separate, immiscible liquid phases called NAPLs. NAPL substances comprise molecules that are nonpolar or weakly polar and therefore have only weak attraction to polar water molecules, but relatively strong London force attraction to other NAPL molecules (see Chapter 2, Section 2.9). Therefore, NAPLs are not very soluble in water; if mixed into water, they separate into a distinct liquid phase with a well-defined, usually visible boundary between the NAPL and the water. The boundary is a visible interface physically dividing the bulk phases of the NAPL and water.

Many NAPLs are mixtures of different hydrocarbon compounds with varying low solubilities. The phase boundary does not prevent compounds within an NAPL mixture from dissolving into the water. The more-soluble compounds in an NAPL mixture will preferentially diffuse across the interface and dissolve into the surrounding water phase. Over time, this changes the composition and physical properties of NAPL mixtures, as well as polluting the water.

NAPL contaminants include a wide range of industrial compounds, such as solvents, heating oil, gasoline, chlorinated hydrocarbons, coal tars, and creosote. NAPL compounds are typically stored in underground and aboveground tanks and transported by pipelines, trains, and tank trucks, all of which are susceptible to leaks and spills.

Due to their low solubility, NAPLs in contact with environmental waters dissolve slowly, acting as potential long-term sources of water contamination. Because of their typically high toxicity, small amounts of NAPL can pollute very large volumes of soil and groundwater. Remediation of NAPL-polluted sites is made difficult by the low solubility and uneven distribution of NAPLs in the subsurface, and the generally low flow rates of groundwater. The accidental release of NAPL liquids into soils and aquifers is a widespread and serious environmental problem.

NAPLs may be further subdivided into light nonaqueous phase liquids (LNAPLs) and dense nonaqueous phase liquids (DNAPLs). LNAPLs are those liquid hydrocarbon compounds or mixtures that are less dense than water, such as gasoline and diesel fuels and their individual components. DNAPLs are liquid hydrocarbon compounds or mixtures more dense than water, such as creosote, polycyclic aromatic hydrocarbons (PAHs), polychlorinated biphenyls (PCBs), coal tars, and most chlorinated solvents (chloroform, methylene dichloride, tetrachloroethene, etc.).

The distinction between LNAPLs and DNAPLs is important because of their different behavior in the subsurface. When an NAPL spill occurs in the vadose zone, LNAPLs will travel downward through soils only to the water table, where they remain floating on the water table surface, whereas DNAPLs can sink through the water-saturated zone to impermeable soil structures such as bedrock, where they collect in bottom pools and enter bedrock fractures*. Different remediation methods

---

* Wherever LNAPLs and DNAPLs are in contact with groundwater, their more soluble components begin to dissolve into the groundwater, creating a dissolved plume (Section 7.3.2).

are required for LNAPLs and DNAPLs. This chapter discusses LNAPLs, and Chapter 8 discusses DNAPLs.

7.2   GENERAL CHARACTERISTICS OF PETROLEUM LIQUIDS: THE MOST COMMON LNAPL

Petroleum liquids are complex mixtures of hundreds of different hydrocarbons, with minor amounts of nitrogen, oxygen, sulfur, and some metals. Nearly all petroleum compounds are nonpolar and not very soluble in water. The behavior of these compounds in a groundwater environment depends on the physical and chemical nature of the particular hydrocarbon blend as well as the particular soil environment. For example, the partition coefficients and migration potential of each individual compound in a mixture depend on the overall composition of the petroleum mixture in which it is found, on the properties of the pure compound, and on the characteristics of the surrounding soil. Furthermore, the composition and properties of petroleum contaminants change with time as the petroleum ages and weathers.

Many nonfuel organic pollutants, such as chlorinated hydrocarbons and pesticides, are more soluble in petroleum than in water. Therefore, if an oil spill occurs where previous organic contamination already exists, the older pollutants tend to concentrate from soil surfaces and pore space water into the fresh oil phase. An oil spill into an already-contaminated soil can mobilize other pollutants that have been immobilized there by sorption and capillarity. As freshly spilled oil moves downward through the soil, immobilized pollutants sorbed to the soil can dissolve into the moving liquid oil and be carried along with it. Analysis of spilled petroleum products will often detect other organic compounds that were previously sorbed to the soil.

RULE OF THUMB

Because many organic pollutants are soluble in NAPLs, analysis of spilled petroleum products will often detect other organic compounds, such as pesticides, that were previously sorbed to the soil but were not originally present in the NAPL spill being investigated.

7.2.1   TYPES OF PETROLEUM PRODUCTS

The first step in refining crude oil into petroleum products is usually fractional distillation, a process that separates the crude oil components according to their boiling points. The resulting products are groups of mixtures, or fractions, each of which has boiling points within a specified range. All but the lightest fractions (lowest boiling temperature ranges) can contain hundreds of different hydrocarbon compounds. The fractions are often classified into the general groups described in Table 7.1. In addition, several pure petrochemicals may be produced during fractional distillation, such as butane, hexane, benzene, toluene, and xylene, for use as solvents, production of plastics and fibers, and reblending into fuel mixtures. Refined petroleum products are further modified by catalytic cracking, blending, and reformulation processes to enhance desirable properties.

TABLE 7.1
Principal Petroleum Fractions from Fractional Distillation

Boiling Range (°C)	Dominant Composition Range	Fraction	Uses
−160 to +30	C1–C4	Gases	LPG, methane, gaseous fuels, feedstock for plastics
30–60	C5–C7	Petroleum ether	Solvents, gasoline additives
90–130	C6–C9	Ligroin, naphtha	Solvents
40–200	C4–C12	Gasoline	Motor fuel
60–200	C7–C12	Mineral spirits	Solvents
150–300	C10–C16	Kerosene	Jet fuel, diesel fuel, lighter fuel oils
300–350	C16–C18	Fuel oil	Diesel oil, heating oil, cracking stock
>350	C18–C24	Lubricating stock	Lubricating oil, mineral oil, cracking stock
Solid residue	C25–C40	Paraffin wax	Candles, toiletries, wax paper
Solid residue	>C40	Residuum	Roofing tar, road asphalt, waterproofing

Note: The notation used here gives the range of carbon atoms in the fraction compounds. For example, C5–C7 means the petroleum fraction that contains mostly hydrocarbon compounds containing between 5 and 7 carbon atoms. This table indicates that as the number of carbon atoms in a hydrocarbon molecule increases, so do its boiling temperature and its viscosity. Volatility decreases as the number of carbon atoms in a compound increases. The relations between molecular size and physical properties are discussed in Chapter 2.

RULE OF THUMB

The larger the hydrocarbon compound and the more carbon atoms it contains, the higher are its boiling point and viscosity, and the lower is its volatility.

7.2.2 Gasoline

Gasoline is among the lightest liquid fractions of petroleum and consists mainly of aliphatic and aromatic hydrocarbons in the carbon number range C4–C12.

Aliphatic hydrocarbons consist of the following:

- *Alkanes*: Saturated hydrocarbons in which all carbons are connected by single bonds. They may have linear, branched, or cyclic carbon-chain structures, such as pentane, octane, decane, isobutane, and cyclohexane.
- *Alkenes*: Unsaturated hydrocarbons in which there are one or more double bonds between carbon atoms. They also may have linear, branched, or cyclic carbon-chain structures, such as ethylene (ethene), 1-pentene, and 1,3-cyclohexadiene.
- *Alkynes*: Unsaturated hydrocarbons in which there are one or more triple bonds between carbon atoms. They also may have linear, branched, or cyclic carbon-chain structures, such as acetylene (ethyne), propyne, and 1-butyne.

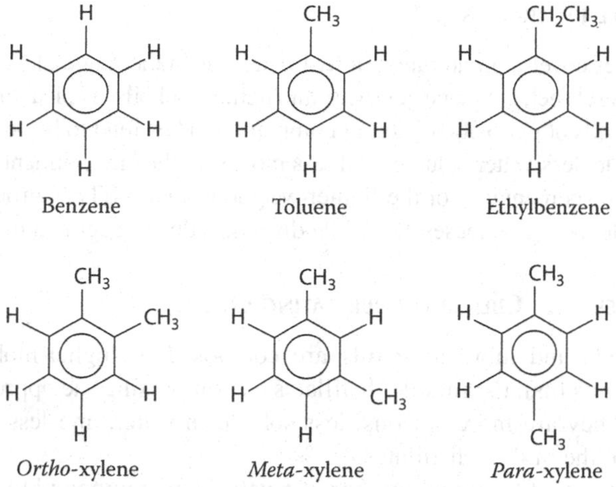

FIGURE 7.1   The BTEX group of aromatic hydrocarbons.

Aromatic hydrocarbons (also called arenes) are hydrocarbons based on the benzene ring as a structural unit. They include monocyclic hydrocarbons such as benzene, toluene, ethylbenzene, and xylene (the BTEX group; see Figure 7.1) and polycyclic hydrocarbons such as naphthalene and anthracene.

RULE OF THUMB

Gasoline mixtures are volatile and include some moderately soluble components, which makes them more mobile in the groundwater environment than heavier (higher carbon number range) petroleum fractions.

Gasolines contain a much higher percentage of the BTEX group of aromatic hydrocarbons (benzene, toluene, ethylbenzene, and the *ortho-* and *para*-xylene isomers) than do other fuels, such as diesel. They contain lower concentrations of heavier aromatics like naphthalene and anthracene than do diesel and heating fuels. Therefore, the presence of BTEX in appropriate concentration ratios* is often a useful indicator of gasoline contamination.

Oxygenated compounds such as alcohols (methanol and ethanol) and ethers (methyl-*tert*-butyl ether [MTBE]) are added as octane boosters and to provide extra oxygen for helping gasoline burn more completely, reducing air pollution. Until recently, MTBE was the most commonly used of these. Gasoline formulations between 1980 and 2003 typically contained around 15% MTBE by volume. The EPA began to phase out MTBE around 2003, replacing it with ethanol, because of concerns about the health effects of MTBE groundwater contamination. The use of MTBE is now banned in the United States, although it continues to be used in some other countries.

---

* The BTEX concentration ratios that are characteristic of fresh and weathered gasoline are discussed in Section 7.7.3.

7.2.3 MIDDLE DISTILLATES

Middle distillates cover a broad range of hydrocarbons in the range of C6 to about C25. They include diesel fuel, kerosene, jet fuels, and lighter fuel oils. Typical middle distillate products are blends of up to 500 different compounds. They tend to be denser, more viscous, less volatile, less water-soluble, and less mobile in the environment than gasoline. They contain low percentages of the lighter-weight aromatic BTEX group, which may not be detectable in older releases of middle distillates due to degradation or transport.

7.2.4 HEAVIER FUEL OILS AND LUBRICATING OILS

Heavier fuel oils and lubricating oils are composed of higher-molecular-weight (MW) compounds than the middle distillates, encompassing the approximate range of C15–C40. They are more viscous, less soluble in water, and less mobile in the subsurface than the middle distillates.

Figure 7.2 relates the carbon number of a petroleum compound to its properties, uses, and the usual instrumental methods used for its analysis.

7.3 BEHAVIOR OF PETROLEUM HYDROCARBONS IN THE SUBSURFACE

7.3.1 SOIL ZONES AND PORE SPACE

As illustrated in Figure 7.3, subsurface soil may be divided into a water-unsaturated zone (also called the vadose zone) extending from the soil surface down to just above the water table, and a water-saturated zone extending from the water table down to bedrock. Capillary forces extend the saturated zone slightly above the water table, establishing a capillary zone of transition between the unsaturated and saturated zones. The capillary zone depends on soil pore sizes and can range from a fraction of an inch in coarse-grained sediments to several feet in fine-grained sediments such as clay.

Each zone contains soil particles with pore spaces between them. In permeable soils, most of the pore spaces are continuous and connected, allowing movement of water and liquid contaminants through them. In the absence of contaminants, pore spaces in the unsaturated zone contain air with some water sorbed to the soil particles. Pore spaces in the saturated zone contain mainly water.

When contaminants enter the subsurface region as spilled liquid petroleum (free product)*

- Volatile compounds vaporize from the free product mixture to the atmosphere and to the air in the soil pore spaces.
- The more-soluble compounds in the free product begin to dissolve into water contained on soil particle surfaces, into water percolating down from the ground surface, and into groundwater in the saturated zone.

---

* Free product is that part of the original or aged NAPL liquid mixture that is still in a nonaqueous liquid state and has not dissolved in water, sorbed to solid surfaces, or volatilized. Loss over time of some free product mixture components by dissolution, sorption, and volatilization generally causes some physical properties of the free product to change with time; viscosity and density increase and volatility decreases.

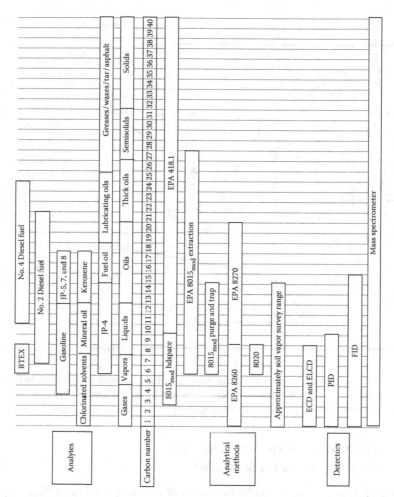

FIGURE 7.2 Hydrocarbon ranges, corresponding uses, and analytical methods. Analytical methods constantly evolve and current method numbers may differ from those in this figure. Detectors are electron capture detector (ECD), electrolytic conductivity detector (ELCD), photoionization detector (PID), and flame ionization detector (FID).

- A small fraction of the free product is taken up and metabolized by microbiota.
- A part of the remaining free product (the least-soluble components) sorbs to soil particles.
- Some liquid free product is trapped and retained in small pore spaces by capillary forces.
- All the above are continuing processes that steadily diminish the total mass of remaining free product as it migrates downward through the soil. If the distance to groundwater is sufficiently large and the amount of spilled free product sufficiently small, all the mobile free product may be lost or immobilized before it reaches the groundwater table.

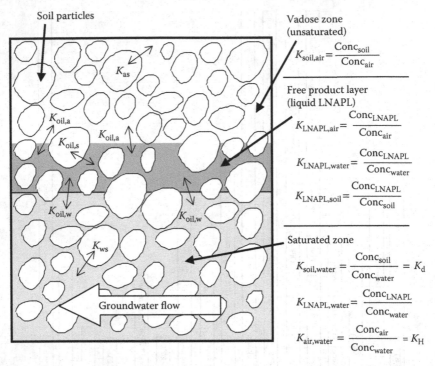

FIGURE 7.3 Soil zones and partitioning behavior of a free product (original bulk liquid) LNAPL pollutant. All the Ks in the equations are partition coefficients. They quantitatively describe how the LNAPL pollutant distributes itself among water, soil, air, and free product.

7.3.2 Partitioning of Light Nonaqueous Phase Liquids in the Subsurface

Because of their low water solubilities, most of the compounds classified as petroleum hydrocarbons are generally considered NAPLs. As described in Section 7.1, when mixed with water, NAPLs separate into a distinct liquid phase with a well-defined boundary between the NAPL and the water. However, all petroleum NAPLs contain some compounds of sufficient solubility to represent a pollution hazard if brought into contact with water. In addition, volatile compounds can escape into the atmosphere, causing air pollution; the insoluble, nonvolatile compounds that remain trapped in the soil can render the soil unfit for many uses (Freeze and Cherry 1979).

In clean soil, the voids of vadose zone earth materials are filled with air and water. After a petroleum release, some soil voids contain immobile petroleum held by capillary forces and sorbed to soil surfaces. There may also be liquid petroleum moving downward through the pore interstices under gravity. If LNAPL reaches the water table, its buoyancy prevents further downward movement and it spreads out horizontally over the water table to form a layer of free product floating above the saturated zone. The individual components of the petroleum then become partitioned into subsurface air spaces, water, and solid soils that are in contact with the free product.

7.3.3 PROCESSES OF SUBSURFACE MIGRATION

After part of spilled petroleum has partitioned from the free product into other phases, hydrocarbons are present in solid, liquid, dissolved, and vapor phases.

- *Solid phase* hydrocarbons are sorbed on soil surfaces or diffused into soil micropores and mineral grain lattices. They are immobile and degrade very slowly.
- *Liquid phase* hydrocarbons exist in the subsurface as
  - Immobile residual liquids held by capillary forces and as a thin layer sorbed to sediments in the unsaturated and capillary zones
  - Free mobile liquids in the pore spaces of the unsaturated zone above the capillary zone
  - Immobile residual liquids trapped in soil pore spaces and sorbed to soil surfaces below the water table in the saturated zone
- *Dissolved phase* hydrocarbons are found in
  - Water infiltrating downward through the unsaturated zone and in residual films of water sorbed to sediments in the unsaturated zone
  - Pore water and residual films of water sorbed to sediments in the capillary zone and elsewhere in the hydrocarbon plume
  - Groundwater in the saturated zone
- *Vapor phase* hydrocarbons are found
  - Mostly in void spaces of the unsaturated zone not occupied by water or liquid hydrocarbons (they are mobile, moving between connected void spaces, sometimes to the surface and into the atmosphere)
  - As small bubbles trapped in the hydrocarbon plume and in the water-bearing zone below the plume, where they generally are immobile
  - Dissolved in the groundwater of the saturated zone, where they move with the groundwater

7.3.4 PETROLEUM MOBILITY THROUGH SOILS

The concept of NAPL partitioning into other phases in the subsurface is nothing new to oil field workers. Liquid petroleum fields are typically found in rock formations of 10–30% porosity. Up to half the pore space contains water. Primary recovery of oil, which relies on pumping out the portion of oil that is mobile and will accumulate in a well, collects only 15–30% of the oil in the formation. Secondary recovery techniques force water under pressure into the oil-bearing rocks to drive out more oil. Primary and secondary techniques together typically extract somewhat less than 50% of the oil from a formation. Tertiary recovery techniques use pressurized carbon dioxide to lower oil viscosity along with detergents to solubilize the oil. Even using tertiary techniques, producers expect 40% of the oil to remain immobile and unrecoverable, unless the porosity of the rock formations can be increased by methods like hydrofracturing, in which water solutions are injected under pressure to widen and lengthen the existing rock fractures.

Similar difficulties are encountered when trying to clean up soils contaminated by a petroleum spill. There is always a fraction of the oil that is strongly sorbed on soil particles or trapped in soil pore spaces, which is not easily removed by water flushing or air sparging. Because liquid NAPL is toxic to microbes (see Chapter 9), biodegradation is negligible, occurring only near the boundary between NAPL and groundwater, where dissolved NAPL components have been strongly diluted.

In the subsurface environment, a significant portion of oil pollution must be regarded as permanent, with a lifetime of well over 25 years, unless deliberate efforts are made to degrade it or physically remove it by excavation (Bredehoeft 1992; MacDonald and Kavanaugh 1994). Immobilized petroleum contaminants in the subsurface act as a long-term source of groundwater pollution, as the more-soluble components continue to diffuse to the oil–water interface and dissolve into the groundwater.

7.3.5 BEHAVIOR OF LNAPL IN SOILS AND GROUNDWATER

LNAPL movement in the subsurface is a continual process of partitioning different components of a mixture among different phases that are present in the subsurface matrix. Spilled LNAPL at or near the soil surface penetrates and thoroughly saturates the soil because there is little trapped water or air to block its movement. Under the influence of gravity, the LNAPL sinks vertically downward, leaving behind in the soil a trail of residual LNAPL trapped by sorption and capillary forces. Capillary forces cause the LNAPL to spread horizontally as well as vertically downward, creating an inverted-funnel-shaped zone of soil contamination (Figure 7.4).

As LNAPL moves downward through soil, a significant portion becomes immobilized by sorption to soil particle surfaces and capillary entrapment in soil pore space. This continually reduces the amount of mobile contaminant. When mobile LNAPL encounters an impermeable soil structure, bedrock, or the water table, its downward movement is halted and it spreads laterally under its internal pressure or down the slope of the retarding surface under gravity. If the water table is deep enough and the LNAPL layer is not replenished by a continuing leak, it may eventually be completely depleted by entrapment in the soil and become essentially immobilized. However, even when immobilized, the trapped free product continues to lose mass into water percolating downward in the vadose zone, into a vapor phase in pore space air and by biodegradation around the LNAPL boundary.

After entering the subsurface, an LNAPL front continues downward, leaving behind an ever-widening inverted funnel of contaminated soil that contains residual immobile LNAPL sorbed to soil surfaces and trapped in pore spaces. The term "immobile" is used loosely and really means that, although some mobility may still occur, it will be very slow compared to the remediation time frame of interest.

A spill may or may not reach the water table. If the groundwater table is far enough below the surface or if the amount of spilled LNAPL is small enough, the mobile LNAPL can be completely depleted by entrapment in the soil before it reaches the groundwater. If the spill is large enough or the groundwater table is shallow, mobile LNAPL, commonly called free product, will contact groundwater and start to spread out over its surface, tending to move more in the downgradient flow

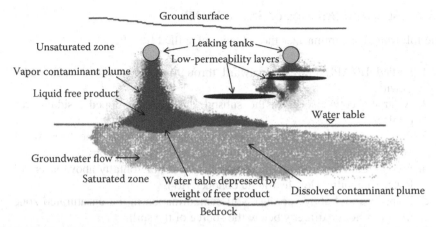

FIGURE 7.4   LNAPL leaking from underground fuel storage tanks migrate downward under gravity. Enough fuel free product has leaked from the left tank to reach the saturated zone and spread out above the water table, moving in the direction of groundwater flow. The smaller spill from the right tank is insufficient to reach the water table and has become immobilized within the unsaturated zone by sorption and capillary forces. The more-soluble components of the free product are present in the dissolved plume, which extends below the free product plume into the saturated zone and moves downgradient with groundwater flow. There also is a vapor plume in the unsaturated zone consisting of the most-volatile components. The vapor plume migrates away from the liquid free product in the vadose zone in all directions independent of gravity. It may enter underground cavities such as sewers and basements, and may escape through the ground surface into the atmosphere.

direction. The weight of the free product depresses the water table locally below the free product column (Figure 7.4).

At the top of the saturated zone, free product will continue to spread laterally as a layer over the water table, leaving a trail of residual LNAPL entrapped in the soil, until it has spread out to a saturation level so low that it all becomes immobile. Lateral spreading of the free product is influenced by a viscous frictional interaction at the water–LNAPL interface, which tends to move the free product preferentially in the direction of groundwater movement, along the hydraulic gradient. The relative downgradient velocities of water and free product depend on their relative viscosities and densities, as well as the soil conductivities for the different liquids.

When the water table rises and falls, the floating free product is moved vertically, buoyed above the changing water level. A rising water table can entrap some LNAPL by sorption and capillary forces within the newly saturated zone below the water table. A falling water table allows floating LNAPL to move downward into soils formerly saturated with water and exposes LNAPL formerly below the water table, allowing some of it to resume partitioning into air and, perhaps, to continue its slow migration downward under gravity. This up-and-down movement smears LNAPL into a region thicker than the free product thickness. With each up-and-down cycle, still more residual LNAPL becomes immobilized in this smear zone. The result is that the smear zone of entrapped LNAPL extends above and below the average level of the water table.

7.3.6 SUMMARY: BEHAVIOR OF SPILLED LNAPL

The following list summarizes the behavior of spilled LNAPL:

1. Spilled LNAPL moves downward through the unsaturated zone under gravity.
2. A large fraction sorbs to the subsoil surfaces as trapped residual free product.
3. Some horizontal spreading occurs in the unsaturated zone because of attractive forces to mineral surfaces and capillary attractions.
4. Free product tends to accumulate and spread horizontally above layers of low permeability (low hydraulic conductivity).
5. Points 3 and 4 above produce soil contamination in the unsaturated zone that does not lie directly below the source of the spill.
6. At the water-bearing region of the capillary zone, the free liquid phase floats on the water and begins to move laterally. The weight of floating LNAPL displaces water underneath it, depressing the water table in the region below the LNAPL.
7. If the spill is small enough, all the LNAPL may become trapped as residual liquid in the vadose zone, so that no LNAPL reaches the water table. However, any portion of the residual LNAPL that dissolves in downward percolating water from precipitation can be carried to the water table and contaminate it.
8. The vapor phase spreads widely in the unsaturated zone and can escape to the atmosphere and accumulate in cellars, sewers, and other underground air spaces.

7.3.7 WEATHERING OF SUBSURFACE CONTAMINANTS

With time, the composition of immobilized oil changes in the following ways:

- Less-viscous components move downgradient through the soils.
- Components that are more viscous remain trapped as residual LNAPL in the vadose zone.
- Volatile components are lost into the atmosphere.
- Soluble components are lost into the groundwater.
- Biodegradable components are degraded by bacterial activity.

However, the total mass of immobilized oil decreases slowly because the loss processes are usually slow, unless they are artificially enhanced as part of a remediation program. The natural rate of free product depletion becomes progressively slower with time, as the remaining contaminants are increasingly enriched in those components that resist the loss mechanisms. The remaining oil becomes more firmly fixed in the subsurface soil, continually releasing its more-soluble components in slowly decreasing concentrations to the groundwater.

RULES OF THUMB

1. Less than 1% of the total mass of a gasoline spill will dissolve into water in the vadose and saturated zones.
2. Since more than 99% of a fuel spill remains as sorbed or free product LNAPL, it is impossible to clean up groundwater fuel contamination simply by pump-and-treat without eliminating the source residual and free product remaining in the soil.

7.3.8 PETROLEUM MOBILITY AND SOLUBILITY

The environmental impact of a contaminant release is determined mainly by the mobility and water solubility of the different components of the contaminant. The most important soil and liquid parameters determining the mobility of LNAPL free product are the following:

- Average soil pore size.
- Percent of soil pore space (soil porosity).
- Density and viscosity of the moving LNAPL. Density is mass per unit volume. Most petroleum products have a density less than the density of water, which is 1 g/mL. Viscosity measures resistance of fluid to flow. Gasoline is less viscous than water and can flow through pores and fissures more easily than water. The heavier petroleum fractions, such as diesel fuel and fuel oils, are more viscous than water and flow less readily and more slowly. Values of density and viscosity for several fuel products are listed in Table 7.2.
- Capillary attraction for the liquid contaminant to soil particles.
- Soil zone (vadose, capillary, or saturated), in which LNAPL is present, determines whether the pore space contains air, water, or contaminant.
- Magnitude of pressure and concentration gradients acting on the liquid free product.

TABLE 7.2
Densities and Viscosities of Selected Fuels Compared to Water

Fluid	Density (g/mL)			Viscosity (centipoise)		
	0°C	15°C	25°C	0°C	15°C	25°C
Water	1.000	0.998	0.996	1.8	1.14	0.9
Automobile gasoline	0.76	0.73	0.68	0.8	0.62	—
Automotive diesel fuel	0.84	0.83	—	3.9	2.7	—
Kerosene	0.84	0.84	0.83	3.4	2.3	2.2
No. 5 jet fuel	0.84	—	—	—	—	—
No. 2 fuel oil	0.87	0.87	0.84	7.7	—	4.0
No. 4 fuel oil	0.91	0.90	0.90	—	47	23
No. 5 fuel oil	0.93	0.92	0.92	—	215	122
No. 6 fuel oil or Bunker C	0.99	0.97	0.96	7.4×10^7	—	3200

TABLE 7.3
Solubility Variability of Gasoline Components from Different Fuel Mixtures

	Concentration Dissolved in Water (mg/L)			
Compound	Regular Leaded	Regular Unleaded	Super Unleaded	Pure Compound
Benzene	30.5	28.1	67.0	1,740–1,860
Toluene	31.4	31.1	107.0	500–627
Ethylbenzene	4.0	2.4	7.4	131–208
1,2-Dichloroethane	1.3	—	—	8,524
Methyl-*tert*-butyl ether (MTBE)	43.7	35.1	966.0	48,000
t-Butyl alcohol	22.3	15.9	933.0	Miscible
m-Xylene	13.9	10.9	11.5	134–196
o-, *p*-Xylene	6.1	4.8	5.7	157–213
1,2-Dibromoethane	0.58	—	—	4,300

LNAPL solubility in water is variable and depends on the chemical mixture. Literature data for solubility of pure compounds can be misleading because the solubility of a specific compound decreases when it is part of a blend (Table 7.3).

RULE OF THUMB

The aqueous solubility of a particular compound in a multicomponent NAPL can be approximated by multiplying the mole fraction of the compound in the NAPL mixture by the aqueous solubility of the pure compound.

Solubility of component i in an NAPL mixture is

$$S_i = X_i S_i^0 \tag{7.1}$$

where

S_i is the solubility of component i in the mixture
X_i is the mole fraction of component i in the mixture
S_i^0 is the solubility of pure component i

7.4 FORMATION OF PETROLEUM CONTAMINATION PLUMES

In the subsurface soil environment, petroleum compounds can be present in four phases, each of which can create its own contaminant plume. The four phases are

1. Liquid petroleum free product (LNAPL)
2. Petroleum compounds sorbed to soil particles

3. Dissolved petroleum components
4. Vaporized petroleum components

Each phase behaves differently and poses different remediation problems. The liquid free product originates directly from the contamination source and initially has the same composition. The sorbed, dissolved, and vapor phases are extracted from the liquid free product as it contacts soil, water, and air in the soil pore spaces. Each phase moves independently in its own distinct contaminant plume.

In general, the vapor contaminant plume moves most rapidly. The dissolved plume moves more slowly, at groundwater velocity or less, depending on its retardation factor. Depending on whether its viscosity is greater or less than water, the free product plume may move slower or faster than the dissolved plume. The sorbed plume may be immobilized, or, in the saturated zone, part of it may be sorbed to mobile colloids and move at approximately the groundwater velocity.

RULES OF THUMB

1. With respect to the total mass of fuel contaminant in the subsurface soil environment, the floating free product* and immobilized oil (trapped by capillary forces and sorbed to soil particles) are generally more than 99% of the total mass.
2. When free product is present, the dissolved phase in the groundwater is generally less than 1% of the total mass. The dissolved plume is just the tip of the contamination iceberg.

Example 1: Comparing Dissolved and LNAPL Free Product Masses

A leaking underground storage tank released 1000 gal of gasoline (density about 0.7 g/mL) to the subsurface. After 1 year, the resulting dissolved-phase plume was about 1000 feet long, 100 feet wide, and averaged 10 feet deep. The average concentration of hydrocarbons in the plume was 2.20 mg/L (estimated by measuring and adding the total volatile hydrocarbon and total extractable hydrocarbon concentrations). The porosity of the aquifer was 0.30. If no hydrocarbon was lost due to volatilization or biodegradation, how much of the original release was in the dissolved phase and how much was in the LNAPL phase?

ANSWER

Total mass released = (1000 gal) (3.78 L/gal) (1000 mL/L) (0.7 g/mL) = 2.646×10^6 g

---

* The mole fraction of compound a in a mixture of several compounds is written X_a, where

$$X_a = \frac{\text{moles of } a}{\text{total moles of all compounds in mixture}}$$

For a mixture containing 1 mole of CCl_4 and 3 moles of $CHCl_3$, $X_{CCl4} = 1/4 = 0.25$ and $X_{CHCL3} = 3/4 = 0.75$.

Note that the sum of all mole fractions must equal unity. The mole fraction of any pure substance equals unity.

Volume of contaminated groundwater = (1000 ft) (100 ft) (10 ft) (0.30)
  (28.3 L/ft^3) = 8.490 × 10^6 L
Mass of dissolved hydrocarbons = (8.490 × 10^6 L) (2.20 mg/L) (1 g/1,000 mg) =
  18,680 g
Mass of LNAPL free product = (2.646 × 10^6 g) − 18,680 g = 2.627 × 10^6 g
Percent of total mass that is dissolved = (18,680 g/2.646 × 10^6 g) × 100 = 0.7%
Percent of total mass that is LNAPL = (2.627 × 10^6/2.646 × 10^6) × 100 = 99.3%

7.4.1 DISSOLVED CONTAMINANT PLUME

Water solubility is the most important chemical property for assessing the impact of
a contaminant on the non-atmospheric parts of the environment. Dissolved contaminants arise when the free product comes in contact with water. The water may be
moisture retained in the soil, precipitation percolating downward through the soil,
groundwater flowing through contaminated soil, or groundwater lying under a layer of
free product. Both crude and refined petroleum products contain hundreds of different
components with different water solubilities, ranging from slightly soluble to insoluble.

RULES OF THUMB

1. In general, lightweight aromatics such as the BTEX group (benzene,
   toluene, ethylbenzene, and xylenes) are the most-soluble components
   of fuel mixtures. If MTBE or ethanol additives are present, they are
   the most-soluble components by far.
2. The overall water solubility of commercial gasoline without additives is
   between 50 and 150 mg/L, depending on its exact composition. When
   free product gasoline is present, the dissolved portion generally accounts
   for less than 1% of the total contaminant mass present in the subsurface.
3. The overall solubility of fresh No. 2 diesel fuel in water is around
   0.4–8.0 mg/L, again depending on its composition. When free product diesel fuel is present, the dissolved portion generally accounts for
   less than 0.1% of the total contaminant mass present in the subsurface.
4. Nevertheless, because of their typically high toxicity, dissolved contaminants can greatly exceed concentrations where water is regarded
   as seriously polluted.

The compositions of the free product and dissolved fractions are very different, as
indicated in Figure 7.5, because the more-soluble compounds partition out of the free
product and become concentrated in the water-soluble fraction.

Dissolved contaminants become a part of the water system and move with the
groundwater, but they usually move at a lower velocity because of retardation by
sorption processes. Sorption to soil and desorption back into the dissolved phase is a
continual process that retards the movement of dissolved contaminants. The amount
of retardation for any particular contaminant depends mainly on the organic content
of the soil; retardation is greater in soils with more organic matter. Because their water

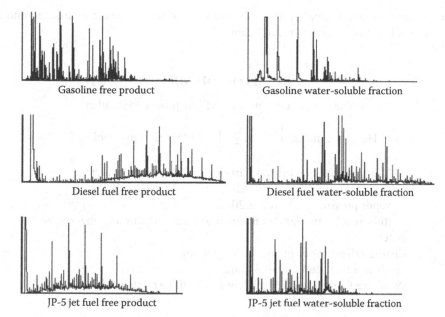

FIGURE 7.5   Gas chromatograph/flame ionization detector (GC/FID) chromatograms of gasoline, diesel, and JP-5 fuels and their respective water-soluble fractions. Time of elution, which corresponds roughly to the number of carbons in the eluted compound, increases from left to right. Thus, peaks corresponding to heavier compounds appear farther to the right in each figure. The composition of free product and dissolved fractions are very different in each type of product because the more-soluble compounds become distributed preferentially into the water-soluble fraction. The water-soluble fractions are composed mainly of 1-, 2-, and 3-ring aromatic hydrocarbons.

solubilities are low, dissolved fuel contaminants continue to partition between the dissolved phase and soil particle surfaces, especially in soils with a high organic content.

RULE OF THUMB

Typical retardation factors for BTEX in sandy soil range from 2.4 for dissolved benzene (groundwater moves 2.4 times faster than benzene) to 6.2 for the dissolved xylene isomers.

7.4.2 Vapor Contaminant Plume

Vapor phase contaminants arise from the volatile components of the free product escaping into adjacent air. Low-mass hydrocarbon components commonly associated with the gasoline fraction are the most volatile. Vapor movement is not influenced by groundwater motion, but is weakly influenced by gravity. It follows the most conductive pathways through the subsurface, from regions of higher to lower pressure. Vapors denser than air can collect in low spots like basements and sewers. Much of the vapor remains trapped in soil near its origin, slowly escaping to the surface atmosphere. A small portion of vapor phase contaminants may dissolve into soil–water, but it is generally insignificant.

RULES OF THUMB

1. A measurable vapor concentration will be produced if either

$$\text{Henry's constant } \left(K_H = \frac{C_a}{C_w} \right) > 0.0005 \text{ atm} \cdot m^3 \cdot mol^{-1}$$

   (this produces significant partitioning from water to air)
   or
   vapor pressure >1.0 torr at 20°C
   (this results in significant diffusion upward through the vadose zone).
2. Characteristic vapor pressures for gasoline:
   Fresh gasoline: 260 torr (0.34 atm)
   Weathered gasoline (2–5 years old): 15–40 torr (0.02–0.05 atm)

7.5 ESTIMATING THE AMOUNT OF LNAPL-FREE PRODUCT IN THE SUBSURFACE

The initial steps in the remediation of a site where an LNAPL spill has occurred are the following:

1. Try to limit the movement of contaminant plumes.
2. Remove from the subsurface as much free product as possible.

As long as free product is present, it continues to partition into the sorbed, dissolved, and vapor contaminant plumes, continually feeding their growth. Only after the mobile LNAPL has been removed from above the water table can remediation of the contaminant plumes be effective.

LNAPL in the subsurface is generally detected and measured by its accumulation in wells. To design a program for removing free product, one must obtain a reliable estimate of the volume of free product that is present. However, the relationship between the thickness of free product that accumulates in a well and the thickness of free product distributed above the water table is easily misinterpreted. This LNAPL thickness in a well is influenced not only by the thickness of LNAPL in the subsurface adjacent to the well but also by soil texture and fluctuations of the water table level.

Figure 7.6 illustrates some of the factors that affect free product accumulation in a well. In the soil subsurface away from a well, liquids are influenced by capillary attractions that draw them into small pore spaces and interstices. Where no LNAPL free product is present, three forces determine the aquifer water table elevation:

1. Gravity pulls water downward.
2. Water pressure in the aquifer acts upward against gravity.

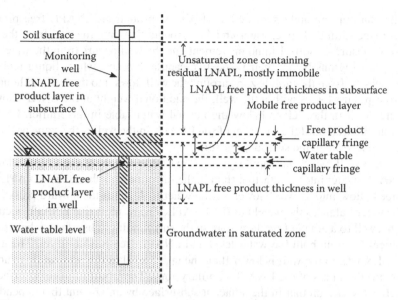

FIGURE 7.6   Thickness of LNAPL accumulated in a well compared to thickness in adjacent subsurface.

3. Capillary forces within the porous subsurface at the interface between the saturated and unsaturated zones also act upward against gravity.
4. The water table rises to the level where the downward force of gravity is balanced by the two upward forces of water pressure and capillary attractions.

Within a well, there are no upward-acting capillary forces affecting the liquid levels. Only the balance between gravity and water pressure in the aquifer determines the water level in a well. The result is that the water level in a well is lower than the top of the water table in the subsurface around the well, at the bottom of the capillary zone in the adjacent soil.

If LNAPL is present floating on the water table, it also develops a capillary zone at its interface with the unsaturated zone. Capillary zones occur at the upper boundaries of both the water table and the free product layer. In the capillary zones, liquids are drawn upward against gravity and are largely immobile, especially in horizontal directions. The thickness of the capillary zones depends on the soil texture. In coarse soils and sands with few capillary-size pore spaces and interstices, capillary zone layers may be only a few millimeters thick; in fine soils and sands, they may be several meters thick.

Where LNAPL free product lies on the water table, the water level is lowered by the weight of LNAPL (see Figure 7.6). Where the LNAPL free product layer is thin, it lies largely above the water capillary zone because the weight of LNAPL cannot easily displace water from this region. Where the LNAPL layer is thick, its excess weight makes it penetrate farther into, or even through, the water capillary zone, moving the free product–water interface still lower.

When an appropriately screened well passes through an LNAPL free product layer into the saturated zone, water and free product flow into the well from the surrounding subsurface soils. Liquid movement into the well occurs from the water and free product regions below their respective capillary zones, where liquid mobility exists. LNAPL from the mobile zone around the well flows into the well; without any upward capillary forces within the well, the additional weight of the LNAPL lowers the water level in the well to below the normal water table in the aquifer. LNAPL flows into the well until the top level of the LNAPL in the well is the same as the top of the mobile zone in the surrounding soil.

Within a well, where no capillary forces exist, the weight of the LNAPL lowers the LNAPL–water interface farther than in the surrounding subsurface. LNAPL will continue to flow into the well, lowering the water table, until the upward pressure of aquifer water balances the weight of the LNAPL. The result is that LNAPL accumulates in a well to a greater thickness than in the surrounding subsurface, where capillary forces buoy up both the water level and LNAPL free product layer. The upper level of LNAPL in the well is lower than the upper level in the surrounding subsurface by the thickness of the LNAPL capillary zone. The LNAPL–water interface in the well is lower than that in the adjacent subsurface by an amount that depends on the soil texture and the thickness of the subsurface layer of the mobile LNAPL. This behavior is illustrated in Figures 7.7 and 7.8.

Equation 7.2 may be used to calculate the water table level in the subsurface adjacent to a well from measurements in a well where LNAPL is present. Equation 7.2 is useful for evaluating and plotting groundwater elevations when LNAPL is present in the wells.

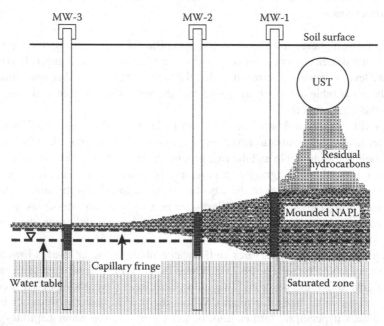

FIGURE 7.7 Comparison of LNAPL thickness in three wells (MW-1, MW-2, and MW-3).

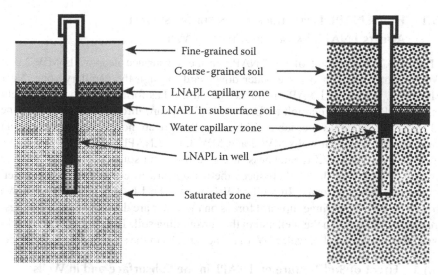

FIGURE 7.8   Effect of soil texture on LNAPL thickness in a well.

$$\text{WTE} = \text{WE}_{\text{well}} + (\text{LNAPL density} \times \text{LNAPL thickness in well}) \qquad (7.2)$$

where

WTE is the water table elevation in subsurface adjacent to the well
WE_{well} is the water elevation at the water–LNAPL interface in the well

An estimate of LNAPL thickness in the adjacent subsurface, ignoring soil properties and capillarity, can be made from (Figure 7.6):

$$t_{\text{subsurface}} \approx \frac{t_{\text{well}}(\text{water density} - \text{LNAPL density})}{\text{LNAPL density}} \qquad (7.3)$$

where

$t_{\text{subsurface}}$ is the thickness of LNAPL in the subsurface adjacent to the well (cm)
t_{well} is the thickness of LNAPL in the well (cm)
Water density is 1.0 g/cm^3
LNAPL density is 0.7–0.8 g/cm^3 for gasoline and diesel fuels

Equation 7.3 is useful for estimating the total volume of LNAPL in a plume whose area has been measured. Details for calculating more accurately the recoverable volume of LNAPL free product in the subsurface from well measurements have been published (Farr et al. 1990; Lenhard and Parker 1990; Parker et al. 1996). Computer programs are also available for this and related calculations. However, Equation 7.3 is often sufficient for the initial evaluation of a remediation program.

7.5.1 How LNAPL Layer Thickness in the Subsurface Affects LNAPL Layer Thickness in a Well

In Figure 7.7, the weight of the LNAPL in the subsurface near wells MW-2 and MW-3 is not sufficient to force water downward through the capillary zone. Near well MW-1, where the LNAPL is thicker because it has formed a dome, the weight of the LNAPL is great enough to force water downward through the capillary zone and below the original water table. Within a well, there are no capillary forces acting upward on the water. In wells MW-2 and MW-3, the LNAPL thickness in the wells is greater than the LNAPL thickness in the surrounding soils because the absence of capillary forces in the wells reduces the net upward forces acting on the water, so the water level is lower. In well MW-1, where the LNAPL has pressed down through the capillary fringe, upward forces on the water are due only to aquifer pressure and are the same in the well and in the surrounding soil. Thus, the LNAPL thickness and the water level in well MW-1 are the same as in the surrounding subsurface.

7.5.1.1 Effect of Soil Texture on LNAPL in the Subsurface and in Wells

Soil texture determines the magnitude of the upward capillary forces that act on subsurface water and LNAPL. Capillary forces are much larger in fine-grained soil than in coarse-grained soil. Consequently, the difference between LNAPL thickness in a well and LNAPL thickness in the adjacent subsurface is greater in fine-grained soil. The effect of soil texture is illustrated in Figure 7.8.

RULES OF THUMB

1. Measured LNAPL thickness in a well often exceeds the corresponding LNAPL thickness in the surrounding subsurface by a factor of 2–10, because LNAPL above the water table flows into the well and depresses the well-water-level.
2. The LNAPL thickness ratio, $h_{well}/h_{subsurface}$, generally increases with decreasing soil particle size, increasing capillary zone thickness, and increasing LNAPL density.
3. A crude estimate, ignoring soil properties, of LNAPL thickness in the adjacent subsurface can be made with Equation 7.4, obtained by rearranging Equation 7.3.

$$\frac{h_{well}}{h_{subsurface}} \approx \frac{\text{LNAPL density}}{\text{Water density} - \text{LNAPL density}} \tag{7.4}$$

a. Since fuel LNAPL (gasoline and diesel) generally has a density of 0.7–0.8 g/cm^3 and the density of water is 1.0 g/cm^3, Equation 7.4 gives, for fuel LNAPLs,

$$\frac{h_{well}}{h_{subsurface}} \approx \text{between } \frac{0.7}{1.0-0.7} \text{ and } \frac{0.8}{1.0-0.8}, \text{ or } 2.3 \text{ and } 4.0$$

7.5.1.2 Effect of Water Table Fluctuations on LNAPL in the Subsurface and in Wells

Water table fluctuations promote vertical spreading of LNAPL in the subsurface and can influence the thickness of LNAPL that collects in monitoring wells.

When the groundwater table rises,

- Some floating LNAPL free product is driven up into the unsaturated zone. Sorbed residual LNAPL in the formerly unsaturated zone can be remobilized by dissolving into the free product, causing further lateral spreading.
- Some free product remains trapped in pore spaces below the water level within the saturated zone.
- Mobile free product layer above the water table becomes thinner.

When the groundwater table falls,

- LNAPL free product floating on the water table moves downward as the water table drops, leaving behind an immobilized fraction of free product as sorbed residual LNAPL retained in the newly unsaturated zone above the lowered water table.
- Mobile free product layer floating on the water table may become thicker because free product formerly trapped below the water table is free to migrate downward to the new water table.

The rising and falling of the water table leaves behind a smear zone of contamination that lies partially in the saturated zone and partially in the unsaturated zone. This behavior is illustrated in Figure 7.9.

7.5.1.3 Effect of Water Table Fluctuations on LNAPL Measurements in Wells

LNAPL spills are often first detected by the appearance of a free product layer above the water in downgradient wells. If the well free product layer diminishes during a remediation program, it is tempting to believe that the cleanup effort is working successfully. An increase in the well free product thickness may initiate a search for new LNAPL sources. However, unless fluctuations in groundwater depth are taken into account, basing such conclusions on changes in the free product layer thickness in wells can lead to serious errors.

When groundwater rises, the thickness of the free product layer in wells generally decreases because a portion of the mobile free product becomes trapped below the water table and becomes immobile, thinning the mobile free product layer. When the water table falls, free product formerly trapped in the saturated zone becomes mobile again and can accumulate in the free product layer over the water table, where it is free to flow into wells. This behavior is illustrated in Figure 7.10.

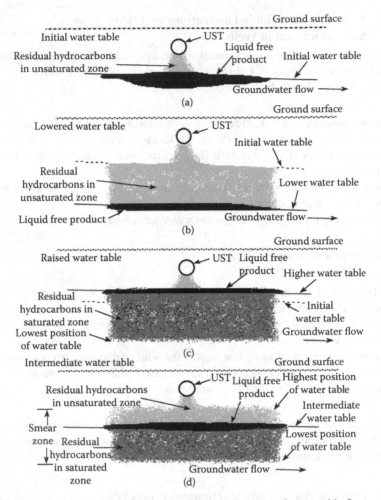

FIGURE 7.9   Spreading of LNAPL into a smear zone because of water table fluctuations.

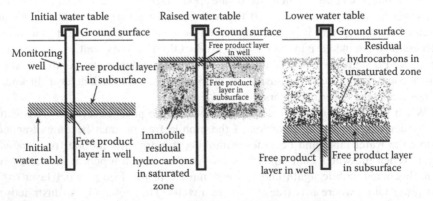

FIGURE 7.10   Effect of fluctuating water table on LNAPL accumulation in a well.

7.6 ESTIMATING THE AMOUNT OF RESIDUAL LNAPL IMMOBILIZED IN THE SUBSURFACE

Residual LNAPL in the subsurface is the portion that will not flow into a well. It is the part of an LNAPL spill that cannot be removed by pumping to the surface. Residual LNAPL must be remediated by processes such as biodegradation, chemical degradation, soil washing, volatilization with heat, vapor extraction, or excavation. Residual LNAPL is retained in the unsaturated zone by adsorption and capillary forces. Therefore, small soil particles and large surface area both increase the amount of residual LNAPL retained. The soil retention factor (volume of LNAPL per volume of soil) depends mainly on the soil pore size distribution, soil wettability, LNAPL viscosity, and LNAPL density.

Usually, more LNAPL is immobilized in the saturated zone than in the unsaturated zone because part of the residual LNAPL in larger pores of the unsaturated zone eventually drains down to the water table. In the unsaturated zone, LNAPL is the wetting fluid and tends to spread into the smaller pores, where it can be retained by capillary forces. However, it can drain downward from the larger pores. In the saturated zone, water is the wetting fluid and LNAPL is the nonwetting fluid. Here, if the LNAPL were mobile, its buoyancy would drive it upward to the water table, but, instead, it is trapped in larger soil pore spaces by immobile water.

Figure 7.11 shows soil retention factors for several kinds of LNAPL in soils of different textures. The retention of LNAPL in soils above the water table usually ranges

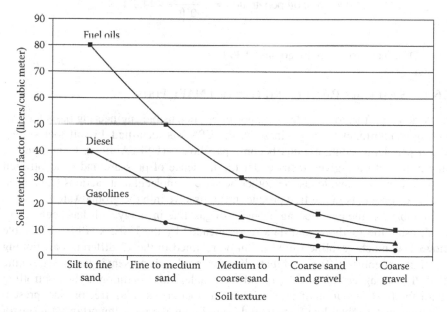

FIGURE 7.11 Soil retention factors for LNAPL fuels in different soils above the water table. Calculations assume a soil bulk density of 1.85 g/cm³ and LNAPL densities of 0.7, 0.8, and 0.9 g/cm³ for gasolines, diesel fuel, and fuel oils, respectively. (Adapted from Mercer, J. and Cohen, R. A., *J. Contamin. Hydrol.*, 6, 107, 1990.)

between about 80 L/m³ of soil for fuel oil in silt, to about 2.5 L/m³ for gasoline in coarse gravel. LNAPL in the unsaturated zone can often be remediated without excavation by some combination of soil washing, volatilization, or bioremediation.

Example 2: Using Soil Retention Factors

One thousand gallons of fuel oil were spilled on a soil consisting mostly of medium to coarse sand. What volume of soil is required to immobilize the 1000 gal? If the spill area was confined by a berm to 100 ft², how deep into the soil will the oil penetrate? Could it endanger a shallow aquifer 35 feet below the surface?

ANSWER

From Figure 7.11, the soil retention factor is about 30 L/m³ for fuel oil in medium to coarse sand. The volume of soil needed to contain the entire spill is

$$V_{soil} = \left(\frac{1000 \text{ gal}}{30 \text{ L/m}^3} \right) \left(\frac{3.785 \text{ L}}{1 \text{ gal}} \right) \left(\frac{35.3 \text{ ft}^3}{1 \text{ m}^3} \right) = 4454 \text{ ft}^3$$

Assume that the oil plume travels downward without spreading until it is all retained and immobilized, so that its cross-sectional area remains 100 ft², a worst case assumption. Then, the depth of oil penetration will be the length of a column with a cross section of 100 ft² and a volume of 4454 ft³.

$$\text{Depth of oil penetration} = \frac{4454 \text{ ft}^3}{100 \text{ ft}^2} = 44.5 \text{ ft}$$

Oil is likely to reach the aquifer at 35 feet.

7.6.1 Subsurface Partitioning Loci of LNAPL Fuels

As part of an effort to provide an authoritative, defensible engineering basis for predicting contaminant behavior in soils, the EPA has identified 13 soil loci among which petroleum contaminants become partitioned (USEPA 1991). Contaminants may move within a given locus under the influence of pressure and concentration gradients, or from one locus to another because of molecular attractions (e.g., when soluble components in the free product layer dissolve into the groundwater).

In Table 7.4, the partitioning behavior of gasoline in sandy soils has been calculated, using the methods and data recommended by the USEPA (1991). The table shows how much gasoline LNAPL can be retained in the 13 different partitioning loci. It is evident from Table 7.4 that when free product is present in locus 7 (gasoline LNAPL floating on top of the groundwater table), this condition is the controlling factor for the distribution of contaminants in other zones. With free product present above the water table, loci 2, 5, 6, and 7 are by far the most important in terms of mass and account for 99.9% of the total soil and groundwater contamination.

The mass of gasoline in the vapor and dissolved states (loci 1, 3, 8, and 12) is insignificant compared to that in the free product above the water table and in the

TABLE 7.4
Relative Importance of Different Subsurface Loci in Sandy Soils for Retention of Gasoline Contamination

	Loci of Subsurface LNAPL Retention	Average Gasoline Retention in Sandy Soils (mg/cm³)	Percent of Total Retention in Sandy Soils
1	Gasoline vapors in soil pores in the unsaturated zone.	0.095	<0.1%
2	Liquid gasoline sorbed to dry soil particles in the unsaturated zone. Locus 2 is especially important in the soil volume immediately below a spill, but not downgradient of the spill.	36	9.88
3	Gasoline dissolved in water on wet soil particles in the unsaturated zone.	0.0010	<0.1%
4	Liquid gasoline sorbed to wet soil particles in the saturated and unsaturated zones.	0.076	<0.1%
5	Liquid gasoline in soil pore spaces within the saturated zone. Locus 5 contaminants may generally be regarded as immobile.	38	10.4
6	Liquid gasoline in soil pore spaces in the unsaturated zone. Contaminants enter locus 6 mainly from free product floating on the groundwater table when the table rises and then falls.	110	30.2
7	LNAPL gasoline free product floating on top of the groundwater table. The most important loss mechanism from locus 7 occurs when a fluctuating water table moves contaminant into loci 5 and 6, where some of it remains trapped.	180	49.4
8	Gasoline dissolved in groundwater.	0.020	<0.1%
9	Gasoline sorbed to colloidal particles in water in the saturated and unsaturated zones.	0.00013	<0.1%
10	Liquid gasoline diffused into mineral grains in the saturated and unsaturated zones.	0.000060	<0.1%
11	Gasoline sorbed onto or into microbiota in the saturated and unsaturated zones.	0.010	<0.1%
12	Gasoline dissolved into the mobile pore water of the unsaturated zone.	0.030	<0.1%
13	Liquid gasoline in rock fractures in the saturated and unsaturated zones.	0.21	0.17

Source: Data from USEPA, *Assessing UST Corrective Action Technology: A Scientific Evaluation of the Mobility and Degradability of Organic Contaminants in Subsurface Environments,* EPA/600/2–91/053, 1991.

smear zone created by a rising and falling water table (loci 2, 5, 6, and 7). Of course for remediation purposes, all loci are important and those with relatively small amounts of contaminant may be the most difficult to remediate to a regulated level. Because it is the most mobile and can flow into wells, the first goal of remediation should be to remove the LNAPL floating on the water table.

Example 3: Calculation of the Contaminant Plume Volume Required to Immobilize 1 Million Gallons of Gasoline

Consider a site where LNAPL from a point source of contamination has leaked downward into sandy soil to the water table, where it has spread out above the saturated zone, moving downgradient in the direction of groundwater flow, as in Figure 7.4 for the left tank. For this case, we can assume that only loci 5 and 6, which comprise the smear zone caused by a fluctuating water table, are effective for immobilizing LNAPL contaminants. Locus 2 lies primarily under the area of the initial leak and probably is smaller than the total free product plume volume. LNAPL in locus 7 may be regarded as mobile. As LNAPL in locus 7 flows downgradient and is subjected to vertical movement caused by a fluctuating water table, it continually loses mass into loci 5 and 6, where it becomes immobile.

Assume an average sandy soil with a hydraulic gradient of 0.009 ft/ft. Using the methods laid out by the USEPA (1991), the estimated flow velocity for LNAPL gasoline in locus 7 is about 1.3 ft/day.* This may be compared with a groundwater velocity of about 10 ft/day for the same conditions. Further assume that the seasonal groundwater table fluctuations average around ±1.5 ft, giving a free product smear zone 3 ft in thickness. Taking the smear zone to be of uniform thickness and assuming that, on average, half of the smear zone is in the saturated zone and half in the unsaturated zone, we can say that loci 5 and 6 are each 1.5 ft thick everywhere adjacent to the free product plume.

ANSWER

Locus 5 RETENTION

Take the density of aged gasoline LNAPL to be 0.74 g/cm³ or 2800 g/gal. From Table 7.4, locus 5 retains 38 mg of gasoline per cubic meter of soil, or about

$$\frac{38 \text{ mg}}{\text{cm}^3} \times \frac{1 \text{ g}}{10^3 \text{ mg}} \times \frac{2.83 \times 10^4 \text{ cm}^3}{\text{ft}^3} = 1.08 \times 10^3 \text{ g of gasoline per cubic foot of soil}$$

The gallons of gasoline stored per cubic foot in locus 5 is

$$\frac{1.08 \text{ g}}{\text{ft}^3} \times \frac{1 \text{ gal}}{2.78 \times 10^3 \text{ g}} = 0.387 \text{ gal/ft}^3$$

The quantity of gasoline stored in locus 5 (1.5 ft thick) below 1 ft² of surface area is

$$\frac{0.387 \text{ gal}}{\text{ft}^3} \times 1.5 \text{ ft} = 0.58 \text{ gal/ft}^2$$

---

* Because of its greater viscosity, diesel fuel would move about one-third as fast.

Locus 6 retention

From Table 7.4, locus 6 retains 110 mg of gasoline per cm^3 of soil, or about 3.11 × 10^3 g of gasoline per cubic feet of soil. Calculations parallel to those above give a result of 1.11 gal of gasoline per cubic feet of soil, and 1.67 gal/ft^2 of surface area in locus 6.

The total quantity of gasoline retained in loci 5 and 6 below 1 ft^2 of surface area is (0.58 + 1.67) gal/ft^2 = 2.25 gal/ft^2, or about 98,000 gal per acre of surface area.

Thus, the soil contaminant plume from a 1-million-gallon gasoline spill in sandy soil, where the water table fluctuates ±1.5 ft annually (creating a 3-ft smear zone), could become immobilized after it had spread into approximately 30 acre-ft, beneath a surface area of about 10 acres.

7.7 CHEMICAL FINGERPRINTING OF LNAPLS

Environmental professionals often need to do more than locate and clean up pollutants. They may be asked to help identify the sources of the contamination. Someone always has to pay for a remediation effort. The high costs that are often involved mean that any prudent potentially responsible party will want convincing proof that it must accept responsibility for the pollution. Polluted sites often have a history of different owners and uses, any or all of which may have contributed to the present site contamination. Off-site sources may also be responsible for all or part of the current pollution problems. Allocating responsibility for the expenses involved in remediation often becomes a legal issue. The methods used in building a legally defensible case that assigns responsibility for soil and groundwater pollutants have become part of a branch of environmental investigations known as environmental forensics.

One of the most common applications of environmental forensics is identifying the sources of fuel contamination. The use of petroleum fuels has been increasing for the past 100 years, and fuel contaminants are pervasive wherever they have been used. There are many potential sources of fuel pollutants, from numerous small gasoline stations to large refinery operations, roadway accidents to criminal acts of disposal, pipeline breaks to tank corrosion. It is no wonder that the origin of any particular site contamination may often be legitimately questioned. Chemical fingerprinting of fuels is one tool in the toolbox of chemical forensics.

In chemical fingerprinting, the chemical characteristics of fuel contaminants are used to help distinguish among different possible sources of the pollutants. For example, knowing that only gasoline was ever used at a certain site, but that the groundwater hydrocarbon contaminants found there are from diesel fuel, clearly points to an off-site source.

7.7.1 First Steps in Chemical Fingerprinting of Fuel Hydrocarbons

Successful chemical fingerprinting of hydrocarbon contamination frequently consists of the sequential application of several investigative steps:

- Identify the fuel type of the contaminants. Are they gasoline, diesel fuel, or fuel oils? Each of these classes has unique chemical characteristics.
- If possible, distinguish between the ages of different contaminated samples. Weathering and aging processes often introduce predictable changes in the chemical makeup of fuels. If samples have weathered differently because of different exposure times, they will have different chemical profiles. Certain

compounds in fuel mixtures are more readily biodegraded, solubilized, or volatilized than others. This causes changes in the overall fuel composition as time passes. The exact nature of the changes is always site specific, but even if approximate ages cannot be assigned, sometimes the ages of different samples can be clearly demonstrated to be significantly different. In some cases, the presence of discontinued additives or certain degradation products may be a useful indicator of age.

*   Also, fuel compositions have changed historically, as more efficient refinery production practices were developed and as clean air legislation and automotive engine development mandated changes. For example, in the United States, gasoline containing organic lead-based octane boosters was first marketed in 1923 and the elimination of lead compounds began in the early 1980s and was complete by 1995. The presence of lead in fuel-contaminated soil or water may indicate an old rather than recent spill, especially if accompanied by dichloroethane and/or dibromoethane, which were lead scavenger gasoline additives introduced around 1927. Table 7.5 chronicles the history of lead and lead scavenger additives in the United States.

TABLE 7.5
Timeline of Lead and Lead Scavenger Additives to Gasoline

1923—Leaded gasoline first marketed using tetraethyl lead (TEL) as an antiknock agent.

1926—U.S. Surgeon General recommends maximum lead content of 3.17 g of lead per U.S. gallon of gasoline (g Pb/gal).

1927–1928—Lead scavengers 1,2-dibromoethane (EDB, ethylene dibromide[a]; 1,2-EDB, $C_2H_4Br_2$) and 1,2-dichloroethane (EDC, ethylene dichloride; 1,2-DCA, $C_2H_4Cl_2$) introduced for use with lead alkyl gasoline additives. Only EDB, not EDC, was used in piston engine aviation gasoline to avoid corrosion of aluminum parts. The amount of scavenger added was designed to react completely with all the lead to form volatile lead halides that would be removed with the exhaust gases. Ratios of EDC–EDB ranged from 1:1 to about 2:1. Characteristic composition of a lead additive package with a 1:1 scavenger ratio during the 1980s was 62 wt% TEL, 18 wt% EDB, 18 wt% EDC, and 2 wt% of other ingredients such as dye, antioxidants, and stability improvers.

1950—Most gasolines in the United States were leaded by this time.

1959—Maximum permitted lead level peaked at 4.23 g Pb/gal. This value is well above the point of diminishing returns on the lead response curve, and it is unlikely that much gasoline was ever produced with this much lead. More characteristic lead concentrations were between 2.25 (for regular) and 3 g Pb/gal (for premium).

1960—Alternative antiknock agent tetramethyl lead (TML) introduced by Chevron (then Standard Oil). After 1960, various mixtures of TEL, TML, and other organic lead additives have been used.

1975—The EPA proposed a scheduled reduction of lead to 1.7 g Pb/gal in 1975, 1.4 g Pb/gal in 1976, 1.0 g Pb/gal in 1977, 0.8 g Pb/gal in 1978, and 0.5 g Pb/gal in 1979.

1980—The EPA set an overall maximum lead limit in all gasolines to 0.5 g Pb/gal, for large refiners.

1982—The EPA set a maximum lead concentration for leaded gasoline, averaged over a 3-month manufacturing period (called a pool standard), at 1.10 g Pb/gal for large refiners. The actual level of lead in any two batches of gasoline could vary.

TABLE 7.5   (*Continued*)
Timeline of Lead and Lead Scavenger Additives to Gasoline

1985—Maximum permitted pool standard for lead in leaded gasoline lowered to 0.5 g Pb/gal for all refiners. Lead credits were allowed. Many states began phasing out leaded gasoline during the middle to late 1980s.

1986—The EPA scaled a pool standard decrease in lead to 0.1 g Pb/gal, to occur from 1986 to 1988.

1987—The EPA eliminated lead credits.

1992—Manufacture of leaded gasoline eliminated in California.

1995—The EPA eliminated lead in all U.S. gasoline produced after 1995, per Section 211(n) of the Clean Water Act.

[a] Despite their names, ethylene dibromide (EDB) and ethylene dichloride (EDC) do not contain double bonds. Their proper chemical names are 1,2-dibromoethane ($BrCH_2CH_2Br$) and 1,2-dichloroethane ($ClCH_2CH_2Cl$). Here, the use of the term "ethylene" is common in the petroleum trade and is based on the use of ethylene as feedstock in their manufacture.

- Look for unique chemical compounds that can serve as markers, which might be present in contamination from one source but not from another. An example might be the presence of MTBE or ethylene dichloride (EDC) in some fuel-contaminated samples but not in others. Different production practices at different refineries sometimes produce subtle but distinctive differences in their fuel products, which can help to distinguish between possible sources. This approach generally requires extensive knowledge about the chemical composition of contamination at the site and, sometimes, knowledge about the chemicals previously and currently used at the site.

Chemical fingerprinting is sometimes fairly simple and sometimes very complicated, even impossible. A few approaches are described here to indicate the possibilities. Complicated cases will require the help of experienced forensic chemists.

Example 4: Estimating the Amounts of Lead, Ethylene Dichloride, and Ethylene Dibromide Contained in a Leaded Gasoline Release

Use Table 7.5 to estimate amounts of lead, EDC, and ethylene dibromide (EDB) contained in 1000 gal of leaded gasoline release believed to have occurred before 1980.

ANSWER

A reasonable estimate for the concentration of lead in leaded gasoline before 1980 may be taken as around 2.0 g/gal. Assuming a typical additive package that was 62 wt% tetraethyl lead ($C_8H_{20}Pb$; TEL), 18 wt% EDC ($C_2H_4Cl_2$), and 18 wt% EDB ($C_2H_4Br_2$), the corresponding concentrations of EDC and EDB can be calculated as follows.
    Assume the following:

    Amount of lead in gasoline = 2.0 g/gal.
    Additive package contained 62% TEL, 18% 1,2-dichloroethane, 18% 1,2-dibromoethane, and 2% other (EDC:EDB = 1:1).

Calculation: all percentages are weight percentages.
Formula for TEL is $C_8H_{20}Pb$; MW = 323.45 g/mol.
C: 29.70%; H: 6.23%; Pb: 64.06%.
2.0 g Pb/gal = 0.64 × (grams of TEL).
TEL = 2.0/0.64 = 3.1 g TEL/gal.
TEL is 62% of additive package.
Therefore, total grams of additive per gal = 3.1 g TEL/0.62 = 5.0 g additive/gal.
EDC = 18% of additive.
Therefore, EDC = 0.18 (5.0 g additive/gal) = 0.9 g EDC/gal.
EDB = 18% of additive.
Therefore, EDB = 0.18 (5.0 g additive/gal) = 0.9 g EDB/gal.

Using these values, a spill of 1000 gal of leaded gasoline that occurred before 1980 could release about 900 g (2.0 lb) each of EDC and EDB to the environment. Both EDC and EDB are moderately soluble and would become dissolved in groundwater, the water concentration depending on the length of contact time, extent of cosolvency with other contaminants, and dilution by water.

7.7.2 IDENTIFYING FUEL TYPES

The most useful analytical tool for identifying different fuel types is the gas chromatograph with a mass spectrometer detector (GC/MS). When fuels are analyzed in a GC, the retention times of different hydrocarbon compounds closely correspond to their boiling points; the higher the boiling point, the longer the retention time. Thus, lower boiling point (more volatile) gasoline components elute from the GC earlier than higher boiling point (less volatile) diesel components.

The physical and chemical characteristics of different fuels were described in Section 7.2. Table 7.1 and Figure 7.2 show that carbon number ranges characteristic of different fuel types correspond to different boiling point ranges; in general, the more carbons in a petroleum molecule, the higher is its boiling point (see the discussion of London forces in Chapter 2, Section 2.8.6). In a gas chromatogram, the longer the retention time of a peak, the higher the boiling point of the corresponding compound and the more carbons in the molecule. Although the presence of structural differences, such as the presence of carbon side chains, introduces some variability, this general principle remains useful. Gas chromatograms of different types of fuel are different from one another and are useful for identifying fuel types.

In Figure 7.12, note that the fresh gasoline GC signature contains more lightweight components (peaks farther to the left, indicating fewer carbon atoms) than do diesel fuel or lubricating oil. When fuel hydrocarbons are weathered by volatilization, dissolution, and biodegradation, the lighter components on the left side of the fresh gasoline and diesel fuel signatures are lost first from the free product LNAPL. Thus, the gas chromatogram of weathered diesel does not resemble that of gasoline, because it lacks the lightweight components that are characteristic of gasoline. Likewise, weathered gasoline does not resemble diesel fuel because it lacks the heavier components that are characteristic of diesel fuel. The components of lubricating oil are all heavier than those found in gasoline or diesel fuel.

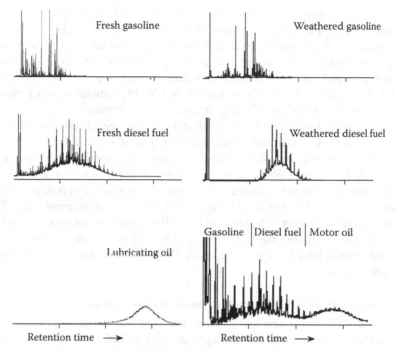

FIGURE 7.12 GCs showing the differences in chromatographic signatures between different types of fresh and weathered petroleum hydrocarbon free product. The graph at lower right is a GC of free product containing a mixture of gasoline, diesel fuel, and motor oil. Humps, where the chromatogram rises above the baseline, are due to hundreds of different hydrocarbon compounds that are not chromatographically resolved.

7.7.3 AGE-DATING FUEL SPILLS

7.7.3.1 Gasoline

As a gasoline additive, benzene increases the octane rating and reduces engine knocking. As a result, gasoline often contained 2–5% by volume of benzene before the 1950s, when TEL replaced it as the most widely used antiknock additive. However, with the global phase-out of leaded gasoline, benzene has made a comeback as a gasoline additive in some nations. In the United States, Canada, and Europe, concern over benzene's negative health effects and its contamination of groundwater has led to the regulation of gasoline's benzene content. In the United States, current gasolines are regulated to 1% benzene on a regional average.* Canadian and European gasoline specifications now contain a similar 1% limit on benzene content.

---

* The 1% by volume limit for benzene in gasoline is a regional average. EPA has established a system of benzene credits that can be traded among regional refiners to allow some production flexibility while keeping the regional benzene average at or below 1%. The intent is to control regional benzene emissions from all sources. The result is that many individual gasolines will have levels less than 1% benzene and a few might exceed 1%.

The gasoline content in fresh gasoline of the other BTEX compounds, toluene, ethyl benzene, and the xylene isomers (see Figure 7.1), is also variable, depending on the manufacturer, octane requirements, seasonal formulations, and geographic regions. Each BTEX compound also weathers at a different rate. Because benzene is more soluble and volatile than other common gasoline constituents, it is depleted more rapidly than other BTEX components in LNAPL samples of older gasoline contamination because of partitioning into air and groundwater.

Several simple, but approximate, indicators of the age of gasoline spills are related to the BTEX constituents. The concentrations of single BTEX compounds in a weathered sample are seldom useful for estimating how long ago the sample was released, because gasolines have never been produced with consistent BTEX concentrations. Furthermore, after a spill or leak, the different components of gasoline are lost by volatilization, dissolution, and degradation at different rates. Thus, knowing the concentration of benzene in a gasoline sample is not particularly helpful by itself for estimating the age of the gasoline. However, certain ratios of the BTEX compounds have been shown empirically to be useful for estimating the age of gasoline releases.

7.7.3.2 Changes in BTEX Ratios Measured in Groundwater

Groundwater BTEX ratio age-dating techniques are based on the different loss rates by volatilization, dissolution, and biodegradation of BTEX compounds from gasoline LNAPL. In groundwater, the sequence of BTEX loss from LNAPL generally begins with benzene, because it is lost most rapidly from gasoline LNAPL by volatilization and partitioning into groundwater; benzene loss is followed by toluene, ethylbenzene, and xylenes, in that order. Although the biodegradation rates of the separate BTEX compounds are very site specific, biodegradation of benzene and toluene is generally faster than that of ethylbenzene and the xylene isomers. Their different loss rates from LNAPL suggest that ratios of BTEX compounds might be useful for estimating the time that has elapsed since a gasoline release.

However, the uncertainty of initial BTEX content in gasoline-contaminated groundwater limits the usefulness of simple ratios such as B/T, B/E, and T/X for contaminant age dating. Nevertheless, it has been found empirically from numerous site studies that a quantity known as the cumulative BTEX ratio (R_{BTEX}) can be a useful dating parameter (Kaplan et al. 1997). R_{BTEX} is defined as

$$R_{BTEX} = \frac{B + T}{E + X} \tag{7.5}$$

where B, T, E, and X are the groundwater concentrations (mg/L) of benzene, toluene, ethylbenzene, and xylene, respectively.

Use of R_{BTEX} appears to minimize the uncertainties related to variations in initial BTEX concentrations. Field and laboratory measurements (Kaplan et al. 1997) show that R_{BTEX} values between 1.5 and 7.0 in gasoline-contaminated groundwater* generally indicate a release that occurred within the last 1–5 years.

---

* The value of R_{BTEX} is also influenced by the amount of gasoline LNAPL in contact with groundwater.

The ratio decreases exponentially with time with an estimated half-life of 2.3 years, and a value of R_{BTEX} less than 0.5 may be taken to indicate a release greater than 10 years old. These relations are more accurate for samples taken close to the release point because additional uncertainty is introduced by spatial separation of the BTEX components during migration through the subsurface (see Chapter 6, Section 6.7). Morrison (2000a,b) discusses additional sources of uncertainty in age dating with R_{BTEX}.

RULES OF THUMB

In the United States, older gasolines (before 1950) typically contained about 2–5% by volume of benzene. Current gasolines are regulated to 1% on a regional average for health reasons. The ratio (by weight concentration) $R_{BTEX} = \frac{B+T}{E+X}$ can be used to obtain an approximate age of dissolved gasoline. Because of the large variability in initial gasoline composition and the site-specific nature of weathering processes, all dating methods using BTEX ratios should be considered only first-order approximations. Whenever possible, additional lines of age-dating evidence should be compared.

Whenever BTEX ratios are used for age dating a fuel release, it is important to consider whether releases of other fuels or substances containing BTEX compounds might contribute to the samples collected.

In groundwater

1. If $R_{BTEX} = \dfrac{B+T}{E+X} = 1.5 - 6.0$, it generally indicates a release less than 5 years old.

2. If $R_{BTEX} = \dfrac{B+T}{E+X} < 0.5$, it generally indicates a release more than 10 years old.

3. The value of R_{BTEX} in groundwater is assumed to decrease exponentially with time after a gasoline release according to

$$R_{BTEX} = 6.0e^{-0.308t} \tag{7.6}$$

where t is the time in years since initial release.

Equation 7.6 assumes that $R_{BTEX} = 6.0$, close to the source shortly after the release (whereas actual ratios immediately after a release were found to range between 1.5 and 6.0 [Kaplan et al. 1997]) and follows first-order decay kinetics with a half-life of 2.3 years.

Other clues to the age of gasoline contamination are based on the presence or absence of TEL, organic manganese compounds, EDB, and EDC, which were common additives to gasoline before 1980 (see Table 7.5). Therefore, the presence of these compounds is supporting evidence for gasoline contamination originating before 1980. However, EDB and EDC are present in some agricultural chemicals and must be used as an indicator with caution.

TABLE 7.6
Approximate Composition of Fresh Diesel Fuel (C6–C24)

Class of Compound	Percent
n-Alkanes (degrade fastest; smaller alkanes degrade faster than larger)	40
iso- and cyclo-Alkanes (degrade slower)	36
Isoprenoids (degrade very slowly)	3–4
Aromatics (most soluble; mainly parent and alkylated benzenes, naphthalenes, phenanthrenes, and fluorenes)	20
Polar (water-soluble sulfur, nitrogen, and oxygen compounds)	1

7.7.3.3 Diesel Fuel

Normal-alkanes (linear single-bonded hydrocarbons) in the approximate carbon number range C6–C24 are the most abundant components of fresh diesel oils, although many other types of hydrocarbons are also present (see Table 7.6). The table also shows that the different types of hydrocarbons in diesel oils biodegrade at different rates. Isoprenoids are branched, unsaturated hydrocarbons that biodegrade more slowly than linear alkanes with similar masses, because their chemical structure inhibits biodegradation.

The solubilities and volatilities of oil range n-alkanes and isoprenoids are quite low and similar for the two chemical classes. Therefore, they have similar rates of weathering by nonbiological processes. However, since isoprenoids biodegrade much more slowly than n-alkanes, the abundance ratio of n-alkanes to isoprenoids changes over time where biodegradation is occurring. Because the passive biodegradation rates of diesel in soils are fairly uniform at similar sites (Christensen and Larsen 1993), the corresponding changes in composition can be used for estimating the age of the diesel contamination.

Christensen and Larsen (1993) found, as had others, that by comparing GC peak-height ratios of pristane (a C19 isoprenoid) with those of n-heptadecane (a linear C17 alkane) an estimate can be made of the degree of biodegradation that the sample has undergone. The ratios of phytane (C20 isoprenoid) and n-octadecane (linear C18 alkane) can be compared in a similar way. Using the relative extents of biodegradation, the relative ages of fuel contamination in different samples from similar sites may be estimated.

When the composition of fresh diesel fuel is compared with weathered fuels that have biodegraded in the subsurface environment, certain composition changes are apparent.

In particular, n-alkanes dominate the composition of fresh fuels and isoprenoids dominate the composition of highly degraded fuels. The C17/pristane peak-height ratio falls from about 2 (for fresh diesel) to 0 (after about 20 years). Figure 7.13 compares GCs of fresh and biodegraded No. 2 diesel oil. Changes in the relative peak heights of n-alkanes and isoprenoids are readily apparent.

Christensen and Larsen (1993) found a linear relation between the age of diesel oils and the C17/pristane peak-height ratio. From their data, it appears possible to

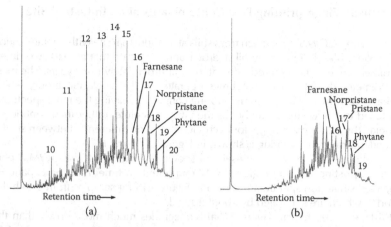

FIGURE 7.13   (a) A GC of typical fresh No. 2 diesel oil. The numbered peaks are linear *n*-alkanes, where the number represents the carbon number (e.g., C17) of the alkane. The *n*-alkanes are the most abundant compounds in fresh diesel fuels and dominate the composition. The peaks of several isoprenoids are labeled. The peak-height ratio of C17/pristane is about 2:1. (b) A GC of biodegraded No. 2 diesel oil. Isoprenoids are more abundant than *n*-alkanes. The peak-height ratio of C17/pristane is about 0.8, indicating an age of about 13 years according to Equation 7.7.

determine the age of a diesel spill to within about 2 years if the following criteria are met:

- It is between 5 and 20 years old.
- It was created by a single sudden spill event.
- It has not been weathered significantly except by biodegradation.

A linear best-fit to their data yields Equation 7.7:

$$\text{Age of diesel fuel (in years)} = -8.3\left(\frac{\text{C17 peak height}}{\text{pristane peak height}}\right) + 19.5 \qquad (7.7)$$

RULES OF THUMB

1. Fresh diesel contains more *n*-alkanes than isoprenoids, such as pristane or phytane. Therefore, if relative depletion of the *n*-alkanes is observed, it indicates that biodegradation has taken place.
2. In a fresh diesel fuel, the C17/pristane and C18/phytane ratios are close to 2. Any ratio less than about 1.5 indicates that biodegradation has occurred.

Example 5: Fingerprinting Fuel Contaminants at an Industrial Site

The following GC/MS data, taken from soils at an industrial site with contamination by gasoline, diesel, and heavy oil hydrocarbons, indicate that at least two different diesel spill events occurred, separated by approximately 6–10 years. The mass spectrometer detector was sometimes operated in the single-ion mode, which increases sensitivity. Single-ion monitoring consists of leaving the mass spectrometer tuned to a fixed mass number as the GC peaks elute, rather than continually scanning the entire mass range for each GC peak. The difference between single-ion and full-range monitoring is shown in Figure 7.14a and b.

To estimate the age of the diesel fuel shown in Figure 7.14, the GC/MS spectra in the pristine/phytane region (15–17-min retention time) must be expanded. Expanded single-ion spectra are shown in Figure 7.15 for samples from Boreholes A and B, which are separated by about 200 yd.

Pristane (16.06 min in Figure 7.15a) biodegrades much more slowly than the C17 n-alkane (16 min). The same is true for phytane (16.92 min) and the C18 n-alkane (16.83 min). The retention times for these peaks are slightly different in Figure 7.15b. In the Borehole A sample, the C17/pristane peak-height ratio is 0.33 and the C18/phytane ratio is 0.26, indicating significant biodegradation. Although site-specific conditions will determine how much time this amount of degradation would require, the diesel in this sample is at least 15–20 years old. Equation 7.7 suggests an age of about 19 years.

In the Borehole B sample, the C17/pristane ratio is 1.2, which indicates a more recent diesel spill. The diesel fuel from Borehole B is about 9 years old by Equation 7.7.

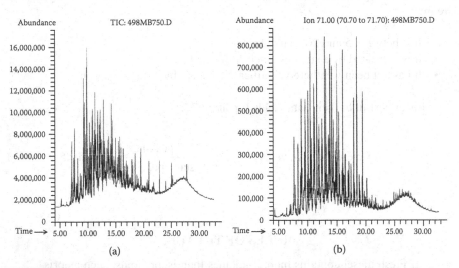

(a)   (b)

FIGURE 7.14 Full-range and single-ion GC/MS spectra of a contaminated soil sample from Borehole A. (a) GC/MS full-range spectrum of fuel contamination extracted from soil at Borehole A. The distribution of compounds indicates that the soil contains gasoline, diesel, and heavy oil hydrocarbons. (b) Single-ion spectrum at mass 71 of the same sample. Mass 71 is a fragment ion common to many hydrocarbon compounds of C7 and larger. It is useful for diesel fingerprinting because the mass 71 ion is produced abundantly in the fragmentation of pristane and phytane.

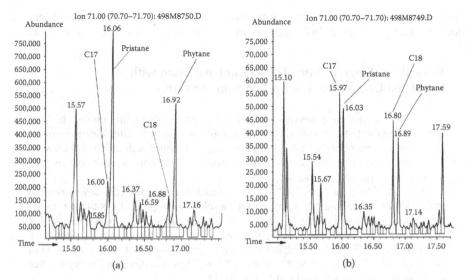

FIGURE 7.15 Expansion of mass 71 single-ion spectra in the pristane/phytane region of contaminated soil from (a) Borehole A and (b) Borehole B.

7.8 SIMULATED DISTILLATION CURVES AND CARBON NUMBER DISTRIBUTION CURVES

A simple approach to chemical fingerprinting is to generate simulated distillation curves (SDCs) and carbon number distribution curves (CNDCs). These plots allow a relative comparison of the concentrations of volatile and semivolatile compounds present in a sample, without requiring a detailed analysis of each compound present. The curves can be generated automatically from computerized GC data without identifying the individual peaks.

An SDC shows the percentage of the sample that would be volatilized at various temperatures. Since the volatility of a petroleum compound is closely related to the size and number of carbon atoms in that compound (compounds with fewer carbon atoms are lighter in weight and boil at lower temperatures than compounds with more carbon atoms), one can estimate the general chemical makeup of petroleum contaminants from an SDC. A CNDC gives similar information, but is often easier to interpret.

SDC and CNDC curves allow a visual comparison of the mass distribution of chemical compounds that are present in the analytical samples, based on their boiling points and masses. The shapes of these curves are distinctly different for different types of hydrocarbon mixtures. Gasoline, for example, contains relatively high concentrations of lightweight hydrocarbons such as benzene, while diesel fuel normally has very low concentrations of these lightweight compounds, but higher concentrations of heavier compounds. Weathering of organic compounds produces predictable changes in the shapes of SDCs and CNDCs.

When CNDCs and SDCs have distinctive shapes, they can be used as a fingerprint for determining whether contamination at one location is different from or similar to contamination at another location. Sometimes this preliminary analysis is sufficient for identification purposes, as in Example 6. In other cases, it serves to guide further study.

Example 6: Fingerprinting Diesel Contamination with Simulated Distillation and Carbon Number Curves

Mr. A, the owner of a newly purchased mountain home, frequently, but not always, detected strong fuel odors in his basement shortly after diesel fuel was delivered to a neighbor's UST. Mr. B, the neighbor, was cooperative and allowed Mr. A to take samples from the UST for analysis and comparison with soil samples from around Mr. A's new home.

In Figure 7.16a, the SDC for No. 2 diesel fuel from Mr. B's UST lies somewhat to the left of the laboratory standard for diesel, showing that it contains a higher percentage of lower-boiling (lighter-weight) compounds. This could indicate that the UST fuel was contaminated slightly with gasoline (perhaps from a tanker truck used to carry both types of fuel) or that it was specially formulated for cold weather use. In Figure 7.16b, the CNDC for the UST fuel has a distinctive fingerprint that includes two prominent peaks at C11 and C13.

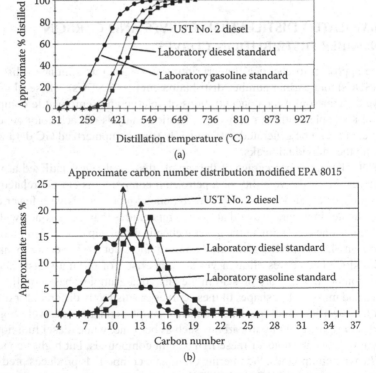

FIGURE 7.16 (a) Simulated distillation curves and (b) carbon number curves comparing No. 2 diesel fuel from an underground storage tank with laboratory standards of diesel fuel and gasoline.

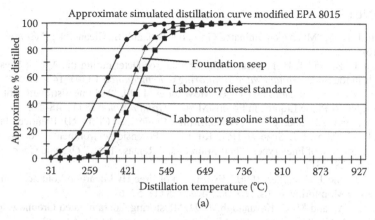

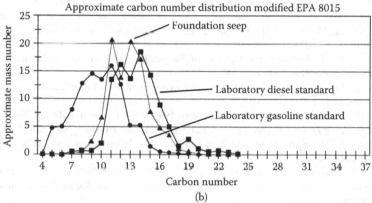

FIGURE 7.17 (a) Simulated distillation curves and (b) carbon number curves comparing free product from a seep adjacent to a house foundation with laboratory standards of diesel fuel and gasoline.

Figure 7.17a and b shows SDC and CNDCs for a soil sample collected from a free product seep adjacent to the foundation of Mr. A's home. Not only is the SDC for the foundation seep sample very similar to the SDC for the UST sample, but also its CNDC contains the two distinctive peaks at C11 and C13 that were also seen in the UST sample.

Mr. B acknowledged that these data strongly implicated his UST as the source of contamination at Mr. A's home, even if it did not explain why the leak appeared to be erratic. When his tank was leak tested, it was found that the filler pipe had cracked where it was fastened to the tank, below the soil surface. It leaked only when the tank was overfilled and diesel rose into the fill pipe above the crack.

EXERCISE

1. Discuss how fluctuations in the groundwater level can influence the distribution of LNAPL hydrocarbon contamination in the subsurface.

REFERENCES

Bredehoeft, J. 1992. "Much Contaminated Groundwater Can't Be Cleaned Up." *Ground Water* 30 (6): 30.

Christensen, L. B., and T. H. Larsen. 1993. "Method for Determining the Age of Diesel Oil Spills in the Soil." *Groundwater Monitoring & Remediation* 13 (4): 142–49.

Farr, A. M., R. J. Houghtalen, and D. B. McWhorter. 1990. "Volume Estimation of Light Nonaqueous Phase Liquids in Porous Media." *Ground Water* 28 (1): 48–56.

Freeze, R. A., and J. A. Cherry. 1979. *Groundwater.* Englewood Cliffs, NJ: Prentice-Hall.

Kaplan, I. R., G. Yakov, T. L. Shan, and P. L. Ru. 1997. "Forensic Environmental Geochemistry: Differentiation of Fuel Types, Their Sources and Release Time." *Organic Geochemistry* 27 (5/6): 287–317.

Lenhard, R. J., and J. C. Parker. 1990. "Estimation of Free Hydrocarbon Volume from Fluid Levels in Monitoring Wells." *Ground Water* 28 (1): 57–67.

MacDonald, J. A., and M. C. Kavanaugh. 1994. "Restoring Contaminated Groundwater: An Achievable Goal?" *Environmental Science & Technology* 13 (4): 142–49.

Mercer, J., and R. A. Cohen. 1990. "A Review of Immiscible Fluids in the Subsurface: Properties, Models, Characterization and Remediation." *The Journal of Contaminant Hydrology* 6: 107–63.

Morrison, R. D. 2000a. "Critical Review of Environmental Forensic Techniques: Part I." *Environmental Forensics* 1: 157–73.

Morrison, R. D. 2000b. "Critical Review of Environmental Forensic Techniques: Part II." *Environmental Forensics* 1: 175–95.

Parker, J. C., D. W. Waddill, and J. A. Johnson. 1996. *UST Corrective Action Technologies: Engineering Design of Free Product Recovery Systems.* EPA/600/SR-96/031, Order No. PB96–153556. Springfield, VA: National Technical Information Service, National Risk Management.

USEPA. 1991. *Assessing UST Corrective Action Technology: A Scientific Evaluation of the Mobility and Degradability of Organic Contaminants in Subsurface Environments.* EPA/600/2–91/053.

USEPA. 2011. *UST Program Facts: Data about the Underground Storage Tank (UST) Program.* Washington, DC, United States Environmental Protection Agency, Office of Solid Waste and Emergency Response, June 2011, http://www.epa.gov/oust/pubs/ustfacts.pdf.

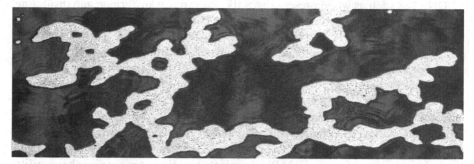

Moving through the subsurface, DNAPLs leave a trail of soil-sorbed and immobile liquid NAPL in the form of disconnected blobs and ganglia of free product that remain behind the trailing end of the downward-moving DNAPL body.

Chapter 8, Section 8.2

8 Behavior of Dense Nonaqueous Phase Liquids in the Subsurface

8.1 DNAPL PROPERTIES

Dense nonaqueous phase liquids (DNAPLs) are liquids, denser than water, that are only slightly soluble in water and therefore exist in the subsurface as a separate fluid phase immiscible with both water and air.* The density of DNAPLs is greater than water (DNAPL density > 1 g/cm³ at 4°C), and their mobility in the subsurface is governed more by gravity and the properties of the DNAPL and surrounding soil than by groundwater movement.

Unlike light nonaqueous phase liquids (LNAPLs) such as gasoline, diesel fuel, and heating oil (which are less dense than water), DNAPLs released into soils can sink below the water table where, in the same manner as LNAPLs, their more-soluble components can slowly dissolve into flowing groundwater and form dissolved contaminant plumes. However, unlike LNAPLs, they may be difficult to locate and remove by pump-and-treat remediation. A release of DNAPL at the ground surface can therefore lead to long-term contamination of both the vadose and saturated zones at a site.

---

* See Chapter 7 to review the properties of NAPLs in general, along with LNAPLs.

DNAPLs such as wood preservatives like creosote, transformer, and insulating oils containing polychlorinated biphenyls (PCBs), polynuclear aromatic hydrocarbons (PAHs), coal tar, and a variety of chlorinated solvents such as trichloroethene (TCE) and tetrachloroethene (PCE) have been widely used in industry since the beginning of the twentieth century. However, their importance as soil and groundwater contaminants was not recognized until the 1980s, mainly because of the limitations of early analytical methods. As a result, chemical material safety data sheets (MSDSs) distributed as late as early 1970 sometimes recommended that waste chlorinated solvents can be discarded by spreading them onto dry ground and allowing them to evaporate. These early MSDSs acknowledged the volatile nature of many DNAPL chemicals but did not recognize their ability to infiltrate rapidly into the subsurface, causing soil and groundwater pollution. It is not surprising that DNAPLs are the contaminants of greatest concern at many Superfund and other hazardous waste sites.

Table 8.1 lists many of the chlorinated DNAPLs commonly found at Superfund sites, along with their chemical formulas, some alternative names, and common abbreviations. Properties of DNAPL liquids and surrounding soils that are useful for predicting DNAPL mobility are described in Table 8.2. Some important DNAPL compounds often found at Superfund sites and their properties are included in Table 8.3.

8.2   DNAPL-FREE PRODUCT MOBILITY

In a DNAPL release, the free product sinks vertically downward through the vadose zone under gravitational forces, spreading laterally under capillary forces and leaving behind a trail of residual soil-sorbed DNAPL. In the vadose zone, DNAPL behaves similarly to LNAPL, moving downward while spreading laterally and leaving a trail of soil-sorbed and immobile liquid NAPL in the form of disconnected blobs and ganglia of free product that remain behind the trailing end of the downward-moving DNAPL body.

8.2.1   DNAPL in the Vadose Zone

Like LNAPL, DNAPL in the vadose zone will partition into solid, liquid, and vapor phases so that different portions are present as free product, pore space vapor, dissolved in water, and sorbed to soil (see Figure 7.3). Because of continual losses to other phases, the downward-moving free product is continually diminished in mass and volume. It also undergoes changes in composition as the more-volatile and -soluble components preferentially leave the free product mixture. A point may be reached at which the remaining DNAPL free product no longer holds together as a continuous phase but rather is present as immobile isolated globules and ganglia, held in place by sorption and capillary forces. Only DNAPL present as a continuous, immiscible, liquid phase is mobile. If sufficient DNAPL was originally present, liquid free product will eventually reach the water table interface between the vadose and saturated zones.

TABLE 8.1

Chlorinated DNAPL Contaminants of Concern at Many Hazardous Waste Sites

Chemical Abstracts Service (CAS) Name	Abbreviation	CAS Number	Other Names	Molecular Formula	Structural Formula
Chloromethane	Artie; R40	74-87-3	Methyl chloride; monochloromethane	CH_3Cl	CH_3Cl
Dichloromethane	Methylene chloride; MC	75-09-2	Methylene dichloride	CH_2Cl_2	CH_2Cl_2
Trichloromethane	CF	67-66-3	Chloroform; methane trichloride	$CHCl_3$	$CHCl_3$
Tetrachloromethane	CT	56-23-5	Carbon tetrachloride	CCl_4	CCl_4
Chloroethane	CA	75-00-3	Ethyl chloride	C_2H_5Cl	Cl_3C-CH_3
1,1-Dichloroethane	1,1-DCA	75-34-3	Ethylidene dichloride	$C_2H_4Cl_2$	Cl_3C-CH_3
1,2-Dichloroethane	1,2-DCA, EDC	107-06-02	Ethylene dichloride	$C_2H_4Cl_2$	Cl_3C-CH_3
1,1,1-Trichloroethane	1,1,1-TCA	71-55-6	Methyl chloroform, chlorothene, methyltrichloromethane	$C_2H_3Cl_3$	Cl_3C-CH_3
1,1,2-Trichloroethane	1,1,2-TCA	79-00-5	Vinyl trichloride β-trichloroethane	$C_2H_3Cl_3$	Cl_2HC-CH_3
Chloroethene	VC	75-01-4	Vinyl chloride; chloroethylene	C_2H_3Cl	$ClHC=CH_2$
1,1-Dichloroethene	1,1-DCE	75-35-4	1,1-Dichloroethylene; vinylidine chloride	$C_2H_2Cl_2$	$Cl_2C=CH_2$
(E)-1,2-Dichloroethene	trans-1,2-DCE	156-60-5	trans-1,2-Dichloroethene; trans-1,2-dichloroethylene; acetylene dichloride	$C_2H_2Cl_2$	t-ClHC=CHCl
(Z)-1,2-Dichloroethene	cis-1,2-DCE	156-59-2	cis-1,2-Dichloroethene cis-1,2-dichloroethylene; acetylene dichloride	$C_2H_2Cl_2$	c-ClHC=CHCl

(Continued)

TABLE 8.1 (Continued)

Chlorinated DNAPL Contaminants of Concern at Many Hazardous Waste Sites

Chemical Abstracts Service (CAS) Name	Abbreviation	CAS Number	Other Names	Molecular Formula	Structural Formula
Trichloroethene	TCE	79-01-6	Trichloroethylene	C_2HCl_3	$Cl_2C = CHCl$
Tetrachloroethene	PCE	127-18-4	Perchloroethylene; tetrachloroethylene	C_2Cl_4	$Cl_2C = CCl_2$
Chlorobenzene	CB	108-90-7	Monochlorobenzene, benzene chloride, phenyl chloride	C_6H_5Cl	C_6H_5Cl
1,2-Dichlorobenzene	1,2-DCB	95-50-1	o-Dichlorobenzene	$C_6H_4Cl_2$	$C_6H_4Cl_2$
1,3-Dichlorobenzene	1,3-DCB	541-73-1	m-Dichlorobenzene	$C_6H_4Cl_2$	$C_6H_4Cl_2$
1,4-Dichlorobenzene	1,4-DCB	106-46-7	p-Dichlorobenzene	$C_6H_4Cl_2$	$C_6H_4Cl_2$
1,2,3-Trichlorobenzene	1,2,3-TCB	87-61-6	vic-Trichlorobenzene	$C_6H_3Cl_3$	$C_6H_3Cl_3$
1,2,4-Trichlorobenzene	1,2,4-TCB	120-82-1	Trichlorobenzol	$C_6H_3Cl_3$	$C_6H_3Cl_3$
1,3,5-Trichlorobenzene	1,3,5-TCB	108-70-3	sym-Trichlorobenzene	$C_6H_3Cl_3$	$C_6H_3Cl_3$
1,2,3,5-Tetrachlorobenzene	1,2,3,5-TECB	634-90-2	1,2,3,5-TCB	$C_6H_2Cl_4$	$C_6H_2Cl_4$
1,2,4,5-Tetrachlorobenzene	1,2,4,5-TECB	95-94-3	s-Tetrachlorobenzene, sym-tetrachlorobenzene	$C_6H_2Cl_4$	$C_6H_2Cl_4$
Hexachlorobenzene	HCB	118-74-1	Perchlorobenzene	C_6Cl_6	C_6Cl_6
1,2-Dibromoethane	EDB	106-93-4	Ethylene dibromide; dibromoethane	$C_2H_4Br_2$	$C_2H_4Br_2$
Polychlorinated biphenyls	PCBs	—	Aroclor; Phenoclor; Pyralene; Clophen; Kaneclor	—	See Section 8.4

TABLE 8.2
DNAPL Properties Important for Predicting Mobility in Environment

Properties of DNAPL/Soil	Definition/Typical Units	Comments
Density (d)	d = mass/volume d = $g \cdot cm^{-3}$; $lb \cdot ft^{-3}$	Density distinguishes between LNAPLs ($d_{LNAPL} < d_{water}$) and DNAPLs ($d_{DNAPL} > d_{water}$). It depends on temperature, pressure, molecular weights of components, intermolecular forces, and bulk liquid structure.
Dynamic viscosity (μ)	μ = fluid internal resistance to flow or shear. The CGS unit is poise (P); SI unit is $N \cdot s \cdot m^{-2}$. 1 P = 100 centipoise (cp) 1 P = 1 $g/cm \cdot s$ = 0.1 $Pa \cdot s$	Dynamic viscosity is a measure of the force required to move a liquid at a constant velocity. The common unit of μ is the cP because water at 20.2°C has a convenient viscosity of 1.000 cP. Viscosity decreases with increasing temperature (note water in Table 8.3). Intermolecular attractions are the main cause of viscosity. The lower the viscosity, the more fluid the liquid and the more easily it will flow through soils. The reciprocal of dynamic viscosity is called fluidity.
Kinematic viscosity (v)	v = dynamic viscosity/density. The CGS unit is stokes (St) or centistokes (cSt). SI units are $m^2 \cdot s^{-1}$; St = P/density; 1 St = 100 cSt = $10^{-4}\ m^2 \cdot s^{-1}$	When the force causing a liquid to move is only due to gravity, as in NAPL movement in the environment, the fluid density, as well as the dynamic viscosity, affects the rate of movement. Using kinematic viscosity includes density in its definition and eliminates the force term (N or Pa). Kinematic viscosity is convenient for calculating hydraulic conductivity, which is inversely proportional to v. Since the density of water at 20.2°C is 0.998 g/cm^3, the kinematic viscosity of water at 20.2°C is, for most practical purposes, equal to 1.0 cSt.
Solubility in water (S)	S = mass of dissolved substance per unit volume of water, in equilibrium with the undissolved substance. For environmental pollutants in water, the common units are mg/L or μg/L.	Solubility measures a compound's tendency to partition from the bulk compound into water. For a single-component NAPL, the solubility is the concentration of dissolved component in equilibrium with the NAPL. For NAPLs that are mixtures, each component of the mixture has its own characteristic solubility, which is generally lower than the solubility of the pure component (see Section 6.3.8). Thus, the overall solubility of an NAPL mixture is variable, depending on its composition, and changes with time as the more-soluble components leave the NAPL by partitioning into the water. Solubility can vary with temperature, pH, total dissolved solids (TDS), and the presence of cosolvents (e.g., detergents, EDTA, etc.). In general, the greater the molecular weight (high polarizability) and symmetry (low polarity) and the fewer hydrogen-bonding atoms, the lower the solubility; see Section 2.9.

(Continued)

TABLE 8.2 *(Continued)*
DNAPL Properties Important for Predicting Mobility in Environment

Properties of DNAPL/Soil	Definition/Typical Units	Comments
Vapor pressure (P_v)	P_v = pressure exerted by a vapor in equilibrium with the liquid or solid phase of the same substance. There are many different units for pressure. The more common units are millimeters of mercury (mm Hg), torr, and atmosphere (atm). The SI unit is the pascal (Pa). $$1 \text{ mm Hg} = 1 \text{ torr} = 760^{-1} \text{ atm}$$ $$= 1.333 \text{ mbar} = 133.3 \text{ Pa}$$ $$= 1.934 \times 10^{-2} \text{ psi } 1 \text{ Pa}$$ $$= 1 \text{N/m}^2 = 10^{-5} \text{ bar}$$ $$= 7.50 \times 10^{-3} \text{ torr}$$ $$= 1.450 \times 10^{-4} \text{ psi}$$	Vapor pressure indicates an NAPL's volatility, or tendency to vaporize, at a given temperature. For a given NAPL, it depends only on the temperature and increases exponentially with increasing temperature. On a molecular level, vapor pressure is an indication of the strength of intermolecular attractive forces; see Section 2.8.6. The vapor pressure of DNAPLs ranges from very high to very low; for example, compare 1,1-dichloroethylene and chrysene in Table 8.2.
Henry's law volatility	The Henry's law volatility of a compound is a measure of the transfer of the compound from being dissolved in the aqueous phase to being a vapor in the gaseous phase.	The transfer process from water to the gaseous phase in the atmosphere is dependent on the chemical and physical properties of the compound, the presence of other compounds, and the physical properties (velocity, turbulence, depth) of the water body and atmosphere above it. The factors that control volatilization are the solubility, molecular weight, vapor pressure, and the nature of the air–water interface through which it must pass. The Henry's constant is a valuable parameter that can be used to help evaluate the propensity of an organic compound to volatilize from water. The Henry's law constant is defined as the vapor pressure divided by the aqueous solubility. Therefore, the greater the Henry's law constant, the greater the tendency to volatilize from the aqueous phase; refer to Table 8.1.

TABLE 8.3
Values for Important Properties of DNAPL Contaminants Commonly Found at U.S. Superfund Sites

Chemical Compound	Density (g/cm3)	Water Solubility (mg/L)	Vapor Pressure (torr)	Henry's Law Constant (atm-m3/mol)	Dynamic Viscosity[a] (cp)	Kinematic Viscosity[a] (cSt)
Water (for comparison)	0.9991 (15°C) 0.9982 (20°C)	—	12.8 (15°C) 17.5 (20°C)	—	1.145 (15°C) 1.009 (20°C)	1.146 (15°C) 1.011 (20°C)
Halogenated semivolatiles						
Aroclor[b] 1242	1.3850	0.45	4.06×10^{-4}	3.4×10^{-4}		
Aroclor[b] 1254	1.5380	0.012	7.71×10^{-5}	2.8×10^{-4}		
Aroclor[b] 1260	1.4400	0.0027	4.05×10^{-5}	3.4×10^{-4}		
Chlordane	1.6	0.056	1×10^{-5}	2.2×10^{-4}	1.104	0.69
1,4-Dichlorobenzene	1.2475	80	0.6	1.58×10^{-3}	1.258	1.008
1,2-Dichlorobenzene	1.3060	100	0.96	1.88×10^{-3}	1.302	0.997
Dieldrin	1.7500	0.186	1.78×10^{-7}	9.7×10^{-6}		
Pentachlorophenol	1.9780	14	1.1×10^{-4}	2.8×10^{-6}		
2,3,4,6-Tetrachlorophenol	1.8390	1,000				
Halogenated volatiles						
Carbon tetrachloride	1.5947	790	91.3	0.020	0.965	0.605
Chlorobenzene	1.1060	490	8.8	3.46×10^{-3}	0.756	0.683
Chloroform (trichloromethane)	1.4850	7,920	160	3.75×10^{-3}	0.563	0.379
1,1-Dichloroethane	1.1750	5,500	182	5.45×10^{-4}	0.377	0.321
1,2-Dichloroethane	1.2530	8,690	63.7	1.1×10^{-3}	0.840	0.67
cis-1,2-Dichloroethylene	1.2480	3,500	200	7.5×10^{-3}	0.467	0.364
trans-1,2-Dichloroethylene	1.2570	6,300	265	5.32×10^{-3}	0.404	0.321
1,1-Dichloroethylene	1.2140	400	500	1.49×10^{-3}	0.330	0.27
1,2-Dichloropropane	1.1580	2,700	39.5	$3.6 \times 10{-3}$	0.840	0.72

(Continued)

TABLE 8.3 (*Continued*)
Values for Important Properties of DNAPL Contaminants Commonly Found at U.S. Superfund Sites

Chemical Compound	Density (g/cm3)	Water Solubility (mg/L)	Vapor Pressure (torr)	Henry's Law Constant (atm-m3/mol)	Dynamic Viscosity[a] (cp)	Kinematic Viscosity[a] (cSt)
Ethylene dibromide	2.1720	3,400	11	3.18×10^{-4}	1.676	0.79
Methylene chloride	1.3250	13,200	350	2.57×10^{-3}	0.430	0.324
1,1,2,2-Tetrachloroethane	1.6	2,900	4.9	5.0×10^{-4}	1.770	1.10
1,1,2-Trichloroethane	1.4436	4,500	0.188	1.17×10^{-3}	0.119	0.824
1,1,1-Trichloroethane	1.3250	950	100	4.08×10^{-3}	0.858	0.647
PCE	1.620	200	14	0.0227	0.890	0.54
TCE	1.460	1,100	58.7	8.92×10^{-3}	0.570	0.390
Trichloromethane (chloroform)	1.4850	7,920	160	3.75×10^{-3}	0.563	0.379
Nonhalogenated semivolatiles						
2-Methyl naphthalene	1.0058	25.4	0.0680	0.0506		
o-Cresol	1.0273	31,000	2.45×10^{-1}	4.7×10^{-5}		
p-Cresol	1.0347	24,000	1.08×10^{-1}	3.5×10^{-4}		
2,4-Dimethylphenol	1.0360	6,200	0.098	2.5×10^{-6}		
m-Cresol	1.0380	23,500	1.53×10^{-1}	3.8×10^{-5}	21.0	20
Phenol	1.0576	84,000	5.293×10^{-1}	7.8×10^{-7}		
Naphthalene	1.1620	31	2.336×10^{-1}	1.27×10^{-3}		
Benzo(a)anthracene	1.1740	0.014	1.16×10^{-9}	4.5×10^{-6}		
Fluorene	1.2030	1.9	6.67×10^{-4}	7.65×10^{-5}		
Acenaphthene	1.2250	3.88	0.0231	1.2×10^{-3}		3.87
Anthracene	1.2500	0.075	1.08×10^{-5}	3.38×10^{-5}		

Dibenzo(a,h)anthracene	1.2520	2.5×10^{-3}	1×10^{-10}	7.33×10^{-8}
Fluoranthene	1.252	0.27	7.2×10^{-5}	11×10^{-6}
Pyrene	1.2710	0.148	6.67×10^{-6}	1.2×10^{-5}
Chrysene	1.2740	6.0×10^{-3}	6.3×10^{-9}	1.05×10^{-6}
2,4-Dinitrophenol	1.6800	6.0×10^{-3}	1.49×10^{-5}	6.45×10^{-10}
Miscellaneous				
Coal tar (45°F)	1.028			18.98
Creosote	1.05			~1.08 (15°C)

Source: Adapted from USEPA, *Dense Nonaqueous Liquids*, S.G. Huling and J.W. Weaver, Ground Water Issue, EPA/540/4-91-002. Washington, DC, Office of Research and Development, Office of Solid Waste and Emergency Response, March 1991.

[a] Dynamic viscosity measures a liquid's resistance to flow. Kinematic viscosity is the ratio of dynamic viscosity to density; see Table 8.2.

[b] Aroclor is the trade name for PCBs manufactured by Monsanto. See Section 8.3.4.

The fraction of liquid hydrocarbon that is retained by sorption and capillary forces in the pores of soils is referred to as residual saturation and is relatively immobile.* Percent residual saturation (%RS) is defined by Equation 8.1:

$$\%RS = 100 \times \left(\frac{\text{volume of NAPL trapped in subsurface pore spaces}}{\text{total volume of pore spaces in the contaminated zone}} \right) \quad (8.1)$$

The amount of residual DNAPL retained in a typical soil such as silt, sand, or gravel is generally between 5% and 20% of the soil pore space.

In the vadose zone, only DNAPL that is in a vapor, dissolved, or liquid free product phase has significant mobility; DNAPL sorbed to soil surfaces or trapped in pores is immobile unless it partitions again into one of the three mobile phases. DNAPL in the vapor phase is generally denser than air and tends to sink. However, it spreads laterally wherever the subsurface is less permeable, often moving far beyond the region of residual saturation, where the vapors can contaminate soils and groundwater distant from the region of the spill.

Because the vapor pressure of many DNAPL compounds is relatively high, the lifespan of residual DNAPL in the unsaturated zone, where vaporization occurs, can be much less than the lifespan of residual DNAPL below the water table, where vaporization cannot occur. The vaporization process can deplete residual DNAPLs having high vapor pressures, such as the solvents TCE and PCE, within 5–10 years in relatively warm and dry climates. This will not entirely eliminate the presence of adsorbed phase and aqueous phase contamination in the unsaturated zone, but it can lead to an absence of DNAPL vapor. The absence of DNAPL vapor in the unsaturated zone, as measured by a photoionization detector (PID) used to detect volatile organic compounds (VOCs) in soils, does not necessarily imply that no DNAPL was ever released at that site in the past, or that past releases of DNAPL have failed to reach the water table.

Water percolating downward through the vadose zone will preferentially leach the more-soluble components of DNAPL from the free product and residual saturation that it contacts, eventually transporting dissolved DNAPL to the saturated zone and contaminating groundwater there. Partitioning of residual saturation into the dissolved phase is facilitated further by the rise and fall of the water table.

8.2.2 DNAPL AT THE WATER TABLE

At the water table interface, DNAPL behaves very differently from LNAPL. Being denser than water, it does not float above the water table but tends to continue downward through the capillary zone of the water table into the saturated zone, where partitioning into the dissolved phase is maximized. To continue moving downward

---

* A common operational definition of NAPL mobility is that mobile NAPL can drain under gravity and pressure gradient into a monitoring well, while immobile NAPL (residual saturation) cannot.

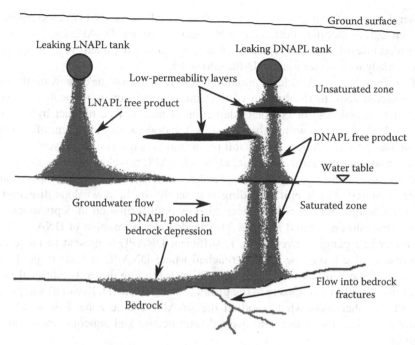

FIGURE 8.1 Comparison of DNAPL and LNAPL movement in the subsurface after a spill. When mobile NAPL encounters stratigraphic units of low permeability, such as a clay lens or bedrock, it spreads out laterally until it can enter a preferential pathway of greater permeability that allows it to continue downward. DNAPL entering fractured rock systems may follow a complex pattern of preferential pathways.

in the saturated zone, DNAPL must displace water held in the soil pore spaces by capillary forces. Consequently, at the water table interface, downward movement slows while DNAPL piles up and spreads laterally. If a sufficient weight of DNAPL accumulates, it presses downward through the capillary zone and continues down through the saturated zone (see Figure 8.1). Because soil surfaces in the saturated zone are already wetted by water, DNAPL movement below the water table does not leave a trail of soil-sorbed DNAPL, although some DNAPL can become trapped as residual saturation where water is held strongly by capillary forces and not readily displaced.

8.2.3 DNAPL in the Saturated Zone

In the saturated zone, DNAPL can exist only in three phases: the continuous liquid free product, dissolved, and residual saturation phases. The vapor phase is absent. In the saturated zone, residual saturation DNAPL is in continual contact with water and, therefore, continually partitions its more-soluble components into the dissolved phase. This causes the DNAPL properties to change progressively, generally toward greater density and higher viscosity. In most saturated zone soils, hydraulic gradients large enough to mobilize horizontal movement of residual DNAPL are unrealistic. Therefore, investigation and remediation activities involving intensive well pumping are not likely to draw residual DNAPL into wells.

If the initial release was large enough, DNAPL will continue downward through the saturated zone to the bottom of the aquifer. Only an impermeable obstruction, such as bedrock, or complete depletion of mobile free product by sorption and capillary retention within the soil stops the downward movement of DNAPL mobile free product. A decrease in soil permeability, such as a clay layer,* whether in the unsaturated or saturated zone, affects DNAPL travel by slowing the downward movement and causing lateral spreading until soils that are more permeable are encountered. The lateral spreading is generally in the downslope direction of the stratigraphic unit but is influenced also by pool formation in depressions and penetration into cracks and fissures. This leads to the formation of DNAPL pools and finger-like ganglia. Eventually, if sufficient DNAPL is present to move past the impermeable layers, bedrock is reached where DNAPL collects in pools and fractures. If the bedrock is slanted, DNAPL may migrate down the physical slope even if the direction is opposite to the groundwater movement. Residual and pooled DNAPL together form what is called the DNAPL source zone. It is within the source zone that dissolution into groundwater occurs and aqueous phase plumes originate.

DNAPL solubilities are generally low, so DNAPL in the saturated zone will continue to dissolve slowly into the groundwater without significant mass diminution over many years. At typically slow groundwater velocities, even a small DNAPL release can persist for decades or longer under natural conditions before a significant part of the DNAPL has dissolved or degraded. Once in the subsurface, it is difficult or impossible to recover all the trapped residual DNAPL, except by excavation. DNAPL that remains trapped in the soil/aquifer matrix will act as a continuing source of groundwater contamination for many years.

DNAPLs with low viscosity (e.g., methylene chloride, perchloroethylene, 1,1,1-TCA, TCE) can infiltrate into soil faster than water. The relative values of DNAPL viscosity and density, with respect to water, indicate how fast it will flow downgradient through the saturated zone compared to water. For example, several low-viscosity chlorinated DNAPLs (refer to Table 8.3) will flow 1.5–3.0 times faster than water, whereas higher-viscosity compounds, including LNAPLs like light heating oil, diesel fuel, jet fuel, and crude oil, will flow 2–10 times slower than water. Both coal tar and creosote typically have a density greater than one and a viscosity greater than water. Note that the viscosity of a NAPL will change with time as different components partition to other phases. As a fresh NAPL loses the lighter volatile components by evaporation, the NAPL becomes more viscous because the remaining heavier, more viscous components comprise a larger fraction of the NAPL mixture.

---

* Also called a *clay lens* or *low-permeability lens*.

RULES OF THUMB

1. DNAPL movement is affected by gravity far more than groundwater movement. It moves with the slope of the bedrock below the aquifer, independently of the direction of groundwater movement, forms pools in bedrock depressions, and enters bedrock fractures.
2. Chlorinated hydrocarbons are generally denser than water (DNAPL). They sink to the bottom of the water table.
3. In Table 8.3, many chlorinated hydrocarbons, including TCE, PCE, 1,1,1-TCA, methylene chloride, chloroform, and carbon tetrachloride, have viscosities less than water. They flow through the saturated zone 1.5–3.0 times faster than water and can penetrate small fractures and micropores, becoming inaccessible to in situ remediation.
4. The percent residual DNAPL retained as immobile liquid in a typical soil such as silt, sand, or gravel is generally between 5% and 20% of the soil pore space.
5. DNAPLs with high vapor pressure can totally evaporate from the DNAPL phase in the vadose zone in a relatively short time. Therefore, the absence of vapor phase DNAPL in the unsaturated zone at a site does not necessarily imply that no DNAPL was ever released at that site in the past or that past releases of DNAPL have failed to reach the water table. Sorbed phase, and dissolved phase contamination may still be present, even though photoionization detector (PID) measurements of the soil not detect any volatile organic compounds (VOCs).
6. In most saturated zone soils, intensive well pumping cannot create a large enough hydraulic gradient to move residual DNAPL into the well.

8.3   TESTING FOR THE PRESENCE OF DNAPL

It is very difficult to locate DNAPL free product with monitoring wells. First, DNAPL remains at the bottom of the monitoring well and may go unnoticed. Second, DNAPL free product may be present in locations seemingly unrelated to the spill location, such as perched on low-permeability layers in pools and cracks or upgradient of the spill at the bottom of the aquifer in pools and fractures in the bedrock. There are often no obvious guidelines as to where a well should be placed or how it should be screened to collect free product.

In addition, there are risks of enlarging the contaminated volume when trying to locate and determine the extent of a DNAPL source zone. Unlike residual DNAPL, pooled DNAPL is relatively easy to mobilize by increasing the hydraulic gradient. An exploratory well can inadvertently be drilled through DNAPL perched on a clay lens or pooled on bedrock, resulting in vertical mobilization into previously uncontaminated regions. It is often prudent to use a "from outside toward inside" approach to delineating DNAPL sites in order to minimize the chances of directly encountering pooled DNAPL during site characterization.

For these reasons, dissolved concentrations of DNAPL-related chemicals in groundwater wells distant from the source zone are often the only evidence that DNAPL free product is present at a site. The EPA has recommended an empirical approach for determining whether DNAPL free product is near a monitoring well where dissolved DNAPL-related compounds have been detected (USEPA 1992). In order to use this approach, one must

1. Measure the concentrations of DNAPL-related compounds dissolved in groundwater samples from the well.
2. Know the composition of the suspected DNAPL. See Example 3 for a useful procedure when the composition of the DNAPL is not known.
3. Calculate the effective solubility (S_{eff}) of the measured DNAPL components.
4. Apply the guidelines of Section 8.3.1.

The effective solubility is the theoretical solubility in water of a single component of a DNAPL mixture. It may be approximated by multiplying the component's mole fraction* in the mixture by the solubility of its pure phase.

$$S_{eff}(a) = X_a S_{pure}(a) \tag{8.2}$$

where

$S_{eff}(a)$ = effective solubility, in mg/L, of component a in a DNAPL mixture
X_a = mole fraction of compound a in the mixture
$S_{pure}(a)$ = pure-phase solubility of compound a, in mg/L

8.3.1 CONTAMINANT CONCENTRATIONS IN GROUNDWATER AND SOIL THAT INDICATE THE PROXIMITY OF DNAPL

If any of the following conditions exist in groundwater, there is a high probability that DNAPL free product is near the sampling location.

• Groundwater concentrations of DNAPL-related chemicals are >1% of either their pure-phase solubility (S_{pure}) for a single-component DNAPL or the effective solubilities (S_{eff}) for components of a DNAPL mixture. The factor of 1% of the solubility is intended to roughly account for the expected concentration decrease due to dilution, dispersion, and degradation of the DNAPL component while moving from the source zone to a monitoring well that is "near" the source. The higher the percentage factor, the closer the well is likely to be to the source zone.

---

* The mole fraction of compound a in a mixture of several compounds is written X_a, where:

$$X_a = \frac{\text{moles of } a}{\text{total moles of all compounds in mixture}}$$

For a mixture containing 1 mole of CCl_4 and 3 moles of $CHCl_3$, $X_{CCl_4} = 1/4 = 0.25$ and $X_{CHCl_3} = 3/4 = 0.75$. Note that the sum of all mole fractions must equal unity. The mole fraction of any pure substance equals unity.

- Soil concentrations of DNAPL-related chemicals are >10,000 mg/kg (1% of soil mass).
- Groundwater concentrations of DNAPL-related chemicals increase with depth or appear in anomalous upgradient/crossgradient locations with respect to groundwater flow.
- Groundwater concentrations of DNAPL-related chemicals calculated from water–soil partitioning relationships are greater than their pure-phase solubility or effective solubility.

8.3.2   CALCULATION METHOD FOR ASSESSING RESIDUAL DNAPL IN SOIL

1. Measure the DNAPL compounds in the soil.
2. Calculate $S_{eff}(a)$ from Equation 8.2 for each compound.
3. Find K_{oc}, the organic carbon–water partition coefficient, in Chapter 6, Table 6.5 or from published literature. Otherwise, estimate it from $\log K_{oc} = \log K_{ow} - 0.21$.
4. Determine f_{oc}, the fraction of organic carbon (OC), in the soil by laboratory analysis. Values for f_{oc} typically range from 0.03 to 0.00017 (mg OC)/(mg soil). Convert values reported in percent (mg OC/100 mg soil) to (mg OC)/(mg soil).
5. Determine or estimate the dry bulk density of the soil (d_b). Typical values range from 1.8 to 2.1 g/cm³ (kg/L).
6. Determine or estimate the water-filled porosity (p_w) of the soil.
7. Determine K_d, the soil–water partition coefficient, from $K_d = K_{oc} \times f_{oc}$, Chapter 6, Equation 6.16.
8. If the soil sample is collected from a source zone, DNAPL free product is present in the soil and the concentrations of DNAPL compounds dissolved in the pore water will be close to their calculated effective solubilities, S_{eff}. Therefore, calculate from Equation 8.3 the minimum DNAPL concentration in soil, $C_{soil}^{min}(a)$, that would result in a pore water concentration equal to S_{eff}.

$$C_{soil}^{min}(a) = \frac{S_{eff}(a) \times (K_d d_b + p_w)}{d_b} \tag{8.3}$$

9. If measured soil concentrations of compound $a > C_{soil}^{min}(a)$, DNAPL free product is likely to be present in the soil sample.
10. If measured soil concentrations of compound $a < C_{soil}^{min}(a)$, DNAPL free product is likely to not be present in the soil sample.

Example 1: Using Groundwater Concentrations to Estimate the Proximity of Residual Single-Component DNAPL

Analysis of a water sample from a monitoring well indicated 6.4 mg/L of PCE (also called PERC). PCE was a target contaminant because a dry cleaning establishment had once been on the site near the well. Is residual PCE DNAPL likely to be in the subsurface upgradient near the well? Use data from Table 8.3.

ANSWER

Since the observed DNAPL is a pure solvent (PCE) and not a mixture, its mole fraction, X, equals unity and $S_{eff} = S_{pure}$. From Table 8.3, the solubility of pure PCE is 200 mg/L. By the guideline in Section 8.3.1, if the measured concentration of a single-component DNAPL in a well is 1% or more of its pure-phase solubility, it is likely that a DNAPL source zone is near the well.

One percent of 200 mg/L is 2.0 mg/L. The measured concentration of PCE in the well is 6.4 mg/L. Because this is significantly larger than 2.0 mg/L, it is likely that a source zone of PCE DNAPL is quite close to the well.

Example 2: Using Groundwater Concentrations to Estimate the Proximity of Residual Multicomponent DNAPL Mixtures, Where the Initial Composition Is Known

A remediation project was being planned for a site that had contained a metal degreasing facility. The degreaser solution that was used consisted of 70 wt% trichloromethane, 15 wt% TCE, and 15 wt% PCE. A matrix of monitoring wells was drilled to try to locate subsurface source zones of DNAPL releases. A water sample from well SW-4 contained 88 mg/L trichloromethane (MW = 119.37), 1.6 mg/L PCE (MW = 165.82), and 4.2 mg/L TCE (MW = 131.37). Is this well likely to be close to an upgradient DNAPL source zone? Use data from Table 8.3.

ANSWER

Convert the weight percentages of each DNAPL component to mole fractions.

100 g of solvent contains 70 g trichloromethane, 15 g TCE, and 15 g PCE.

$$70 \text{ g trichloromethane} = \left(\frac{70 \text{ g}}{119.37 \text{ g/mol}}\right) = 0.586 \text{ mol}$$

$$15 \text{ g trichloroethylene} = \left(\frac{15 \text{ g}}{131.38 \text{ g/mol}}\right) = 0.114 \text{ mol}$$

$$15 \text{ g tetrachloroethylene} = \left(\frac{15 \text{g}}{165.82 \text{ g/mol}}\right) = 0.090 \text{ mol}$$

Total moles of DNAPL in 100 g solvent = 0.586 + 0.114 + 0.090 = 0.790 mol

$$\text{Mole fractions : X(trichloromethane)} = \left(\frac{0.586}{0.790}\right) = 0.742$$

$$\text{X(trichloroethylene)} = \left(\frac{0.114}{0.790}\right) = 0.144$$

$$\text{X(tetrachloroethylene)} = \left(\frac{0.090}{0.790}\right) = 0.114$$

Sum of mole fractions = 0.742 + 0.144 + 0.114 = 1.000

Calculate S_{eff} from Equation 8.2 and Table 8.3.

- S_{eff} (trichloromethane) = 0.742 × 7920 mg/L = 5877 mg/L
- S_{eff} (TCE) = 0.144 × 1100 mg/L = 158.4 mg/L
- S_{eff} (PCE) = 0.114 × 200 mg/L = 22.8 mg/L

By the guideline in Section 8.3.1, if the measured concentration in a well of a multicomponent DNAPL mixture is 1% or more of the effective solubilities of its components, it is likely that a DNAPL source zone is near the well. The measured concentrations

- C_{meas} (trichloromethane) = 88 mg/L
- C_{meas} (TCE) = 4.2 mg/L
- C_{meas} (PCE) = 1.6 mg/L

are all greater than 1% of their respective effective solubilities. Therefore, the sampled well is likely to be close to an upgradient DNAPL source zone.

Example 3: Using Groundwater Concentrations to Estimate the Proximity of Residual Multicomponent DNAPL Mixtures, Where the Initial Composition Is Not Known

When the composition of the source DNAPL mixture is not known, a variation of the method used in Example 2 can be applied. From Equation 8.2, the mole fraction of component a is

$$X_a = \frac{S_{eff}(a)}{S_{pure}(a)}$$

and the sum of mole fractions of all components of the mixture must equal unity:

$$\sum_i X_a = \sum_i \frac{S_{eff}(a)}{S_{pure}(a)} - 1$$

In the absence of any dilution, dispersion, or degradation effects, $S_{eff}(a)$ will be equal to the measured well concentration of component a, $C_{meas}(a)$. Using the EPA 1% guideline to account for loss effects, $C_{meas}(a) = 0.01\ S_{eff}$. This gives Equation 8.4, which uses only the measured concentrations of DNAPL components in a well and their pure compound solubilities to describe the conditions where a DNAPL source zone is likely to be near the monitoring well.

$$\sum_i \frac{(0.01)S_{eff}(a)}{S_{pure}(a)} = \sum_i \frac{C_{meas}(a)}{S_{pure}(a)} \geq 0.01 \tag{8.4}$$

Suppose, in Example 2, we did not know the initial composition of the DNAPL solvent. Assume that the only data available were the measured well concentrations:

- C_{meas} (trichloromethane) = 288 mg/L
- C_{meas} (TCE) = 4.2 mg/L
- C_{meas} (PCE) = 1.6 mg/L

The use of Equation 8.4 is illustrated in Table 8.4.

The sum of column 4 is greater than 0.01, and, therefore, it is likely that a source zone of a DNAPL mixture is upgradient near the monitoring well.

TABLE 8.4
Estimating the Proximity of Residual Multicomponent DNAPL Mixtures of Unknown Composition, Using Equation 8.4

Compound	Measured Concentration in Monitoring Well, C_{meas} (mg/L)	Solubility of Pure Compound, S_{pure} (mg/L)	$\dfrac{C_{meas}}{S_{pure}}$
CF	88	7920	0.0111
TCE	4.2	1100	0.00382
PCE	1.6	200	0.008
$\displaystyle\sum_i \dfrac{C_{meas}}{S_{pure}}$			0.02292

Example 4: Using Soil Concentrations Below the Water Table to Estimate the Proximity of Residual Single-Component DNAPL

TCE was measured to be 452 mg/kg in a soil sample from the saturated zone. No other DNAPL compounds were detected. Measured soil data are

Porosity, $p_w = 0.27$
Dry bulk density, $d_b = 1.9$ kg/L
Fraction of organic carbon (OC), $f_{oc} = 0.003$

Is a TCE DNAPL free product phase likely to be present in the soil sample?

ANSWER

Since only TCE was present, $S_{eff} = S$, the pure compound water solubility. Use Table 5.5 to obtain the following data for TCE:

$S = 1100$ mg/L
$K_{oc} = 166$ L/kg

Calculate C_{soil}^{min} (TCE) from Equation 8.3.

$$C_{soil}^{min}(a) = \frac{S_{eff}(a) \times (K_d d_b + p_w)}{d_b}$$

$$= \frac{1100 \text{ mg/L} \times ((166 \text{L/kg})(0.003)(1.9 \text{ kg/L}) + 0.27)}{1.9 \text{ kg/L}}$$

$$C_{soil}^{min}(a) = 704 \text{ mg/kg}$$

Since the measured TCE soil concentration of 452 mg/kg < 704 mg/kg, it most likely is residual TCE rather than free product DNAPL.

Example 5: Using Soil Concentrations Below the Water Table to Estimate the Proximity of Residual Multicomponent DNAPL, Where the Initial Composition Is Not Known

Even though the initial composition of a DNAPL mixture is not known, the sum of its mole fractions must equal unity. Under conditions of DNAPL saturation, the sum of measured soil concentrations of the DNAPL components divided by their saturated values, C_{soil}^{min}, is equivalent to the sum of their mole fractions. Therefore,

$$\sum_a \frac{C_{meas}(a)}{C_{soil}^{min}(a)} \geq 1$$

Table 8.5 demonstrates an example calculation for determining if DNAPL free product was present in a soil sample found to contain four different DNAPL compounds. The soil had the following properties:

Porosity, $p_w = 32\%$
Dry bulk density, $d_b = 1.7$ g/cm³
fraction of organic carbon, $f_{oc} = 0.003$ (0.3%)
Values for K_{oc} are from Chapter 6, Table 6.5 and K_d is calculated from Chapter 6, Equation 6.16: $K_d = K_{oc}f_{oc}$

C_{soil}^{min} is calculated from Equation 8.3: $C_{soil}^{min}(a) = \dfrac{S_{eff}(a) \times (K_d d_b + p_w)}{d_b}$

Since the sum of the estimated mole fractions, $\sum_a \dfrac{C_{meas}(a)}{C_{soil}^{min}(a)}$, is greater than unity, DNAPL was present in the soil sample.

TABLE 8.5
Example Soil Concentration Calculation for Multicomponent DNAPL

Compound	C_{meas} (mg/kg)	Pure Compound Solubility, S (mg/L)	K_{oc} (L/kg)	K_d	C_{soil}^{min} (mg/kg)	$\dfrac{C_{meas}}{C_{soil}^{min}}$
PCE	109	200	155	0.465	131	0.834
TCE	368	1100	166	0.498	755	0.487
Chlorobenzene	84	490	219	0.657	399	0.211
1,1,1-Trichloroethane	188	950	110	0.330	689	0.273
					$\sum \dfrac{C_{meas}}{C_{soil}^{min}} =$	1.805

8.4 POLYCHLORINATED BIPHENYLS

8.4.1 BACKGROUND

PCBs are a family of stable synthetic organic compounds containing 209 individual compounds known as congeners (see Figure 8.2). PCBs were manufactured and sold under various trade names (Aroclor, Pyranol, Phenoclor, Pyralene, Clophen, and Kaneclor) as complex mixtures differing in their average chlorination level. They have no smell or taste and may take the form of oily liquids or solids.

In the United States, production of PCBs was banned in 1977 due to their toxicity and persistence in the environment. There are no known natural sources of PCBs in the environment.

Since 1929, about 1.4 billion pounds of PCBs have been commercially produced, the majority in the United States. It is estimated that several hundred million pounds have been released to the environment. The world's primary producer was Monsanto, who produced PCBs under the trade name Aroclor from 1930 to 1977. General Electric had a rival product called Pyranol. As shown in Figure 8.2, individual PCB compounds are formed by substituting between 1 and 10 chlorine atoms onto the biphenyl aromatic structure. These substitutions can produce 209 different congeners (homologues and isomers).

PCBs have many desirable properties for commercial applications: very high chemical, thermal, and biological stability; low water solubility; low vapor pressure; high dielectric constant; and high flame resistance. It is not surprising that PCBs found wide application as coolant and insulation fluids in transformers and capacitors and as lubricants, flame retardants, plasticizers, solvent extenders, organic diluents, additives to epoxy paints, heat transfer fluids, in hydraulic fluids, in pesticides, and in printing inks. PCBs are also by-products of many industrial processes, such as the manufacturing of chlorinated solvents and chlorinated benzenes.

Industrial-grade PCBs are mixtures of PCB compounds blended to give particular overall properties, such as viscosity, electrical resistance, boiling point, and so on. For example, Aroclor-1242 (also called PCB-1242) (see Table 8.1), is actually a mixture of more than 60 different PCB congeners with varying degrees of chlorination.

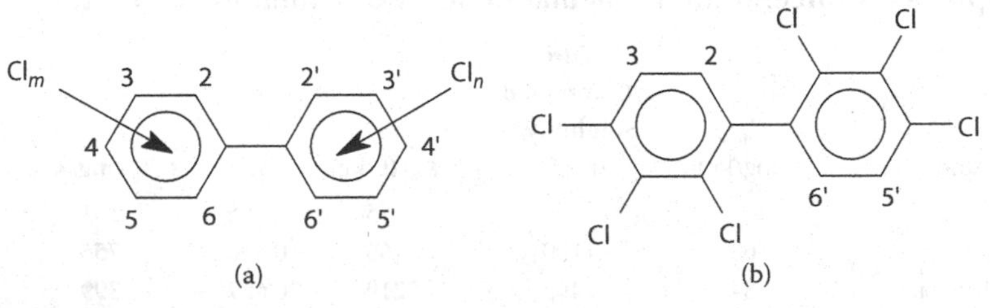

FIGURE 8.2 (a) General structure of PCBs. (b) One particular 6-chlorine PCB congener out of the 209 different types possible. The general formula for a PCB is $C_6Cl_mH_{5-m}C_6Cl_nH_{5-n}$, where m and n each can be any integer between 1 and 5. Individual PCB compounds are formed by substituting between 1 and 10 chlorine atoms onto the biphenyl aromatic structure. This substitution can produce 209 different congeners (homologues and isomers).

It naturally has a complicated gas chromatogram. A four-digit numbering system was assigned to the mixtures. The first two numbers indicate the number of carbon atoms and the third and fourth numbers give the weight percent of chlorine; Arochlor-1242 contains 12 carbon atoms and 42% chlorine by weight.

8.4.2   Environmental Behavior

PCBs are very stable species and do not degrade readily in the environment. Most of the released PCBs are believed to remain in mobile environmental reservoirs (Alder et al. 1993). They even survive ordinary incineration and can escape as vapors up the smokestack.

The wide use of PCBs has resulted in their common presence in soil, water, and air. PCB dispersion from source regions to global distribution occurs mainly through atmospheric transport and subsequent deposition. Because of their low vapor pressure and water solubility, PCBs typically have very high partition coefficients to abiotic and biotic particles. In surface water aquatic systems, sediments are potentially an important PCB reservoir. In the absence of PCB contamination from a nearby source of contamination, such as a hazardous waste site, PCBs are unlikely to be present in most groundwaters because of their low solubility and strong attraction to soils and minerals; they are not very mobile through subsurface soils.

Environmental contamination was first reported in 1966, when high levels of PCBs were found in fish. PCBs in wastes released into Lake Michigan were found in the fatty tissue of fish and in the breast milk of nursing mothers who ate the fish. Children nursed by these mothers showed higher rates of development and learning disorders than those of nursing women in the same region who had not eaten the fish. Similar developmental effects were reported from Japan and Taiwan, where children of women who had eaten PCB-contaminated rice products were underdeveloped physically and mentally. Adults working with PCBs were susceptible to a skin condition called chloracne, which produces pustules and cysts.

Eventually it was discovered that PCBs did not easily biodegrade and their use was restricted. In 1976, PCBs became regulated under the Toxic Substances Control Act and safe disposal became a major concern. Between 1974 and 1979, PCBs were used only in the production of capacitors and transformers. Monsanto stopped producing Aroclors in October 1977. In 1986, an international agreement was signed to ban most uses of PCBs and phase out the rest.

Although the production of PCBs was banned in 1977, they are still prevalent in our water today. PCBs released into the ground years ago are still cycling through the water systems and into drinking water. They leak from old electrical devices, including power transformers, capacitors, television sets, and fluorescent lights, and are released from hazardous waste sites and historic and illegal refuse dumps. They persist in fatty foods, such as certain fish, meat, and dairy products. Currently, new contamination arises mainly from old PCB sources cycling from soil to the atmosphere and back to soil again, stormwater runoff from landfills, and releases associated with obsolete and junked industrial equipment, such as PCB-insulated electrical transformers. One of the largest environmental cleanup projects in the United States, expected to finish in 2015, involves dredging PCBs from the Hudson River, where over one million pounds of PCBs were deposited between 1947 and 1977 in the bottom sediments of the upper Hudson River by two General Electric Co. capacitor-manufacturing plants.

8.4.3 HEALTH EFFECTS

The toxicity of PCBs is a complicated issue since each congener differs in its toxicity. All PCBs are listed by the EPA as known carcinogens. When they are incinerated, they can produce dioxins (see Appendix A), which are rated by the EPA among the most toxic substances.

- Occasional external exposure to PCBs at any concentration can cause skin irritations, such as acne and rashes.
- Long-term exposure, as in a work place, may cause irritation of the nose and lungs.
- Short-term exposures above the EPA maximum contaminant level (MCL) can potentially cause acne-like eruptions and pigmentation of the skin, hearing and vision problems, and muscle spasms.
- Long-term ingestion or inhalation above EPA air and water MCLs can have effects similar to acute poisonings: irritation of nose, throat, and gastrointestinal tracts, changes in liver function, problems with the thymus gland, and immune deficiencies or reproductive or nervous system difficulties.
- PCBS are endocrine disruptor chemicals, possibly harmful to the human endocrine system.
- Lifetime exposure above EPA MCLs leads to increasing cancer risk.

8.4.4 REGULATION

In public water supplies, the EPA has listed PCBs as a primary drinking water contaminant, with an enforceable MCL of 0.0005 mg/L. The maximum contaminant level goal (MCLG) for PCBs in drinking water set by the EPA is zero.

8.4.5 WATER TREATMENT

PCBs are removed from water most effectively with sorption to activated or catalytic carbon or alumina or filtration through membranes at the separation level of reverse osmosis or nanofiltration (see Chapter 3, Section 3.11.6.7). Removal by conventional coagulation, flocculation, and sedimentation, followed by filtration can also be used because PCBs have low solubility and bind readily to particulates. Also, some types of advanced oxidation are known to decompose the PCB molecule.

8.4.6 ANALYSIS OF PCBs

The gas chromatograph/mass spectrometer detector (GC/MS) patterns of the different PCB mixtures show considerable overlap. Common petroleum products, such as motor oil, also generate peaks in the PCB region. For these reasons, unambiguous identification of particular PCBs requires meticulous laboratory technique, especially if other organic compounds are present that have peaks in the PCB GC/MS regions.

8.4.7 CASE STUDY: MISTAKEN IDENTIFICATION OF PCB COMPOUNDS

A metal recycling company shredded automobile bodies, large appliances, industrial components such as power transformers and manufacturing equipment, etc. The nonmetallic residue from the shredding operation is called fluff and consists of shredded solid plastics, foamed plastic, rubber, glass, wood, and so on. The fluff was oily, having absorbed much of the residual oil remaining in the original metal components.

Fluff was disposed off by transport to a landfill. Acceptance by the landfill operators was conditional on a chemical analysis that showed the fluff did not contain excessive levels of toxic materials. High toxicity would require the fluff to be classified as hazardous waste, with more stringent disposal conditions. PCBs were a toxic substance of concern. If PCB levels exceeded 50 mg/kg, the fluff would be classified as a hazardous waste, requiring special and expensive disposal methods.

For several years, the recycling company had never had PCB analyses from their fluff that were higher than about 15 mg/kg. Then, although they had no reason to believe their mix of shredded materials had changed significantly, the laboratory analyses were suddenly showing greater than 50 mg/kg of PCBs. Were these results accurate or not? Because PCB mass spectra overlapped the motor oil GC/MS spectral range, it was possible that oil compounds were being mistaken for PCBs.

Arrangements were made with a knowledgeable laboratory director to be especially careful in sample cleanup and preparation. PCBs are very stable and can withstand strong acid and base extractions that will decompose most oils. Figure 8.3a and b compares the GC/MS spectra from an inadequate and a satisfactory cleanup procedure on similar fluff samples containing PCBs. By modifying the sample cleanup to more completely decompose any petroleum oils present, it was

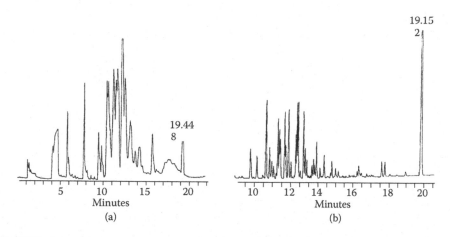

FIGURE 8.3 GC/MS spectra of oily waste samples containing PCBs. (a) Incomplete removal of oils results in an overlap of oil and PCB spectra, causing poor resolution of the PCB components and an overestimation of PCB concentrations in the sample. (b) Better cleanup preparation of the sample decomposes most of the oil contaminants. This spectrum shows an expanded portion of (a) in the 10–20 minute region. Individual PCB compounds show much better separation, and the measured PCB concentration is much lower than in (a).

shown that the PCB concentrations were well below the hazardous waste threshold and the fluff need not be treated as a hazardous waste.

REFERENCES

Alder, A. C., M. M. Haggblom, S. R. Oppenheimer, and L. Y. Young. 1993. "Reductive Dechlorination of Polychlorinated Biphenyls in Anaerobic Sediments." *Environmental Science Technology* 27: 530–8.

USEPA. March 1991. *Dense Nonaqueous Liquids*, Scott G. Huling and James W. Weaver, Ground Water Issue, EPA/540/4-91-002. Washington, DC, Office of Research and Development, Office of Solid Waste and Emergency Response.

USEPA. January 1992. *Estimating Potential for Occurrence of DNAPL at Superfund Sites*, Publication 9355.4-07FS, NTIS PB92-963338. Washington, DC, Office of Emergency and Remedial Response.

(Photo by Eugene Weiner)

Biodegradation is often the most economical and practical approach to remediation of NAPLs and DNAPLs.

9 Biodegradation and Bioremediation of LNAPLs and DNAPLs

9.1 BIODEGRADATION AND BIOREMEDIATION

Biodegradation is the chemical breakdown of organic contaminants into smaller compounds through metabolic or enzymatic processes of living organisms in the environment, primarily bacteria, yeast, and fungi. It differs from chemical and physical degradation processes (e.g., chlorine oxidation, reduction by metallic iron particles, hydrolysis, photolysis, and catalysis on reactive surfaces) in being caused by the action of living organisms. Some chemical structures are more susceptible to microbial breakdown than are others; vegetable oils, for example, will biodegrade more readily than petroleum oils, which, in turn, biodegrade more readily than polycyclic aromatic hydrocarbons (PAHs). Bioremediation* of soil and groundwater at

---

* Bioremediation is one of the several approaches for the remediation of sites contaminated with nonaqueous phase liquids (NAPLS). It is, however, the only remediation method treated with some depth in this book. Other methods, such as soil vapor extraction, air sparging, chemical oxidation, and low-temperature thermal desorption, are more appropriate for specialized reports. However, the fact that bioremediation processes are occurring does not preclude the additional use of more "active" remediation methods. A good starting reference for other remediation options is the EPA's manual (USEPA 2004).

sites contaminated with total petroleum hydrocarbons (TPH) and benzene, toluene, ethylbenzene, and xylene (BTEX) is a well-established technology compared to sites contaminated with PAHs, chlorinated VOCs, pesticides and herbicides, and explosives, which are more difficult to biodegrade.

When biodegradation converts hazardous contaminants into less-hazardous or benign substances (sometimes the products are more hazardous), it is called bioremediation. Microbial metabolism is a series of biological reactions, predominantly oxidation–reduction (also called "redox") reactions, which convert organic compounds into energy and carbon to sustain microbial growth. In a typical metabolic redox reaction, an organic carbon compound serves as an electron donor that microbes use as a food source. The transfer of an electron from the donor to an electron acceptor proceeds through multiple reaction steps that generate energy and materials, carbon, and other elements for microbial cell growth.

Organic pollutants are often toxic because of their chemical structure that allows them to interfere with the normal functions of living organisms. Changing their structure in any way, which always occurs when an electron is lost or gained, will change their properties and may make them less toxic or even, in a few cases, more toxic. Eventually, usually with many reaction steps, biodegradation breaks organic pollutants into smaller and smaller molecules, finally ending with carbon dioxide, water, and mineral salts if the process is not interrupted. Although these final products represent the destruction of the original pollutant, some of the intermediate steps may temporarily produce compounds that are also pollutants, sometimes more toxic than the original. An example of a more toxic biodegradation product is vinyl chloride, formed as an intermediate in the anaerobic biodegradation chain of the common solvent trichloroethene (TCE) (see Section 9.6).

For organic contaminants of low solubility and low volatility, for which most of the remediation processes described in Chapters 2 and 3 (e.g., chemical oxidation and reduction, volatilization, sorption, coagulation, precipitation, filtration, and photolysis) are inefficient or impossible, biodegradation is the ultimate form of contaminant removal for soil and groundwater cleanup. Because some fraction of contaminants always becomes nearly irreversibly sorbed to soil and cannot readily be removed by desorption treatments (see Chapter 6, Section 6.7.3), pollutants can remain in place for many years, serving as a continual source of groundwater contamination. In such cases, there are only three practical approaches to achieve a complete remediation:

1. Excavating the contaminated soil and sending it to a landfill or treating it on the surface (e.g., by incineration).
2. Isolating the contaminated soil by capping its surface with a membrane or other impervious cover and diverting surface and groundwater away from the area.
3. Allowing or assisting biodegradation to transform the contaminants into nonhazardous substances.

Biodegradation is often the most economical and practical approach to remediation. When biodegradation becomes bioremediation, that is, when it is used as the main site cleanup process, it may require some assistance. If the rate of natural bioremediation (intrinsic bioremediation) is too slow, adding nutrients, oxygen, and appropriate microbes (engineered bioremediation) can often accelerate it. Much progress has been made in recent years in understanding the different processes of biodegradation and adapting them to successfully degrading types of organic compounds once thought to be resistant to biodegradation, such as chlorinated hydrocarbons (National Research Council 2000). The EPA publishes many reports related to this new technology. The EPA website at http://www.epa.gov/gateway/science/land.html provides the latest references.

9.2   BASIC REQUIREMENTS FOR BIODEGRADATION

Biodegradation is a redox process. This means that energy for biodegradation arises from electron transfer reactions. There are six basic requirements that must be present for biodegradation to occur:*

1. Suitable degrading organisms, generally bacteria or fungi.
2. Electron donors that are the energy source (food) for the organisms. The electron donors are generally organic carbon compounds, which are mostly converted to CO_2, water, and mineral salts in the redox reactions that comprise the metabolism of the organisms.
3. Electron acceptors, generally O_2, NO_3^-, SO_4^{2-}, Fe^{3+}, and CO_2.
4. Carbon for cell growth, which comes from organic carbon. About 50% of bacterial dry mass is carbon.
5. Nutrients, including nitrogen, phosphorus, calcium, magnesium, iron, and trace elements. Of these, nitrogen and phosphorus are needed in the largest quantities. Bacterial dry mass is about 12% nitrogen and about 2–3% phosphorus.
6. Acceptable environmental conditions: pH, salinity, hydrostatic pressure, solar radiation, toxic substances, oxygen, and so on must all be within acceptable limits for the particular bioprocesses. Microbes capable of degrading petroleum hydrocarbons generally prefer a pH between 6 and 8 and temperatures between 5°C and 25°C. Temperatures lesser than 5°C tend to inhibit biodegradation in general.

Microbial reactions can be grouped into two classes: aerobic and anaerobic. Aerobic microbial reactions, such as the breakdown of fuel oils to carbon dioxide and water, require the presence of oxygen as an electron acceptor. Anaerobic microbial reactions, such as the reductive dechlorination of chlorinated solvents, utilize electron acceptors other than oxygen. When sufficient oxygen is present, it will be utilized in preference to alternative electron acceptors and only aerobic reactions will occur. When sufficient oxygen is absent, only anaerobic reactions

---

* These requirements are similar to those for metabolic processes in animal life forms. People, for example, degrade food (electron donors) to obtain energy, breathe air to obtain oxygen (the main electron acceptor), require carbon and other nutrients for cell growth, and produce waste (degraded food), cell structures, and energy.

will occur. Aerobic and anaerobic reactions generally require different kinds of bacteria, with aerobic reactions usually being considerably faster than anaerobic reactions.

In the list of six basic requirements on the previous page, items 1–5 must all be present in sufficient quantities for biodegradation to proceed. However, different pollutants can have very different quantity requirements. For example, the redox processes involved in the biodegradation of fuel hydrocarbons are very different from the biodegradation processes of chlorinated aliphatic hydrocarbons. In the biodegradation of fuel hydrocarbons, especially BTEX, aerobic reactions are far more efficient than anaerobic reactions. Since carbon electron donors can include the contaminant fuel molecules themselves, electron donors will generally be in excess, and these aerobic redox reactions are generally limited by electron acceptor availability. Under aerobic conditions where oxygen is the main electron acceptor, an adequate supply of oxygen generally means that biodegradation will proceed until all the contaminants accessible to the microbes are degraded. As aerobic biodegradation is the only natural process that actually reduces the mass of petroleum hydrocarbon contamination, it is the most important (and preferred) remediation mechanism for petroleum hydrocarbons.

On the contrary, the most highly chlorinated aliphatic hydrocarbons, such as the solvents tetrachloroethene (PCE) and TCE, typically are biodegraded by reductive dechlorination, an anaerobic process where the electron acceptors are the chlorinated solvent molecules themselves. Thus, as long as chlorinated contaminant remains, there are electron acceptors available and the abundance of electron donors (e.g., fuel hydrocarbons, landfill leachate, or natural organic carbon) will always be the limiting factor. If the subsurface environment is depleted of electron donors before all the chlorinated aliphatic hydrocarbons are biodegraded, biological reductive dechlorination will cease.

9.3 BIODEGRADATION PROCESSES

The most important metabolic processes in biodegradation are redox reactions, classified according to the species serving as the electron acceptor:

- Aerobic respiration
  - Microbes use oxygen as electron acceptors to transform organic carbon into carbon dioxide.
  - Electrons are transferred from the contaminant to oxygen. The contaminant is sequentially oxidized and the oxygen reduced, ultimately forming water and carbon dioxide.
  - The key requirement is adequate oxygen.
- Anaerobic respiration
  - Microbes use an "oxygen substitute" to serve as electron acceptors (usually nitrate, sulfate, Mn^{4+}, Fe^{3+}, or CO_2). Organic carbon is chemically transformed, often to carbon dioxide but sometimes to methane.

TABLE 9.1

Normal Sequence of Biodegradation Reactions

Oxygen Reduction	$\{CH_2O\}+O_2 \rightarrow CO_2+H_2O$	(Aerobic)	(9.1)
NO_3^- reduction to N_2 (denitrification)	$5\{CH_2O\}+4NO_3^-+4H^+ \rightarrow 5CO_2+2H_2+7H_2O$	(Anaerobic)	(9.2)
Mn^{4+} reduction	$\{CH_2O\}+2MnO_2+4H^+ \rightarrow CO_2+2Mn^{2+}+3H_2O$	(Anaerobic)	(9.3)
Fe^{3+} reduction	$\{CH_2O\}+4Fe(OH)_3+8H^+ \rightarrow CO_2+4Fe^{2+}+11H_2O$	(Anaerobic)	(9.4)
SO_4^{2-} reduction	$2\{CH_2O\}+SO_4^{2-}+H^+ \rightarrow 2CO_2+HS^-+2H_2O$	(Anaerobic)	(9.5)
Methane production	$2\{CH_2O\} \rightarrow CH_4+CO_2$	(Anaerobic)	(9.6)
	For organic matter with element ratios different from the model compound $\{CH_2O\}$, especially those containing chlorine, reduction leads to the generation of H_2, which then undergoes a redox reaction with dissolved CO_2 as follows: $CO_2+4H_2 \rightarrow CH_4+2H_2O$	(Anaerobic)	(9.7)

Note: In the stoichiometric equations, organic pollutants being degrading are approximated by the model compound CH_2O, which has element ratios similar to typical hydrocarbon contaminants.

- Electrons are transferred from the contaminant (oxidizing it) to an electron acceptor (reducing it). The products formed by the anaerobic electron acceptors (CO_2, NO_3^-, SO_4^{2-}, and Fe^{3+}) are shown in Table 9.1 (Equations 9.1 through 9.7).
- The key requirement is an adequate supply of electron acceptors and an absence of oxygen.
- Cometabolism
  - Enzymes produced by microbes during the degradation of organic matter fortuitously react chemically to transform a contaminant that resists biodegradation. For example, during biodegradation of methane, some bacteria produce an enzyme that breaks down chlorinated solvents such as TCE.
  - The key requirement is the presence of a substance that, when metabolized by microbes, produces the right enzymes to transform the contaminants.

Each of these three biodegradation processes requires an electron donor that is oxidized and an electron acceptor that is reduced. The overall electron transfer process provides metabolic energy for the microbes. Favorable conditions for biodegrading various organic compounds are listed in Tables 9.2. and 9.3, which identify some of the site-specific factors that favor successful bioremediation.

Upon accepting electrons from an energy source, electron acceptors are converted to the products indicated in Equations 9.1 through 9.7. Only Equation 9.1 is

TABLE 9.2
Biodegradation Processes for Some Organic Compounds

Hydrocarbons (HCs)

Gasoline, diesel, fuel oil	Readily biodegradable under aerobic conditions; more slowly degradable under anaerobic conditions
PAHs	Aerobically biodegradable under a narrow range of conditions
Creosote	Readily biodegradable under aerobic conditions
Alcohols, ketones, esters	Readily biodegradable under aerobic conditions
Ethers	Biodegradable under a narrow range of conditions using aerobic or nitrate-reducing microbes

Chlorinated aliphatic HCs

Highly chlorinated (e.g., PCE)	Biodegraded and cometabolized by anaerobic microbes, primarily by reductive dechlorination; cometabolized by aerobic microbes in special cases
Less chlorinated (e.g., DEC)	Aerobically biodegradable under a narrow range of conditions; cometabolized by anaerobic microbes

Chlorinated aromatic HCs

Highly chlorinated (e.g., pentachlorophenol)	Aerobically biodegradable under a narrow range of conditions; cometabolized by anaerobic microbes
Less chlorinated (e.g., chlorobenzene)	Readily biodegradable under aerobic conditions

Polychlorinated biphenyls (PCBs)

Highly chlorinated	Cometabolized by anaerobic microbes
Less chlorinated	Aerobically biodegradable under a narrow range of conditions

Nitroaromatics

Trinitrotoluene, nitrobenzene, etc.	Aerobically biodegradable; converted to innocuous VOCs under anaerobic conditions

Metals

Cr, Cu, Ni, Hg, Cd, Zn, etc.	Solubility and reactivity can be changed by a variety of microbial processes

aerobic; all the others are anaerobic. Equations 9.1 through 9.7 are listed in the order of energy release; Equation 9.1 releases the most energy and Equation 9.7 the least. Thus, when oxygen is available, Equation 9.1 will occur before any of the others. If all the oxygen is consumed and nitrate is available, Equation 9.2 becomes the preferred process, and so on.

TABLE 9.3
Hydrogeologic Factors Favoring In Situ Bioremediation

Type of Bioremediation	Important Site Characteristic	Favorable Indicators
Engineered	Transmissivity of subsurface to fluids	Hydraulic conductivity >10^{-4} cm/s (if system circulates water)
		Intrinsic permeability >10^{-9} cm^2 (if system circulates air)
	Relative uniformity of subsurface medium	Common in river delta deposits, floodplains of large rivers, and glacial outwash aquifers
	Low residual concentrations of NAPL contaminants on subsurface solids	NAPL concentration <10,000 mg/kg
Intrinsic/passive	Consistent groundwater flow (velocity and direction)	Seasonal variation in depth to water table <1 m
		Seasonal variation in regional flow path <25°
	Presence of pH buffers	Carbonate minerals (limestone, dolomite, and shell material)
	High concentration of electron acceptors	Oxygen, nitrate, sulfate, Fe^{3+}
	Presence of elemental nutrients	Nitrogen and phosphorus

9.3.1 CASE STUDY: PASSIVE (INTRINSIC) BIOREMEDIATION OF FUEL LNAPLs, CALIFORNIA SURVEY

Sometimes the best approach to treat soil contaminated with fuel hydrocarbons is to do nothing. If passive biodegradation rates are high enough, a pollutant plume may shrink or become immobile within an acceptably short time frame.

Passive, or intrinsic, bioremediation is becoming increasingly acceptable as a treatment alternative, especially for fuel hydrocarbons (light nonaqueous phase liquids, or LNAPLs), if there is no immediate threat to water uses and if natural site conditions are favorable (see Table 6.11). With favorable site conditions, passive remediation may be expected to eventually stabilize a contaminant plume's length and mass, even if an active source such as free product is present in the subsurface, continually dissolving contaminants into the plume. If all mobile active sources are removed to the point of residual saturation, biodegradation will often reduce the plume mass to the point of completing the cleanup.

A survey of fuel hydrocarbon contamination in California by the Lawrence Livermore National Laboratory, commissioned in 1994 by the UST Program of the California State Water Resources Control Board (Rice et al. 1995), shows the value of a passive approach to remediation. A detailed analysis of 271 cases involving leaking underground fuel tanks (LUFTs) showed that

- In general, fuel hydrocarbon plume lengths change slowly and tend to stabilize at short distances from the source.
- Plume boundaries, defined by a contaminant concentration of 10 ppb, extended no farther than 76 m in 90% of the cases.
- Plume mass decreased more rapidly than plume length.
- Residual fuel in the soil degraded more slowly than dissolved fuel and continued to dissolve contaminants into the groundwater. Thus, the length of the plume is defined mainly by the extent of soils containing residual adsorbed fuel.
- In 50% of sites with no actively engineered remediation, groundwater benzene concentrations decreased about 70% as fast as where pump-and-treat plus excavation treatment was applied.
- After removing contaminant sources, degradation generally removed 50–60% of the remaining pollutant mass per year.

RULE OF THUMB

Once a fuel hydrocarbon (e.g., gasoline, diesel, and heating oil) source is removed, passive remediation requires 1–3 years to reduce the dissolved plume mass by a factor of 10.

9.4 NATURAL AEROBIC BIODEGRADATION OF NAPL HYDROCARBONS

As discussed in Chapters 7 and 8, liquid hydrocarbons of low water solubility (oils, many solvents, gasoline, etc.) are called NAPLs. NAPLs are further divided into hydrocarbons that are less dense than water, LNAPLs, and hydrocarbons that are denser than water, DNAPLs. If NAPLs are mixed with water, they separate from water into a separate immiscible liquid phase. LNAPL floats on the water surface and DNAPL sinks to the bottom of the water. Generally, most of the NAPL (>90%) is in the immiscible phase and a small fraction dissolves into the water (<10%).

When natural aerobic biodegradation of NAPLs occurs in the saturated zone, indigenous aerobic bacteria react with dissolved oxygen (DO) to consume some of the immiscible-phase NAPL directly. These bacteria also release a biosurfactant that helps to increase the rate of NAPL dissolution into groundwater, enhancing their food supply.

RULES OF THUMB

1. Within the unsaturated zone, pore space contains mostly air, and in situ aerobic biodegradation of NAPL is often limited by the availability of nitrogen nutrients. Addition of nitrogen, as nitrate or ammonia, usually enhances biodegradation.

2. Within the saturated zone, DO is usually the limiting factor for in situ aerobic biodegradation of NAPL.

3. For gasoline, the contaminants of regulatory interest are usually the most-toxic and most-soluble components, known as BTEX (benzene, toluene, ethylbenzene, and xylene isomers).

4. It takes about 1 mg of O_2 to biodegrade 0.32 mg of BTEX.

5. The most easily biodegraded hydrocarbons are low-molecular-weight unbranched alkanes (smaller than C30–C40) and aromatics (smaller than C10). About 95% of their mass is converted to CO_2 and water in a few months.

6. The remainder, unbranched alkanes larger than C40 and branched alkanes, alkenes, and aromatics larger than C10, can resist degradation for many years.

7. Polar hydrocarbons containing S, O, N, Cl, or Br are often resistant to biodegradation.

8. Highly water-soluble chemicals biodegrade more readily than do those with low water solubility.

9. Chemicals that sorb weakly to soils biodegrade more readily than do those that sorb strongly. Strongly sorbed chemicals are less available to microbes.

10. Chemicals with small K_{ow} values biodegrade more readily than do those with large K_{ow} values.

11. Chemicals that leach easily from soils biodegrade more readily than those that are not easily leached. These chemicals are more soluble and less strongly sorbed.

12. In the saturated zone, rates of hydrocarbon biodegradation roughly double for every 10°C increase in groundwater temperature, in the range of 5–25°C.

Example 1

How long will it take to naturally biodegrade the BTEX contained in 250 kg of gasoline (an LNAPL) immobilized within the saturated zone, given the following conditions?

Depth of LNAPL zone into saturated zone = 2 m
Width of LNAPL zone in saturated zone = 10 m
Groundwater Darcy velocity = 1 m/day
Background upgradient DO concentration = 5 mg/L
Oxygen–hydrocarbon consumption ratio = 1 mg O_2/0.32 mg BTEX (from Rules of Thumb).

Assume that gasoline LNAPL is immobilized by sorption in the soil matrix and that BTEX is 25% of the LNAPL weight. Also assume that aerobic biodegradation is essentially instantaneous compared to normal groundwater movement. In the presence of excess oxygen, aerobic bacteria can degrade 1 mg/L of BTEX in about 8 days, essentially instantaneous compared to the years often required for flowing groundwater to replenish a plume area with oxygen. Under these conditions, the rate of biodegradation is equal to the rate at which sufficient DO can be brought into the residual gasoline LNAPL zone by groundwater flow. Approach the problem by calculating how long it would take enough water to pass through the plume cross-sectional area to supply 1 mg O_2 per 0.32 mg of BTEX.

ANSWER

$$\text{Time to degrade} = (250\,\text{kg NAPL}) \times \left(\frac{0.25\,\text{g BTEX}}{1\,\text{g NAPL}}\right) \times (10^6\,\text{mg/kg}) \times \left(\frac{1\,\text{mg O}_2}{0.32\,\text{mg BTEX}}\right)$$

$$\times \left(\frac{1\,\text{L}}{5\,\text{mg O}_2}\right) \times \left(\frac{1}{(2 \times 10)\text{m}^2}\right) \times \left(\frac{1\,\text{day}}{1\,\text{m}}\right) \times \left(\frac{1\,\text{m}^3}{10^3\,\text{L}}\right)$$

Time to degrade = 1950 days or about 5.4 years.

This approach of calculating how rapidly DO can be supplied to the plume area does not work for DNAPL composed of chlorinated compounds because they are resistant to aerobic biodegradation. It also neglects the fact that part of the LNAPL becomes increasingly less unavailable to microbes because of stronger sorption and penetration into cracks and soil pores with time, which increases the time needed for remediation.

9.5   DETERMINING THE EXTENT OF BIOREMEDIATION OF LNAPL

Spilled fuel LNAPL, a major type of subsurface contamination, is present in the subsurface as

- Mobile LNAPL-free product, which will drain into a well under gravity
- Residual LNAPL held by adsorption and capillarity in soil pore spaces, which is immobile and unable to drain into a well under gravity
- Dissolved LNAPL compounds in water
- Volatile LNAPL vapors

In an LNAPL spill, the most-soluble and-volatile components are lost first from the LNAPL-free product by water leaching (washing and volatilization). Nevertheless, months later, the remaining LNAPL still contains about

- 90% of the benzene
- 99% of the total BTEX
- 99.9% of the TPH

Since most of the LNAPLs have low water solubility, it seems clear that the first remediation step should be to remove physically as much LNAPL as possible. However, frequently less than 10% of the total LNAPL can be removed by recovery of mobile LNAPL. The remaining part stays trapped in the soil by sorption and capillarity. For this remaining part of the contamination, the best choice for corrective action may be bioremediation, especially if excavation is not practical. The next remediation step should be to determine if intrinsic bioremediation is occurring at a sufficient rate that no other action is required.

The U.S. Air Force Center for Environmental Excellence has published a technical document that describes a protocol for data collection and analysis that can be used for judging whether intrinsic bioremediation is occurring at a useful rate (Wiedemeier et al. 1995). This report is notable for comprehensively discussing the current state of knowledge and for its thorough list of references. Much of the following material is adapted from the Wiedemeier report.

9.5.1 USING CHEMICAL INDICATORS OF THE RATE OF INTRINSIC BIOREMEDIATION

Certain water quality parameters change because of biodegradation. By measuring how these parameters change with time and location within a contaminant plume in the saturated zone, the occurrence and rate of active biodegradation can be determined. The most important parameters that change are presented in Table 9.4 and include the pollutants being degraded, the waste products that are formed, and the redox reactants. For aerobic respiration, the electron acceptor is oxygen. Anaerobic electron acceptors include nitrate, sulfate, ferric iron, manganese, and carbon dioxide.

TABLE 9.4
Water Quality Parameters That Indicate Biodegradation Activity

Chemical Indicators in Groundwater for Biodegradation	Trend in Indicator Concentration during Biodegradation	Processes Responsible for Trend
Hydrocarbon concentrations	Decreases	Biodegradation
Dissolved oxygen (DO)	Decreases	Aerobic respiration
Nitrate	Decreases	Denitrification
Manganese(II)	Increases	Manganese(IV) reduction
Iron(II)	Increases	Iron(III) reduction
Sulfate	Decreases	Sulfate reduction
Methane	Increases	Methanogenesis
Alkalinity	Increases	Increased by aerobic respiration, denitrification, iron(III) reduction, and sulfate reduction; not affected much by methanogenesis
ORP (pE)	Generally decreases toward plume center	Serves as a crude indicator of which redox reactions may be operating at a given time
Volatile fatty acids	Increases	Metabolic byproducts of biodegradation

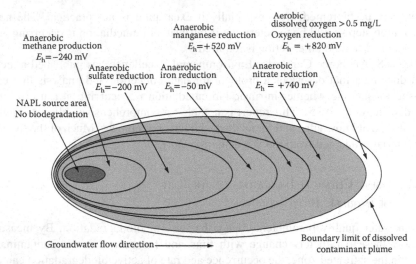

FIGURE 9.1   Idealized LNAPL plume in the saturated zone, showing aerobic (white) and anaerobic (shaded) zones of biodegradation in a dissolved NAPL plume. The distribution of different electron acceptors according to redox potential is also shown.

Figure 9.1 shows how the redox reactions associated with these electron acceptors are distributed in an idealized LNAPL plume undergoing active biodegradation. The processes that establish this distribution are

1. Aerobic biodegradation consumes available oxygen resulting in anaerobic conditions in the core of the plume and a zone of oxygen depletion along the outer margins.
2. The progressive lowering of the redox potential from the outer plume boundary toward the source zone, as preferred anaerobic electron acceptors become depleted according to Equations 9.1 through 9.6, establishes successive zones where different electron acceptors are dominant.
3. As aerobic biodegradation is relatively rapid and the rate of DO replacement is slow in the saturated zone, DO becomes depleted where LNAPL concentrations are high. Thus, anaerobic redox reactions occur in most of the plume, where the abundance of anaerobic electron acceptors is large relative to DO. Since the anaerobic zone is typically more extensive than the aerobic zone, anaerobic biodegradation is usually the dominant process overall.
4. Water carrying DO diffuses into the plume from outside the plume boundary, enabling aerobic redox reactions to occur around the plume periphery.
5. It is likely that all biodegradation is inhibited in the LNAPL source area because of pollutant concentrations high enough to be toxic to microorganisms.
6. For both aerobic and anaerobic processes, the rate of contaminant degradation is controlled by the concentration of electron acceptors, not the rate that microorganisms consume the electron acceptors. As long as there is a sufficient supply of the electron acceptors, the rate of metabolism does not make any practical difference in the length of time required to achieve remediation objectives.

The first step in determining if biodegradation is occurring at a rate fast enough to be useful for remediation is to look for appropriate changes in these chemical indicators (Table 9.4), which are discussed in more detail in Sections 9.5.2 through 9.5.9.

9.5.2 HYDROCARBON CONTAMINANT INDICATOR

If significant LNAPL biodegradation is occurring, hydrocarbon concentrations will diminish with time and distance from the spill source. However, because this could occur due to dilution, confirmatory evidence is always required. Seek confirmatory evidence by measuring hydrocarbon concentrations in the groundwater plume close to the flow centerline and near the spill source, where dilution has less effect. Analyze the groundwater for hydrocarbon compounds of regulatory concern. These are usually BTEX and trimethylbenzenes, but total volatile hydrocarbons (TVH), total extractable hydrocarbons (TEH, also called semivolatiles), and PAHs often also should be measured. The determination of whether biodegradation is occurring at a useful rate hinges on the rate at which these parameters are disappearing.

Based on solubilities, the highest combined dissolved concentrations of BTEX plus trimethylbenzenes should not be greater than about 30 mg/L for JP-4 jet or diesel fuel or about 135 mg/L for gasoline. If these concentrations are exceeded, sampling errors have possibly occurred, such as collecting some free product. This error is likely if emulsification of LNAPL has occurred in the water sample.

Figures 9.2a and 9.2b are idealized representations of a dissolved total BTEX groundwater plume caused by a gasoline spill, measured twice, 1 year apart. Comparing total BTEX plumes at the start of the study and 1 year later shows that,

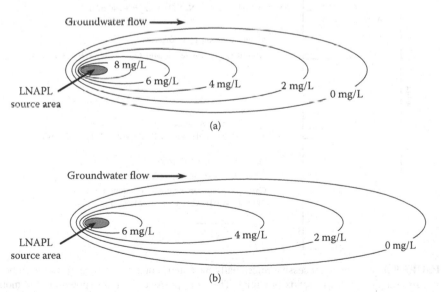

FIGURE 9.2 Idealized dissolved total BTEX isopleths (lines of equal concentration) for a groundwater plume. (a) Initial total BTEX measurement. Contour interval is 2 mg/L. Indicated isopleth values represent the concentration in mg/L of total BTEX. (b) Measurement of total BTEX isopleths 1 year later.

although plume area has increased a little, the total mass of BTEX in the plume has decreased. Data from other chemical indicators may provide evidence that the decrease is primarily due to intrinsic bioremediation and not dilution.

9.5.3 ELECTRON ACCEPTOR INDICATORS

Subsurface microbes utilize different redox reactions in the order of decreasing energy-yielding value (from top to bottom in Figure 9.3). If oxygen is available, using it as an electron acceptor will always yield the greatest energy. Therefore, aerobic biodegradation reactions always occur first whenever sufficient DO is available. The DO level within the plume will be below background levels outside the plume wherever aerobic degradation is occurring.

Anaerobic processes will begin when DO has been depleted sufficiently, depending on which electron acceptors are available. Reduction of nitrate is the second highest energy-yielding process and occurs next if nitrate is present. However, if DO is present in concentrations greater than 0.5 mg/L, it is toxic to anaerobic-only (obligate) bacteria. Therefore, nitrate denitrification cannot begin until most of the DO has been consumed.

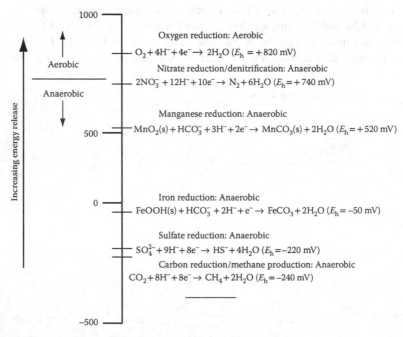

FIGURE 9.3 Order of successive microbially mediated redox reactions is from top to bottom, from higher to lower redox potential. The more positive the redox potential, the more energy is released per electron transferred. (Adapted from Wiedemeier, T.H. et al., *Technical Protocol for Implementing Intrinsic Remediation with Long-Term Monitoring for Natural Attenuation of Fuel Contamination Dissolved in Groundwater*, U.S. Air Force Center for Environmental Excellence, San Antonio, 1995.)

Electron acceptors available in the groundwater determine the sequence in which biodegradation reactions will occur. As the redox potential changes from positive to increasingly negative values, different electron acceptors are used in the biodegradation process. As reduction of electron acceptors progresses, the redox potential (E_h) of the groundwater becomes increasingly negative and the energy obtained per electron transfer decreases.

Figure 9.1 shows how aerobic and anaerobic biodegradation zones develop in an LNAPL plume undergoing active biodegradation. Figure 9.3 illustrates the normal sequence of the most important biodegradation redox reactions and the potential at which they are initiated. Table 9.1 gives the stoichiometry of the biodegradation reactions.

9.5.4 DISSOLVED OXYGEN INDICATOR

If significant aerobic biodegradation is occurring, DO will diminish with time in the plume zone. DO will be consumed first before other electron acceptors can be used. Each 1.0 mg/L of O_2 consumed by microbes will destroy approximately 0.32 mg/L of BTEX, based only on the production of CO_2 and H_2O. This is a conservative estimate that ignores the conversion of carbon to cell mass. If cell mass production is included, each 1.0 mg/L of O_2 consumed by microbes can destroy as much as 0.97 mg/L of BTEX under ideal conditions. Because other factors such as nutrient availability affect respiration, it is best to use the more conservative 0.32 mg/L value as the amount of BTEX destroyed per 1.0 mg/L of O_2 consumed. The overall redox reaction of oxygen with benzene is

$$C_6H_6 + 7.5O_2 \rightarrow 6CO_2 + 3H_2O \tag{9.8}$$

Example 2

Suppose the background DO in groundwater at a remediation site is 6 mg/L. Then, excluding cell mass production, the groundwater conservatively will have the capacity to aerobically biodegrade up to 1.9 mg/L of BTEX:

$$6 \text{ mg/L DO} \times \frac{0.32 \text{ mg/L BTEX}}{1.0 \text{ mg/L DO}} = 1.9 \text{ mg/L BTEX}$$

FIELD TEST FOR DO CONSUMPTION BY MICROBES

- Analyze groundwater for DO at different locations in the BTEX plume.
- Areas with elevated BTEX concentrations should have DO concentrations that range from less than background to zero.
- This is strong evidence for the occurrence of aerobic biodegradation.

9.5.5 NITRATE AND NITRITE DENITRIFICATION INDICATOR

If significant biodegradation is occurring and DO has been depleted to less than 0.5 mg/L, anaerobic reactions that utilize nitrate (NO_3^-) and nitrite (NO_2^-) as electron acceptors can commence. If nitrate/nitrite is present in the plume, their concentrations will diminish with time in the plume zone. Each 1.0 mg/L of dissolved NO_3^- will destroy about 0.21 mg/L of BTEX. The final reaction products are CO_2, H_2O, and N_2.

The denitrification steps are

$$NO_3^- \rightarrow NO_2^- \rightarrow NO \rightarrow N_2O \rightarrow NH_4^+ \rightarrow N_2 \qquad (9.9)$$

The overall redox reaction with benzene is

$$6NO_3^- + 6H^+ + C_6H_6 \rightarrow 6CO_6 + 6H_2O + 3N_2 \qquad (9.10)$$

Requirements for denitrification are

- Dissolved nitrate/nitrite
- Organic carbon
- Denitrifying bacteria
- Reducing conditions (DO < 0.5 mg/L)

Denitrification is favored when

$$6.2 < pH < 10.2$$
$$-200\,mV < redox\ potential\,(E_h) < +665\,mV$$

Nitrate reduction is rapid. The rate at which nitrate and nitrite are supplied by groundwater to the reduction zone limits the reaction rate. Under denitrifying conditions, biodegradation of BTEX occurs in the following order:

$$toluene > p\text{-xylene} > m\text{-xylene} > ethylbenzene > o\text{-xylene} \gg benzene$$

FIELD TEST FOR NITRATE + NITRITE CONSUMPTION BY MICROBES

- Analyze groundwater for nitrate plus nitrite.
- Areas with elevated BTEX concentrations should have nitrate plus nitrite concentrations that range from less than background to zero.

9.5.6 METAL REDUCTION INDICATORS: MANGANESE(IV) TO MANGANESE(II) AND IRON(III) TO IRON(II)

The reduction of oxidized forms of iron and manganese (Fe^{3+} and Mn^{4+}) results in the production of reduced species that are water-soluble. Elevated levels of these reduced metals (Fe^{2+} and Mn^{2+}) in the plume relative to background is indicative of anaerobic biodegradation.

If significant biodegradation is occurring and DO and nitrate/nitrite have been depleted, reduction of manganese from Mn(IV) to Mn(II) and iron from Fe(III) to Fe(II) will be initiated. Figure 9.3 shows that as the redox potential decreases, manganese will be reduced first, followed by iron. As the reduced forms of both metals are more soluble than the oxidized forms, dissolved manganese(II) and iron(II) concentrations will increase with time in the plume zone if there are available forms of manganese(IV) and iron(III) minerals to be used as electron acceptors. Aquifer sediments often contain large quantities of manganese(IV) and iron(III), frequently in the form of amorphous iron and manganese oxyhydroxides, so the reduction reaction, Equation 9.11, is a common occurrence.

$$Mn(IV), Fe(III)(insoluble) \rightarrow Mn^{2+}, Fe^{2+} (soluble) \tag{9.11}$$

The presence of increasing concentrations of Mn^{2+} and Fe^{2+} within the BTEX plume is a strong evidence for anaerobic biodegradation reactions.

FIELD TEST FOR METAL REDUCTION BY MICROBES

- Analyze groundwater for Mn^{2+} and Fe^{2+}.
- Areas with elevated BTEX concentrations should have elevated (relative to background) Mn^{2+} and Fe^{2+} concentrations.

9.5.7 SULFATE REDUCTION INDICATOR

After available DO, nitrate, manganese(IV), and iron(III) are consumed, available sulfate can be used as an electron acceptor. Sulfate reduction to sulfide is favored at pH 7 and E_h −200 mV. Sulfate-reducing microorganisms are sensitive to temperature, inorganic nutrients, pH, and redox potential. Small imbalances in environmental conditions can severely limit the rate of BTEX degradation through sulfate reduction. Each 1.0 mg/L of BTEX that is biodegraded requires the reduction of about 4.7 mg/L of sulfate. The overall reaction for benzene oxidation by sulfate reduction is

$$7.5H^+ + 3.75SO_4^{2-} + C_6H_6 \rightarrow 6CO_2 + 3.75CH_2S + 3H_2O \tag{9.12}$$

FIELD TEST FOR SULFATE REDUCTION BY MICROBES

- Analyze groundwater for sulfate (SO_4^{2-}), and perhaps sulfide (S^{2-}).
- Depleted sulfate concentrations and increased sulfide concentrations (relative to background) within the BTEX plume indicates active biodegradation by SRB.

9.5.8 METHANOGENESIS (METHANE FORMATION) INDICATOR

After available DO, nitrate, manganese(IV), iron(III), and sulfate are consumed, the redox reactions of organic carbon to form CH_4 (C^{4-}) can be used to biodegrade BTEX, resulting in an increase in CH_4 concentrations in the plume zone.

Methanogenesis generates less energy for microbes than the other reducing reactions and always occurs last, after other electron acceptors have been depleted. Methanogenesis causes the redox potential to fall below −200 mV at pH 7. The presence of elevated levels of methane in the presence of elevated levels of BTEX indicates that BTEX biodegradation is occurring as a result of methanogenesis.

Because methane is not present in LNAPL fuels, the presence of methane above background in groundwater adjoining LNAPL fuels indicates microbial degradation of fuel hydrocarbons. The overall reaction for benzene oxidation by methanogenesis is

$$C_6H_6 + 4.5H_2O \rightarrow 2.25CO_2 + 3.75CH_4 \tag{9.13}$$

This reaction occurs in a minimum of four steps, at least one of which involves CO_2 accepting electrons and reacting with H^+ to form CH_4. In the process, C_6H_6 is oxidized to form additional CO_2. The biodegradation of 1 mg/L of BTEX by methanogenesis produces about 0.78 mg/L of methane.

FIELD TEST FOR METHANOGENESIS

- Analyze groundwater for methane (CH_4).
- High methane concentrations (relative to background) within the BTEX plume indicate active biodegradation by methanogenesis.

9.5.9 Redox Potential and Alkalinity as Biodegradation Indicators

Groundwater redox potential and alkalinity also undergo measurable changes in regions where significant biodegradation of fuel hydrocarbons occurs.

9.5.9.1 Using Redox Potentials to Locate Anaerobic Biodegradation within the Plume

Each successive biodegradation redox reaction involving the electron acceptors in Table 9.1 from DO to methane lowers the redox potential of the groundwater in which it occurs. Thus, a decrease with time of groundwater redox potential should serve as an indication of biodegradation activity.

FIELD TEST FOR OCCURRENCE OF REDOX REACTIONS

- Map groundwater redox potentials at the site. Include at least one location upgradient of the plume. It is important to avoid aeration of well samples for these measurements.
- Locations where groundwater redox potentials are lower than background are where electron acceptor species are being reduced, a sign of biodegradation.

- Redox potentials within the plume can help to indicate which electron acceptors are active in different locations.
- In regions of biodegradation activity, the zone of low redox potential (reducing zone) will become larger with time, as diminishing DO concentrations move farther and farther from the spill region.

9.5.9.2 Using Alkalinity to Locate Anaerobic Biodegradation within the Plume

Groundwater alkalinity increases during aerobic respiration, denitrification, iron(III) reduction, and sulfate reduction and is unchanged during methanogenesis. The two main processes that increase alkalinity are

1. All of the biodegradation redox reactions produce carbon dioxide (see Table 9.1; Equations 9.1 through 9.7). Addition of CO_2 increases the total carbonate and alkalinity of the groundwater.
2. Redox reactions involving nitrate, manganese(IV), iron(III), and sulfate as electron acceptors all consume acidity as H^+ (see Table 9.1; Equations 9.2 through 9.5). This also increases alkalinity in groundwater where these reactions take place.

A measurement of alkalinity within a hydrocarbon plume can be used to infer the amount of petroleum hydrocarbons destroyed. For every 1 mg/L of alkalinity (as $CaCO_3$) produced, 0.13 mg/L of BTEX is destroyed (Wiedemeier et al. 1995).

FIELD TEST FOR USING ALKALINITY AS AN INDICATOR OF BIODEGRADATION

- Map groundwater alkalinity concentrations at the site. Include at least one location upgradient of the plume.
- Locations where groundwater alkalinity is higher than background are where CO_2 is being produced and H^+ is being consumed, which are signs of biodegradation.
- Alkalinity levels within the plume can help to indicate the amount of petroleum hydrocarbons that have been destroyed.
- In regions of biodegradation activity, the zone of higher alkalinity will become larger with time.
- Increased alkalinity levels within the plume can be used to infer the extent of biodegradation occurring.

9.6   BIOREMEDIATION OF CHLORINATED DNAPLS

Chapter 8 contains a general discussion of chlorinated DNAPLs. Chlorinated solvents are among the most frequently encountered organic groundwater contaminants in the United States (Morrison 2000; USEPA 2000). Although these compounds have been known since before 1900, large-scale production and use of chlorinated solvents began around 1950 and continued until around 1978, when regulations were enacted controlling their production and use. The largest uses of chlorinated solvents are vapor degreasing of metal parts and dry cleaning of clothing (Morrison 2000).

The most important process for natural biodegradation of highly chlorinated solvents such as PCE, TCE, and 1,1,1-TCA is reductive dechlorination, where a chlorine atom is removed from the solvent molecule and replaced with a hydrogen atom. The chlorine atom acquires an electron from the hydrogen atom and enters the solution as a chloride anion (Wiedemeier et al. 1996, 1998). The major requirements for reductive dechlorination are the presence of electron donors to serve as a source of reducing power and metabolic energy and a population of dehalorespiring microorganisms. The required electron acceptor is the chlorinated solvent molecule itself. Electron donors commonly found in contaminated groundwater include petroleum-derived aromatic hydrocarbons found in fuels (BTEX and the trimethylbenzene isomers) or other types of organic carbon (landfill leachate, aliphatic fuel hydrocarbons, or natural organ carbon) (Wiedemeier et al. 1996, 1998; Wilson et al. 2001).

9.6.1   Reductive Dechlorination of Chlorinated Ethenes

There are six chlorinated ethenes:

1. Tetrachloroethene (PCE: $CCl_2 = CCl_2$)
2. Trichloroethene (TCE: $CCl_2 = CHCl$)
3. cis-1,2-DEC ($CHCl = CHCl$)
4. trans-1,2-DEC ($CHCl = CHCl$)
5. 1,1-DEC ($CCl_2 = CH_2$)
6. Vinyl chloride ($CHCl = CH_2$)

Starting with PCE, reductive dechlorination proceeds by sequential removal of one chlorine atom after another, replacing each with a hydrogen atom, in the following series of steps:

$$PCE(CCl_2 = CCl_2) \rightarrow TCE(CCl_2 = CHCl) \tag{9.14}$$

$$TCE \rightarrow cis\text{-}1,2\text{-}DCE(CHCl = CHCl);$$
$$trans\text{-}1,2\text{-}DCE(CHCl = CHCl); 1,1\text{-}DCE(CCl_2 = CH_2) \tag{9.15}$$

$$DCE \rightarrow vinyl\,chloride\,(CHCl = CH_2) \tag{9.16}$$

$$Vinyl\,chloride \rightarrow ethene\,(CH_2 = CH_2) \tag{9.17}$$

Equations 9.14 through 9.17 are illustrated by the flow diagram of Figure 9.4. Of the three DCE isomers, cis-1,2-DCE is usually the most abundant product, while

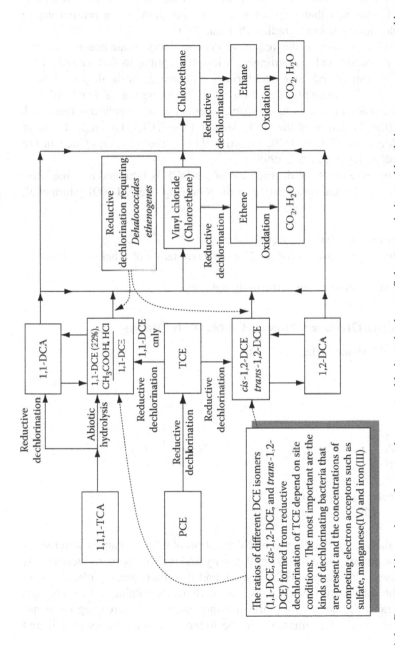

FIGURE 9.4 Decomposition pathways for several common chlorinated solvents. Solvents are designated by their common names or abbreviations. Chemical names and formulas may be found in Chapter 8 (Table 8.1).

trans-1,2-DCE and 1,1-DCE are formed in lesser amounts (Wiedemeier et al. 1996, 1998; Wilson et al. 2001). The presence of *cis*-1,2-DCE as a reaction product is an indication that reducing conditions exist. Even if aerobic conditions prevail, detection of 1,1-DCE indicates that microsites of anaerobic activity are present where reductive dechlorination is biodegrading PCE and TCE.

Each successive product in the reductive dechlorination sequence is less oxidized than its precursor and, accordingly, is less susceptible to further reduction. Consequently, the rate of reductive dechlorination decreases as the degree of chlorination decreases. For example, Equation 9.16 (the formation of vinyl chloride from the DCE isomers) requires more strongly reducing site conditions than does Equation 9.15 (the formation of the DCE isomers from TCE). The degradation of maximally chlorinated PCE to TCE, Equation 9.14, is the most rapid step in the sequence (Wiedemeier et al. 1996, 1998).

It has been observed that the degradation of *cis*-DCE is inhibited at some sites, apparently because of some combination of the following (Newell 2001; Shim et al. 2001):

- Particular essential bacteria were absent.
- Inhibiting enzymes were produced by the degradation of other organics at the site.
- Redox conditions were not sufficiently reducing.

9.6.2 Reductive Dechlorination of Chlorinated Ethanes

There are nine chlorinated ethanes (CAs):

1. CA
2. 1,1-DCA
3. 1,2-DCA
4. 1,1,1-TCA
5. 1,1,2-TCA
6. 1,1,1,2-TeCA
7. 1,1,2,2-TeCA
8. Pentachloroethane (PCA)
9. Hexachloroethane (HCA)

In general, increasing the degree of chlorination of ethanes causes water solubility and vapor pressure to decrease and density and melting point to increase. At room temperature, CA is a gas, HCA a solid, and the others are liquids. CA is an LNAPL and the others are DNAPLs. All are sufficiently soluble to be a concern as water pollutants. Toxicity to aquatic life increases with increasing chlorination. All of the chlorinated ethanes may be found as contaminants in soil and groundwater.

Although chlorinated ethanes (see Table 8.1) are similar to chlorinated ethenes in that they biodegrade by reductive dechlorination, two additional characteristics of CAs are important to note:

1. They also degrade abiotically at environmentally significant rates, often at rates much greater than biodegradation.
2. Both biological and abiotic degradation of CAs can produce chloroethenes (Wiedemeier et al. 1996, 1998).

In the case of 1,1,1-TCA, a common industrial solvent and degreaser, abiotic chemical transformation is the most prevalent environmental degradation process, producing 1,1-DCE (about 22%) by elimination reactions and acetic acid (about 78%) by hydrolysis (Smith 1999).

9.6.3 Case Study: Using Biodegradation Pathways for Source Identification

Chlorinated contaminants found in soils and groundwater in the subsurface of an urban Superfund site included PCE, TCE, 1,1-DCE, cis-1,2-DCE, trans-1,2-DCE, and CA.

From Figure 9.4, it is evident that the presence of cis-1,2-DCE, trans-1,2-DCE, 1,1-DCE, and 1,2-DCA are indicative of reductive dechlorination reactions of PCE and TCE. However, 1,1-DCE can result from both the reductive dechlorination of PCE and TCE and the abiotic dechlorination of 1,1,1-TCA. Although 1,1,1-TCA was not detected, it could be possible that it had been present but was fully degraded, leaving only the degradation product 1,1-DCE.

The presence of CA, which is also a degradation product of 1,1,1-TCA but not of PCE and TCE, is supporting evidence that 1,1,1-TCA was also once present at the site. This result might possibly implicate a new PRP for sharing in the remediation costs if they were known to have used 1,1,1-TCA as a solvent in their activities.

REFERENCES

Morrison, R.D. 2000. *Environmental Forensics: Principles and Applications*. Boca Raton, FL: CRC Press

National Research Council. 2000. *Natural Attenuation for Groundwater Remediation*, Chapter 3. Washington, DC: The National Academy of Sciences Press.

Newell, C.J. 2001. *The BIOCHLOR Natural Attenuation Model*, Powerpoint presentation, Headquarters U.S. Air Force, January 31, 2001.

Rice, D.W. et al. 1995. *California Leaking Underground Fuel Tank (LUFT) Historical Case Analyses*. UCRL-AR-122207, Lawrence Livermore National Laboratory, Environmental Protection Department Environmental Restoration Division, University of California Livermore, California, Submitted to the State Water Resources Control Board Underground Storage Tank Program and the Senate Bill 1764 Leaking Underground Fuel Tank Advisory Committee, November 16, 1995, California State Water Resources Control Board.

Shim, H. et al. 2001. "Aerobic Degradation of Mixtures of Tetrachloroethylene, Trichloroethylene, Dichloroethylenes, and Vinyl Chloride by Toluene-o-Xylene Monooxygenase of Pseudomonas stutzeri OX1." *Applied Microbiology and Biotechnology* 56: 265–9.

Smith, J. 1999. "The Determination of the Age of 1,1,1-Trichloroethane in Groundwater." In *Conference Abstracts, Second Executive Forum on Environmental Forensics*, p. 1. Southboro, MA: International Business Communications.

USEPA. 2000. *Engineered Approaches to In Situ Bioremediation of Chlorinated Solvents: Fundamentals and Field Applications.* Solid waste and emergency response (5102G), EPA 542-R-00–008, July (revised), http://www.epa.gov/tio/download/remed/ engappinsitbio.pdf, accessed September 25, 2012.

USEPA. 2004. *How to Evaluate Alternative Cleanup Technologies for Underground Storage Tank Sites: A Guide for Corrective Action Plan Reviewers.* EPA 510-R-04–002, May 2004, www.epa.gov/oust/pubs/tums.htm, accessed September 25, 2012.

Wiedemeier, T.H. et al. 1995. *Technical Protocol for Implementing Intrinsic Remediation with Long-Term Monitoring for Natural Attenuation of Fuel Contamination Dissolved in Groundwater.* San Antonio, TX: U.S. Air Force Center for Environmental Excellence.

Wiedemeier, T.H. et al. 1996. "Overview of the Technical Protocol for Natural Attenuation of Chlorinated Aliphatic Hydrocarbons in Ground Water Under Development for the U.S. Air Force Center for Environmental Excellence." In *Symposium on Natural Attenuation of Chlorinated Organics in Ground Water,* Dallas, TX, September 11–13, EPA/540/R-96/509.

Wiedemeier, T.H. et al. 1998. *Technical Protocol for Evaluating Natural Attenuation of Chlorinated Solvents in Groundwater.* USEPA, EPA/600/R-98/128, National Risk Management Research Laboratory, Office of Research and Development, U.S. Environmental Protection Agency, Cincinnati, Ohio.

Wilson, J.T. et al. 2001. *Evaluation of the Protocol for Natural Attenuation of Chlorinated Solvents: Case Study at the Twin Cities Army Ammunition Plant.* EPA/600/R-01/025, March 2001, U.S. Environmental Protection Agency, Office of Research and Development, Washinton, D.C.

Uranium is the only radionuclide for which the chemical toxicity has been found to be greater or equal to its radiotoxicity and for which its drinking water standard is expressed in terms of a reference dose (RfD), rather than a sample's activity (pCi/L).

Chapter 10, Section 10.5.2

10 Behavior of Radionuclides in the Water and Soil Environment

10.1 INTRODUCTION

This chapter is intended to give the nonspecialist a helpful understanding of how radionuclides behave in water and soil environments. Another purpose is to assemble information used for evaluating environmental radionuclide measurements into a form that is useful for a nonnuclear environmental professional. For example, the drinking water maximum contaminant level (MCL) for gross beta (β) emissions is 4 mrem/year, but laboratory results are generally given in terms of pCi/L. Tables and rules of thumb for many conversions between units used in radiochemistry are found in this chapter. A third purpose, less important perhaps than the first two, is to offer a concise introduction to the basics of radioactivity and the properties of radiation. The nuclear processes of fission and fusion, which few environmental professionals become involved with, are not covered. Section 10.2, which comprises the introduction to nuclear structure, is not essential for the rest of the chapter. However, reading it might help to remove some of the mystery that often surrounds a layman's perception of radionuclides and radioactivity.

10.2   RADIONUCLIDES

A radionuclide is an atom that has a radioactive nucleus. A radioactive nucleus is an atomic nucleus that emits radiation in the form of particles or photons, thereby losing mass and/or energy and changing its internal structure to become a different kind of nucleus, perhaps radioactive, perhaps a different element, or perhaps a different isotope of the same element. All radionuclides have finite lifetimes, ranging between billions of years to less than nanoseconds; each time a particle is emitted, the original radionuclide is transformed into a different species. The emitted particles can possess enough energy to penetrate into solid matter, altering and damaging the molecules with which they collide. Radionuclides cannot be neutralized by any chemical or physical treatment; they can only be confined and shielded until their emission activity dies to a negligible level. Radionuclides are unique in being the only pollutants that can act at a distance, harming life forms and the environment without physical contact.

This section addresses two basic questions about atomic nuclei:

1. What are atomic nuclei made of?
2. Why are some nuclei radioactive and some not radioactive?

Sections 10.2.1 through 10.2.10 discuss the fundamentals of radioactivity and radionuclides. Sections 10.3 through 10.4 discuss the conditions under which radiation is a hazard to the environment and to human health and offer a guide through the "maze" of different units (becquerels, curies, rads, rems, and more) used to measure the amount of radiation and doses received.

Finally, in Sections 10.5 and 10.6, the fate and transport behavior of radionuclides in the environment is discussed. This behavior is mainly dependent on their chemical properties rather than their nuclear properties.

10.2.1   A Few Basic Principles of Chemistry

10.2.1.1   Matter and Atoms

All matter is composed of atoms. There are about 118 different kinds of atoms with different chemical and physical properties. Ninety-two exist naturally and the others are all radioactive with such short half-lives that, if they had been present on Earth initially, they have since disappeared by radioactive decay. We know about them from theory and by experimentally creating them in high energy physics laboratories with particle colliders.

These different atoms are called the elements* and comprise the entries in the periodic table, reproduced on the inside front cover of this book. For a working definition of an atom, regard it as the smallest particle of matter that can chemically react with other matter; it is identified as a particular element by measuring its

---

* Many of the 92 naturally occurring elements have radioactive isotopes (Section 10.2.3) with short lifetimes. These, however, have not disappeared because they are formed continually as radioactive "daughter" products in the decay chains of the long-lived, naturally occurring radioactive elements uranium and thorium (Section 10.2.10).

properties. Atoms can combine to form larger units of two or more atoms called molecules. A molecule is the smallest bit of matter that is recognizable as any chemical substance other than an element. Molecules can assemble into the still larger entities that make up the world of rocks, life forms, soils, oceans, and atmosphere that we perceive with our normal five senses.

Atoms themselves have a substructure; they are assembled from subatomic particles called protons, electrons, and neutrons. This is the deepest level of subdivision needed for interpreting chemical behavior. However, a description of nuclear structure and radioactivity requires that we consider the next deeper level of substructures where still smaller units of matter, called quarks, leptons, and gluons combine to make protons, electrons, and neutrons. These are discussed briefly in Section 10.2.5. More substructures beyond quarks, leptons, and gluons enter the realms of quantum physics and theoretical speculation.

The structure of an atom is determined by the properties of its component parts, the protons, electrons, and neutrons. The attractive and repulsive forces associated with protons, electrons, and neutrons play important roles in how an atom interacts with other matter.

- There are two different electrical charges, called positive and negative. An electrical charge exerts a force on other electrical charges.
- Like charges (e.g., two positive charges or two negative charges) repel one another, while unlike charges (e.g., a positive and a negative charge) attract one another.
- A single positive charge is equal in magnitude to a single negative charge, but opposite in sign. One positive charge can attract and neutralize one negative charge, resulting in zero net charge.
- An electron carries a single negative charge and has insignificant mass (about 9.109×10^{-28} g) compared to a proton or a neutron.
- A proton carries a single positive charge and is about 1800 times heavier than an electron (about 1.673×10^{-24} g).
- A neutron has zero electric charge and a mass nearly the same as a proton, but just a little heavier (1.675×10^{-24} g). We will see that a neutron may be regarded as a combination of one proton with one electron.

Every atom has a small positively charged nucleus in its center containing both protons and neutrons. The positive charge of the nucleus electrically attracts electrons from its surrounding space until the nucleus is encircled by a "cloud" of electrons (see Figure 10.1). The electrons are not drawn into the nucleus itself because of short-range repulsive forces. The number of electrons in the cloud is equal to the number of positive charges in the nucleus so that the atom is electrically neutral overall.* The nucleus contains essentially all the mass of the atom in its protons and neutrons, while the electron cloud is virtually weightless by comparison.

---

* Electrical neutrality is always a property of an isolated, stable atom. However, in normal matter, where every atom is generally in the company of many others, there are interactions that distort the electron clouds around the nuclei, sometimes resulting in polar or ionized atoms that are electrically charged; see Chapter 2.

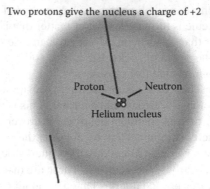

Two protons give the nucleus a charge of +2

Proton   Neutron

Helium nucleus

Two electrons form a spherical cloudlike orbital
around the nucleus with a charge of −2

FIGURE 10.1 A representation of a helium atom with two protons and two neutrons in its nucleus. The two positive charges of the protons attract and hold two negatively charged electrons depicted as a spherical cloud of negative charge surrounding the nucleus. In an actual helium atom, the diameter of the electron cloud is approximately 100,000 times larger than the diameter of the nucleus.

10.2.1.2   Elements

Originally, a chemical element was defined as a substance that cannot be decomposed by chemical means into simpler substances. The test was whether any of its chemical properties could be changed by chemical decomposition processes. Elements identified by such tests could be combined into new substances called compounds having new properties, but decomposition of the compounds always brought back the original set of starting elements. However, there were many different elements, each with a unique set of properties. There had to be reasons for the differences among elements.

Eventually, in the early 1900s, the internal nuclear structure of atoms was revealed and the properties of elements were shown to depend on the number of protons in the nucleus. Each different element has a different number of protons in its nucleus. The periodic table arranges the elements from left to right in rows so that each successive element contains one more proton in its nucleus than the preceding element, beginning with hydrogen, which has one proton. Each element is numbered with an atomic number equal to the number of protons in its nucleus and each element has a unique set of chemical and physical properties.

For example, the element carbon, with atomic number 6, has six protons in its nucleus and the element nitrogen, with atomic number 7, has seven protons in its nucleus. The carbon nucleus can attract six electrons to itself before the atom becomes neutral and does not attract additional electrons. The electrical forces within the atom hold the electrons to the vicinity of the nucleus, and because the electrons repel one another, the six carbon electrons become distributed around the nucleus in a pattern unique to carbon atoms. Likewise, the electron pattern around a nitrogen atom with seven protons in its nucleus and seven electrons distributed around it is unlike that of carbon and unique to nitrogen.

One of the most obvious differences between carbon and nitrogen is that a large quantity of carbon atoms forms a solid at room temperature, whereas a large quantity of nitrogen atoms forms a gas. The reasons for this have to do with their different electron distributions. When atoms come near one another, they interact with their electron clouds; the nuclei remain separated by relatively large distances. The attractions that form different compounds or cause a substance to be a gas, liquid, or solid at a particular temperature depend on the nature of the electron distributions around the interacting atoms.

In summary, the chemical properties of an element are primarily determined by the number of electrons it contains, and the number of electrons is equal to the number of positively charged protons in the nucleus.

10.2.2 Properties of an Atomic Nucleus

We have seen that an element is defined by the number of protons in its nucleus. What about the neutrons in the nucleus, what do they do? Since neutrons are not charged, they cannot attract or repel electrons and therefore do not affect the number or distribution of electrons around the nucleus. Because the chemical properties of each element are determined by that element's electron distribution, neutrons in the nucleus do not influence the chemical properties of the element. Thus, two different nuclei with the same number of protons but different numbers of neutrons have the same chemical properties and are the same element. However, they differ in their masses because of their different number of neutrons. Such atoms are called different isotopes of the same element. We will see that neutrons serve to hold protons together in a nucleus against the repulsive forces between the positive electrical charges of protons.

Most of the naturally occurring elements are mixtures of several isotopes, for example, a gram of pure carbon is a collection of carbon atoms, each of which contains exactly 6 protons but different numbers of neutrons. While most of the carbon atoms contain 6 neutrons, others may contain anywhere between 3 to 10 neutrons. There are 8 known isotopes of carbon, of which 6 are radioactive. The term "nuclide" refers to the nucleus of a particular isotope. Collectively, all the isotopes of all the elements form the set of nuclides. The distinction between the terms "isotope" and "nuclide" is somewhat blurred, and they are often used interchangeably. Isotope is best used when referring to several different nuclides of the same element and when the chemistry of the element is of interest as well as its isotope-specific nuclear properties. Nuclide is more generic and is used when referencing only one nucleus or several nuclei of different elements and the emphasis is mainly on nuclear properties.

10.2.2.1 Nuclear Notation

- The number of protons in a nucleus is called either the atomic number or the proton number and is designated by Z.
- The number of neutrons in a nucleus is called the neutron number and is designated by N.
- The sum of protons and neutrons in a nucleus is called the mass number* and is designated by A.

---

* The sum of neutrons and protons is called the mass number because, in an atom, neutrons and protons are generally considered to be of equal mass, 1 u each, while the mass of the electrons is neglected. See Section 10.2.3 for additional discussion of using relative masses rather than absolute masses for protons and neutrons.

The symbolic representation of the nucleus of an element is $^A_Z X$, where X is the chemical symbol of the element. The number of neutrons in X is found from $N = A - Z$. Note that there is some redundancy in this notation because only Z or X is needed to define an element, but not both. For this reason, Z is sometimes omitted and the nucleus may be written as $^A X$. If needed, Z can be obtained from the periodic table.

Examples of nuclear notation:

- $^4_2 He$ is the nucleus of the most abundant* isotope of the element helium (He), with two protons and two neutrons ($N = A - Z = 4 - 2 = 2$).
- $^{56}_{26} Fe$ is the nucleus of the most abundant isotope of the element iron (Fe), with 26 protons and 30 neutrons ($N = A - Z = 56 - 26 = 30$).
- $^{58}_{26} Fe$ is the nucleus of a less abundant isotope of the element iron (Fe), with 26 protons (Fe) and 32 neutrons ($N = A - Z = 58 - 26 = 32$).

RULES OF THUMB

1. All nuclei are composed of two types of particles: protons and neutrons.
2. The number of electrons in an atom equals the number of protons in its nucleus, making the atom electrically neutral.
3. The atomic number (or proton number), Z, equals the number of protons in the nucleus.
4. The neutron number, N, equals the number of neutrons in the nucleus.
5. The mass number, A, equals the total number of nucleons (protons plus neutrons) in the nucleus.
6. The nuclei of all atoms of a particular element must have the same number of protons but can contain different numbers of neutrons.
7. Isotopes are different forms of the same element, having the same number of protons (same atomic number: Z) but different numbers of neutrons.

10.2.3 ISOTOPES

Different isotopes of the same element have different mass numbers ($A = Z + N$) because they have the same number of protons but different numbers of neutrons. A stable isotope is one that does not spontaneously decompose into a different nuclide. With two exceptions, hydrogen-1 ($^1_1 H$) and helium-3 ($^3_2 He$), the number of neutrons is equal or greater than the number of protons in the stable nuclides.

---

* The percent natural abundance (PNA) of a particular isotope in a given sample of an element is

$$PNA = \frac{\text{Number of atoms of the particular isotope in the sample}}{\text{Total number of atoms of all isotopes of that element in the sample}} \times 100\%$$

The sample being measured must be a naturally occurring sample of the element as found on Earth. Natural abundances can vary over a wide range. For example, the natural abundances of the stable isotopes of oxygen are 99.759% for $^{16}_8 O$ (oxygen-16), 0.037% for $^{17}_8 O$ (oxygen-17), and 0.204% for $^{18}_8 O$ (oxygen-18).

For convenience, when comparing relative atomic masses, as when analyzing mass spectral data, an atomic mass unit, or u,* is defined to be exactly one-twelfth of the mass of a single atom of the most abundant isotope of carbon, $^{12}_{6}C$. This definition is used because it results in the most precise mass spectrometer determination of the relative masses of other isotopes. Since carbon-12 contains 6 protons, 6 neutrons, and 6 electrons, the definition of a u implies that protons and neutrons are considered to be of equal mass, 1 u each, while the mass of the electrons is neglected. The mass of any nucleus in atomic mass units (u) is equal to its mass number, $A = N + P$.

There are about 118 different elements, while the number of different isotopes identified so far is about 3000, of which only about 266 are stable.[†] Clearly, most elements are a mixture of several isotopes. Most elements with proton numbers between 1 and 82 have at least two stable isotopes, a few have only one, and there are many with more than two (e.g., tin has 10 stable isotopes). All isotopes with proton numbers greater than 82 are unstable and radioactive. If the number of neutrons N is plotted as a function of the number of protons Z in the nuclei of each of the approximately 266 stable isotopes, Figure 10.2 results. Thousands of unstable (radioactive) isotopes are not included in Figure 10.2.

Figure 10.2 shows a dependence of the stability of a nucleus on its neutron-to-proton ratio (N/Z). Figure 10.2 also reveals some interesting relationships between the numbers of protons and neutrons in a stable nucleus and the abundance of the corresponding isotope. This data has been used to develop theoretical models for the internal structure of nuclides.

- There is a zone of stability (shaded area in Figure 10.2) within which all stable nuclei lie. If a nucleus has an N/Z ratio too large or too small, so that it falls outside the stable zone, it will be unstable and radioactive.
- For the lighter stable elements, from $Z = 1$ (hydrogen, $^{1}_{1}H$) to about $Z = 20$ (calcium, $^{40}_{20}Ca$), the number of neutrons in the most abundant isotope is approximately equal to the number of protons, that is, the slope of a best-fit line through the lighter stable isotopes is close to unity.
- For stable elements heavier than calcium, the best-fit line bends noticeably upward, away from the $N = Z$ line. As the number of protons increases, the ratio of neutrons to protons needed to produce a stable nucleus also increases to a maximum of about 1.5–1.

---

* The atomic mass unit (u, shortened from amu) is also called the dalton (Da); see footnotes in Chapter 1, Section 1.3.4, *,[†] on page 16 and * on page 17, for more information.

[†] The number of identified stable isotopes depends on how stability is defined because experimental methods are currently capable of measuring radioactive decay half-lives as long as 10^{19} years, which is about a billion times longer than the current estimated age of the universe, 13.7×10^9 years. Several isotopes once thought to be completely stable have been shown in recent years to be slightly radioactive with very long half-lives. An example is $^{209}_{83}Bi$ (bismuth-209), traditionally regarded as the element with the heaviest stable isotope. However, in 2003, bismuth-209 was shown to be an α emitter with a half-life of 19×10^{18} years (de Marcillac et al. 2003). Although such long-lived isotopes may be regarded as stable for any practical purpose, their instability is of great theoretical interest. Bismuth-209 and several other isotopes with comparably long half-lives are often treated as stable and are still included within the stable zone in Figure 10.2.

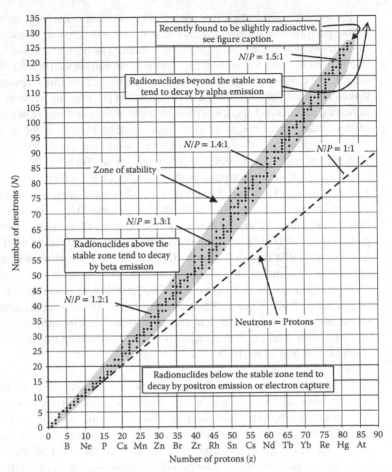

The figure contains the following labels:

- Recently found to be slightly radioactive, see figure caption.
- $N/P = 1.5:1$
- Radionuclides beyond the stable zone tend to decay by alpha emission
- $N/P = 1.4:1$
- $N/P = 1:1$
- Zone of stability
- $N/P = 1.3:1$
- Radionuclides above the stable zone tend to decay by beta emission
- $N/P = 1.2:1$
- Neutrons = Protons
- Radionuclides below the stable zone tend to decay by positron emission or electron capture
- Number of neutrons (N)
- Number of protons (z)
- 0 5 10 15 20 25 30 35 40 45 50 55 60 65 70 75 80 85 90
- B Ne P Ca Mn Zn Br Zr Rh Sn Cs Nd Tb Yb Re Hg At

FIGURE 10.2 Plot of the neutron/proton ratio (N/Z) for 266 stable nuclei. Unstable nuclei are not shown. For nuclei with 20 or fewer protons, stable nuclei have N/Z close to unity. As Z increases beyond 20, nuclei require an increasing N/Z ratio to be stable. There are no known stable nuclei with more than 82 protons. (Bismuth, with 83 protons, was once thought to be stable but has recently been found to be slightly radioactive; see footnote on the previous page). For nuclei with 82 or fewer protons, the envelope of the dots (shaded area) represents a zone of stability. Nuclei with an N/Z ratio either too large or too small to lie within the zone of stability are unstable (radioactive).

- The maximum number of protons in a stable nucleus appears to be 82 (in the three stable isotopes of lead, $^{206}_{82}Pb$, $^{207}_{82}Pb$, $^{208}_{82}Pb$; see footnote † on page 389). All nuclei with 83 or more protons are unstable (radioactive).
- If we classify all nuclides by whether their numbers of protons and neutrons are even or odd, four groups are evident:
  - Even Z and even N (e.g., $^{32}_{16}S$, $^{12}_{6}C$); this group contains more than half of all stable nuclides.
  - Odd Z and odd N (e.g., $^{14}_{7}N$, $^{2}_{1}H$); this group contains the fewest stable nuclides, and $P = N$ in all of the odd Z–odd N stable nuclides.

- Even Z and odd N (e.g., $^{13}_{6}C$, $^{67}_{30}Zn$); this group contains about 20% of the stable nuclides.
  - Odd Z and even N (e.g., $^{19}_{9}F$, $^{63}_{29}Cu$); this group contains about 16% of the stable nuclides.
- The natural abundances of isotopes (see footnote * on page 388) can vary over a wide range. Nuclides with an even number of protons or neutrons, or both, are the most abundant, indicating that even numbers of nucleons impart an increased probability of formation.
- Certain numbers of protons and neutrons are especially favored combinations for forming a nuclide. These numbers, all even, are called magic numbers. They are 2, 8, 20, 26, 28, 50, 82, and 126. The analogy between the nucleon magic numbers and the unusual stability of elements with filled electron shells (the noble gases with 2, 10, 18, 36, 54, and 86 electrons) has led to the development of theories of a nuclear shell structure for nuclides. The most abundant nuclides have Z or N numbers that correspond to the magic numbers. Nuclei that have both Z and N equal to one of the magic numbers are called "doubly magic" and are especially abundant. Some examples of doubly magic isotopes are helium-4 ($^{4}_{2}He$), oxygen-16 ($^{16}_{8}O$), calcium-40 ($^{40}_{20}Ca$), calcium-48 ($^{48}_{20}Ca$), tin-100 ($^{100}_{50}Sn$), and lead-208 ($^{208}_{82}Pb$). Helium-4 and oxygen-16 are the second and the third most abundant isotopes in the universe, after hydrogen-1 ($^{1}_{1}H$, with zero neutrons).
- The zone of stability seen in Figure 10.2 contains all of the stable nuclides. However, some nuclides that lie within the stable zone are not stable and all these have an odd number of protons, an odd number of neutrons, or both. Examples are technetium (Tc, $Z = 43$) and promethium (Pm, $Z = 61$), which have no stable isotopes. There are also nuclides like argon and potassium that have both stable and unstable nuclides within the zone of stability. Argon ($Z = 18$) has an even number of protons and has stable isotopes with even numbers of neutrons (Ar-36, Ar-38, and Ar-40), but its isotopes with odd numbers of neutrons (Ar-37 and Ar-39) are unstable. Potassium ($Z = 19$) has an odd number of protons and has stable isotopes with an even number of neutrons (K-39 and K-41), but its isotopes with odd numbers of neutrons (K-38, K-40, and K-42) are unstable. Two other potassium isotopes with an even number of neutrons (K-37 and K-45), which nevertheless are unstable, lie just outside the zone of stability.

10.2.4 Nuclear Forces

Why do nucleons remain assembled together in a nucleus at all? The existence of both stable and radioactive nuclides is evidence that sometimes they do and sometimes they do not. Some nuclides appear to be completely stable, some have very long half-lives (they hold together for long periods of time; thousands to billions of years), and some have very short half-lives (they hold together for very short periods of time; days to fractions of a second).

Protons are packed so closely together in an atomic nucleus that the coulombic repulsive force between them, which varies inversely with the square of the distance between charges of the same sign, is very strong. Another force must be present that is attractive and strong enough to hold the nucleus together. This very strong force is called the nuclear force. The nuclear force is a strong attraction between all nucleons, whether they are protons or neutrons. It is neither electrical nor gravitational in nature, is always attractive, and is a short-range force, acting only over very small distances (about 10^{-13} cm). When protons or neutrons are within about 10^{-13} cm of each other, the nuclear force binds them together strongly, overcoming the electrostatic repulsion between protons.

The nuclear force has the following important properties:

- It is extremely strong, much stronger than gravitational or electrical forces.
- It has a very short range, about 10^{-13} cm, and can become saturated; one nucleon can only exert its nuclear force on a limited number of other nucleons.
- It is always attractive and is charge independent; for nucleons within the 10^{-13} cm effective range, the force is equally as strong between two neutrons, two protons, or a proton and a neutron. However, although two neutrons or a proton and a neutron can only attract each other because they experience only nuclear and not coulombic forces, two protons also have a coulombic repulsion, which can negate the attraction of their nuclear force under certain conditions.
- Although the nuclear force is much stronger than electrostatic forces (coulombic attraction and repulsion) at very small distances, electrostatic forces are effective over much longer distances. In the case of two protons alone, the nuclear force cannot hold them together against their coulombic repulsion. There are no stable nuclei consisting of two or more protons with no neutrons.
- Because each neutron in a nucleus adds additional forces of attraction to every nucleon within its attractive range without adding any electrostatic repulsion, their presence in the nucleus is very important for holding the protons together. Add one neutron to an unstable two-proton nucleus (item 4 above) and the stable helium-3 nuclide is formed, although its natural abundance ($1.3 \times 10^{-4}\%$) is very low. Add two neutrons and the very stable helium-4 nuclide is formed, with almost 100% abundance.
- The fact that electrostatic proton–proton repulsive forces are long range and influence all the protons in the nucleus, while nuclear forces are short range and saturate with only a few nucleons, gives rise to the important observation that as the number of protons in a nucleus becomes greater, a relatively greater number of neutrons are needed to stabilize the nucleus. This can be seen in Figure 10.2, where the zone of stability curves upward with increasing proton number.

RULES OF THUMB

1. Light nuclei (up to about $Z = 20$, $^{40}_{20}\text{Ca}$) are stable with approximately an equal number of protons and neutrons.
2. Heavier nuclei require more neutrons than protons to be stable because the attractive nuclear force is short range and saturates, while repulsive coulombic force between protons is long range and does not saturate. As the number of protons increases, the coulombic repulsion increases rapidly and more and more neutrons are needed to hold the nucleus together.
3. At $Z = 83$, the repulsive force of 83 protons cannot be negated by adding more neutrons. All nuclei with $Z = 83$ or greater are unstable (radioactive). The maximum number of protons in a stable nucleus appears to be 82. The nuclide $^{208}_{82}\text{Pb}$ has the distinction of being the stable nuclide with the largest mass number and the largest atomic number. All nuclei with $Z \geq 83$ or $N \geq 126$ are unstable (radioactive).
4. All elements with atomic numbers between 83 (bismuth) and 92 (uranium) are naturally occurring unstable radionuclides (on Earth). Many elements with $Z > 82$ have been created in laboratory experiments (up to $Z = 118$ at the time of publication). All of these have half-lives too short to have persisted to the current time if they were present during the formation of Earth (see Section 10.2.10).
5. There are three causes of radioactivity related to the neutron/proton ratio in an atomic nucleus. A nucleus will be radioactive if:
   a. There are more than 82 protons or more than 126 neutrons in the nucleus.
   b. There are 82 or fewer protons in the nucleus but the neutron/proton ratio is too low or too high to lie within the zone of stability.
   c. A few isotopes have neutron/proton ratios within the zone of stability but are unstable because they contain odd numbers of both neutrons and protons.

10.2.5 QUARKS, LEPTONS, AND GLUONS

This section offers just a taste of the ever-deepening complexity of nuclear physics. It is not needed for using any other part of this book and is included only as a reminder that complete, in-depth knowledge of a topic is seldom needed to accomplish something useful in engineering and science.

For about 30 years after their discovery in the early 1900s, protons and neutrons, along with electrons, were believed to be the fundamental particles of matter.

However, studies of radioactivity and high-energy particle physics eventually revealed that matter could be subdivided still further. Two early observations started the search for an inner structure of protons and neutrons.

1. Certain radioactive nuclides were observed to emit negatively charged particles identical to electrons, called β particles. How could negative particles come from a nucleus consisting of protons and neutrons?
2. Neutrons in the free state, after being emitted from nuclei in nuclear reactions, are not stable. A free neutron decays into an electron and a proton, with a half-life of about 13 minutes.

The emission of an electron from a nucleus could be (and was at first) regarded as the decay of a confined neutron into a proton and an electron. The electron is released from the nucleus while the proton is retained in the nucleus. Emission of an electron from a radioactive nucleus effectively changes a nuclear neutron into a proton, making the nucleus more positive by one charge unit. This addition of a proton to the nucleus increases an atom's atomic number by one unit without changing its mass number. For a time, it was proposed that the neutron was actually a combination of an electron and a proton. However, further studies revealed greater complexity.

There appear to be three families of more fundamental particles, called quarks (six different kinds), leptons (six different kinds), and force carriers (the photon is the force carrier particle that carries electromagnetic forces). The strong nuclear force is carried by a force carrier particle called a gluon. The electron is the most familiar lepton. Protons and neutrons contain three quarks each, of just two different kinds (named *up quark*, or *u*, and *down quark*, or *d*). The three quarks in a neutron are a *udd* combination and the three quarks in a proton are a *uud* combination. When a neutron decays, one of its *down quarks* is transformed into an *up quark*. In this process, the neutron becomes a proton and conservation of charge and momentum is preserved by the creation of an electron and an antineutrino.

10.2.6 RADIOACTIVITY

An unstable nuclide cannot hold all of its nucleons together indefinitely. Eventually, the nucleus will change its internal structure by losing energy in the form of high-energy photon radiation or losing mass and energy by releasing one or more nucleons as energetic particles, a process called radioactive decay or radioactivity. After a decay event, the new nuclear structure may or may not be stable; if not, the release of photons and nucleons will continue. The timing of each decay event is unpredictable but, after many events, the rate of decay is statistically predictable and consistent for each radionuclide. The photons and nucleons released are collectively called radiation or emissions. Radioactivity is the result of an unstable nucleus rearranging its nucleons by emitting radiation, a process that continues until a stable nuclear configuration is achieved.

The most common forms of radionuclide emissions are named after the first three letters of the Greek alphabet—alpha (α), beta (β), and gamma (γ).* These emissions accompany the process of transmutation, a nuclear reaction in which an unstable isotope of one element, called the parent isotope, is transformed into an isotope of a different element (possibly still radioactive, possibly not), called a daughter isotope.

All known nuclides with $Z > 82$ (Pb) are radioactive and most of these undergo α decay, which is the radiation that lowers the value of Z most efficiently. Radioactive nuclides with $Z \leq 82$ are those with a neutron/proton ratio that does not fall within the zone of stability (Figure 10.2) or, in a few cases, nuclides with an odd number of both neutrons and protons. Isotopes with too many neutrons decay by β emission, which converts a neutron into a proton, decreasing the N/Z ratio. Isotopes with too few neutrons decay by positron emission or electron capture, which changes a proton into a neutron, increasing the N/Z ratio.

10.2.6.1   Alpha (α) Emission

Alpha (α) emission is ejection of an α particle from a nucleus of an atom. The α particle is a nuclide unit consisting of two protons and two neutrons. After emitting an α particle, the nucleus has lowered its atomic number by two units and its mass number by four units. For example, after emission of an α particle, the nuclide uranium-238, $^{238}_{92}U$, becomes thorium-234, $^{234}_{90}Th$.

An α particle is identical to a helium-4 nucleus, $^{4}_{2}He$, and will become a helium atom when it comes to rest and acquires two electrons from its surroundings. It carries two positive charges and has a mass number of four, the largest and heaviest particle emitted by natural radioactivity. The fact that these four nucleons are emitted as a single unit from a radioactive nucleus testifies to the unusual stability of the nuclear combination of two protons and two neutrons.

10.2.6.2   Beta (β) Emission

Beta (β) emission is the ejection of a β particle (an electron) and an antineutrino from a nucleus. Beta decay changes a neutron into a proton. Since the simplest model of a nucleus states that a nucleus contains only protons and neutrons, β decay was one of the first indications that neutrons could be regarded as being a combination of an electron and a proton. The term "beta particle" is a historical term used in the early description of radioactivity. A nucleus that has too many neutrons for stability can decrease the N/Z ratio by emitting an electron in β decay.

---

* Other less common forms of radioactivity are not regulated by the EPA because they normally are not hazards to health or the environment. They include

- Electron capture, where a parent nucleus captures one of its orbital electrons and emits a neutrino. This converts a nuclear proton to a neutron and lowers the nuclide's atomic number by one unit.
- Positron emission, also called positive β ($\beta+$) decay, where a positron (antielectron) is emitted. Positron emission has the same effect on the nucleus as electron capture, decreasing the atomic number by one unit.
- Internal conversion, where electric fields within the nucleus interact with orbital electrons, resulting in the ejection of an orbital electron from an outer shell of the atom.

TABLE 10.1

Nuclear Changes Caused by Radioactive Emissions

Type of Emission	Symbol	Mass Number (A) Change	Atomic Number (Z) Change	Example
Alpha, α	^4_2He	Decreases by 4	Decreases by 2	$^{226}_{88}\text{Ra} \rightarrow {}^{222}_{86}\text{Rn} + {}^4_2\text{He}$
Beta, β	$^0_{-1}\text{e}$	No change	Increases by 1	$^{14}_6\text{C} \rightarrow {}^{14}_7\text{N} + {}^0_{-1}\text{e}$
Gamma, γ	$^0_0\gamma$	No change	No change	$^{12}_5\text{B} \rightarrow {}^{12}_6\text{C}^* + {}^0_{-1}\text{e}$, followed by $^{12}_6\text{C}^* \rightarrow {}^{12}_6\text{C} + {}^0_0\gamma$
Positron	$^0_{+1}\text{e}$	No change	Decreases by 1	$^{38}_{19}\text{K} \rightarrow {}^{38}_{18}\text{Ar} + {}^0_{+1}\text{e}$
Electron capture	EC	No change	Decreases by 1	$^{123}_{52}\text{Te} + {}^0_{-1}\text{e} \rightarrow {}^{123}_{51}\text{Sb}$
Neutron	^1_0n	Decreases by 1	No change	$^{13}_4\text{Be} \rightarrow {}^{12}_4\text{Be} + {}^1_0\text{n}$

Notes:
1. C* signifies a carbon nucleus in an excited energy state.
2. The principles of balancing nuclear equations, as in the Example column, are explained in Section 10.2.7.
3. Electron capture creates a vacancy in the inner orbital electron orbital, leaving the orbital electrons in an excited energy state. As orbital electrons cascade downward to return to the ground energy state, electromagnetic photons, including x-rays, are emitted.

10.2.6.3 Gamma (γ) Emission

Gamma (γ) emission usually occurs when the prior emission of an α or β particle leaves the nucleus in an excited energy state. It can then relax to the more stable ground state by emitting a high-energy γ photon (see Table 10.1). Gamma radiation is the highest energy form of electromagnetic radiation because it results from transitions between widely spaced nuclear energy levels. Next on the electromagnetic energy scale are x-rays and ultraviolet light, which result from transitions between more closely spaced energy levels of the orbital electrons.

10.2.7 BALANCING NUCLEAR EQUATIONS

A nuclear equation describes the nuclear changes that occur because of radioactivity. The examples in Table 10.1 are all balanced nuclear equations. The rules for balancing nuclear equations are very simple:

- In a nuclear equation, the sum of the mass numbers (A) on both sides of the equation must be equal, to establish conservation of mass.
- The sum of the charge numbers for nuclides, electrons, neutrons, and gammas must be equal on both sides of the equation to establish conservation

of charge. Note that the atomic number (Z) is actually the nuclear charge number so that balancing the atomic numbers is the same as balancing the nuclear charges.

- For each particle in the equation, the chemical symbol (X), mass number (A), and atomic number (Z) are used in the form $^A_Z X$. For nonnuclides like electrons, protons, neutrons, and gammas, the charge number is the same as the atomic number for nuclides and is the lower-left subscript.

Example 1

Write the balanced equation for a nuclear reaction in which uranium-238 emits an α particle to form thorium-234.

ANSWER

Uranium has atomic number 92 and thorium has atomic number 90. Therefore,

$$^{238}_{92}U \rightarrow\ ^{234}_{90}Th + ^4_2He$$

The mass numbers balance: 238 on the left $= 234 + 4$ on the right.
The charge numbers balance: 92 on the left $= 90 + 2$ on the right.

Example 2

When nitrogen-14 in the upper atmosphere absorbs a neutron that enters the atmosphere from outer space, the nuclear reaction forms carbon-14. What other particle must also be formed?

ANSWER

$$^{14}_{7}N + ^1_0n \rightarrow\ ^{14}_{6}C + ?$$

The unknown particle must have a mass number of 1 so that the mass numbers on both sides of the equation are equal to 15 and a charge number of 1 so that the charge numbers on both sides of the equation are equal to 7. The particle with a mass number of 1 and a charge number of 1 is the proton, written ^1_1H (or ^1_1p). The balanced equation is

$$^{14}_{7}N + ^1_0n \rightarrow\ ^{14}_{6}C + ^1_1H$$

Example 3

The carbon-14 formed in Example 2 is radioactive, emitting a β particle. Its use in carbon age dating is discussed in Example 7. Write the balanced equation that identifies the product nuclide of carbon-14 decay.

ANSWER

$$^{14}_{6}C \rightarrow \, ^{14}_{7}? + \, ^{0}_{-1}e$$

The unknown product must have an atomic number of 7 so that $-1 + 7$ (on the right) $= 6$ (on the left). The element with atomic number 7 is nitrogen and the balanced equation is

$$^{14}_{6}C \rightarrow \, ^{14}_{7}N + \, ^{0}_{-1}e$$

Example 4

Complete and balance the following nuclear reaction, which could occur in a particle accelerator.

$$^{250}_{98}Cf + \, ^{11}_{5}B \rightarrow \, ? + 5\,^{1}_{0}n$$

ANSWER

The sum of mass numbers on the right must equal 261, to be the same as the sum of mass numbers on the left. The sum of charge numbers on the right must equal 103, to be the same as the sum of charge numbers on the left. Since the five neutrons on the right have a total mass number of 5 and a total charge number of 0, the unknown nuclide must have a mass number of 256 and a charge, or atomic, number of 103. The element with an atomic number of 103 is Lawrencium so that the unknown nuclide is $^{256}_{103}Lr$. The balanced equation is

$$^{250}_{98}Cf + \, ^{11}_{5}B \rightarrow \, ^{256}_{103}Lr + 5\,^{1}_{0}n$$

Example 5

Write a nuclear reaction for the α decay of polonium-210.

ANSWER

The charge number is 84 for polonium and 2 for an α particle ($^{4}_{2}He$). The mass number is 210 for the polonium isotope and 4 for an α particle. Therefore, the other product nucleus has a mass number of $210 - 4 = 206$ and a charge number of $84 - 2 = 82$. The product is lead-206. The balanced equation is

$$^{210}_{84}Po \rightarrow \, ^{206}_{82}Pb + \, ^{4}_{2}He$$

10.2.8 RATES OF RADIOACTIVE DECAY

The rate of radioactive decay of a nuclide can only be determined by counting the number of decay events occurring in a given time. Decay rates are most easily measured with pure samples to ensure that only one kind of emitting nuclide is present. However, even a sample that begins pure will become contaminated by radioactive daughter products. When more than one kind of emitter is present, their total emissions have to be sorted out by identifying the different kinds of particles, their characteristic energies, and their different rates of emission. In the discussion that follows, it is assumed that a pure nuclide is the source of all emissions. Under these conditions, the rate of radioactive decay is the number of disintegrations per unit time, which is equal to the decrease in the number of parent radioactive nuclei per unit time.

Radioactive decay follows a first-order rate law, which means that the rate of decay of a given radionuclide at any time is directly proportional to the number of radioactive nuclei remaining at that time. This means, of course, that the radioactivity of an isolated nuclide must diminish over time. Mathematically, the rate equation is written

$$\text{Rate} = \frac{dN}{dt} = -kN \tag{10.1}$$

where

N = the number of radioactive nuclei present.
t = time.
k = a constant proportional to the rate of decay, known as the decay rate constant. k has a unique value for each radioisotope, which is determined by measuring that isotope's half-life (see Section 10.2.8.1). The minus sign indicates that N decreases with time.

Integrating Equation 10.1 gives

$$\ln\left(\frac{N}{N_0}\right) = -kt \tag{10.2}$$

where

N_0 = initial number of radioactive nuclei at the start of a measurement
N = the number of radioactive nuclei remaining after a time t.

10.2.8.1   Half-Life

For radioactive decay, it is usual to express the rate in terms of the half-life. The half-life is the time required for half of the radioactive nuclei initially present at any time to undergo disintegration.

For example, if there were 10,000 radioactive nuclei present in a sample at the start of a measurement, the half-life is the time it takes for the sample to decay until there are only 5000 of the original radioactive nuclei remaining. Equation 10.2 can be used to derive a simple expression for the rate constant in terms of half-lives. Let $t_{1/2}$ be the time required for half of the initial nuclei to decay, that is, the half-life. When one half-life of the nuclide being measured has elapsed, the number of remaining nuclei, N, will equal half of the initial number of nuclei present at the start of the measurement, N_0, or $N = (1/2)N_0$.

The integrated Equation 10.2 then becomes, $\ln\left(\dfrac{\frac{1}{2}N_0}{N_0}\right) = \ln \frac{1}{2} = -0.693 = kt_{1/2}$ so that

$$k = \frac{0.693}{t_{1/2}} \tag{10.3}$$

and

$$\ln\left(\frac{N}{N_0}\right) = -\left(\frac{0.693}{t_{1/2}}\right)t \tag{10.4}$$

The use of Equations 10.3 and 10.4 is shown in the examples that follow.

Example 6

Measurements on a sample containing iodine-131 indicate an initial activity of 3153 disintegrations per minute (dpm). After 52.5 hours, the activity has fallen to 2613 dpm. What is the half-life of $^{131}_{53}I$? The activity of a sample is directly proportional to N, the number of radioactive nuclei present at the time of measurement.

ANSWER

Insert the values into Equation 10.4:

$$\ln\left(\frac{N}{N_0}\right) = -\left(\frac{0.693}{t_{1/2}}\right)t$$

$$\ln\left(\frac{2613}{3153}\right) = \ln(0.8287) = -\left(\frac{0.693}{t_{1/2}}\right)(52.5 \text{ hours})$$

$$t_{1/2} = -\frac{0.693 \times 52.5 \text{ hours}}{\ln(0.8287)} = 194 \text{ hours}$$

Example 7

$^{14}_{6}C$ is a β emitter with a half-life of 5730 years (y). It is formed when $^{14}_{7}N$ in the upper atmosphere absorbs a neutron from outer space. Radioactive carbon dating assumes that formation and decay of carbon-14 are in equilibrium and that the concentration of carbon-14 in the atmosphere has been constant, proportional to 15.3 dpm, for the past several 100,000 years. Plants incorporate carbon (mainly ^{12}C, ^{13}C, and ^{14}C) from atmospheric CO_2 into their body structures and animals eat the plants. While plants and animals are alive, the ^{14}C disintegration rate in their tissues is the same as in the atmosphere (15.3 dpm) and remains constant because there is continual carbon exchange with the environment. But when living species die, there is no further carbon exchange with the environment and the ^{14}C concentration in their tissues begins to decrease because of ^{14}C decay. The ^{14}C activity in the tissues of once living species can be used to estimate the elapsed time since death.

A piece of wood found in an archeological digging site had a ^{14}C count rate of 9.8 dpm. Estimate how long this wooden sample has been buried.

ANSWER

Let N be the current ^{14}C count rate of 9.8 dpm. Assume that the wood originally had a ^{14}C count rate, N_0, of 15.3 dpm, the current atmospheric ^{14}C disintegration rate. Then, the age of the sample is found from Equation 10.3:

$$\ln\left(\frac{N}{N_0}\right) = -\left(\frac{0.693}{t_{1/2}}\right)t$$

$$\ln\left(\frac{9.8}{15.3}\right) = \ln(0.641) = -0.445 = -\left(\frac{0.693}{5730 \text{ y}}\right)(t)$$

$$t = -\frac{5370 \text{ y} \times (-0.445)}{0.693} = 3446 \text{ y}^{-1}$$

Example 8

It is a rule of thumb that a radioactive sample is effectively "gone" after about 15 half-lives. What fraction of the original activity is left after 15 half-lives?

ANSWER

$$\text{Let } t = 15t_{1/2}. \text{ Then, } \ln\left(\frac{N}{N_0}\right) = -\left(\frac{0.693}{t_{1/2}}\right)t = -\left(\frac{0.693}{t_{1/2}}\right)(15t_{1/2}) = -10.4$$

N/N_0 = antiln(−10.4) = $e^{-10.4}$ = 3.06 × 10^{-5} or about 0.003% of the original activity remains.

Example 9

The half life of ^{14}C is 5720 years (y). Calculate the weight in grams, W, of ^{14}C that would have an activity (emission rate) of 1 mCi (3.700 × 10^7 dps).

ANSWER

By definition, activity (A) equals number of disintegrations per second, or dN/dt. Set A equal to Equation 10.1.

$$A = \frac{dN}{dt} = -kN$$

Step 1: Use the known half-life of ^{14}C with Equation 10.3 to solve for k, the decay rate constant.

$$t_{1/2} = \frac{0.693}{k} = 5720 \text{ y}$$

$$k = \frac{0.693}{t_{1/2}} = \frac{0.693}{5720 \text{ y}} = 1.21 \times 10^{-4} \text{ y}^{-1}$$

Step 2: Solve for N, the number of ^{14}C atoms present.

$$N = \frac{\text{grams of } ^{14}C}{\text{mass number of } ^{14}C} \times \text{Avogadro's number} = \frac{W}{14} \times 6.02 \times 10^{23}$$

Step 3: Solve for A.

$$A = \frac{1.21 \times 10^{-4} \text{y}^{-1}}{3.15 \times 10^{7} \text{s y}^{-1}} \times \frac{W}{14} \times 6.02 \times 10^{23} = 1.65 \times 10^{12} \text{s}^{-1} \times W$$

Step 4: Set the activity to 1 mCi and solve for W.

$$W = \text{grams of } ^{14}C \text{ emitting 1 mCi} = \frac{3.70 \times 10^{7} \text{s}^{-1}}{1.65 \times 10^{12} \text{s}^{-1} \text{g}^{-1}} = 2.24 \times 10^{-5} \text{g}$$

10.2.9 RADIOACTIVE DECAY SERIES

A radioactive nucleus seeks greater stability by spontaneous emission of nuclear particles. After a particle is emitted, the original nuclide has changed the neutron and proton content of its nucleus and has become a different nuclide.

- If the new nuclide resulting from a radioactive emission has a stable neutron/proton ratio (see Section 10.2.4), the new nuclide is not radioactive and lies within the zone of stability, as given in Figure 10.2.
- If the new nuclide resulting from a radioactive emission has an unstable neutron/proton ratio, it is also radioactive and eventually will emit another nuclear particle. Each new radioactive nuclide will emit additional radiation until a stable nuclide is finally formed and the decay series ends.
- The particle (α, β neutron, etc.) with the highest probability of being emitted is one that changes the neutron/proton ratio in a way that moves the nuclide most efficiently toward the zone of stability (in Figure 10.2).
- Nuclides with mass numbers that are too large ($A > 208$) can decrease the mass number most efficiently by emitting α particles, the heaviest emission. Therefore, most heavy radioisotopes ($A > 208$) are α emitters. Emission of

an a particle ($^4_2$He) makes Z decrease by 2, N decrease by 2, and A decrease by 4 (e.g., $^{238}_{92}$U $\rightarrow$ $^{234}_{90}$Th + $^4_2$He).

- Nuclides with too many neutrons can decrease the N/Z ratio by emitting an electron in β decay ($_{-1}^0$e), which decreases N by 1 and increases Z by 1 without changing (e.g., $^{14}_6$C $\rightarrow$ $^{14}_7$N + $_{-1}^0$e).

- Nuclides with too few neutrons can increase the N/Z ratio by emitting a positive electron, or positron ($_{+1}^0$e), which increases N by 1 and decreases Z by 1 without changing A (e.g., $^{20}_{11}$Na $\rightarrow$ $^{20}_{10}$Ne + $_{+1}^0$e).

- Often, after radioactive decay, a nucleus is left in an excited energy state. It can relax to the stable ground state by emitting a high-energy γ-ray photon, for example, $^{12}_5$B $\rightarrow$ $^{12}_6$C* + $_{-1}^0$e, followed by $^{12}_6$C $\rightarrow$ $^{12}_6$C + γ.

10.2.10 NATURALLY OCCURRING RADIONUCLIDES

Before 1940, the periodic table ended with uranium, $^{238}_{92}$U. No elements with $Z > 92$ (called transuranic elements) were known to exist because they are all radioactive with half-lives that are short compared with the lifetime of the Earth, currently believed to be about 4.5×10^9 years. Although transuranic elements might have existed during the early years of Earth's history, they have long since decayed away to stable elements with $Z \leq 82$. There are just three radioisotopes found naturally on Earth with half-lives long enough to have persisted since Earth's creation. They are uranium-238 $\left(^{238}_{92}\text{U}, t_{1/2} = 4.67 \times 10^9 \text{ years} \right)$, uranium-235 $^{235}_{92}$U, $t_{1/2} = 7.13 \times 10^8$ years), and thorium-232 $\left(^{232}_{90}\text{Th}, t_{1/2} = 1.39 \times 10^{10} \text{ years} \right)$ All the other naturally occurring radioisotopes found on the Earth today are daughter isotopes in the decay series of these three parent radioisotopes.

Therefore, there are just three naturally occurring radioactive decay series* (Figures 10.3 through 10.5). Each series starts with one of the long-lived parent radioisotopes whose half-life exceeds that of any of its daughter products. A series continues forming one radioactive daughter after another by emitting α and β particles until a stable isotope is finally made. The half-life of each daughter is an indication of its stability; a longer half-life means a more stable nuclide. The three decay series, starting with $^{238}_{92}$U, $^{235}_{92}$U, and $^{232}_{90}$Th, terminate in the stable isotopes of lead $^{206}_{82}$Pb, $^{207}_{82}$Pb, and $^{208}_{82}$Pb respectively.

A series sometimes follows alternative paths in which a main sequence of, say, α emission followed by β emission is reversed to be a β emission followed by α emission. If undisturbed by natural chemical separation processes, such as groundwater dissolving a soluble mineral containing a daughter isotope and carrying it away from the original rock formation, a parent radioisotope and its daughters reach a secular equilibrium. Secular equilibrium occurs when each radioisotope in the decay series is at a concentration where its rate of formation equals its rate of decay. This condition requires that the activity (A, dpm) of each isotope in the series be identical.

---

* See footnote † on page 389. Although bismuth-209 is naturally occurring and radioactive, its half-life is too long (19 × 10^{18} years) for it to produce a detectable series of daughter products. Therefore, it is not included among the naturally occurring decay series.

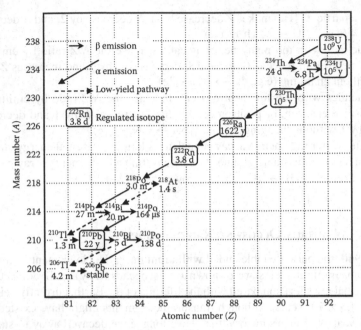

FIGURE 10.3   Uranium-238 natural decay series.

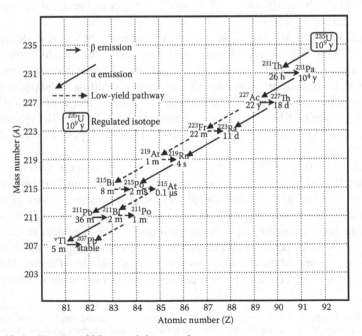

FIGURE 10.4   Uranium-235 natural decay series.

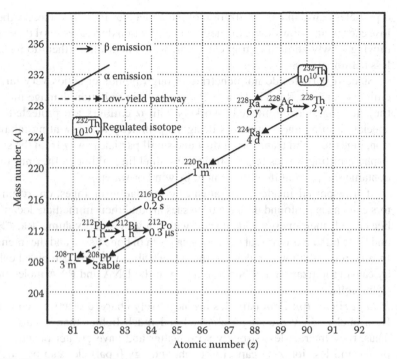

FIGURE 10.5 Thorium-232 natural decay series.

10.3 EMISSIONS AND THEIR PROPERTIES

The health and environmental hazards associated with different radioisotopes are based primarily on the type and energy of their radiation emissions. Except for radon, most of the commonly encountered radioisotopes are heavy metals and these represent an additional risk based on their chemical toxicity.

When radioisotope radiation passes through human tissue, it interacts with the molecules it encounters, losing energy by forcing electrons out of their normal molecular orbitals, resulting in broken bonds, ionization, and configuration changes in the molecules. This can cause damage or death to cells in the tissue. Damage to a cell's reproductive mechanisms can result in abnormal reproductive behavior such as cancer.

Three types of radiations present the greatest environmental concerns because they are produced by radionuclides that are commonly found in minerals and waste products and have a long enough half-life to allow dangerous quantities to accumulate.

1. *Alpha particles* (*α*): Alpha particles are doubly charged helium nuclei. Because they are charged and are relatively heavy (more than 7000 times heavier than β particles), they interact strongly with orbital electrons in matter and lose their kinetic energy over very short distances of travel. In air, a typical α particle may travel about 10 cm, whereas in water or tissues, its range is about 0.05 cm. They are completely stopped by a single sheet of paper or the outer layer of skin. In their passage through matter,

α particles cause intense ionization of the molecules in their wake. As they lose energy and slow down, the degree of ionization decreases until they are moving slowly enough to trap electrons, which neutralizes them to harmless helium atoms.

Since α particles do not penetrate through liquids or solids very far, α emitters (such as Ra, Th, U, and Pu) cause little radiation damage unless they are ingested or inhaled. Workers with α emitters are protected by special clothing and respirators designed primarily for preventing inhalation, ingestion, and transport of dust and small particles away from the site. Alpha-emitting wastes require little or no shielding, although they must be contained to prevent their movement by wind or water.

If transported inside an organism by inhalation or ingestion, α emitters can cause profound damage to tissues around their immediate location because all their energy is deposited within a very small volume. Ra, ^{90}Sr, and ^{133}Ba (all α emitters) substitute for calcium in bone tissue and the intense localized α radiation can destroy the tissue's ability to produce blood cells by causing ionization and bond-breaking in the DNA and RNA molecules of the cells.

2. *Beta particles (β)*: Beta particles are negatively charged electrons emitted from nuclei. They are singly charged and much lighter than α particles. Thus, they interact less strongly with matter and have greater penetrating power and less ionizing capacity. Higher-energy β particles can penetrate skin and travel up to 9 m in air. Although their damage is less localized, their cumulative effects can be as serious as those of α particles. Because their range in matter is so long, β-emitting wastes require shielding by a minimum of 5 mm of aluminum or 2 mm of lead.

3. *Gamma rays (γ) and x-rays*: Gamma and x-rays are photons (electromagnetic radiation), are uncharged, and have no mass. They interact with matter relatively weakly by quantum mechanical processes rather than by collisional impact. They have a much greater penetrating power than α or β particles and deposit their energy over much longer path lengths. Gamma- and x-ray sources require extensive shields to block their emissions. Unlike charged particle radiation, the attenuation of γ radiation by shielding is exponential, and, in principle, there is always some probability that a percentage of γ particles penetrate any thickness of shielding. For this reason, γ shielding is usually described in terms of the thickness required to attenuate γ radiation by a certain factor. For example, 4 in of lead shielding will attenuate 1-MeV γ-rays by a factor of 3200 and 2 MeV γ-rays by a factor of 175.

The damage produced in matter by any of these particles depends on the activity (number of particles emitted per second) and on their energy. A given activity of tritium causes less damage than the same activity of ^{14}C because the β particles from tritium are lower energy than β particles from ^{14}C. Tritium β particles have a maximum energy of 0.0179 MeV, whereas those from ^{14}C are 0.156 MeV. In a sufficient mass of tissue to absorb all the β particles completely, ^{14}C will deposit almost 10 times more energy than the same activity of tritium.

10.4   UNITS OF RADIOACTIVITY AND ABSORBED RADIATION

Concentrations of radionuclides in the environment are typically expressed in terms of activity of the radionuclide per unit of volume of water (e.g., picocuries per liter, or pCi/L), per unit volume of air (e.g., picocuries per cubic meter, or pCi/m^3), or per unit mass of soil of solid (e.g., picocuries per kilogram, or pCi/kg). Activity measures the rate of disintegration of a radionuclide per unit mass or volume. Because the carcinogenic effect of a radionuclide is due to its emissions, concentrations of radionuclides are generally measured in terms of activity for health evaluation purposes. An exception is uranium, which is regulated for both its chemical toxicity and its radioactivity. Uranium has a drinking water limit expressed as a mass concentration (30 µg/L, see Section 10.5.2).

This section summarizes the three basic kinds of radioactivity measurements: (1) emission rate or activity, (2) absorbed dose, and (3) dose delivering a given biological effect. Tables 10.2 through 10.4 provide conversion factors among the most commonly used quantities. Table 10.6 gives calculated conversion factors, for many β-and photon-emitting radionuclides, between the emission rate (becquerels, curies) and the dose delivering a given biological effect (sieverts, rems), for which there are no simple direct conversions. Table 10.7 helps to compare measurements commonly reported in pCi/L with drinking water standards for β and photon emitters commonly given in terms of rems.

10.4.1   ACTIVITY

Activity is the number of disintegrations or emissions per second in a radioactive sample. Its International Standard (SI) unit is the becquerel (Bq) and the conventional unit more commonly used in the United States is the curie (Ci).

TABLE 10.2
Radioactivity Parameters

Quantity	SI Units	Special SI Name/Symbol	Conventional Name/Symbol	Conversion: From Conventional to SI Units
Activity (dps)	s^{-1}	becquerel (Bq)	curie (Ci)	3.7 × 10^{10} Bq/Ci
Absorbed dose (kinetic energy absorbed per unit mass of absorbing matter)	J/kg	gray (Gy)	rad (rad)	0.01 Gy/rad
Dose equivalent	J/kg	sievert (Sv)	rem (rem)	0.01 Sv/rem
Photon exposure	C/kg	C/kg	roentgen (r, R)	2.58 × 10^{-4} C/kg/r

Note: C = coulombs.

TABLE 10.3
Unit Conversions That Apply to Disintegration Rate (Activity)

To Convert From	To	Multiply By
curies (Ci)	picocuries (pCi)	10^{12}
curies (Ci)	nanocuries (nCi)	10^9
curies (Ci)	microcuries (µCi)	10^6
curies (Ci)	millicuries (mCi)	10^3
curies (Ci)	becquerels (Bq)	3.7×10^{10}
curies (Ci)	disintegrations per second (dps)	3.7×10^{10}
picocuries (pCi)	nanocuries (nCi)	10^{-3}
picocuries (pCi)	microcuries (µCi)	10^{-6}
picocuries (pCi)	millicuries (mCi)	10^{-9}
picocuries (pCi)	curies (Ci)	10^{-12}
picocuries (pCi)	becquerels (Bq)	3.7×10^{-2}
picocuries (pCi)	disintegrations per second (dps)	3.7×10^{-2}
picocuries (pCi)	disintegrations per minute (dpm)	2.22
becquerels (Bq)	curies (Ci)	2.7×10^{-11}
becquerels (Bq)	picocuries (pCi)	27
becquerels (Bq)	disintegrations per second (dps)	1
disintegrations per second (dps)	becquerels (Bq)	1

Notes: milli (m) = 10^{-3}; micro (µ) = 10^{-6}; nano (n) = 10^{-9}; pico (p) = 10^{-12}.

TABLE 10.4
Unit Conversions That Apply to Dose and Exposure

To Convert From	To	Multiply By
gray (absorbed dose of 1 J/kg)	rad (absorbed dose of 100 erg/g)	100
gray (absorbed dose of 1 J/kg)	roentgen (R) (exposure dose; 1 R = radiation dose depositing 84 erg/g of air or 93 erg/g of water)	107
rad (absorbed dose of 100 erg/g)	gray (Gy) (absorbed dose of 1 J/kg)	0.01
rad (absorbed dose of 100 erg/g)	roentgen (R) (exposure dose; 1 R = radiation dose depositing 84 erg/g of air or 93 erg/g of water)	1.07
rem (dose equivalent = rad × RBM factor)	sievert (Sv) (dose equivalent = gray × RBM factor)	0.01
sievert (Sv) (dose equivalent = gray × quality factor)	rem (total absorbed dose = rads × quality factor)	100

(Continued)

TABLE 10.4 (*Continued*)
Unit Conversions That Apply to Dose and Exposure

To Convert From	To	Multiply By
roentgen (r, R) (exposure dose; 1 r = radiation dose depositing 84 erg per g of air or 93 erg per g of water)	rad (absorbed dose of 100 erg/g)	0.93
roentgen (r, R) (exposure dose; 1 r = radiation dose depositing 84 erg/g of air or 93 erg/g of water)	gray (Gy) (absorbed dose of 1 J/kg)	0.0093

$$1 \text{ Bq} = 1 \text{ disintegration/s}$$
$$1 \text{ Ci} = 3.700 \times 10^{10} \text{ disintegrations/s}$$

The activity is a measure of the rate of nuclear disintegrations and, therefore, the half-life of the radionuclide; longer half-lives mean lower activity. The activity does not give any information about the kinds of particles emitted or their effects in the environment.

The definition of a curie originally was based on the disintegration rate of 1 g of radium, which is 3.700×10^{10} disintegrations/s. However, now, it is simply defined as the above quantity and is independent of any experimentally determined value.

RULE OF THUMB

$1 \text{ pCi} = 10^{-12} \text{ Ci} \approx 2.2$ disintegrations/min ≈ 1 disintegration every 27 seconds.

10.4.2 ABSORBED DOSE

Absorbed dose measures the amount of energy actually deposited per kilogram of a receiving body, regardless of the type of radiation. Its SI unit is the gray (Gy) and the conventional unit is the rad.

$$1 \text{ Gy} = \text{an absorbed dose of 1 J/kg}$$
$$1 \text{ rad} = \text{an absorbed dose of 100 erg/g} = \text{absorbed dose of 0.01 J/kg } (1 \text{ erg} = 10^{-7} \text{ J})$$

Note that there is no time period specified. Every 100 ergs absorbed per gram of mass is a dose of 1 rad. Thus, rads, which are a dose and not a rate, are cumulative with time and cannot be directly related to curies, which are a constant rate in time.

The number of rads per unit time that correspond to a specified activity depends on the nature of the particles emitted, their energy, and the absorbing, or stopping, power of the matter in which the particles deposit their energy.

The difference between the units of rad (or gray) and curie (or becquerel) is that rads indicate the amount of energy absorbed by matter, whereas curies indicate the number of nuclei disintegrating per second.

The roentgen, abbreviated as r or R, is a dose unit related to photon emissions. It measures the ionizing ability of x-rays and γ-rays. One roentgen is the amount of photon activity that produces 2×10^9 ion pairs in 1 cm^3 of air. It represents an absorbed dose of photon energy where air is the absorbing matter instead of human tissue. In general, the exposure to radiation in roentgen units is numerically approximately equal to the absorbed dose in rads (see Table 10.4).

RULES OF THUMB

1. Alpha particles are heavier and more highly charged than β particles. Therefore, they lose their energy faster to their surroundings and penetrate shorter distances. Beta particles lose their energy over longer paths.
2. Alpha particles produce approximately 20 times more tissue damage per unit length of travel than β particles. Compared to β particles, the relative biological effectiveness (RBE) for α particles = 20, taking RBE for β particles = 1 (see Table 10.5).

10.4.3 DOSE EQUIVALENT

Dose equivalent measures the amount of energy that produces a certain biological effect. It is an empirical quantity that attempts to quantify the fact that the biological hazard from radiation depends on two factors: the amount of energy absorbed by tissues and the type of radiation. The dose equivalent is the absorbed dose multiplied by a quality factor, called the RBE factor. Its SI unit is the sievert (Sv) and the conventional unit is the rem (roentgen equivalent man).

TABLE 10.5

Relative Biological Effectiveness Values for Several Types of Radiations, Based on Whole-Body Exposure

Radiation	RBE
x- and γ-Rays	1
β-Rays and electrons	1
Thermal neutrons	2
Fast neutrons	10
High-energy protons	10
α-Particles	20
Fission fragments, heavy particles of unknown charge	20
Heavy ions	20

Note: RBE values are also called quality factors or weighting factors.

Dose equiv. (Sv) = Absorbed dose in SI units × RBE = grays × RBE

Dose equiv. (rem) = Absorbed dose in conventional units × RBE = rads × RBE

When a dose of radiation is expressed in sieverts or rems, it does not matter what type of radiation it is since it has already been adjusted by an RBE factor. RBE factors (Table 10.5) are weighting factors for different kinds of radiation but do not consider different sensitivities of various organs. All internal organs or organ systems are treated the same as the whole body.

Another dose equivalent term that is sometimes used is the rem EDE (effective dose equivalent), which is a dose adjusted for different radiation types and by an organ weighting factor to account for organ sensitivity to radiation. The EPA continues to reevaluate the complicated issues involved in setting dose equivalent standards based on risks determined by rem EDE models (EPA 2000a). At present, a 4 mrem/year maximum exposure to critical organs remains the drinking water standard. As Table 10.5 indicates, 1 rad of particles absorbed causes about 20 times the biological damage as 1 rad of β particles, 20 rem versus 1 rem.

Sieverts and rems are the product of two quantities: the first is the energy absorbed, as given by grays or rads; the second is the quality factor, which depends on the type of radiation. The quality factor is an empirical quantity that relates a tissue dosage unit (sievert or rem) to an energy absorption unit (gray or rad) by

Number of sieverts or rems = Number of grays or rads × RBE

An allowable rem dose depends on the tissues irradiated and the amount of time over which the radiation is received. A given amount of radiation damage in a hand or foot is less significant than the same damage in the thyroid gland. Since many tissues can regenerate, a low rate of radiation energy deposition in tissues may be less significant than an equal total energy deposition occurring in a much shorter time.

Note again that there is no time factor and that sieverts and rems, like grays and rads, are dose units and not rate units. Rems per hour would indicate a dose rate.

10.4.4 Unit Conversion Tables

10.4.4.1 Converting between Units of Dose Equivalent and Units of Activity (Rems to Picocuries)

There is no direct conversion factor between rems and picocuries. Rems must be calculated from rads using equations that consider the particular radionuclides present and the type and magnitude of body exposure. Table 10.6 gives calculated concentrations (in pCi/L) of β, γ, and x-ray emitters in drinking water that result in an absorbed dose of 4 mrem/year to the total body or to any critical organ (EPA 2002), However, a provision is made for monitoring gross β activity without calculating a rem value. The gross β standard for drinking water is 4 mrem/year, but a β screening standard is set at 50 pCi/L. An analysis of the major radionuclides and a calculation of the rem value are required only if the sample gross β activity exceeds 50 pCi/L.

TABLE 10.6

Calculated Concentrations (pCi/L) of β and Photon Emitters in Drinking Water Yielding a Dose of 4 mrem/year to the Total Body or to Any Critical Organ as Defined in NBS Handbook 69

Nuclide	pCi/L	Nuclide	pCi/L	Nuclide	pCi/L	Nuclide	pCi/L
H-3	20,000	Sr-85m	21,000	Sb-124	60	Er-169	300
Be-7	6,000	Sr-85	900	Sb-125	300	Er-171	300
C-14	2,000	Sr-89	20	Te-125m	600	Tm-170	100
F-18	2,000	Sr-90	8	Te-127	900	Tm-171	1,000
Na-22	400	Sr-91	200	Te-127m	200	Yb-175	300
Na-24	60	Sr-92	200	Te-129	2,000	Lu-177	300
Si-31	3,000	Y-90	60	Te-129m	90	Hf-181	200
P-32	30	Y-91	90	Te-131m	200	Ta-182	100
S-35 inorg	500	Y-91m	9,000	Te-132	90	W-181	1,000
Cl-36	700	Y-92	200	I-126	3	W-185	300
Cl-38	1,000	Y-93	90	I-129	1	W-187	200
K-42	900	Zr-93	2,000	I-131	3	Re-186	300
Ca-45	10	Zr-95	200	I-132	90	Re-187	9,000
Ca-47	80	Zr-97	60	I-133	10	Re-188	200
Sc-46	100	Nb-93m	1,000	I-134	100	Os-185	200
Sc-47	300	Nb-95	300	I-135	30	Os-191	600
Sc-48	80	Nb-97	3,000	Cs-131	20,000	Os-191m	9,000
V-48	90	Mo-99	600	Cs-134	80	Os-193	200
Cr-51	6,000	Tc-96	300	Cs-134m	20,000	Ir-190	600
Mn-52	90	Tc-96m	30,000	Cs-135	900	Ir-192	100
Mn-54	300	Tc-97	6,000	Cs-136	800	Ir-194	90
Mn-56	300	Tc-97m	1,000	Cs-137	200	Pt-191	300
Fe-55	2,000	Tc-99	900	Ba-131	600	Pt-193	3,000
Fe-59	200	Tc-99m	20,000	Ba-140	90	Pt-193m	3,000
Co-57	1,000	Ru-97	1,000	La-140	60	Pt-197	300
Co-58	300	Ru-103	200	Ce-141	300	Pt-197m	3,000
Co-58m	9,000	Ru-105	200	Ce-143	100	Au-196	600
Co-60	100	Ru-106	30	Ce-144	30	Au-198	100
Ni-59	300	Rh-103m	30,000	Pr-142	90	Au-199	600
Ni-63	50	Rh-105	300	Pr-143	100	Hg-197	900
Ni-65	300	Pd-103	900	Nd-147	200	Hg-197m	600
Cu-64	900	Pd-109	300	Nd-149	900	Hg-203	60
Zn-65	300	Ag-105	300	Pm-147	600	Tl-200	1,000
Zn-69	6,000	Ag-110m	90	Pm-149	100	Tl-201	900
Zn-69m	200	Ag-111	100	Sm-151	1,000	Tl-202	300
Ga-72	100	Cd-109	600	Sm-153	200	Tl-204	300
Ge-71	6,000	Cd-115	90	Eu-152	200	Pb-203	1,000
As-73	1,000	Cd-115m	90	Eu-154	60	Bi-206	100
As-74	100	In-113m	3,000	Eu-155	600	Bi-207	200
As-76	60	In-114m	60	Gd-153	600	Pa-230	600

(Continued)

TABLE 10.6 (*Continued*)

Calculated Concentrations (pCi/L) of β and Photon Emitters in Drinking Water Yielding a Dose of 4 mrem/year to the Total Body or to Any Critical Organ as Defined in NBS Handbook 69

Nuclide	pCi/L	Nuclide	pCi/L	Nuclide	pCi/L	Nuclide	pCi/L
As-77	200	In-115	300	Gd-159	200	Pa-233	300
Se-75	900	In-115m	1,000	Tb-160	100	Np-239	300
Br-82	100	Sn-113	300	Dy-165	1,000	Pu-241	300
Rb-86	600	Sn-125	60	Dy-166	100	Bk-249	2,000
Rb-87	300	Sb-122	90	Ho-166	90		

Source: From EPA, Implementation Guidance for Radionuclides, United States Environmental Protection Agency, Office of Ground Water and Drinking Water (4606M), EPA 816-F-00-002, March 2002, www.epa.gov/safewater.

Example 10

The radioactive isotope ^{40}K (odd Z and odd N) is naturally present in biological tissues and subjects humans to a certain amount of unavoidable internal radiation. The body of an adult weighing 70 kg contains about 170 g of total potassium, mostly in intracellular fluids. The relative natural abundance of ^{40}K is 0.0118% by weight, its half-life is 1.28×10^9 years (y), and it emits β particles with an average kinetic energy in electron volts of 1.40 MeV (1.40×10^6 eV; 1 eV = 1.60×10^{-19} J).

1. Calculate the total activity, A, of ^{40}K, in millicuries, for a 70-kg adult.
2. Find the radiation dose rate in mrem/y.

Note: Potassium-40 is not included in Table 10.6 because human exposure to it is governed entirely by the body's metabolic processes and its natural abundance. Therefore, it cannot be controlled by regulation.

ANSWER

1. First calculate the decay rate constant, k, in units of s^{-1}:

$$k = \frac{0.693}{t_{1/2}} = \frac{0.693}{(1.28 \times 10^9 \text{ y})(3.16 \times 10^7 \text{ s/y})} = 1.71 \times 10^{-17} \text{ s}^{-1}$$

Let the number of ^{40}K atoms = N.

$$N = \left(\frac{170 \text{ g K}_{total}}{40.0 \text{ g/mol}}\right)\left(\frac{1.18 \times 10^{-4} \text{ mol } ^{40}\text{K}}{\text{mol of K}_{total}}\right) \times 6.02 \times 10^{23} \text{ atoms/mol}$$

$$= 3.02 \times 10^{20} \text{ atoms of } ^{40}\text{K}$$

$$\text{Activity} = A_{dps} = \frac{dN}{dt} = kN$$

$$A_{dps} = \left(1.71 \times 10^{-17} \text{s}^{-1}\right)\left(3.02 \times 10^{20} \text{ atoms}\right)$$

$$= 5.16 \times 10^3 \text{atoms/s (dps)}$$

Converting to curies gives

$$A_{Ci} = \frac{5.16 \times 10^3 \text{ s}^{-1}}{3.7 \times 10^{10} \text{ s}^{-1}/\text{Ci}} = 1.4 \times 10^{-7} \text{ Ci} = 0.14 \text{ } \mu\text{Ci}$$

2. To find the dose in rems, find the radiation energy (E) absorbed per kilogram of body tissue and convert to rems.

$$A_{dps}/\text{kg} = \frac{5.16 \times 10^3 \text{ s}^{-1}}{70 \text{ kg}} = 73.7 \text{ s}^{-1} \text{ kg}^{-1}$$

Energy per β particle $= 1.40 \times 10^6$ eV $= (1.40 \times 10^6 \text{ eV})(1.60 \times 10^{-19} \text{ J/eV}) = 2.24 \times 10^{-13}$ J.

$$\text{Total energy/kg body tissue} = \text{Energy/}\beta \text{ particle} \times A_{dps}/\text{kg}$$
$$= (2.24 \times 10^{-13} \text{ J})(73.7 \text{ s}^{-1} \text{ kg}^{-1})$$
$$\text{Total energy/kg body tissue} = 1.65 \times 10^{-11} \text{ J s}^{-1} \text{ kg}^{-1}$$

The number of rems is the number of centijoules (10^{-2} J) of radiation energy absorbed per kilogram of tissue weight multiplied by the RBE factor. From Table 10.5, RBE = 1.

Since ^{40}K is distributed throughout all body tissues, we can assume that essentially all the β radiation energy is absorbed within the body.

Therefore, rems = total energy absorbed/kg = 1.65×10^{-11} J/s/kg.

Rems/y = $(1.65 \times 10^{-11} \text{ J/s/kg})(3.16 \times 10^7 \text{ s/y})(100 \text{ cJ/J}) = 5.22 \times 10^{-2}$ rem/y.

Radiation dose rate = 52.2 mrem/y.

Table 10.7 shows the latest estimate (EPA 2011) of the average annual dosage from all radiation sources per person in the United States, which is about 620 mrem/year. About 6% of our radiation exposure is from internal body sources, almost entirely ^{40}K, and is unavoidable because it is part of our body's chemical makeup. As shown in Table 10.7, about 56% of the total average annual dose is from natural sources of radiation, and of that, most is from radon. Of the other 44%, the majority is from medical diagnosis and treatments.

10.5 NATURALLY OCCURRING RADIOISOTOPES IN THE ENVIRONMENT

Naturally occurring radionuclides become widely distributed in low concentrations in soil and water by weathering and erosion of rocks followed by transport dissolved in groundwater or as a gas. The geological distribution of the parent radionuclide determines where the first daughter elements are formed, but because many daughters are more mobile than the parent radionuclide, they can move long distances from their parent dissolved in groundwater or as a gas (radon). The radionuclides most commonly found dissolved in groundwater are

TABLE 10.7

Sources of Radiation Exposure in the United States with Estimated Average Total Annual Dose Contributions to Individuals

Sources	Approx. Av. Annual Dose to a Person in the United States (mrem/year)	Approx.% of Total Dose
Natural background sources: About 56% of total exposure (347 mrem/year)		
Inhaled radon	200	32
Ingested with food	40	6
Cosmic radiation	24 (sea level) to 50 (Denver)	4–8
Terrestrial radiation	28	4
Internal body	40	6
Man-made sources: About 44% of total exposure (273 mrem/year)		
Medical procedures (x-ray, tomography, radiation treatment, etc.)	267	43
Consumer products	12	2
Occupational exposure	<0.6	<0.1
Weapons testing fallout	<1.2	<0.2
Nuclear fuel cycle	<0.6	<0.1
Miscellaneous	<0.6	<0.1

Note: The total annual dose is about 620 mrem/year.

- From the uranium-238 decay series: uranium-238, uranium-234, radium-226, and radon-222.
- From the thorium-232 decay series: radium-228.*

Other radionuclides of these two decay series and all isotopes of the uranium-235 decay series are usually not present in significant amounts in groundwater because they form highly insoluble compounds or have half-lives short enough to preclude the buildup of large concentrations. Naturally occurring radionuclides that form insoluble compounds and have long enough half-lives to be detectable will be found in surface waters as particulates associated with suspended and bottom sediments.

Naturally occurring radionuclides with $Z > 82$ are those in the three radioactive decay series of uranium-238, uranium-235, and thorium-232 (see Figures 10.3 through 10.5). Unless there have been releases from nearby facilities that make or use enriched uranium, the uranium-235 series is much less important than the others because of the low natural abundance of U-235 (Table 10.8).

---

* Although thorium-232 has a long half-life, it generally is not found in groundwater because it normally is in the Th(IV) oxidation state and forms insoluble compounds. The only daughter product of thorium-232 with a long enough half-life to be of concern is radium-228, with a half-life of 6 years; see Figure 10.5.

TABLE 10.8
Major Natural and Artificial Radionuclides Found in Surface Water and Groundwater

Radionuclide	Half-Life	Average Concentration in Natural Waters	Emissions (MeV)	Origins (Nuclear Reaction, Environmental Pathways, etc.)
Naturally occurring from cosmic radiations				
Carbon-14 (^{14}C)	5730 years	6 pCi/g-carbon	β	Thermal neutrons from cosmic radiation or nuclear weapons testing reacting with atmospheric ^{14}N, $^{14}_{7}N + ^{1}_{0}n \rightarrow ^{14}_{6}C + ^{1}_{1}H$
Potassium-40 (^{40}K)	1.28×10^9 years	<4 pCi/L	β, γ	0.0119% of natural K
Silicon-32 (^{32}Si)	104 years		β	Nuclear fission of atmospheric ^{40}Ar by protons from cosmic sources
H-3, Tritium (^3H)	12.43 years	10–25 pCi/L	β	Thermal neutrons from cosmic radiation or nuclear weapons testing reacting with atmospheric ^{14}N, $^{14}_{7}N + ^{1}_{0}n \rightarrow ^{12}_{6}C + ^{3}_{1}H$
Naturally occurring from ^{238}U and ^{232}Th decay series				
Uranium-238 (^{238}U)	4.5×10^9 years	0.1–10 μg/L ($^{238}U + ^{234}U$)	α	Parent of decay chain of ^{238}U; migration from point of generation
Uranium-234 (^{234}U)	2.5×10^5 years	0.1–10 μg/L ($^{238}U + ^{234}U$)	α	In decay chain of ^{238}U; migration from point of generation
Radium-223 (^{223}Ra)	11.4 days		α, γ	In decay chain of ^{238}U; migration from point of generation
Radium-224 (^{224}Ra)	3.7 days		α, γ	In decay chain of ^{232}Th; migration from point of generation
Radium-226 (^{226}Ra)	1620 years	0.01–0.1 pCi/L	α, γ	In decay chain of ^{238}U Migration from point of generation
Radium-228 (^{228}Ra)	6.7 years		β	In decay chain of ^{232}Th; migration from point of generation
Radon-219 (^{219}Rn)	4.0 seconds		α	In decay chain of ^{238}U; migration from point of generation
Radon-220 (^{220}Rn)	56 seconds		α	In decay chain of ^{232}Th; migration from point of generation
Radon-222 (^{222}Rn)	3.8 days	1 pCi/L surface, >10^3 pCi/L groundwater	α	In decay chain of ^{238}U; migration from point of generation

Thorium-230 (^{230}Th)	7.54×10^4 years		α, γ	In decay chain of ^{238}U; migration from point of generation
Thorium-234 (^{234}Th)	24.1 days		β, γ	In decay chain of ^{238}U; migration from point of generation
Thorium-232 (^{232}Th)	1.4×10^{10} years	0.01–1 μg/L	α	Parent of decay chain of ^{232}Th; migration from point of generation
From fission reactions in reactors and weapons				
Strontium-90 (^{90}Sr)	28.5 years	0.5 pCi/L	β	Fission products with highest yields and biological activity
Iodine-131 (^{131}I)	8.04 days		β, γ	
Cesium-137 (^{137}Cs)	30.0 years	0.1 pCi/L	β, γ	
Barium-140 (^{140}Ba)	12.75 days		β, γ	
Zirconium-95 (^{95}Zr)	64.02 days		β, γ	
Cerium-141 (^{141}Ce)	32.5 days		β, γ	
Strontium-89 (^{89}Sr)	50.55 days		β, γ	Fission products of lesser yields; listed in order of decreasing yield
Ruthenium-103 (^{103}Ru)	39.254 days		β, γ	
Krypton-85 (^{85}Kr)	10.72 years		β, γ	From nonfission neutron reactions in reactors
Cobalt-60 (^{60}Co)	5.271 years		β, γ	From nonfission neutron reactions in reactors
Manganese-54 (^{54}Mn)	312.20 days		γ	From nonfission neutron reactions in reactors
Iron-55 (^{55}Fe)	2.73 years		x-ray	From high-energy neutrons acting on stable ^{56}Fe in weapon hardware

Naturally occurring radionuclides with $Z < 82$, like hydrogen-3, carbon-14, potassium-40, rubidium-87, and so on, either have long enough half-lives to have persisted since the Earth's formation or are formed continuously by the action of radioisotope and cosmic radiation on stable atoms in the Earth's atmosphere and geosphere.

Table 10.8 summarizes some important information for the major radionuclides found in surface water and groundwater. Table 10.9 contains data concerning radionuclides commonly used for hydrological studies of natural waters.

10.5.1 CASE STUDY: RADIONUCLIDES IN PUBLIC WATER SUPPLIES

The widespread distribution of radionuclides in groundwater is a serious problem. Approximately 80% of public water systems (PWSs) in the United States get all or most of their water from underground sources. Many of these groundwater sources have concentrations of radionuclides above EPA MCLs for drinking water (Table 10.8). Ever since the EPA first proposed drinking water standards for radionuclides in 1976, there continue to be many PWSs with exceedances of EPA MCLs. In 1992, 415 PWSs (about 0.7% of the U.S. total) serving nearly 2 million people reported exceedances of the EPA radionuclide MCLs in their delivered water. The number of PWSs reporting exceedances dropped to 200 by 1998 (EPA 1999). The main problem was that the treatment technologies most effective for removing dissolved radionuclides, ion exchange, and reverse osmosis (RO) were not commonly used by PWSs, especially those serving small communities.

Even today, many PWSs continue to operate with radionuclide concentrations above EPA MCLs, even with present-day treatment technologies. A few examples of the widespread existence of systems operating above the radium MCL are listed as follows:

- The 2005 Illinois EPA Annual Compliance Report for Community Water Supplies reports 90 PWS facilities with 287 exceedances of the radium MCL of 5 pCi/L. At year end, 101 PWSs had returned to compliance, with 186 still in violation (IEPA 2006).
- The 2005 New Mexico EPA Annual Compliance Report for Community Water Supplies reports 15 PWS facilities (an increase of 4 over 2004) with 42 exceedances of the gross α and combined uranium MCLs of 15 pCi/L and 30 µg/L (mg/L) respectively (NMED 2006).
- In Colorado in 2007, there were approximately 90–135 water supply systems reported to be operating with radionuclide concentrations in excess of MCLs (Miller and Stanford 2006).

It is important to keep in mind that most PWS exceedances of radionuclide MCLs are not considered to be acute problems requiring immediate correction. Radionuclide MCLs are determined by considering the potential health effects to adults resulting from a 70-year exposure, with a water consumption of 2 L per day. Typical exceedances occurring in a PWS do not pose immediate health risks, any more than do medical x-rays or high altitude airplane trips. The EPA recognizes that,

TABLE 10.9
Radionuclides Commonly Used for Hydrological Studies

Radionuclides	Half-Life	Average Concentration in Natural Waters	Emissions (MeV)	Origins	Environmental States Measured
Hydrogen-3 (^3H, tritium)	12.43 years	10-25 pCi/L	β	Cosmic radiation, weapons testing	Natural surface water and groundwater; organic matter
Carbon-14 (^{14}C)	5730 years	6 pCi/g-carbon	β	Cosmic radiation, weapons testing	Carbon dioxide, carbonate minerals, organic matter
Chlorine-36 (^{36}Cl)	301,000 years	—	β	Cosmic radiation	Chloride minerals, Cl-(aq)
Argon-39 (^{39}Ar)	269 years	—	β	Cosmic radiation	Argon gas
Krypton-85 (^{85}Kr)	10.72 years	—	β	Nuclear fuel processing	Krypton gas
Krypton-81 (^{81}Kr)	210,000 years	—	ec	Cosmic radiation	Krypton gas
Iodine-129 (^{129}I)	1.6×10^7 years	—	β	Cosmic radiation, nuclear reactors	
Radon-222 (^{222}Rn)	3.8 days	1 pCi/L surface, >10^3 pCi/L groundwater	α	In decay chain of ^{238}U	Migration from point of generation
Radium-226 (^{226}Ra)	1620 years	0.01-0.1 pCi/L	α, γ	In decay chain of ^{238}U	Migration from point of generation
Thorium-230 (^{230}Th)	7.54×10^4 years		α, γ	In decay chain of ^{238}U	Migration from point of generation
Uranium-234 (^{234}U)	2.46×10^5 years	0.1-10 μg/L ($^{238}U + {}^{234}U$)	α	In decay chain of ^{238}U	Migration from point of generation
Uranium-238 (^{238}U)	4.47×10^9 years	0.1-10 μg/L ($^{238}U + {}^{234}U$)	α	Parent of decay chain of ^{238}U	Migration from point of generation

TABLE 10.10

EPA National Drinking Water Standards for Radionuclides (December 2011)

Radionuclide	MCL	MCLG
Radium-226 and radium-228 combined	5 pCi/L	0
Gross alpha emitters (including radium-226 but excluding radon and uranium)[a]	15 pCi/L	0
All beta particles and photon emitters[a]	4 millirem (mrem)/year; any organ or whole body	0
Uranium	30 μg/L (equals activity level of 20 pCi/L)[b]	0
Radon[c]	300 pCi/L (final rule pending)	—

Note: MCLs are based on a 70-year exposure, with a water consumption of 2 L per day.

[a] Measurement of gross alpha and beta activity is used primarily as screening tests. If gross alpha or beta activity measurements exceed their MCLs, a more complete analysis is required to identify specific radioactive isotopes.

[b] If uranium concentration is reported in pCi/L, multiply by 1.49 μg/pCi to get mass in μg/L.

[c] See Section 10.5.4.1 for information about current (December 2011) radon regulations.

because of background radiation, risks from exposure to radiation cannot be eliminated; the goal is to minimize them.

In view of the 70-year, 2 L per day drinking water consumption basis for determining radionuclide MCLs, EPA regulations do not require treatment to the MCL limit within any specified time frame (Table 10.10). The action required for most radionuclide exceedances is not rejection of the water supply but continued acceptance, while corrective actions are undertaken. The exact manner of correction is decided by individual states. A common approach used by many states for most exceedances of naturally occurring radionuclides is to require that the water supplier investigates corrective measures, such as best available technology (BAT) treatment or alternative supplies, while continuing a stringent monitoring program. Unless the annual average radionuclide levels are deemed high enough to be of immediate concern, continued use of the water supply is considered acceptable (EPA 1976).

10.5.2 URANIUM

Uranium (U, atomic number 92) has 18 isotopes with atomic masses ranging from 222 to 242. All are radioactive. Only ^{234}U, ^{235}U, and ^{238}U are found naturally. Pure uranium emits only alpha particles accompanied by a low level of gamma radiation. The mass differences among the uranium isotopes are relatively small and the isotopes do not normally fractionate through natural physical or chemical processes. Properties of the three naturally occurring uranium isotopes are presented in Table 10.11.

Uranium is the only radionuclide for which the chemical toxicity has been found to be greater or equal to its radiotoxicity and for which its drinking water standard

TABLE 10.11

Properties of the Naturally Occurring Uranium Isotopes

Uranium Isotope	Half-Life (years)	Specific Activity (disint./s/ms)	Decay Constant, k (years⁻¹)	Abundance in Natural U (Atom%)	Abundance in Natural U (Weight%)
U-234	2.48×10^5	2.28×10^5	2.79×10^{-6}	0.0055	0.0054
U-235	7.13×10^8	79.0	9.72×10^{-10}	0.7204	0.7114
U-238	4.51×10^9	12.31	1.537×10^{-10}	99.2741	99.2830
Natural mix of U isotopes	—	25.0	—	—	—

is expressed in terms of a reference dose (RfD).* Since α emission is the most common measurement of uranium, an activity-to-mass conversion is useful. The EPA has assumed that the mix of uranium isotopes commonly found in PWSs will have a conversion factor of 0.9 pCi/μg, which means that a drinking water MCL of 30 μg/L will typically correspond to 27 pCi/L. The EPA considers that 30 μg/L (27 pCi/L) is protective of both chemically caused kidney toxicity and radiation-caused cancer.

10.5.2.1 Uranium Geology

Uranium occurs naturally in the Earth's crust, with an average concentration of around 2–3 mg/kg (Langmuir 1997). Uranium is common in crustal rocks around the world and is found in trace amounts in nearly all rocks and soils. Ninety percent of the world's known uranium sources are contained in conglomerates and sandstone. The Earth's crust contains 2–3 mg/kg uranium on average. It averages about 1.3 mg/kg in sedimentary rocks and ranges between 2.2 and 15 mg/kg in granites and between 20 and 120 mg/kg in phosphate rocks (Langmuir 1997). Igneous rocks with high silicate content, such as granite, have a uranium content above average, while the contents of sedimentary and basic rocks, such as basalts, are below average.

Because of its chemical reactivity, it is not present as free uranium in the environment. In nature, uranium is generally found as an oxide, such as in the olive-green-colored mineral pitchblende, which contains triuranium octaoxide (U_3O_8). Uranium dioxide (UO_2) is the chemical form most often used for nuclear reactor fuel. Both oxide forms are solids that have a low solubility in water and are relatively stable over a wide range of environmental conditions.

U_3O_8 is the most stable form of uranium and is the form usually found in nature. The most common form of U_3O_8 is "yellow cake," a solid named for its characteristic color that is produced during the uranium mining and milling process. UO_2 is a solid ceramic material and is the form in which uranium is most commonly used as

---

* The RfD is an estimate of a daily ingestion exposure to the population, including sensitive subgroups, which is likely to be without an appreciable risk of deleterious effects during a 70-year lifetime. Dissolved uranium is a kidney toxin and the risk of toxic kidney effects from uranium depends on the mass of uranium ingested. The EPA oral RfD for uranium is 0.6 μg/kg body weight/day.

a nuclear reactor fuel. At ambient temperatures, UO_2 will gradually convert to U_3O_8. Uranium oxides are extremely stable in the environment and are thus generally considered the preferred chemical form for storage or disposal.

Under reducing conditions, uranium is mainly found as the oxide UO_2, an insoluble compound found in minerals. With oxidizing conditions, it forms the oxide UO_3, a moderately soluble compound found in surface waters. Its concentration in most natural waters is between 0.1 and 10 µg/L, although concentrations as high as 1–15 mg/L may exist in water in contact with uranium ore deposits (Hem 1992). Uranium is found in about 25% of the country's water supplies.

Uranium can exist in the oxidation states U(III), U(IV), U(V), and U(VI), but only the uranous [U(IV)] and uranyl [U(VI)] oxidation states are commonly found in the environment. Oxidized species [U(VI)], such as the uranyl cation UO_2^{2+} or anionic species formed at high pH, are the most soluble, favoring the wide distribution of uranium in the oxidized portion of the Earth's crust. The reduced species [U(IV)] are only slightly soluble (Hem 1992). Uranium is a very mobile element in the environment because of the solubility of the oxidized form, UO_3.

Important U(IV) minerals include uraninite (UO_2 through $UO_{2.25}$, the main component of pitchblende) and coffinite ($USiO_4$). Aqueous U(IV) is inclined to form sparingly soluble precipitates, adsorb strongly to mineral surfaces, and partition into organic matter, thereby reducing its mobility in groundwater.

Important U(VI) minerals include carnotite [$(K_2(UO_2)_2(VO_4)_2$], schoepite ($UO_3 \cdot 2H_2O$), rutherfordine (UO_2CO_3), tyuyamunite [$Ca(UO_2)_2(VO_4)_2$], autunite [$Ca(UO_2)_2(PO_4)_2$], potassium autunite [$K_2(UO_2)_2(PO_4)_2$], and uranophane [$Ca(UO_2)_2(SiO_3OH)_2$]. Some of these are secondary phases, which may form when sufficient uranium is leached from contaminated wastes or a disposal system and migrates downstream. Uranium is also found in phosphate rock and lignite at concentrations that can be commercially recovered. In the presence of lignite and other sedimentary carbonaceous substances, uranium enrichment is believed to be the result of uranium reduction to form insoluble precipitates such as uraninite.

10.5.2.2 Uranium in Water

Uranium in surface water is often leached from phosphate deposits and mine tailings and transported in runoff from phosphate fertilizers (which can contain up to 150 ppm of uranium) from agricultural land. Greater than 99% of uranium transported by runoff from land to freshwater systems is in suspended particles and remains in the sediment. Dissolved levels in U.S. streams are usually between 0.1 and 10 µg/L (Hem 1992), but can exceed 20 µg/L in irrigation return flows because of phosphates and evaporative concentration. In waters associated with uranium ore deposits, dissolved uranium concentrations may be greater than 1000 µg/L.

Uranium is widely dispersed in groundwater because of its presence in soluble minerals, its long half-life, and its relatively high abundance. The highest concentrations in groundwater are found in granite rock and granitic sediments. Uranyl cation (U^{6+}) forms strong carbonate complexes in most waters above pH 6. Complexation with carbonate anions greatly increases the solubility of uranium minerals, facilitating uranium mobility. The solubility of uranium is also enhanced by complexation with phosphate, sulfate, and fluoride ions and with organic compounds, especially humic substances. Because it is readily dissolved and transported by oxidizing ground-waters, it can be transported to areas far from its original location.

Uranium is less mobile in reducing groundwater, where U(IV) forms solids of low solubility, and the dominant uranus ion (U^{4+}) and its aqueous complexes tend to sorb very strongly to humic materials and mineral surfaces in the aquifer. U(IV) concentrations in reducing groundwater are usually less than 10 μg/L.

Uranium activity as high as 652 pCi/L is observed in both surface water and groundwater. The average uranium concentration in surface water is estimated to be about 1 pCi/L and in groundwater about 3 pCi/L. The population-weighted average uranium concentration in community drinking water supplies ranges between 0.3 and 2.0 pCi/L.

RULES OF THUMB

1. Uranium in surface runoff is mostly (>99%) in sediment form.
2. Uranium is the only radionuclide for which its chemical toxicity risk is greater or equal to its cancer risk from emissions.
3. The drinking water standard for uranium is 30 μg/L. This corresponds to 27 pCi/L in most PWSs. The conversion factor is 1 μg/L = 0.9 pCi/L.
4. Uranium, generally, is least mobile in reducing (anaerobic) environments free of complexing anions and most mobile in oxidizing (aerobic) environments with high concentrations of complexing anions.
5. High alkalinity (carbonate content) and high levels of sulfate, fluoride, and humic materials increase uranium solubility by complexation.
6. Reducing conditions in groundwater and lake sediments promote uranium sorption to solid surfaces and greatly lower concentrations and mobility.
7. Phosphate fertilizers can contain up to 150 ppm of uranium.

10.5.3 RADIUM

All the isotopes of radium between Ra-223 and Ra-230 have been observed. Only Ra-223, Ra-224, Ra-226, and Ra-228 occur naturally (see Table 10.8). Ra-226 and Ra-228 are usually the only radium isotopes of environmental interest because their half-lives (1620 and 6.7 years respectively) are long enough to allow substantial environmental accumulation.

In addition to their own radiation emissions, three of these radium isotopes present additional environmental and health concerns due to the fact that they decay into radon. Radon is a gas at room temperature and is thus more mobile in the environment.

- Ra-223, in the uranium-235 decay series, decays to radon-219 by α emission.
- Ra-224, in the thorium-232 decay series, decays to radon-220 by α emission.

- Ra-226, in the uranium-238 decay series, decays to radon-222 by α emission.
- Ra-228, in the thorium-232 decay series, decays to actinium-228 by β emission.

Radium-226 is of greatest concern because it is the most abundant and decays into the most abundant radon isotope, radon-222.

Radium belongs to periodic table group 2A, the alkaline earth metals, and its chemical properties are therefore most similar to those of barium, strontium, and calcium. In the body, radium behaves like calcium and becomes incorporated into bone structure, where it poses a serious risk of bone cancer. In PWSs with radium problems, radium is often found precipitated as a carbonate or sulfate along with calcium and magnesium deposits in pipes of the distribution system.

Radium in solution exists in only the Ra(II) oxidation state and does not easily complex in water. It forms carbonate and sulfate salts of very low water solubility. Radium salts of chloride, nitrate, and bromide are soluble. Sorption can remove radium from solution by adsorption and coprecipitation by scavengers such as iron hydroxide and barium sulfate.

10.5.3.1 Radium in Soil

Radium-226 and radium-228 are found in soil throughout the world. In the United States, the mean concentration of Ra-226 in 356 surface soil samples collected from 33 states was 1.1 pCi/g (Myrick et al. 1981).

Typical concentrations by rock type are

- Sandstone, 0.71 pCi/g
- Limestone, 0.42 pCi/g
- Shale, 1.1 pCi/g

Coal burning and uranium mining/milling operations have produced elevated levels of radium in soil. The concentration of Ra-226 in soils contaminated by mining and milling activities can be 40,000 pCi/g or more. Using uranium concentrations as an indicator of radium levels, elevated radium levels in soil are expected in the western third of the continental United States, including large areas of California and Idaho, and in Wisconsin, Minnesota, the Appalachian Mountains, and Florida (ATSDR 1990).

10.5.3.2 Radium in Water

The solubility of radium generally is low but increases with decreasing pH. Dissolved radium occurs mainly as Ra^{2+} within the pH range 3–10. Dissolved radium concentrations tend to be highest in reducing waters high in iron and manganese and low in sulfate. Increased concentrations of total dissolved solids also increase dissolved radium concentrations because fewer sediment and soil surface sites are available for sorption of radium species.

Radium-226, a daughter product of ^{238}U, is the most common radium isotope in natural waters because of the abundance and mobility of its parent uranium-238 in groundwater. It is more abundant than ^{228}Ra, a daughter product of ^{232}Th, because uranium is generally more abundant and is more soluble than thorium.

Although ^{228}Ra is chemically similar to ^{226}Ra, its distribution in groundwater is very different for several reasons. The relatively short half-life of ^{228}Ra limits the potential for transport without the parent being present. Consequently, ^{228}Ra cannot migrate far from its source before it decays to another progeny. Thorium-232, the parent of ^{228}Ra, is extremely insoluble and is not subject to mobilization in most groundwater environments. The very low solubility of thorium (much lower than uranium) limits the distribution of ^{228}Ra in groundwater. Areas associated with the presence of ^{228}Ra include the east coastal plain and high plains aquifers.

Radium levels are generally higher in groundwater than in surface water. Typical concentrations for groundwater are 0.5–25 pCi/L and for surface water 0.01–1 pCi/L. Water concentrations for radium seldom exceed 50 pCi/L. In general, shallow groundwater has less radium than deep aquifers, and treated water has less radium than raw groundwater. The radium content of surface water is usually lower than most groundwater supplies. Ra-226/228 average concentration in community drinking water supplies ranges from 0.3 to 0.8 pCi/L but may exceed the EPA drinking water MCL of 5 pCi/L wherever uranium-238 and thorium-232 minerals are abundant.

10.5.4 RADON

Radon (Rn) is a naturally occurring chemically inert gas, in the same family as helium, neon, and argon. Radon is produced in the natural decay chains of ^{238}U, ^{235}U, and ^{232}Th and is the only naturally occurring gaseous element that is radioactive. It has no odor, color, or taste and is about 7.4 times denser than air. All radon isotopes are radioactive and decay by α emission. Radon accounts for about 50% of the average radiation exposure of people in the United States (see Table 10.7).

There are three naturally occurring isotopes of radon:

- ^{219}Rn, half-life = 3.9 seconds; in decay chain of ^{235}U
- ^{220}Rn, half-life = 54.5 seconds; in decay chain of ^{232}Th
- ^{222}Rn, half-life = 3.8 days; in decay chain of ^{238}U

Because of its longer half-life, only ^{222}Rn is environmentally important. Wherever ^{238}U is in the soil, ^{222}Rn is formed continuously as a daughter product. Because radon is an inert gas, it moves freely through soil fissures and pore spaces without reacting to become a less-mobile radionuclide.

Its half-life of 3.8 days is long enough for significant amounts of radon to travel far from their point of origin and dissolve in groundwater, diffuse to the surface atmosphere, and collect in underground voids such as caves, building basements, subways, water wells, and sewers. On the other hand, its half-life is short enough that after 40 days (about 10 half-lives) an isolated sample has decayed to a negligible activity. Some water treatment methods for removing radon are based on simply holding it in one place (e.g., on carbon sorption media or in a holding tank) long enough for it to decay to negligible activity. Radon is sufficiently soluble in water to reach concentrations as high as 300,000 pCi/L where uranium minerals in granites and uranium minerals in pegmatites associated with metamorphic rocks are in the area. However, radon concentrations around 10,000 pCi/L are more typical. As with most gases, its solubility in water varies inversely with water temperature; the colder the water, the greater is radon's solubility.

Because it is an inert gas and unreactive in the environment, it is the most mobile radioisotope and can travel long distances dissolved in groundwater and through voids and fissures in the vadose zone as a gas. Radon's average concentration in groundwater is about 1,000 pCi/L, but the variability is very large.

Radon is found nearly everywhere in soil, air, and water; even outdoor air contains low levels of radon (typically about 0.4 pCi/L). Radon enters building structures through floor and wall cracks, drains, hollow concrete blocks, openings around pipes, windows, and so on. Radon is denser than air (density of radon at STP = 9.73 g/L, density of air at STP = 1.29 g/L) and does not diffuse evenly into the atmosphere but tends to collect in voids and low spots.

Radon in soil under homes is the biggest source of radon in indoor air and presents a greater risk of lung cancer than radon in drinking water. The EPA's National Residential Radon Survey found that over 6% of all homes nationwide have average annual indoor radon levels above 4 pCi/L (Marcinowski et al. 1944). Most indoor radon comes from uranium and radium decay in rocks and soil around a home, although other, usually less significant, sources of indoor radon are water and some construction materials.

Radon is also drawn into the low-pressure zone within a pumping well's cone of depression. The result is that radon is drawn from remote crevices and fractures into the well. For buildings whose water is supplied by a well, radon dissolved in groundwater can enter through the water distribution system, becoming volatilized into the indoor air space as the water is used, especially in showers and washing machines. Public water supplies tend to have lower radon levels than wells because the longer residence times allow more complete radioactive decay. Also, where radon is a known problem, public supplies often treat their water by aeration, venting the radon gas to the atmosphere.

10.5.4.1 Health Issues

Radon is a known human lung carcinogen and was reported to be the second leading cause of lung cancer in the United States (National Academy of Sciences 1999). It is estimated that radon is responsible for 15,000–20,000 lung cancer deaths per year in the United States (National Academy of Sciences 1999).

By far, the primary health risk for radon is exposure through inhalation, not ingestion. Drinking radon-rich water appears to result in only a very minor increase in the risk of stomach cancer (National Academy of Sciences 1999). The primary concern of exposure to radon-rich water is its contribution to radon in indoor air as it volatilizes out of the water phase. This occurs during normal household water use, particularly during showers and during the washing of dishes and clothes.

Most radon that is inhaled is also exhaled. However, radon remaining in the lung decays in several steps to form daughter radioactive isotopes of polonium, lead, bismuth, and tellurium, all with short half-lives (see Figure 10.3). These radon decay products are chemically toxic metals that are readily retained in the respiratory system and, over time, will damage sensitive lung and bronchial tissues. Also, the short half-lives of the daughter products allow them to undergo further decay before the action of mucus in the bronchial tubes can clear them out. Thus, the lungs and bronchia are exposed to additional α and β emissions. The final decay product in the radon decay chain is the stable toxic lead isotope Pb-206.

Even very small exposures to radon can, over time, result in lung cancer, and researchers have not yet been able to determine a safe lower threshold. Smokers exposed to elevated levels of radon are particularly susceptible to contracting lung cancer because of the synergistic relationship between radon, smoking, and lung cancer. Generally speaking, health risks associated with radon are due to long-term exposure, from about 5 to 25 years (National Academy of Sciences 1999).

The EPA has proposed an MCL for radon in water of 300 pCi/L and an alternate MCL of 4000 pCi/L for public water suppliers that have radon mitigation programs for their customer base (EPA 1999). Radon dissolved in public water supply systems nationwide averages about 250 pCi/L. The proposed radon regulation does not affect PWSs or their customers; the EPA must promulgate a final regulation before a federal radon regulation will be enforced and has not yet (as of December 2011) done this. Until a final regulation is promulgated, states may establish their own radon regulations, with EPA approval.

RULES OF THUMB

1. Radon is the largest source of radiation exposure to the public and is considered a serious health risk.
2. The average waterborne radon level in U.S. public groundwater supplies is about 353 pCi/L. However, the range of reported activities is very large, from zero up to at least 300,000 pCi/L.
3. 10,000 pCi/L in water produces about 1 pCi/L in confined air adjacent to the water.
4. The current "action level" for airborne radon is 4 pCi/L. The EPA recommends that action be taken to lower airborne radon levels that exceed 4 pCi/L (40,000 pCi/L in the water) in inhabited spaces.
5. There are no EPA standards for radon in water as of December 2011; an MCL of 300 pCi/L for public water supplies is being considered. Individual states may have established their own standards.
6. Water treatment for radon is usually either aeration to release it to the atmosphere or simply holding it in one place (e.g., on carbon sorption media or in a holding tank) long enough for it to decay to negligible activity, at least 45–60 days. The half-life of radon = 3.8 days.

10.5.5 CASE STUDY: TRITIUM-BASED AGE DATING OF GROUNDWATER

A community in Central America wished to determine the source of nutrient contamination, mainly nitrates, in three water supply wells. To help identify the recharge areas for these wells, it was decided to "age date" the well water using tritium as a tracer. Tritium age dating, although it allows only a qualitative estimate of groundwater age, is relatively simple and fast compared to other age dating methods and may be sufficient if an unambiguous result is obtained.

10.5.5.1 Tritium Age Dating Method

Tritium (^{3}H, also written as T) is a radioactive isotope of hydrogen of very low natural abundance (around 1 part in 10^{17} of total earth hydrogen) and a half-life of 12.3 years (Table 10.9). A natural atmospheric background of tritium is produced from cosmic ray radiation interacting with upper atmosphere gas molecules. Naturally produced atmospheric tritium accounts for a fairly constant background concentration in precipitation of about 1–15 pCi/L, depending on how atmospheric conditions distribute it spatially and temporally. Additional large quantities of tritium were released during the period of atmospheric atomic weapons testing beginning around 1950 and peaking about 1963. Atmospheric atomic weapons testing ceased about 1980 and, since then, atmospheric tritium concentrations have been decreasing due to radioactive decay and losses by precipitation.

Tritium rapidly exchanges with hydrogen atoms in water molecules, becoming a part of the atmospheric water molecules. This tritium enters the hydrological cycle via precipitation. Once tritiated water enters groundwater, it ceases to equilibrate with the atmosphere and its concentration decreases by radioactive decay, dilution with nontritiated groundwater, and dispersion. In the Northern Hemisphere, tritium concentrations in precipitation reached a peak around 3000 pCi/L in the early1960s (Clark and Fritz 1997), dropping to current (2012) concentrations of near prebomb-testing background concentrations (~1–15 pCi/L).

10.5.5.1.1 Groundwater Mean Residence Time and "Age"

Groundwater recharge occurs mainly by infiltration of precipitation and surface runoff into the subsurface, where the water can move by different pathways and times of passage. At any location in an aquifer, the groundwater contains a mixture of waters that were recharged at different times and from different locations. In the vadose (unsaturated) zone and just below the water table, a given volume of groundwater may contain waters that entered the subsurface at times differing by weeks to months or years. Deeper groundwater can include components recharged over a greater time range of months to many years.

The short half-life of tritium (12.3 years) insures that the low baseline precipitation activity (3–15 pCi/L) present in tritiated groundwater that was recharged before atmospheric testing of thermonuclear weapons (between 1950 and 1976) has decayed to below detection at the present time (2012). Therefore, the presence of measurable concentrations of tritium in groundwater is an unambiguous evidence of some degree of active recharge during the time since 1950.

The age of groundwater is interpreted to be a measure of the time since the water sample was last in contact with the atmosphere. When only the groundwater tritium activity is used as a tracer, it is assumed that the collected water sample has traveled from the point of recharge to the point of sampling by "plug" flow, with no dilution by other waters, and that the time of travel can be determined by the extent that the tritium activity has diminished, using the known half-life.

However, groundwater in a well is always a mixture of water molecules that entered the subsurface at different times. The assignment of an age to groundwater is really an integration of the travel times of individual water molecules from their point of entry into the subsurface to the point of sampling.

When this age is to be determined by a tracer molecule like tritium, a further complication is introduced by the fact that the spatial and temporal global distribution of elevated tritium levels in precipitation has been very uneven. The most accurate use of tritium for dating groundwater is near locations where the tritium activity in precipitation has been consistently monitored from the onset of atmospheric nuclear weapons testing to the present. Unfortunately, there are relatively few such locations and most tritium-only-based age dating methods must rely on interpolating an initial precipitation tritium activity between locations with known values.

Uncertainties introduced by spatial and temporal variations in precipitation tritium activity and by groundwater mixing allow only qualitative deductions to be drawn from tritium-only dating procedures. The following general observations are helpful when qualitatively interpreting tritium groundwater data.

- By the year 2012 in most regions, tritium concentrations in precipitation have fallen to (or nearly to) the background levels preceding nuclear weapons testing.
- Bomb tritium activity, found in groundwater recharged since 1950, is greater in the Northern Hemisphere middle latitudes than in the lower northern latitudes, tropics, and Southern Hemisphere.
- Coastal precipitation generally has much smaller tritium concentrations than continental (interior) stations because it contains a larger proportion of water that has not yet mixed with upper atmosphere water, where tritium is more concentrated. This is true for precipitation during the thermonuclear era as well as today.
- Groundwater with large tritium levels (>5 TU) certainly contains some bomb tritium.
- Groundwater with small tritium levels (close to 1 TU) may be mixtures of groundwater recharged recently (since 1980) and groundwater recharged prior to bomb testing (early 1950s).
- Although tritium in precipitation has mostly returned to prebomb testing natural levels, there are still spatial and seasonal variations.

10.5.5.1.2 Application of Tritium Age Dating Principles to this Case Study

Table 10.12 summarizes the tritium content measured in three water supply wells and a nearby surface water stream in the Central American community.

As discussed above, tritium-based age dating requires knowledge of the tritium content of the recharging precipitation. To estimate this quantity, data for tritium concentrations in rainfall between about 1950 and 2000 at the Atomic Energy Agency monitoring station in Ocala, Florida (Katz 2004), a location that should be similar to that of the Central America community with respect to global tritium distribution, both being near the coast (~40 km inland) in the lower northern latitudes (Ocala, Florida: ~29° N latitude; Central America community: ~20° N latitude).

TABLE 10.12
Tritium Content of Three Water Supply Wells and a Surface Water Stream (All Sampled on Same Day in January 2012)

Well Identification	Tritium Content (TU)
A	0.4±0.27
B	0.96±0.25
C	0.95±0.27
Surface water stream	2.04±0.21

TABLE 10.13
Qualitative Interpretation of Tritium-Based Age Dating

Approximate Recharge Period[a]	Typical Tritium Concentration (TU)[b,c] in Rainfall During the Recharge Period at Ocala, Florida	Typical Tritium Concentration Expected in Ocala, Florida, Groundwater Sampled in 2011[d]
Before 1950	~2	Undetectable (<~0.01 TU) (>5 half-lives since recharge)
1950–1962	~2 to 80	Undetectable to ~5 TU (~5 to 4 half-lives since recharge)
1962–1975	~80 (in 1962) to 500 (at bomb-testing peak in 1965) to 10 (in 1975)	~5 (1962) to 31 (1965) to 2 (1975) TU (~4 to 3 half-lives since recharge)
1975–1987	~10 to 5	~2 to 1 TU (~3 to 2 half-lives since recharge)
1987–1999	~5 to 3	~2 to <1 TU (~2 to 1 half-lives since recharge)
1999–2011	~2	~2 to <1 TU (<1 half-life since recharge)

Source: Derived from Katz, B.G., *Environ. Geol.*, 46, 693, 2004, Figure 3.

[a] Recharge period indicates the apparent age of groundwater when estimated from tritium concentrations. It measures the apparent length of time since the water sample was last in contact with the atmosphere. Well water samples are a mixture of water molecules with an age distribution that may span a wide range. A tritium groundwater age is the apparent mean age of the mixed sample.

[b] Tritium concentrations are expressed in tritium units (TU). One TU corresponds to one tritium atom in 1018 hydrogen atoms, which is a count rate in water of 3.2 pCi/L. Groundwater today seldom has more than 50 TU and commonly is in the <1–10 TU range.

[c] At inland locations and more northern latitudes, typical tritium concentrations can be roughly two times larger than that at Ocala, Florida, during the same recharge periods.

[d] Adjusted for half-life attenuation, assuming undiluted plug flow of recharge water.

Table 10.13 provides apparent ages of groundwater when estimated from tritium concentrations.

Tritium concentrations in all three well waters are less than 1 TU, but definitely above the method detection limit. The presence of some low level, but still detectable, tritium activity eliminates the early period before 1962. The presence of activities <1 TU qualitatively "dates" the well water as having been recently recharged since 1987 and most likely since about 1999, for the following reasons:

- The tritium half-life of 12.3 years assures that water recharged with a baseline tritium level of around 2–3 TU will have decayed to an undetectable level (<0.1 TU) after about four half-lives (50 years). The fact that tritium levels near 1 TU were detected in all the samples tested indicates that the sampled water cannot have been recharged prior to bomb testing, which started around 1950. Therefore, it had to have been recharged during a more recent period consistent with a present activity of less than 1 TU.
- Two periods in Table 10.13 meet the above criteria: 1987–1999 and 1999–2011. A choice between these periods requires knowledge of the input activity in rainfall.
- The rainfall input activity for recharge of the wells can be estimated to be the same as the tritium activity measured in the surface water stream (2.04 ± 0.21 TU), since only recent recharge times need to be considered.

From Table 10.13, only the most recent recharge period of 1999–2011 is consistent with both an input activity of 2.04 ± 0.21 TU and a groundwater activity of ~2 to <1 TU. Since the measured groundwater activity of <1 TU indicates some decay of activity from the estimated input activity of ~2 TU, the most likely apparent age of the groundwater is closer to 1999 than to 2011.

10.6   TECHNOLOGIES FOR REMOVING URANIUM, RADIUM, AND RADON FROM WATER

The Radionuclides Rule (EPA 2000b, 2001), lists BATs and Small System Compliance Technologies (SSCTs) for radionuclide treatment, based on their efficiency at removing radionuclides from drinking water. Water suppliers are not required to use a listed BAT or SSCT; any technology that achieves compliance with the MCL and has regulatory acceptance is allowed.

For uranium and radium removal, BATs listed in the Radionuclide Rule consist of RO, ion exchange, coagulation/filtration, and lime softening. Other non-BAT technologies, such as iron/manganese removal, preformed hydrous manganese oxide (HMO) addition with filtration, adsorptive media (such as activated alumina [AA]), and electrodialysis/electrodialysis reversal (ED/EDR), are also capable of removing either radium or uranium. States may adopt their own set of acceptable BATs.

Several of the treatment methods are pH dependent. Table 10.13 shows how pH affects the forms of radium and uranium present in groundwater.

Table 10.14 lists the most common radionuclide removal technologies currently in use, along with useful information about each.

TABLE 10.14

Common Radionuclide Removal Technologies Currently in Use

Treatment Technology	Radionuclides Removed by Treatment					Comments
	Combined Radium-226/Radium-228	Uranium	Radon	Gross Alpha	Beta/Photon	
Ion exchange (IX) (anion, cation or anion/cation exchange in a mixed media column) Exchanges ions weakly bound to media surface sites for more strongly bound contaminant ions	BAT (cation, mixed-bed IX)	BAT (anion, mixed-bed IX)		BAT (mixed-bed IX)	BAT (mixed-bed IX)	Anion exchange resins generally exchange chloride for anionic contaminants (e.g., uranium, arsenic). Cation exchange resins generally exchange hydronium ion, sodium, or potassium for cationic contaminants (e.g., radium). Anion and cation exchange resins can be combined in a single unit to remove cationic and anionic contaminants together. IX can remove up to 99% of contaminants depending on the resin, pH, and competing ions (e.g., sulfate, hardness). Particulates and metals (e.g., iron, manganese, or total organic carbon [TOC]) in the source water can clog IX media or block exchange sites. Pretreatment can help to remove fouling compounds and preserve bed life; however, pretreatment may also remove radionuclides and create a disposal problem. Anion exchange can reduce alkalinity; adding carbonate or caustic may be necessary. The regeneration solution contains high concentrations of the contaminant ions; disposal options should be carefully considered before choosing this technology. Also removes arsenic, sulfate, fluoride, nitrate, magnesium, and calcium.

Technology					Description
RO Pressure-driven membrane separation of contaminants larger than the membrane pores	BAT	BAT	BAT	BAT	RO can remove up to 99% of most contaminants, including radionuclides, but does not remove gaseous contaminants like carbon dioxide and radon. Pretreatment is generally required to limit membrane fouling and scaling from hardness, colloids, and bacteria. Rejected water and used membranes may be radioactive; disposal options should be carefully considered before choosing this technology. Also removes most dissolved contaminants, including arsenic, nitrate, metals, TOC, and microbial contaminants
Lime softening (LS) Lime is added to raise pH and precipitate carbonate sludge, which also incorporates many contaminants. Enhanced LS adds magnesium carbonate to the lime	BAT SSCT	BAT SSCT	BAT SSCT	BAT SSCT	Radium is in the same periodic family as calcium and magnesium and tends to precipitate with them. Divalent uranium can also precipitate as a carbonate at high pH. Single-stage LS (pH > 10) can remove between 50% and 80% of radium. Enhanced LS (pH > 10.6) is needed to remove uranium and can remove up to 90% of radium and uranium. Because calcium and magnesium are also removed, it may be necessary to increase alkalinity or add corrosion inhibitors to treated water for corrosion control. Wastes generated by LS include backwash water, sludge, and aged/ineffective media. The concentration of radionuclides in these residuals may impact disposal options; disposal options should be carefully considered before choosing this technology. Also removes arsenic, iron, manganese, and hardness

(Continued)

TABLE 10.14
Common Radionuclide Removal Technologies Currently in Use

Treatment Technology	Radionuclides Removed by Treatment					Comments
	Combined Radium-226/Radium-228	Uranium	Radon	Gross Alpha	Beta/Photon	
ED/EDR Drives dissolved cations and anions through separate membranes with an electric field. Polarity is reversed to clean the membranes	BAT	Used, but not BAT				The units can be highly automated and only require monitoring of operational parameters and periodic maintenance. ED/EDR may be an effective alternative for small systems that have multiple contaminants. The costs of these systems are relatively high compared to other radionuclide treatment options. Capital costs are high and operating costs are increased by required acid washes for radionuclides and by disposal costs. The reject stream and membranes will contain elevated concentrations of radionuclides and other contaminants. Also removes many dissolved ions, including arsenic, nitrate, perchlorate, and hardness.
Greensand manganese oxide filtration Uses manganese-coated filter media to oxidize and sorb contaminants	BAT					Greensand is a naturally occurring mineral (glauconite) used as an ion exchange and sorption medium. It is activated by adding soluble manganese(II), then treating with permanganate. Radium is removed by sorption; no oxidation occurs. Hardness and pH affect greensand performance. Iron, manganese, and other polyvalent cations compete for sorption sites and limit radium removal. Sorption of radium increases as pH increases from 5 to 9. A high iron-to-manganese ratio may lower radium sorption, but the presence of iron is beneficial to arsenic sorption. Radium removal improves as pH increases from 5 to 9. Arsenic removal by greensand requires pH > 6.8. Also removes arsenic, iron, and manganese (up to 10 ppm).

AA	BAT	AA can remove up to 99% of uranium, depending on pH and competing ion concentrations. AA is also effective at removing other ions such as arsenate, fluoride, sulfate, and selenate. However, optimal sorption for different pollutants is pH dependent. For example, arsenic removal by AA is optimum at a pH between 5.0 and 6.0. At this pH, however, uranium is a neutral molecule and may not be removed as effectively. TDS and sulfate have less surface saturation effect on AA compared to other sorption media. The regeneration solution and exhausted sorption media contain high concentrations of the contaminant ions; disposal options should be carefully considered before choosing this technology.
AA uses a porous, granular, aluminum media to sorb ions from solution		
Preformed hydrous manganese oxide (HMO) filtration	SSCT	Most applicable to small systems that already have filtration in place. HMO filtration can remove up to 90% of radium. Also removes arsenic; high hardness enhances arsenic removal. High iron levels may require pretreatment to remove iron because of iron competition for sorption sites. Wastes include backwash water, sludge, and aged/ineffective filtration media; disposal options should be carefully considered before choosing this technology.
Preformed manganese oxide particles are added to water to sorb radium and then filtered		

(Continued)

TABLE 10.14

Common Radionuclide Removal Technologies Currently in Use

Treatment Technology	Radionuclides Removed by Treatment					Comments
	Combined Radium-226/Radium-228	Uranium	Radon	Gross Alpha	Beta/Photon	
Enhanced coagulation/filtration (EC/F) Coagulants are mixed with influent water to form filterable particles and flocs		BAT				EC/F is appropriate mainly for surface water systems, of which few have significant uranium. EC/F can remove up to 90% of uranium at pH 10, but is generally not effective for radium. At pH levels typically used in treatment plants, removal efficiencies are generally between 50 and 80%. Wastes generated by coagulation/filtration include backwash water, sludge, and aged/ineffective filtration media; disposal options should be carefully considered before choosing this technology. Also removes arsenic, iron, and manganese.
Coprecipitation with barium sulfate	BAT					Adding soluble barium chloride to water containing dissolved radium and sulfate causes coprecipitation of a highly insoluble radium-containing barium sulfate sludge, potentially removing up to 95% of radium. Since static mixing, detention basins, and filtration are required, it is most applicable to the systems that already treat for high sulfate levels and have suitable filtration in place. Wastes include the sludge, filter backwash, and sludge supernatant; disposal options should be carefully considered before choosing this technology. A variation of this technology is to impregnate ion exchange/sorption media with barium sulfate; dissolved radium becomes bound to sorption sites, releasing Ba^{2+} into the reject water.

Treatment Technology	Uranium	Radon	Combined Radium-226/Radium-228	Gross Alpha	Beta/Photon	Comments
Granular activated carbon (GAC)		Used, but not BAT				GAC filter systems can be effective for removing radon from water. Wastes include radioactive filter bed material. However, since the half-life of radon is only 3.8 days, spent GAC media need only be stored long enough for it to decay to a safe level. A retention time of 45–60 days lowers radon activity to about 0.03–0.003% of its original value. A carefully sized system may not accumulate a dangerous level of activity. GAC systems should not be installed in homes or inhabited areas.
Aeration systems for water		BAT				Aeration treatment forces small air bubbles through contaminated water in an atmospheric-vented chamber. After sufficient aeration, the water is essentially free of radon. Because radon is a "noble" gas of very low chemical reactivity, it does not react with water or other constituents and does not sorb to solids. Although it is sufficiently soluble to be a radioactive risk, it is readily displaced from water into the atmosphere by aeration. Aeration systems are effective at removing radon from water and, if the aeration location is well ventilated, have the advantage over GAC treatment of not accumulating radioactive wastes. Aeration systems require periodic cleaning to remove particulates that come from minerals in the water.

Notes: All treatment technologies, BAT or non-BAT, produce radioactive wastes that require appropriate disposal.

EXERCISES

1. Complete the following nuclear equations (in every case, the unknown is just one kind of particle):

    a. $^{218}_{84}Po \rightarrow \, ^{4}_{-2}He + \underline{\quad}$

    b. $^{248}_{98}Cf + \, ^{18}_{8}O \rightarrow \, ^{263}_{106}Sg + \underline{\quad}$

    c. $^{1}_{0}n + \, ^{235}_{92}U \rightarrow \, ^{90}_{38}Sr + 2\, ^{1}_{0}n + \underline{\quad}$

    d. $^{2}_{1}H + \, ^{1}_{1}H \rightarrow \underline{\quad} + \, ^{0}_{0}\gamma$

2. The half-life for a certain radioisotope is 10 hours. If you start with 1 g of the isotope, how long will it take for it to be reduced to 1 mg (1/1000 g)?

3. Briefly, discuss radon gas. How is it formed and why is it hazardous?

4. After a uranium-238 atom emits one particle, what isotope does it become? Use the periodic table inside the front cover, but do not refer to other figures in this chapter.

5. A scientist is studying isotopes of the element thorium (Th), which has the atomic number 90. He isolates two different samples, each of which contains a single pure isotope of thorium. All the atoms in one sample have an atomic weight of 230 mass units and all the atoms in the other sample have an atomic weight of 234 mass units. Which is the correct explanation below?

    a. The lighter sample must have lost weight in a chemical reaction.

    b. Each atom of the heavier sample has four more protons in the nucleus than do the atoms in the lighter sample.

    c. Each atom of the heavier sample has four more neutrons in the nucleus than do the atoms in the lighter sample.

    d. He must have made some mistake because all atoms of the same element must be identical. Both samples cannot be thorium.

6. Using only the periodic table, complete the following table (Z = atomic number; A = isotope mass number):

Chemical Symbol	Z	A	Protons	Neutrons	Electrons
Mo	42			56	
	34	80			34
Ag		109			
	37		17		16

7. Use Figure 10.2 to write an equation for the probable mode of decay of each of the following radionuclides. Your equation should be in the form of the answers to Question 1.

    a. $^{38}_{16}S$

    b. $^{38}_{20}Ca$

    c. $^{239}_{94}Pu$

8. Water containing radioactive waste material with a half-life of 0.5 years is stored in underground tanks. If the waste concentration is 500 pCi/L, how long will it take for it to decay to less than 0.1 pCi/L?

9. Radon-222 has a half-life of 3.8 days. The source water for a drinking water treatment plant contains 1200 pCi/L of radon. Because aeration is not an option at this plant, how long must they store the water to meet the EPA proposed MCL of 300 pCi/L?

10. Tritium $\left(^3_1H\right)$ has a half-life of 12.3 years. It is formed in the atmosphere by reactions of thermal neutrons from cosmic rays reacting with nitrogen-14. Because of its short half-life, a steady state concentration of tritium is maintained in the atmosphere. Tritium becomes incorporated in atmospheric water molecules, replacing a stable hydrogen-1 atom, and falls to earth in precipitation. When the tritium level in groundwater from a spring was measured, it was 0.4 times that of surface water in the area of the spring.
   a. Estimate the subsurface transit time of the groundwater from the recharge point to the spring out flow.
   b. Write the nuclear reaction for the process that forms tritium in the atmosphere. Check your answer with Table 10.9.

REFERENCES

ATSDR. December 1990. *Toxicological Profile for Radium*. Agency for Toxic Substances and Disease Registry, U.S. Public Health Service.

EPA. July 9 1976. "Title 40, Part 141, National Interim Primary Drinking Water Regulations." *Federal Register* 41(133): 28402.

EPA. 1999. *25 Years of the Safe Drinking Water Act: History and Trends*. Washington, DC, EPA Office of Water. EPA 816-R-99-007.

EPA. March 2000a. *Radionuclides Notice of Data Availability*. Technical Support Document, United States Environmental Protection Agency, Office of Ground Water and Drinking Water.

EPA. December 7 2000b. "National Primary Drinking Water Regulations; Radionuclides; Final Rule." *Federal Register*, 76708–53.

EPA. 2001. *Radionuclides Rule: A Quick Reference Guide*. Available at http://www.epa.gov/ogwdw/radionuclides/pdfs/qrg_radionuclides.pdf.

EPA. March 2002. *Implementation Guidance for Radionuclides*. United States Environmental Protection Agency, Office of Ground Water and Drinking Water (4606M), EPA 816-F-00-002. http://www.epa.gov/safewater.

EPA. 2011. *Radiation Doses in Perspective*. Available at http://www.epa.gov/radiation/understand/perspective.html.

Hem, J. D. 1992. *Study and Interpretation of the Chemical Characteristics of Natural Water*. 3rd ed. U.S. Geological Survey Water-Supply Paper 2254, United States Government Printing Office, Washington, DC.

IEPA. 2006. "2005 Illinois EPA Annual Compliance Report for Community Water Supplies." Illinois Environmental Protection Agency.

Katz, B. G. 2004. "Sources of nitrate contamination and age of water in large karstic springs of Florida." *Environmental Geology*, 46:689–706; http://www.swfwmd.state.fl.us/documents/numeric-nutrient/ATTACHMENT_N_Katz_Sources_Nitrate_Contamination.pdf.

Langmuir, D. 1997. *Aqueous Environmental Chemistry*. Upper Saddle River, NJ: Prentice Hall, 600pp.

de Marcillac, P. et al. 2003. "Experimental Detection of α-Particles from the Radioactive Decay of Natural Bismuth." *Nature* 422 (24): 876–8.

Marcinowski, F., R. M. Lucas, and W. M. Yeager. 1944. "National and Regional Distributions of Airborne Radon Concentrations in U.S. Homes." *Health Physics* 66 (6): 699–706.

Miller, J., and D. Stanford. 2006. "Radium and Uranium: Compliance Strategies and Emerging Issues." Presented to Colorado Rural Water Association Annual Conference, February 13–16. Colorado Springs, Colorado.

Myrick, T. E., B. A. Berven, and F. F. Haywood. 1981. "State Background Radiation Levels: Results of Measurements Taken During 1975–1979." Report to the U.S. Department of Energy, Oak Ridge National Laboratory, Oak Ridge, TN. ORNL/TM-7343.

National Academy of Sciences. 1999. *Risk Assessment of Radon in Drinking Water.* Committee on Risk Assessment of Exposure to Radon in Drinking Water, Board on Radiation Effects Research, Commission on Life Sciences, National Research Council. Washington, DC: National Academy Press.

NMED. 2006. "Public Water System Compliance Report." Drinking Water Bureau, New Mexico Environment Department.

(Photo courtesy of Gary Witt)

On most rivers, the fraction of "new" water diminishes with distance from the headwaters, as the water becomes more used and reused.

Chapter 3, Section 3.11.1

11 Selected Topics in Environmental Chemistry

11.1 TREATMENT OF POLLUTANTS IN URBAN STORMWATER RUNOFF

(This topic assumes familiarity with the material in Chapters 3 through 5.)

Stormwater runoff carries solid and dissolved forms of metals as well as other chemical pollutants, including soil sediments and various kinds of debris, such as paper, plastic, garbage, leaf, and plant litter. Because a major constraint on stormwater treatment systems is that they provide passive or near-passive treatment, the biggest challenge for treatment of stormwater pollutants is their removal by means that require minimal or no operator control or external power sources.

The most common water treatment methods available for pollutants in urban stormwater are control of oxidation-reduction potential (ORP) and pH, addition of chemical reactants to aid precipitation, and increasing the length of retention time in vaults or basins (to increase flocculation and settling). For passive or near-passive treatment systems, the methods available for utilizing these treatments are limited.

- Stormwater treatment vaults and basins generally try to maintain positive ORP (aerobic) conditions for the best treatment of the most common stormwater pollutants, while minimizing the accumulation of undecomposed sludge, and avoiding the development of undesirable odors. ORP is mainly regulated by the dissolved oxygen (DO) level. If DO is depleted sufficiently by biodegradation of organic matter, ORP can become negative in value and anaerobic conditions will prevail. Two methods for increasing DO are (passively) by encouraging oxygen diffusion from the atmosphere by inducing turbulence in the feed flow to the treatment basin and (near-passively) by mechanical diffusion via perforated tubing connected to an air pump.

- pH control is usually not appropriate for passive treatment because it normally requires chemical additions. The pH in a stormwater treatment system is usually determined by the prevailing environmental conditions, and normally is in the range of 6–9.
- The detention time of water in the treatment system is important because it determines the amount of time available for chemical reactions to precipitate dissolved pollutants and for suspended solids to flocculate, coagulate, and settle out. Detention time is determined by the magnitude of water flows that must be handled and by the design capacity of the treatment system.
- Adding chemical reactants other than oxygen is a treatment approach that requires regular maintenance and, therefore, is usually not an option in passive treatment systems. A passive treatment system normally will contain only the chemical species that chance to flow into it from its drainage basin. Since water chemistry has a strong influence on the successful operation of a stormwater treatment system, it is helpful to determine the characteristic site-specific chemical composition of water to be treated.

It is evident from this list that a passive/near-passive stormwater treatment system has some control over just two variables: ORP and retention time. Optimal adjustment of these parameters will depend on site-specific conditions that influence the other important parameters, mainly water composition and pH.

11.1.1 Effect of ORP

Oxidizing conditions (greater than about 2 ppm of DO) are optimal for removal of organic materials by biodegradation, and (when pH is high enough) precipitation of ORP-sensitive metals as hydroxides and carbonates. However, ORP-insensitive metals and ORP-sensitive metalloids tend to be present as soluble species under oxidizing conditions (Chapter 5, Section 5.6).

Reducing conditions (less than about 1 ppm of DO) and nonacid pH values are optimal for precipitation of ORP-insensitive metals and ORP-sensitive metalloids. In addition, because reducing conditions slow biodegradation processes markedly, organic sediments and debris accumulate, immobilizing dissolved metals by surface sorption, as evidenced by the efficiency of wetlands in this respect. However, if the organic load is too high, biodegradation can consume DO to the point of causing anaerobic conditions and slowing further biodegradation. This can result in the rapid accumulation of undecomposed sludge, which must be frequently removed, and may cause odor problems because of the generation of ammonia, hydrogen sulfide, and various microbially generated gases (Chapter 4, Section 4.6).

When sufficient sulfate is present in the stormwater, a condition that is not very common, reducing conditions are also suitable for precipitating both ORP-sensitive and ORP-insensitive metals, because sulfate is rapidly reduced to sulfide by sulfate-reducing bacteria and most metals will then precipitate as insoluble metal sulfides. Sulfide precipitation can achieve lower dissolved metal concentrations over a wider pH range than can hydroxide and carbonate precipitation. However, sulfide can react with water to form H_2S, a toxic gas with a very disagreeable odor (Section 4.5). Water conditions promoting the formation of H_2S are a sulfate concentration >60 mg/L, ORP less than −60 mV, and pH <6.

11.1.2 Removing Pollutants from Stormwater

Typical stormwater treatment systems, such as stormwater collection vaults, are designed primarily to remove sediment and debris. They serve only incidentally to remove other pollutants such as metals, accomplishing this mainly by having a long enough retention time, under oxidizing conditions, to precipitate a fraction of the dissolved metals as hydroxides and oxides. However, at many locations, incidental metal removal may not be adequate if nonpoint source contamination by metals is an important secondary (and sometimes primary) cause of water quality impairment.

Ideally, a stormwater vault will collect storm runoff, retain the sediments, and immobilize the pollutants so that they do not enter surface water bodies or groundwater aquifers. Although this goal is never perfectly attained in practice, knowledgeable management of a storm vault's design and operating conditions can go a long way toward assuring that these best management practice (BMP) devices are in fact operating at their best. Unfortunately, the optimal environmental conditions for removing pollutants are often different for different pollutants. For passive treatment, this makes it necessary to find the most efficient compromise for each site. Nevertheless, because most sites have a limited variety of critical pollutants, satisfactory operating conditions are often attainable.

Dissolved pollutants may include metals, nitrogen and phosphorus nutrients, and soluble organic compounds. Although the focus of this case study is the removal of metal compounds from stormwater, it is necessary to understand the behavior of nutrients and organic matter in a stormwater detention basin because the ORP is strongly affected by biodegradation processes.

Dissolved organic substances that biodegrade are removed by biodegradation, which converts them to carbon dioxide, water, and a variety of less objectionable organic substances. They are referred to as BOD because they consume DO in the course of their removal by biodegradation.

Dissolved metals are removed from stormwater by precipitation as insoluble compounds and adsorption to the surfaces of sediments and organic debris.

Pesticides and herbicides are removed by biodegradation and adsorption to the surfaces of sediments and organic debris.

Phosphorus, from fertilizers, cleaning products, and animal wastes, is removed by precipitation as insoluble compounds, adsorption to the surfaces of sediments and organic debris, and utilization as an essential nutrient by biodegrading microbes.

Nitrogen, from fertilizers and animal wastes, may be present as organic nitrogen, ammonia, nitrites, or nitrates. Organic nitrogen (protein, urea, etc.) is converted to ammonia, which is removed by adsorption to solid surfaces and by bacterial conversion to nitrogen gas. If the DO level is high enough, ammonia will be converted to nitrites and nitrates. All nitrite and nitrate compounds have low adsorption potential and high solubility. Although a small fraction of the nitrates and nitrites present may be utilized as nutrients by biodegrading microbes, the much larger dissolved fraction cannot be removed in a stormwater vault or detention basin.

Good operation of a stormwater treatment BMP means maintaining conditions that:

- Trap sediments, so they are not washed out with overflow releases.
- Promote the conversion of dissolved metals to solid forms by precipitation and sorption to sediments.
- Retard dissolution of metals already precipitated and desorption of metals sorbed to sediments.
- Encourage biodegradation of dissolved and solid organic substances, such as pesticides and petroleum products.

Unfortunately, no single set of operating conditions can realize all these goals simultaneously. For example, maintaining adequate oxidizing conditions (greater than about 2 ppm of DO) in a stormwater vault encourages biodegradation of organic matter and precipitation of certain metals, such as iron, manganese, and copper. On the other hand, some pollutants, such as arsenic, selenium, lead, and zinc, tend to be present as soluble species under oxidizing conditions. These pollutants are best retained as solids when reducing conditions exist (<1 ppm of DO). When sufficient sulfate is present, reducing conditions are best for retaining most metals because the sulfate is reduced to sulfide, and metals precipitate as insoluble metal sulfides. However, if the organic load is high, reducing conditions can cause odor problems because of the generation of ammonia and hydrogen sulfide.

The bottom line is that proper operation of a stormwater vault is site specific and may require some compromises. In some cases, it might be necessary to operate one or more stormwater vaults in a manner that provides sequential treatment in separate oxidizing and reducing zones.

11.1.3 CASE STUDY: BEHAVIOR OF TRACE METALS IN STORMWATER VAULTS

The behavior of the trace metals most commonly measured in urban runoff (Pb, Cu, Zn, Cd, Cr, and Ni; plus the metalloid As) is strongly affected by the presence of iron and manganese, two generally more abundant metals not often measured in urban runoff.

For most trace metals (except for Cr, Cu, and Hg), the conditions of greatest solubility are low pH and oxidizing positive ORP. The insoluble elemental form of a metal is usually the most reduced form. However, environmental conditions seldom reach low-enough ORPs to precipitate elemental metals. In general, less severe reducing conditions that do not precipitate elemental metals help to retain trace metal cations sorbed to organic sediments, as evidenced by the efficiency of wetlands in this respect. In addition, under reducing conditions, sulfate is reduced to sulfide, one of the most efficient precipitants of dissolved metals. As long as the pH remains between 6.5 and about 9, none of the trace metals commonly monitored in stormwater will increase its solubility with decreasing ORP.

However, the more abundant iron and manganese behave differently. Both of these metals pass through a soluble reduced state at environmental pH values when water ORP is negative. Therefore, precipitated Fe and Mn may be dissolved from sediments under reducing conditions. Because dissolved trace metals tend to sorb strongly to iron and manganese precipitates, the formation of Fe and Mn precipitates often serves as a sink for dissolved trace metals. If Fe and Mn solids dissolve under reducing conditions, the sorbed trace metals can become mobilized. Thus, if Fe and Mn are present in high concentrations, reducing conditions could mobilize trace metals. If Fe and Mn concentrations are low, reducing conditions are not likely to mobilize trace metals.

Two conditions could serve to retain trace metals on sediments under reducing conditions in the presence of Fe and Mn solids:

1. If the sulfate concentration is high, it will be reduced to sulfide and react to form insoluble Fe and Mn sulfides, which will remain efficient scavengers of dissolved trace metals. In addition, all trace metal sulfides are insoluble.
2. If the organic content of the sediments is high, trace metals will sorb preferentially to most organic solids and be retained under reducing conditions, as in a wetland.

However, if the organic content of trapped sediments is high, the combination of oxidizing conditions with high COD or BOD levels may be of concern. These are the conditions that promote oxidation of organic matter with the concurrent generation of dissolved carbon dioxide, resulting in lowering pH to values as low as 3 or 4. Under conditions in a sediment trap, low pH could be the most significant environmental factor for mobilizing metals. Thus, low DO, which will slow biodegradation processes, may be a benefit in the case of high BOD levels in sediment traps. The use of stormwater vaults for passive treatment of runoff will always represent performance compromises, but by adapting to site conditions, these BMPs can make important improvements to water quality.

11.2   WATER QUALITY PROFILE OF GROUNDWATER IN COAL-BED METHANE FORMATIONS

Drilling for oil and gas often raises concerns about potential detrimental impacts that the drilling and production activities might have on surface waters and water wells in the vicinity. An energy company planning to explore for coal-bed methane in a western state undertook a baseline water quality study of water wells and surface waters in the proposed exploration region. The plan was to collect historical and current water quality data into a database, and then compare water samples from the geologic formations where methane was found with other samples from surface water and groundwater sources in the area. The assumption was that waters with similar chemical profiles might be hydraulically connected and that dissimilar profiles would indicate that the water sources were not strongly connected hydraulically. Waters not hydraulically connected to the coal-bed methane formations were deemed unlikely to be affected by drilling and production activities.

This approach is based on the recognition that the water quality of wells and springs is influenced by the hydrogeology of their water sources. Total dissolved solid (TDS) concentrations in groundwater tend to increase with the length of time that the water is in contact with minerals in the geologic formations. TDS concentrations in wells and springs located in groundwater recharge areas, close to where precipitation first enters an aquifer, are expected to be comparatively low because this groundwater has been in the formation for a relatively short time. Likewise, a well or spring located in a deeper groundwater discharge area is likely to have higher TDS concentrations since this water has had a longer residence time in the subsurface. Overall, groundwater flow in the energy company's study area was generally toward deeper parts of the geologic formations. Thus, TDS concentrations should increase with increasing well depth, as was generally observed.

Because the database included water quality measurements from approximately 500 individual water sources, a visual method utilizing Stiff diagrams (see Figure 11.1) proved convenient for comparing chemical profiles. In a Stiff diagram, each ion concentration (in milliequivalents per liter) is plotted on a separate horizontal axis, extending out on each side from a central vertical axis representing zero concentration. In Figure 11.1, anions are plotted to the right of zero and cations to the left. The plotted points are connected as in Figure 11.1 to give an irregular polygon shape that is distinctive for different water types. It is obviously important to always arrange the ion plot species in the same order.

For these studies, the Stiff diagrams plotted the sample concentrations of six major ions (sodium, calcium, magnesium, chloride, bicarbonate, and sulfate) as shown in Figure 11.1. These major ions normally comprise over 90% of all dissolved solids in natural waters; therefore, their total concentration may be taken to closely represent the TDS concentration of the sample. The shape of a Stiff diagram allows a rapid identification of a water quality profile for the major ions, for example, predominantly sodium carbonate (Na_2CO_3) versus sodium chloride (NaCl). The total area of the polygon gives an approximate indication of the TDS concentration.

The study found that water from exploration wells drilled into the formations where coal-bed methane was expected had a distinct major ion profile compared to streams, springs, and water supply wells in the study area. From this, the energy company concluded that drilling gas wells into the coal-bed methane formations was not likely to affect most existing water supplies because there appeared to be no significant hydraulic connections between them.

Figure 11.1 shows characteristic Stiff diagrams for different water sources obtained from this study.

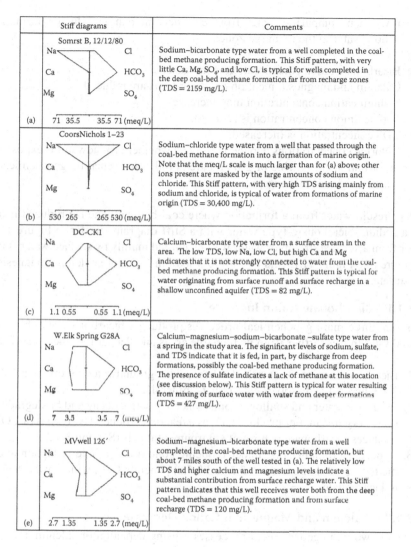

	Stiff diagrams	Comments
(a)	**Somrst B, 12/12/80** Na ⎯ Cl Ca ⎯ HCO$_3$ Mg ⎯ SO$_4$ 71 35.5 35.5 71 (meq/L)	Sodium–bicarbonate type water from a well completed in the coal-bed methane producing formation. This Stiff pattern, with very little Ca, Mg, SO$_4$, and low Cl, is typical for wells completed in the deep coal-bed methane formation far from recharge zones (TDS = 2159 mg/L).
(b)	**CoorsNichols 1–23** Na ⎯ Cl Ca ⎯ HCO$_3$ Mg ⎯ SO$_4$ 530 265 265 530 (meq/L)	Sodium–chloride type water from a well that passed through the coal-bed methane formation into a formation of marine origin. Note that the meq/L scale is much larger than for (a) above; other ions present are masked by the large amounts of sodium and chloride. This Stiff pattern, with very high TDS arising mainly from sodium and chloride, is typical of water from formations of marine origin (TDS = 30,400 mg/L).
(c)	**DC-CK1** Na ⎯ Cl Ca ⎯ HCO$_3$ Mg ⎯ SO$_4$ 1.1 0.55 0.55 1.1 (meq/L)	Calcium–bicarbonate type water from a surface stream in the area. The low TDS, low Na, low Cl, but high Ca and Mg indicates that it is not strongly connected to water from the coal-bed methane producing formation. This Stiff pattern is typical for water originating from surface runoff and surface recharge in a shallow unconfined aquifer (TDS = 82 mg/L).
(d)	**W.Elk Spring G28A** Na ⎯ Cl Ca ⎯ HCO$_3$ Mg ⎯ SO$_4$ 7 3.5 3.5 7 (meq/L)	Calcium–magnesium–sodium–bicarbonate –sulfate type water from a spring in the study area. The significant levels of sodium, sulfate, and TDS indicate that it is fed, in part, by discharge from deep formations, possibly the coal-bed methane producing formation. The presence of sulfate indicates a lack of methane at this location (see discussion below). This Stiff pattern is typical for water resulting from mixing of surface water with water from deeper formations (TDS = 427 mg/L).
(e)	**MVwell 126′** Na ⎯ Cl Ca ⎯ HCO$_3$ Mg ⎯ SO$_4$ 2.7 1.35 1.35 2.7 (meq/L)	Sodium–magnesium–bicarbonate type water from a well completed in the coal-bed methane producing formation, but about 7 miles south of the well tested in (a). The relatively low TDS and higher calcium and magnesium levels indicate a substantial contribution from surface recharge water. This Stiff pattern indicates that this well receives water both from the deep coal-bed methane producing formation and from surface recharge (TDS = 120 mg/L).

FIGURE 11.1   Stiff diagrams of representative water sources.

11.2.1 GEOCHEMICAL EXPLANATION FOR THE STIFF PATTERNS

When precipitation flows over and under the earth's surface, it reacts with the minerals that it contacts, accumulating a characteristic dissolved ion content. The chemical makeup of a collected water sample reflects the history of its prior flow path, particularly the minerals it has contacted, its contact time with these minerals, and the temperature, pH, and redox potential along its flow path. The major dissolved ions in groundwater usually are those used in the Stiff diagrams of Figure 11.1: Na$^+$, Ca^{2+}, Mg^{2+}, Cl$^-$, CO$_3^{2-}$, and SO$_4^{2-}$. As groundwater moves deeper toward bedrock, geochemical processes change its chemical profile, generally in a predictable way.

Typically, when comparing water from coal-bed methane–producing formations with shallow water in the recharge zone:

- Bicarbonate anion concentration is increased.
- Calcium and magnesium cation concentrations are reduced.
- Sodium cation concentration may increase.
- Sulfate anion concentration is reduced.
- TDS concentration is increased.
- Chloride concentrations depend mainly on whether groundwater is exposed to formations of marine origin and are not affected much by geochemical reactions like sorption or precipitation.

As a result, water from a formation where coal-bed methane is produced is usually a sodium-bicarbonate-type water with a Stiff diagram similar to Figure 11.1a. Water from a surface stream or water well recharged mainly by surface water is usually more like Figure 11.1c and d, with less sodium and more calcium, magnesium, and sulfate.

11.2.1.1 Bicarbonate Anion Increase

There are three main geochemical processes producing bicarbonate enrichment as groundwater moves farther from its recharge areas and deeper into the subsurface:

1. Bicarbonate anion is enriched when acidic groundwater (Sections 3.4.2, 3.4.3, and 5.4.1) dissolves bicarbonate minerals.
2. In surface water and shallow groundwater, oxidative decay and biodegradation of organic matter produce CO_2 as a reaction product. Hydrolysis of CO_2 produces bicarbonate at normal environmental pHs (Section 3.4.2).
3. In deeper groundwaters, under reducing conditions, sulfate reduction and methanogenesis (Sections 8.5.7 and 8.5.8) generate CO_2, which hydrolyzes to bicarbonate.

11.2.1.2 Calcium and Magnesium Cation Decrease

There are two main geochemical processes causing depletion of calcium and magnesium as groundwater moves farther from its recharge areas and deeper into the subsurface:

1. The most important depletion process is the precipitation of calcium and magnesium carbonates (calcite and dolomite) that occurs because of the higher concentrations of bicarbonate that develop along the flow path. The solubility of calcium and magnesium carbonates decreases with increasing bicarbonate concentration (Stumm and Morgan 1981).
2. Ion exchange with sodium on clays can also deplete calcium and magnesium concentrations. Ion-exchangeable clays along the flow path generally contain a large fraction of sorbed Na^+. The divalent cations Ca^{2+} and Mg^{2+} sorb more strongly to clays than monovalent Na^+ and, therefore, will exchange with Na^+ on the clay surface, releasing Na^+ into solution.

11.2.1.3 Sodium Cation May Increase

There are three main geochemical processes causing sodium to increase as groundwater moves farther from its recharge areas and deeper into the subsurface:

1. In recharge areas, infiltrating groundwater dissolves soil salts, often accumulating relatively high concentrations of calcium, magnesium, and sodium cations. Processes described previously subsequently deplete calcium and magnesium, but not sodium. Sodium has no important precipitation reactions that control its concentrations.
2. Where groundwater contacts clay minerals, sodium is displaced from clay surfaces by ion exchange with calcium and magnesium and becomes dissolved into the water.
3. Groundwater in contact with formations of marine origin can accumulate extremely high concentrations of sodium (>100,000 mg/L).

11.2.1.4 Sulfate Anion Decrease

There are two important geochemical processes causing depletion of sulfate as groundwater moves farther from its recharge areas and deeper into the subsurface (see Section A.2.38 in Appendix A, and Chapter 4, Section 4.5.2 for additional details):

1. Sulfate salts are abundant in soils where oxidizing conditions exist. Infiltrating water can accumulate sulfate concentrations >1500 mg/L. However, as water moves deeper to where reducing conditions prevail, sulfate is lost by being reduced to sulfide, forming hydrogen sulfide gas or precipitating as low-solubility metal sulfides.
2. Reduction of sulfate to sulfide is greatly accelerated by microorganisms, and sulfate-reducing bacteria are ubiquitous in the subsurface (Chapter 4, Section 4.5 and Chapter 9: 9.5.7). Methane formation in coal beds begins under reducing conditions after sulfate and other electron acceptors have been depleted (Chapter 9: 9.5.8). Therefore, methanogenesis can occur only after sulfate is bioreduced to a low concentration. Thus, biogenic coal-bed methane development is always accompanied by a nearly total depletion of sulfate by bacterial reduction to sulfide.

The unique chemical profile of groundwater associated with coal-bed methane formations has been suggested as a diagnostic tool for identifying promising sites for coal-bed methane extraction (Voast 2003).

11.3 INDICATORS OF FECAL CONTAMINATION: COLIFORM AND STREPTOCOCCI BACTERIA

Detecting and preventing fecal contamination is of prime importance for all drinking water systems and recreation water managers. Fecal wastes may contain enteric pathogens (disease-causing organisms from the intestines of warm-blooded animals) such as viruses, bacteria, and protozoans (which include *Cryptosporidium* and *Giardia*). Fecal-contaminated water is a common cause of gastrointestinal illness,

including diarrhea, dysentery, ulcers, fatigue, and cramps. It also may carry pathogens that cause a host of other serious diseases such as cholera, typhoid fever, hepatitis A, meningitis, and myocarditis.

Testing water directly for individual pathogenic organisms is, so far, impractical
for several reasons:

- There are so many different kinds of pathogens that a comprehensive analysis would be very expensive and time-consuming, whereas time is of the
  essence for pathogen detection.
- Pathogens can be dangerous at small concentrations, which require large
  sample volumes for analysis. This adds to the time and cost of analysis.
- Reliable analytical methods for several important pathogens are difficult
  or not even available. Also, not all waterborne pathogenic microorganisms
  are known.
- A satisfactory alternative is available, namely, the identification of indicator species that are easy to measure and are always present with enteric
  pathogens.

Hence, awareness of possible contamination by enteric pathogens is based on
detecting the more easily identified indicator species, whose presence indicates that
fecal contamination may have occurred.

The five indicator species most commonly used today are total coliforms, fecal
coliforms, *Escherichia coli*, fecal streptococci, and enterococci. All are bacteria
normally present in the intestines and feces of warm-blooded animals, including
humans. All but *E. coli* consist of groups of bacterial species that are similar in
shape, habitat, and behavior. *E. coli* is a single species within the fecal coliform
group. Some strains of *E. coli* are pathogenic, but the other indicators are usually not
pathogens and do not pose a danger to humans or animals. However, if any of the
indicators are present in water, the accompanying presence of a dangerous population of enteric pathogens is a possibility.

All the indicator species are easier to measure than most pathogens, but harder
to kill. Therefore, treatment that satisfactorily destroys the indicator species may be
assumed to have also destroyed enteric pathogens that were present. For example,
in wastewater disinfection, it is assumed that a decrease in fecal coliforms to <200
fc/100 mL (fecal coliforms per 100 mL of sample) will have eliminated the great
majority of pathogens.

Total coliforms and fecal coliforms are the old reliable indicators of fecal contamination, used since the 1920s to protect public health. However, both have limitations that stimulate regulators to continue seeking improved methods (see Sections
11.3.1 through 11.3.5).

11.3.1 TOTAL COLIFORMS

Total coliform bacteria, which include fecal coliforms and *E. coli*, are widespread
in nature. In addition to their animal intestine habitat, they occur naturally in plant
material and soil. Therefore, their presence does not necessarily indicate fecal

contamination. Total coliforms are not recommended as indicators of recreational water contamination, where they are usually present from soil and plant contact. Total coliforms are the standard test for contamination of finished drinking water, where contamination of a water supply or distribution system by fecal, plant, or soil sources is not acceptable. Federal drinking water standards are based on total coliform bacteria. The Environmental Protection Agency's maximum contaminant level (MCL) for drinking water is 0 total coliforms per 100 mL of water for 95% of samples after treatment (chlorination, ozonation, and UV).*

11.3.2 Fecal Coliforms

Fecal coliforms are a more fecal-specific subset of total coliform bacteria. However, even the fecal coliform group contains a genus, *Klebsiella*, with member species not necessarily fecal in origin. *Klebsiella* coliforms are found in large numbers in textile, pulp, and paper mill wastes. Fecal coliforms are widely used to monitor recreational waters and are the only indicator approved for classifying shellfish waters by the U.S. Food and Drug Administration's National Shellfish Sanitation Program.

On the basis of statistical data, the EPA has recently begun to recommend *E. coli* and enterococci as better indicators of health risk from water contact. However, many states still use fecal coliforms for this purpose, in part so that new data can be directly compared with historical data. Natural surface waters usually contain some background level of fecal coliforms, usually <15–20 fc/100 mL MPN (most probable number). In sewage entering a waste treatment plant, the fecal coliform count may be above 10 million fc/100 mL. Satisfactory disinfection of secondary effluent from a waste treatment plant is defined by an average fecal coliform count of <200 fc/100 mL. Fecal coliforms are normally absent after wastewater percolates through 5 ft of soil.

State water standards for fecal coliform levels vary, but typical state standards are (geometric mean values) as follows:

- Class 1 Primary Contact Recreational: 200 fc/100 mL
- Class 1 Secondary Contact Recreational: 2000 fc/100 mL
- Domestic Water Supply (before treatment): 2000 fc/100 mL

11.3.3 Escherichia coli

E. coli is a single species of fecal coliform bacteria that occurs only in fecal matter from humans and other warm-blooded animals. Although *E. coli* bacteria in general are necessary for proper digestion of food and do not cause illness unless

---

* The EPA Total Coliform Rule states: For water systems collecting 40 or more samples per month, no more than 5% can test positive for total coliforms. No more than one sample can be positive for total coliform if fewer than 40 samples per month are collected. Every sample containing total coliforms must be analyzed for fecal coliforms, which must be equal to zero coliforms per 100 mL, i.e., the MCLG = 0 for total coliforms, fecal coliforms, and *E. coli*.

introduced directly into an open wound or the urinary tract, their presence in water indicates fecal contamination and may suggest the presence of harmful bacteria such as *Salmonella* and *Giardia*. However, a few strains of *E. coli*, generally found in poorly prepared food, are disease-causing pathogens. The infamous *E. coli O157:H7* creates a toxic byproduct that can be fatal to children and seniors.

EPA studies (USEPA 1986) indicate that, in fresh water, *E. coli* correlates better with swimming-related illness than do fecal coliforms. Since 1986, the EPA has been recommending that states use *E. coli* as an indicator for fecal-contaminated freshwater recreation areas, instead of fecal coliform. States vary in their adoption of this recommendation.

Concentrations of *E. coli* (like other bacteria) can vary by orders of magnitude, and a single measurement is never a good indicator of health risks. The EPA requires using the geometric mean of a number of measurements, based on site-specific conditions, to minimize bias in reported values. The EPA-recommended action criteria for *E. coli* are as follows:

* For a single sample: 235-mpn colonies/100 mL*
* For the geometric mean: 126-mpn colonies/100 mL

States may set *E. coli* standards more stringent but not more lenient than these EPA criteria.

11.3.4 FECAL STREPTOCOCCI

The normal habitat of fecal streptococci bacteria is the gastrointestinal tract of warm-blooded animals. Because humans differ from most other animals in the relative amounts of coliforms and streptococci normally present in their intestines, it was believed in the past that the ratio of fecal coliforms to fecal streptococci could be used to differentiate human fecal contamination from that of other warm-blooded animals. A ratio >4 was considered indicative of human sources, and a ratio <0.7 suggested animal sources. However, this is no longer regarded as a reliable test because of variable survival half-lives in water of different species of streptococci and the effects of wastewater disinfection and different analytical procedures on the measured coliform/streptococci ratio.

11.3.5 ENTEROCOCCI

Enterococci are a subgroup of fecal streptococci, differentiated by their ability to thrive in saline water over a wide range of temperatures. In this respect, they mimic many enteric pathogens better than do the other indicators. The EPA has recommended enterococci as the best indicator of health risk in salt recreational waters and as a useful indicator in fresh waters as well.

---

* mpn colonies = most probable number of bacterial colonies, based on the counting technique.

11.3.6 CASE STUDY: BACTERIAL SOURCE TRACKING

Segments of a creek in an eastern state were cited on the state's 303d list of impaired waters as contaminated in excess of the whole-body contact standard by fecal coliform from unknown agricultural sources. The state standards for fecal coliform and *E. coli* in streams classified as fishable/swimmable were a geometric mean of 200-mpn colonies/100 mL and 126-mpn colonies/100 mL of water, respectively. These standards were enforced during the swimming season (April–October). Poultry farms were suspected to be the source of the fecal coliform.

The basic approach to identifying particular species as the source of fecal contamination is known as bacterial source tracking, also known as microbial source tracking. Because direct identification of pathogenic microbes in stream water is difficult and expensive, preferred methods estimate the potential presence of waterborne pathogens indirectly by measuring harmless indicator organisms of fecal origin. The advantage of using indicators is that they are present in feces in large numbers and are easily measured.

Methods employing indicators are based on identifying genetic or functional characteristics of the target bacteria that associate them with specific animal hosts. This requires the creation of a reference library of bacterial genetic properties from known host species near the site. Genetic data profiles of bacteria isolated from environmental water samples are then compared to library data for host identification. This study used DNA source tracking to determine the sources of fecal coliform and *E. coli* in the creek watershed and confirmed that poultry feces appeared to be the main source of fecal contamination.

RULES OF THUMB

1. To determine whether recreational water or wastewater meets state water quality requirements, find out which indicator (fecal coliform, *E. coli*, or enterococci) your state uses for recreational or wastewaters and measure that one.
2. For compliance with federal drinking water standards, measure total coliforms, which include fecal coliform and *E. coli*.
3. For a nonregulatory determination of the health risks from recreational water contact, measure *E. coli* or enterococci.

11.4 REUSING MUNICIPAL WASTEWATER: THE PROBLEM OF PATHOGENS

Reuse of municipal wastewater requires special care to minimize hazardous exposure of humans and animals to waterborne pathogens. Wastewater that has received secondary treatment generally contains residual active viruses and other pathogens (Rose and Gerba 1991), which can persist to varying degrees after release to the environment. Removal of microorganisms is accomplished by filtration, adsorption, desiccation, radiation, predation, and exposure to other adverse conditions.

Because of their large size, protozoa and helminths are removed primarily by filtration. Bacteria are removed by filtration and adsorption. Fecal coliforms are normally absent after wastewater percolates through 5 ft of soil. Virus removals are not as well documented (USEPA 1981), but it is generally agreed that viruses are removed from subsurface water flow almost entirely by adsorption to soil particles. As shown below, removal is not the same as inactivation and is generally not permanent. Viruses sorbed to soil particles can be released again still in an active state. Once viruses are mobilized in the environment, they may become inactivated by a variety of factors, such as higher temperatures, pH, loss of moisture, exposure to sunlight, inorganic cations and anions, and the presence of antagonistic soil microflora.

Viral persistence in the environment can be prolonged by low temperatures, adsorption to particulate surfaces, and moist conditions (Bitton 1975). Under appropriate conditions, viruses can remain infective for several months in wastewater sludge and environmental waters (Hurst 1989). In a study of virus occurrences in treated wastewaters in Arizona and Florida, average virus levels ranged from 13 to 130 plaque-forming units per 100 L (pfu/100 L) (Rose and Gerba 1991). In most of the viral-positive samples, viral effluent quality significantly exceeded the standard for unrestricted irrigation of 1 pfu/40 L. The infections dose for viruses is reported to be 1–10 viral units (USEPA 1992a). The authors performed a risk analysis, which found that viral effluent quality in Arizona and Florida would potentially produce two infections per 1000 exposed persons.

A review of the general literature shows that the persistence of pathogens associated with the use of municipal wastewater for aquifer recharge is very site specific. In particular, the behavior of viruses, which are more difficult to remove or inactivate than bacteria or parasites, depends strongly on environmental conditions and the types of viruses present (Bitton 1980). Some studies found that viruses are completely inactivated quickly after introduction to the soil/groundwater environment, whereas other studies found that they can persist for long periods (over 1 year) (Pinholster 1995; USEPA 1981) and over long travel distances (100 m) (USEPA 1981).

There is no question that viruses have a potential for high mobility and persistence in groundwater. A National Research Council report (NRC 1994) emphasizes that there are significant uncertainties associated with predicting the transport and fate of viruses in recharged aquifers, and that these uncertainties make it difficult to determine the levels of risk from any infectious agents still contained in the disinfected wastewater. The NRC report endorses groundwater recharge practices in general, but cautions that current recharge technologies are "especially well-suited to nonpotable uses such as landscape irrigation" and that "potable reuse should be considered only when better-quality sources are unavailable." This report states further that "water quality monitoring and operations management should be more stringent for recharge systems intended for potable reuse."

The state of California has been in the forefront of wastewater-recycling applications because of chronic water shortages and the threat of saltwater incursions into freshwater aquifers. The California water reuse regulations pay particular attention to enteric viruses (viruses that are shed in fecal matter) because of the possibility of contracting disease with relatively low doses and the difficulty of routine examination of wastewater for their presence. California requires essentially virus-free effluent via a full treatment process for wastewater reuse applications with high potential for human exposure (Asano et al. 1992).

11.4.1 TRANSPORT AND INACTIVATION OF VIRUSES IN SOILS AND GROUNDWATER

Viruses are the smallest wastewater pathogens, consisting of a nucleic acid genome enclosed in a protective protein coat called a capsid. A virus capsid contains many ionizable proteins that are subject to protonation and deprotonation reactions in water, depending on the pH and ionic strength of the water. At low pH, virus particles tend to carry a positive charge because of attached H^+ ions. As the pH rises, the positive charge on a virus particle decreases, then passes through zero at the isoelectric point, and becomes negative due to increasing numbers of attached OH^- ions. As a result, viruses can have the ion-exchange characteristics of either cations or anions, depending on the pH.

In groundwater and soils, viruses move as colloidal particles. Due to their small size, it is believed that viruses are not significantly removed from groundwater by membrane filters coarser than reverse osmosis (RO) or nanofiltration (NF), see Chapter 3, Sections 3.11.6.8 and 3.11.6.9. Viruses become attached to soil particles mainly by sorption forces arising from electrostatic interactions, London forces, hydrophobic forces, covalent bonding, and hydrogen bonding (Aronheim 1992; USEPA 1981). As a result, the extent to which viruses are sorbed to soils depends strongly on pH, temperature, ionic strength, and flow velocity of the water, as well as the mineral-organic composition and particle size distribution of the soil and the particular type of virus. Clay soils are more retentive than sandy soils, and finely divided soils retard virus mobility more than coarser soils (USEPA 1981). Two recent summaries of the behavior of viruses and other pathogens in soils and groundwater are found in Gerba and Smith (2005) and Chu et al. (2003).

The isoelectric point of enteric viruses is usually below pH 5, so that in most soils, enteroviruses carry a negative charge, as do most soils. In general, virus adsorption to soil is enhanced at lower pH values (pH < 7), where soil and virus charges are opposite, and reduced at higher pH values (pH > 7) (Wallis et al. 1972). If chemical conditions change or the flow velocity is increased, either locally by microscopic changes or overall by macroscopic changes, sorbed virus particles can be detached from soil surfaces and returned to suspension in the flow. Waters with high TDS concentrations favor adsorption to soils because electrostatic repulsion is minimized in waters with high ionic strength. A rain event can dilute TDS levels near the surface and cause a burst of released viruses. The same burst effect can occur with a release of higher pH water that locally raises the water column pH from 7.2 to 8 or 9 (USEPA 1981). For these reasons, virus adsorption to soils cannot be considered a process of absolute immobilization of the viruses from the water. Infective viruses are capable of release from soil particles after immobilization for long periods. Any environmental change that reduces their attraction to soil particles will result in their further movement with groundwater.

The presence of organic matter, such as humic and fulvic acids, in soils has been shown to inhibit the adsorption of viruses to soil surfaces by competing with viruses for adsorption sites (Burge and Enkiri 1978; Gerba 1975; Lo and Sproul 1977; Scheurman et al. 1979; Sobsey and Hickey 1985). In one study (Scheurman et al. 1979), the presence of organic substances in an aqueous environment reduced the retention of viruses in a soil column from >99% to <1.5%.

Adsorption to soil particles may prolong viral lifetimes in aqueous environments (Bitton et al. 1979; Foster et al. 1980; Sobsey et al. 1980; Stotsky et al. 1980; USEPA 1981). Viruses bound to solids are as infectious to humans and animals as the free viruses (Hurst et al. 1980; USEPA 1981). Virus survival in soil depends on the nature of the soil, temperature, pH, moisture, and the presence of antagonistic soil microflora.

In one study using f2 bacteriophage and poliovirus type 1, 60–90% of the viruses were inactivated at 20°C within 7 days after the initial release to the soil (Lefler and Kott 1974). However, after the first 7 days, the inactivation rate slowed and polioviruses could still be detected at 91 days; f2 viruses survived longer than 175 days. At lower temperatures, up to 20% of the polioviruses survived longer than 175 days. Other studies indicate that virus lifetimes may range from 7 days to 6 months in soils and from 2 days to >6 months in groundwater (USEPA 1981).

11.5　OIL AND GREASE ANALYSIS

When you analyze for oil and grease (O&G), you are actually measuring a group of substances that have similar solubility characteristics in a designated solvent. "Oil and grease" is defined as any substance recovered from an acidified sample by extracting it into a designated solvent, and which is not volatilized during the analysis. The extraction process is called liquid–liquid extraction (LLE). After the sample is extracted, the extract may be measured either gravimetrically (drying and weighing) or by infrared spectroscopy.

The solvents used have the ability to dissolve not only O&G but also other organic substances. Some non-O&G materials commonly included in the determination of O&G are certain sulfur compounds, chlorophyll, and some organic dyes. There is no known solvent that will dissolve selectively only O&G. Some heavier residuals of petroleum (coal tar, used motor oil, etc.) may contain significant amounts of material that do not extract into the solvent. The method is entirely empirical, and duplicate results with a high degree of precision can be obtained only by strict adherence to all details of the analytical procedure.

The designated solvent changed from petroleum ether and *normal*-hexane in 1965 to freon-113 (trichlorotrifluoroethane) in 1976. While freon-113 was still the prescribed solvent, an alternative solvent (80% *n*-hexane and 20% methyl-tert-butyl ether (MTBE)) was developed for gravimetric methods, to help phase out the use of freons. Then, on May 14, 1999, the EPA approved the new Method 1664 for analyzing O&G, which must replace all other methods. It differs mainly in details of the procedure and in the quality assurance and quality control (QA/QC) requirements. Method 1664 uses pure *n*-hexane with either LLE or solid-phase extraction. Because *n*-hexane is less dense than water, LLE will not work well with all sample matrices.

11.5.1 SILICA GEL TREATMENT

The "total oil and grease" designation includes oils and fats of biological origin (animal and vegetable fats and oils) as well as mineral oils (petroleum products). Silica gel has the ability to sorb certain organic compounds known as polar compounds. Since petroleum hydrocarbons are mostly nonpolar, and most hydrocarbons of biological origin are polar, silica gel is used to separate these different types of hydrocarbons. If a solution of nonpolar and polar hydrocarbons is mixed with silica gel, the polar hydrocarbons, such as fatty acids, are removed selectively from the solution. The materials not eliminated by silica gel adsorption are designated as "petroleum hydrocarbons" by this test. Normally, total O&G is measured first and then petroleum hydrocarbons are determined after the animal and vegetable fats are removed by silica gel adsorption. The difference between total O&G and petroleum hydrocarbons is designated as biologically derived hydrocarbons.

11.6 QUALITY ASSURANCE AND QUALITY CONTROL IN ENVIRONMENTAL SAMPLING

QA/QC are the set of principles and practices that, if strictly followed in a sampling and analysis program, will produce data of a known and defensible quality. The effectiveness of any monitoring effort depends on its QA/QC program. A well-designed QA/QC program reduces, as far as possible, the potential for undetected errors appearing in the analytical results. It has many components, ranging from competency certification of personnel to replicate sampling. The overall goal is to minimize or correct all the errors that might occur, by introducing explicit steps into the sampling and analysis protocol for identifying, measuring, and controlling these errors.

11.6.1 FIELD AND LABORATORY QA/QC METHODS

Field QA/QC is intended to collect in the field and deliver to the laboratory a sample where analyte properties have not changed significantly from their values in the environmental waters being tested. An obvious, but not trivial, part of field QA/QC is properly identifying all samples so that measured analyte concentrations can be assigned to a known location and time with confidence.

Laboratory QA/QC is designed to produce a quantitative measure of certain properties of the sample with known limits of uncertainty. There are some overlapping QA/QC practices, such as including field duplicates and blind known samples with a field sample set to test a laboratory's performance.

This section deals only with field QA/QC protocols. Laboratory practices are generally out of the control of a sampling program manager, who ideally will have sufficient experience with the QA/QC standards of the laboratory being used to have confidence in the final data. However, laboratory performance can change and it is prudent to remain alert to possible laboratory errors. Some techniques for checking laboratory performance are included in Section 11.6.4.

11.6.2 ESSENTIAL COMPONENTS OF FIELD QA/QC

11.6.2.1 Sample Collection

Proper sampling is a separate science in its own right. The choice of automated versus hand-sampling equipment—that is, the details of how to get a representative sample into the container, and decisions of whether grab, composite, or time/space-integrated samples are appropriate—is not addressed here.* The importance of obtaining a representative sample for analysis is underscored when one recognizes that the variability of the sampled media is the greatest source of variability in analytical results, far exceeding measurement uncertainties.

Every field performance sampling protocol has three main objectives:

1. To collect stormwater runoff samples that are acceptable representations of environmental conditions at the place and time of sampling.
2. To store and transport these samples in a manner that maintains the important physical and chemical properties of the sample. This generally requires that samples are properly cooled and, if necessary, preservatives added and pH adjusted as soon as possible after collection.
3. To prevent contamination of samples that can result in false analytical results.

To accomplish these objectives, all procedures that involve sample handling must be consistently controlled. This includes selecting, cleaning, and properly using equipment such as sample splitters, sample containers, tubing, gloves, and other materials that may come in contact with the sample, both in the field and during transport to the laboratory.

Quality control checks (equipment and field blanks) are designed to test how well the sampling procedures are executed. However, even if all sampling procedures are performed correctly, most water sources, such as lakes, streams, wetlands, or stormwater runoff, are inherently heterogeneous, and successive samples collected from the same location will generally have some degree of variability in composition. For this reason, when certain comparisons are required, such as comparing analyses of the same sample by different laboratories or seeking correlations among different water quality parameters (e.g., association of dioxins with different kinds of sediments), it is important to make all comparisons from splits of a single sample and not from different samples, even if collected within a short time from the same location. (See Section 11.6.5 for additional discussion of this problem.)

However, certain QA/QC practices are common to all sampling procedures:

- *Sample contamination must be avoided*: All parts of the sampling equipment and containers that contact the sample must be scrupulously clean. Sample containers must be inert to the sample and preservatives.
- *Samples must be properly preserved*: In general, all samples should be stored and transported at 4°C (ice temperature) or less. In addition, samples

---

* Detailed guidance for these procedures can be found at the end of this chapter in the References.

for certain analyses require chemical preservatives, for example, acid addition to less than pH 2 for metals and ammonia, base addition to greater than pH 12 for cyanide, and so on.

- *Samples must be unambiguously identified*: If the sample container is not prelabeled, it must be securely labeled with a unique identification at the time of collection or removal from a sampler. The collector's name, date and time, station designation, form of preservation, and desired analyses should also be on the label. Sample identification must be entered on the report form without delay after labeling the sample container.
- *Samples must be packaged for transportation*: In general, samples are transported in a cooler. The cooler lid must not accidentally open, container lids must not leak, and containers should be protected against breakage.
- *Chain-of-custody and field reporting documentation must be maintained*: Careful completion of chain-of-custody (COC) and sampling report forms are essential parts of field QA/QC. COC form helps to ensure sample integrity from collection to data reporting. Entries on the COC form track the possession and handling of each sample from the time of collection through analysis and final disposition, and demonstrate that all standard steps of sample control have been followed.

The sample report form is needed for interpretation of the reported data and must be designed carefully to require appropriate entries for all relevant information. The report form will always include the name or initials of the sampler and have labeled places to enter sampling location, sample identification, date, and sampling time for each sample. It might also include space for other relevant information such as weather conditions, observations concerning the sampled waters (odor, color, flow, etc.), and sampling procedure (grab, time-integrated, etc.).

Most commercial laboratories provide coolers, ice packs, clean sample containers with preservatives appropriate for requested analyses, container labels, and COC forms and seals as part of their analytical services.

11.6.3 FIELD SAMPLE SET

A field sample set consists of the environmental samples and several kinds of QC samples. The project manager must determine which kinds of QC samples should be collected, based on the purposes of the sampling program.

11.6.3.1 Quality Control Samples

Some field samples should be collected for QC purposes. Sampling procedures can contribute to both systematic and random errors. Many environmental waters are inherently not uniformly mixed. Field blanks, spikes, and duplicates will help to estimate errors caused by sampling bias (usually contamination and/or nonrepresentative samples) and calculate sampling precision. Field QC samples must be handled exactly the same way as the environmental samples, using identical sampling devices, sampling protocol, storage containers, preservation methods, storage times, and transportation methods. For a correct interpretation, QC samples must be subject to the same holding time criteria as the environmental samples.

11.6.3.2 Blank Sample Requirements

There are several types of blank samples, each serving a distinct QC purpose. Wherever a possibility exists for a sample to become contaminated, a blank should be devised to detect and measure the contamination. The most commonly collected blanks include field, trip, and equipment blanks. Although it may be prudent to collect a full set of blanks, most of the blank samples usually do not need to be analyzed. Analysis costs can be reduced if the strategy of blank sampling is understood. First, analyze only the field blanks, which are susceptible to the broadest range of contaminant sources. If these indicate no problems, the other, more specific, blanks can be discarded or stored. If a problem is suspected, other blanks can be analyzed for confirmation of the problem and to discover the source. Holding time limits must be observed with blanks as well as with environmental samples.

- *Field blanks*: At least one clean water sample should be exposed to the same sampling conditions as the environmental samples at each sampling site. The analytical laboratory can generally provide analyte-free distilled water for this purpose. Field blanks are transferred from one container to another, passed through automatic equipment, or otherwise exposed to the conditions at the sampling site.

At a minimum, the field blank container is opened at the sampling site and exposed to the air for approximately the same time as the environmental samples. It is then capped, labeled, and sent to the laboratory with other samples.

Field blanks measure incidental or accidental sample contamination throughout all the steps of transportation, sampling, sample preparation at the laboratory, and analysis. They help to assure that artifacts are recognizable and are not mistaken as real data.

Trip blanks: For each type of container and preservative, at least one container of analyte-free water should travel unopened from the laboratory to the sampling sites and back to the laboratory. Trip blanks serve to identify contamination from the container and preservative during transportation, handling, and storage.

Equipment blanks: These are especially important with automated sampling equipment. Equipment blanks, sometimes referred to as "rinsate blanks," document adequate decontamination of the sampling equipment. These blanks are collected by passing analyte-free water through the sampling equipment after decontamination and before resampling.

The analytical laboratory will run a similar set of blanks to document contamination and errors arising from handling and analysis procedures in the laboratory.

11.6.3.3 Field Duplicates and Spikes

Field duplicates are two separate environmental samples collected simultaneously at an identical source location and analyzed individually. Field duplicates are sensitive to the total sample variability, that is, variability from all sampling, storage, transportation, and analytical procedures. Where the goals of the sampling program warrant it, for example, for permit monitoring requirements, at least one field dupli- cate per day should be collected for each analyte. More than two replicates may be required in cases where it is difficult to obtain representative samples.

Field spikes are environmental samples to which known amounts of the analytes of interest are added. Ampoules containing carefully measured amounts of analytes can be purchased from many chemical supply sources. Field spikes can identify storage, transportation, and matrix effects, such as loss of volatile compounds and analytical interferences caused by certain compounds present in the environmental source. In a spiked sample with no problems, the measured analyte concentration should be equal to the concentration measured in the environmental sample from the same source plus the added spike concentration, within the limits of precision of the method.

11.6.4 UNDERSTANDING LABORATORY-REPORTED RESULTS

When a laboratory reports that a target compound was not detected, it does not mean that the compound was not present. It always means that the compound was not pres- ent in a concentration above a certain lowest reporting limit (RL). There always is the possibility that the compound was present at a concentration below the RL. The laboratory might even have identified the compound below the RL but not reported it because the concentration could not be quantified within acceptable limits of error.

Reported results of analyte concentrations are never exact. There always is some margin of error. The only kind of measurement that can be exact is a tally of discrete objects, for example, dollars and cents or the number of cars in a parking lot. When measuring a quantity capable of continuous variation, such as mass, length, or the concentration of benzene in a sample, there always is some uncertainty. Like an irrational number, measurements capable of continuous variation can always be expressed, in principle if not in experimentally valid numbers, to more significant figures, and any answer with a finite number of digits is always an approximation.

11.6.4.1 Precision: How Scatter in Data Is Reported

Repeated measurements of the same quantity, even under essentially identical condi- tions, will always yield results that are scattered randomly about some average value, because of uncontrollable variations in environmental, experimental, and operator behavior. The person who interprets experimental data must always bear in mind that no reported experimental value necessarily represents the true value. In other words, it is never possible to be completely certain of a result. Nevertheless, we still try to answer questions such as, "Does this sample indicate a violation of a discharge limit?" or "Does the data indicate that my remediation activities are beginning to work?" Our answers to such questions must always acknowledge that there is a range of error in the available data. The purpose of QA/QC is to assure that the reported value lies within a small enough range of error to be useful data.

Part of laboratory QA/QC involves determining how much scatter in the data is to be expected from different analytical procedures. The precision inherent in a par- ticular procedure is a measure of how much deviation may be expected in repeated measurements of the same sample. Standard deviation is a common way of quantitatively expressing the precision, or reproducibility, of a measurement. A more thorough treatment of the standard deviation can be found in elementary statistics texts or in the latest edition of *Standard Methods for the Examination of Water and Wastewater* (Eaton et al. 1995). The procedure for using the standard deviation to express the reliability of measurements is described by the EPA in several documents (e.g., USEPA 1992b, 2007). Here, it is only necessary to understand that, for many measurements of the same quantity, one standard deviation above and below the average value of the measurements will include about 68% of all the individual values if the scatter is due to truly random fluctuations and not biased. For example, suppose the concentration of benzene in a sample is measured many times to yield an average value of 12.3 µg/L and a standard deviation of 2.7 µg/L. Then, 68% of all the individual measurements can be expected to lie within the range of 12.3 ± 2.7 µg/L, that is, between 9.6 and 15.0 µg/L. Two standard deviations will include about 95% of all the values.

The larger the standard deviation, the greater is scatter in the data, that is, the poorer the precision, or reproducibility, of repeated measurements. The standard deviation, and hence the measurement reproducibility, depend strongly on the matrix in which the analyte is being measured. In general, stream and groundwater samples can be measured more precisely than wastewater and soil samples. For this reason, it is not possible to make general statements about the reliability of measurements based only on the concentration values. The sample matrix is taken into account by determining the standard deviation.

11.6.4.2 Terminology of Analytical Measurement Limits

Every analysis has its own unique set of uncontrollable variables that will cause the measured value to differ from some true value in an unknown way. The realities of time and budget limits normally limit sampling programs to rely on fewer sample analyses than would be statistically ideal. For this reason, the EPA uses a concept called quantitation limits to provide safety margins in their regulations. Although the nomenclature for different quantitation limits has varied over time, their function has remained the same: to establish a legally defensible value that can be used to limit the concentration of a pollutant in the environment.

In Figure 11.2, several different kinds of measurement limits are defined in terms of the reported concentration expressed as the number of standard deviations above the instrumental zero. Those and several other common terms that are used to denote the reliability of analytical data are discussed in Sections 11.6.4.2.1 through 11.6.4.5.

11.6.4.2.1 *Method Detection Limit*

Method detection limits (MDLs) are specific to both the technique and the analyte being measured. The MDL is the theoretical method detection capability, defined as the minimum concentration of a substance that can be measured and reported with 99% confidence that the analyte concentration is greater than zero. It is the smallest concentration that can be detected above the noise in a procedure and within a defined confidence level. MDL values are determined statistically by performing the

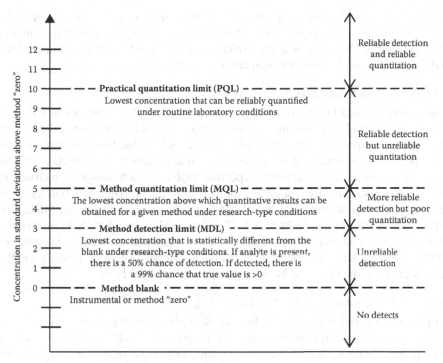

FIGURE 11.2 Statistical measures of data reliability, based on the measured concentration expressed in standard deviations above the instrumental zero. Standard deviation is determined by repeated measurements of a laboratory control sample.

complete analytical procedure (extraction/digestion, cleanup, and instrumental analysis) on a minimum of seven replicate spiked samples in an otherwise clean, interference-free matrix (water, soil, etc.) representative of the environmental matrix being tested. Since MDLs are measured with laboratory standards and not field samples, matrix interferences are not present and similar MDLs should not be expected for field samples, where site-specific matrix interferences are often encountered. Of all the measurement limits discussed here, only the MDL is rigorously defined (USEPA 2007). The statistical calculation of the MDL puts it about three standard deviations above the instrumental zero.

11.6.4.2.2 Method Quantitation Limit

The method quantitation limit (MQL) is an estimate of the lowest concentration above which reliable quantitative results can be obtained under research laboratory conditions. It represents the value at which the laboratory has demonstrated the ability to reliably measure targets within prescribed performance criteria, and it establishes the lowest concentration at which data may be reported without qualification. The MQL can be measured, but is often estimated to be 3 to 10 times the MDL (5–10 standard deviations above the instrumental zero) for a given target analyte. The MQL is often set at the level of the lowest calibration standard for the method, and the lowest calibration standard for each target must be at least three times the MDL. All target analytes detected below the MQL would be flagged as estimates (i.e., J-flagged). The MQL is a fixed reference value based on some multiple of the MDL, and it is not adjusted for sample-specific parameters.

11.6.4.2.3   Practical Quantitation Limit*

The practical quantitation limit (PQL) is an estimation of the lowest concentration above which reliable quantitative results can be obtained under routine laboratory conditions. It represents a practical and routinely achievable limit that can be used with reasonable assurance that any reported value greater than the PQL is reliable and may be used for all regulatory reporting. PQLs for each analytical method are arrived at by statistically evaluating the performance of different laboratories. If performance data are not available, the PQL can be estimated from the MDL. The EPA believes that setting the PQL at 5–10 times the MDL is generally a fair expectation for routine operation at most qualified government and commercial laboratories (USEPA 1994). Many laboratories determine their own routine performance PQLs and often use three times the MDL as the lowest level for a PQL.

11.6.4.2.4   Laboratory Reporting Limit

In addition, many laboratories include an RL (laboratory reporting limit; LRL or RL) in their analytical reports, which represents a threshold concentration above which that laboratory has confidence in their quantitative values and below which the laboratory reports a given result as nondetected.[†] LRLs are laboratory-determined values that may be based on client-requested RLs, regulatory action levels, or multiples of the MDL, but they should never be less than the lowest instrumentation calibration level. Depending on the laboratory, the LRL for an analyte may be anywhere between its MDL and its PQL, but typically is about 3–5 times the MDL for a given analyte.

11.6.4.2.5   Method Blank (Instrumental or Method "Zero")

Even when no sample is being measured, every analytical instrument generates some minimum output signal from its detector, called background noise. The method blank (or method "zero") is based only on the system noise. In the context of analytical detection limits, the lowest concentration of a target analyte that produces a signal that can be reliably distinguished from the background noise is the true detection limit of the instrumental system. This is statistically defined as the minimum analyte concentration that produces a signal 2–3 times the standard deviation of the blank signal (the noise level).[‡]

---

* Other names for practical quantitation limit (PQL) are estimated quantitation limit (EQL), lower limit of quantitation (LLQ), and limit of quantitation (LOQ). Other names for the MQL are LOQ and ML (method limit).

† Although a laboratory may report a result that is below the LRL as ND (nondetect), it is better when they report the measured value and indicate that it is below the LRL by "U-flagging" it; alternatively, they may report it as a "less than" concentration (e.g., <###, where ### is the LRL).

‡ The signal-to-noise ratio (S/N) is often used to gauge instrument sensitivity, and as a rule of thumb, S/N values in the range of 3–5 are considered sufficient to distinguish signal from noise.

RULE OF THUMB

When an official PQL has not been designated for a particular analysis, an unofficial conservative PQL of 10 standard deviations above the instrument zero (or 3.3 times the MDL) is often used.

11.6.5   Environmental Sampling of Low-Solubility Pollutants at Parts-per-Billion and Lower Concentrations

Special sampling and analytical procedures are needed for measuring environmental pollutants quantitatively at parts-per-billion (ppb) concentrations or lower. Even slight accidental sample contamination, which might not be significant in more routine sampling programs, can cause serious errors when sampling for very low concentrations of pollutants. It is important to use sampling protocols that minimize such errors, to follow the protocols carefully and consistently, and to include always the field QC blanks designed to help detect and quantify accidental contamination when it occurs.

Pollutants with low aqueous solubility, such as dioxins, Polychlorinated biphenyls (PCBs), polycyclic aromatic hydrocarbons (PAHs), and many metal compounds, require additional special attention because they tend to partition preferentially from the dissolved state to sorbed and precipitated states on solid surfaces, such as sediments and container walls. Most water sources are inherently heterogeneous and successive samples collected from the same location will generally have some degree of variability in composition. This is especially true for suspended and total solids, which never are uniformly distributed in environmental waters in the same way that dissolved pollutants can be.

When sampling for low-solubility pollutants, which are mostly found associated with the solids in a water sample, the greatest source of variability measured pollutant concentrations in replicate samples arises from spatial and temporal fluctuations in the total solids concentrations of the source. Replicate samples that vary significantly in solids concentration will also have less agreement in chemical pollutant concentrations.

For low-solubility pollutants, any sampling step that requires transferring a sample from the original collection container to other containers has the potential for introducing quantitation errors because sediments and other solids carrying pollutants are seldom transferred in a consistent manner. Sample transfer difficulties are minimized by prerinsing collection and transfer containers with source water before collecting the actual sample and by always using a device, such as a cone splitter for splitting a single sample containing sediment into two or more replicate samples. Using a cone splitter to obtain replicate samples by splitting a single sample has been shown to improve reproducibility among replicate samples split from a single sample containing sediment.*

---

* The cone splitter is a device originally designed by the U.S. Geological Survey (USGS) to improve the accuracy of splitting water samples containing sediment into two or more replicate samples; see the USGS 1997, 2007a,b,c at the end of this chapter for details concerning the use of a cone splitter in environmental sampling.

11.6.5.1　Case Study: Inherent Variability in Parts-per-Billion Analyses of Dioxins and Other Low-Solubility Pollutants

This case study contains a very brief description of certain statistical and technical limitations inherent in the sampling and chemical analysis of ppb concentrations of low-solubility pollutants such as dioxins and certain metals. Although the following discussion is focused on dioxins, the same issues apply to any low-solubility pollutant that must be analyzed at ppb concentrations.

11.6.5.1.1　Analytical Variability

Analytical limitations introduce a significant and unavoidable degree of variability in the reported results. As a current example, the expected variability of repeat measurements on a single sample containing the dioxin 2,3,7,8-tetrachlorodibenzo-p-dioxin (2,3,7,8-TCDD) at 3×10^{-8} mg/L (30 pg/L)* seems to be between ±25% and ±10% (measured as %-relative standard deviation, RSD), depending on the amount of interfering substances in the sample.[†] This is a laboratory-induced variability that must be added to the field sampling variability, which has been reported to be as large as ±30% for particulate sample types.[‡]

As sample concentrations decrease for the same analytical method, precision inevitably decreases and expected variability increases. Achieving acceptable analytical results for ppb samples requires extreme rigor and consistency in all the separate analytical steps of sample extraction, concentration, calibration, and instrumental operation. The following discussion uses dioxins as an example of an analyte with very low water solubility, but it applies to other low solubility analytes as well.

The two main reasons for a large degree of analytical variability in dioxin measurements are as follows:

1. The very low water solubility of dioxins in general.
    a. The water solubility of TCDD is 19×10^{-6} mg/L. In general, all of the dioxin congeners containing four or more chlorine atoms have similar low water solubilities.
    b. Low solubility causes dioxins in environmental water samples to be largely associated with sediments. Although the TCDD solubility of 19×10^{-6} mg/L is greater than the EPA drinking water MCL for TCDD (3×10^{-8} mg/L), it is small enough to result in most waterborne dioxin molecules being bound to sediments.
    c. Because of total suspended sediment (TSS) inhomogenieties, TSS samples from the same source tend to have higher variability in their measured concentrations than would be the case for dissolved analytes. *Standard Methods*[§] reports an analytical precision (standard deviation)

---

* 30 pg/L (30×10^{-12} g/L) is the EPA drinking water standard for 2,3,7,8-dioxin and is a common requirement for analytical samples.
† This value is estimated from information in EPA Method 1613, some reported values from the literature, and a telephone discussion with the chief analyst at Pace Laboratories.
‡ Clement, R. 1997.
§ Single-laboratory analyses of 77 samples containing a known TSS of 293 mg/L (Eaton et al. 1995).

of 21.20 mg/L for TSS analyses of water samples containing 293 mg/L of a TSS standard. This is equivalent to an RSD of ±7.2%.

d. When dioxins are distributed between dissolved and solid phases, extra analytical processing steps are required for filtration, drying the solids, different extraction procedures to separate the analyte from the solid and from water, and recombining the extracted analytes into a single sample—all before concentrating the sample to increase analytical sensitivity.

2. The necessity to measure quantitatively the analyte at a very low concentration.

a. This requires that at least 1 L of collected sample be extracted and concentrated. At least 5% of the concentrate must be injected into the GC/MS instrument to assure that there is enough dioxin present in the injected sample to obtain the required sensitivity. To analyze for 30 pg/L of dioxin, the lab calibrates with 10-pg/L spikes and adjusts instruments for a 1-pg/L detection limit. Measurement errors can be introduced in every step.

11.6.5.1.2 Sampling Variability

Sampling procedures introduce additional measurement variability, which must be added to the analytical variability. Dioxins in water containing solid sediments tend to concentrate on suspended solids carried in the flow, and it is essentially impossible to collect sequential samples from the same source in which the solids and, consequently, the dioxins are equally distributed.

The greatest source of variability in sampling water containing sediment comes from fluctuations in the suspended solids content of different samples collected at the same time from the same source. Replicate samples collected sequentially, whose percent solid levels vary significantly between replicates, will have poor agreement in dioxin concentrations between replicates. The percent standard deviation of dioxin concentrations for samples of equal sediment concentrations depends on the sample size, increasing as sample size decreases.

Thermal mixing assures that the dissolved portion of dioxin is uniformly distributed, but the sediment-sorbed portion will be spatially variable for several reasons:

- Particles with different masses are transported in flowing water at different velocities and settle vertically at different rates.
- Different sediment particles can have different surface areas because they may be of different sizes, have different porosities, and have different weathering histories.
- Dioxins will sorb differently to particles of different chemical origins. In general, for equal weights, organic sediments will contain more dioxin than inorganic sediments.
- Particles from different sources may have different densities and will not be homogeneously distributed throughout the sample volume.

Even if a single large sample were theoretically split precisely into smaller replicates, each containing exactly the same sediment concentrations, there still would be variability in their dioxin concentrations because different sediment particles will contain different numbers of sorbed dioxin molecules. The smallest dioxin variability would only be achieved if all replicates contained identical particle size distributions, identical particle density distributions, identical particle surface area distribution, and identical distributions of organic and mineral sediments, a presently impossible goal.

Now, the cone-splitter appears to be one of the best methods for splitting stormwater samples. It is simple to use and has been shown to split the aqueous samples with significant solids or sediment content more reliably than the more commonly used tilt and pour method, which has poor repeatability and yields split samples with widely varying solids concentrations and particle size distributions.

11.7 DEICING AND SANDING OF ROADS: CONTROLLING ENVIRONMENTAL EFFECTS

Road sanding and deicing, undertaken to enhance winter highway safety, have the potential to contribute significant amounts of sediment and chemicals to the receiving waters of surface runoff. To minimize impacts on surface waters, it is often necessary to incorporate physical and operational controls designed to

- Reduce the application of sand and deicing chemicals as far as practicable.
- Manage surface flow from treated roads and stockpiled materials in a manner that retains sediment and infiltrates dissolved chemicals.

11.7.1 METHODS FOR MAINTAINING WINTER HIGHWAY SAFETY

Snow and ice on roads reduce wheel traction and cause drivers to have less control of their vehicles. Highway departments currently use a site- and event-specific combination of three approaches for mitigating the effects of highway snow and ice:

1. Apply antiskid materials, such as sand or other gritty solids, to road surfaces to improve traction.
2. Apply deicing chemicals that melt snow and ice by lowering the freezing point of water.
3. Plow roads to remove the snow and ice.

Although highway safety is the first concern in the use of snow control measures, environmental impacts are also important. Many highway departments are evaluating the effectiveness of alternative chemicals and operating procedures for minimizing the environmental impacts of sanding, deicing, and snow removal, while not compromising road safety.

11.7.2 ANTISKID MATERIALS

The most commonly used antiskid material is sand, usually derived from either rivers or crushed aggregate. Other abrasives such as volcanic cinders, coal ash, and mine tailings are sometimes used, based on their local availability and cost. River sand is rounded and smooth and is somewhat less effective than crushed aggregate, which is rough and angular. However, river sand is generally cleaner and less contaminated than crushed aggregate. Between 3% and 30% by volume of deicing chemicals are often mixed with sand for increased effectiveness. The amount of sand required is very site and event specific. For example, in the Denver, Colorado metro area, the average amount of sand applied per snow event is 800–1200 lb per lane-mile of treated road, more sand generally being required in the western part of the city than in the eastern part (Chang et al. 1994). In Glenwood Canyon, Colorado, where postevent sand removal is especially difficult, highway maintenance personnel have reduced the use of sand recently from 280 to 60 lb per lane-mile by increasing the use of chemical deicers (Rocky Mountain Construction 1995).

11.7.2.1 Environmental Concerns of Antiskid Materials

Air and water contamination are potential concerns with the use of sand and other antiskid grits. In Denver, fine particulates generated by traffic abrasion of road sand have been found to contribute around 45% of the atmospheric PM_{10} (airborne particulate matter <10 μm in diameter) load during winter. In 1997, EPA standards for PM_{10} were 50 mg/m^3, annual arithmetic mean, and 150 mg/m^3, 24-h arithmetic mean. Efforts to attain compliance with these standards have compelled communities to increasingly use chemical deicers in place of antiskid grits (Chang et al. 1994).

Although airborne particulates from road sand are significant as atmospheric pollution, they represent an insignificant fraction of the total mass of sand applied to the roads. Essentially all the sand applied for traction control becomes potential wash-load that eventually is either flushed to receiving waters (including sewers, streams, and lakes), trapped in sediment control structures, or swept up and deposited in landfills.

11.7.3 CHEMICAL DEICERS

A variety of water-soluble inorganic salts and organic compounds are used to melt snow and ice from roads. The most commonly used road deicer is NaCl, because of its relatively low cost and high effectiveness. Other accepted road deicing agents are potassium chloride, calcium chloride ($CaCl_2$), magnesium chloride, calcium magnesium acetate (CMA), potassium acetate, and sodium acetate.* These chemicals may be used in solid or liquid forms and are frequently combined with one another in various ratios. Different deicer formulations have been rated for overall value based on performance (melting, penetration, and disbonding snow from the road surface), corrosivity, spalling of road surface, environmental impacts, and cost (Chang et al. 1994). Commercial formulations that use chloride salts usually include corrosion inhibitors, generally regarded as effective and worth the additional cost.

---

* Several deicers, ethylene glycol, methanol, and urea, are useful mainly for specialized purposes such as airplane and runway deicing, but seldom find highway use for reasons of poor performance, high costs, toxicity, and difficulty in application.

11.7.3.1 Chemical Principles of Deicing

Water containing dissolved substances always has a lower freezing point than pure water. Any soluble substance will have some deicing properties. How far the freezing point of water is lowered depends only on the concentration of dissolved particles, not on their nature. Given the same concentration of dissolved particles, the freezing point lowering of water will be the same for NaCl, $CaCl_2$, ethylene glycol, or any other solute. The solubility of each deicing substance at the final solution temperature determines how many particles can go into solution, and this is the ultimate limit on the lowest freezing point attainable; ice will melt as long as the outdoor temperature is above the lowest freezing point of the solute–water mixture. Pure NaCl theoretically can melt ice at temperatures as low as $-6°F$, but no lower. $CaCl_2$ is effective down to $-67°F$.

When a salt dissolves to form positive and negative ions, each ion counts as a dissolved particle. Ionic compounds such as NaCl and $CaCl_2$ are efficient deicers because they always dissociate into positive and negative ions upon dissolving, forming more dissolved particles per mole than nonionizing solutes. One NaCl molecule dissolves to form two particles, Na^+ and Cl^-, and one $CaCl_2$ molecule forms three particles, one Ca^{2+} and two Cl^-, whereas the organic molecule ethylene glycol ($C_2H_6O_2$) does not dissociate and dissolves as one particle. Three molecules of dissolved ethylene glycol are needed to lower the freezing point the same amount as one molecule of $CaCl_2$. Another advantage of calcium and magnesium chlorides is that they dissolve exothermically, releasing a significant amount of heat that further helps to melt snow and ice. Conversely, NaCl does not release heat upon dissolving; in fact, the dissolution of NaCl is slightly endothermic and has a small cooling effect.

The differences in effectiveness for different deicing chemicals are related primarily to their different solubilities at environmental temperatures, number of dissolved particles formed per pound of material, and exothermicity of dissolution. Organic deicers (e.g., CMA and ethylene glycol) are said to be more effective than salts at breaking the bond between pavement and snow, allowing for easier plowing and snow removal. Organic deicers are also believed to be stored in surface pores of pavement, aiding in disbonding the snow and possibly prolonging the period of effective deicing.

11.7.3.2 Corrosivity

The main advantage of organic deicers, such as CMA, over inorganic chloride salts, such as NaCl, is their lower corrosivity. Corrosivity results from chemical and electrolytic reactions with solid materials. The chemical corrosivity of chloride salts arises mostly from the chemical reactivity of chloride ions and does not depend strongly on which salt is the source of the chloride ions. Electrolytic corrosivity affects metals, mainly iron alloys, and occurs when dissolved salt ions transfer electrons between zones of the metal surface having slightly different composition, allowing atmospheric or DO to react chemically with the metal. Electrolytic corrosivity depends, in a complicated fashion, on the nature of the metal surface and the nature of the dissolved ions. However, electrolytic corrosion for any surface will always increase as the total ion concentration (often measured as TDS or electrical conductivity) increases.

When chloride is present, chemical and electrolytic corrosion acts synergistically to accelerate the overall corrosion rate. The addition of corrosion inhibitors to commercially formulated salt deicers is reported to reduce salt corrosivity. The main reason for using chloride salts rather than nitrates, fluorides, bromides, and the like, is the relatively low toxicity of chlorides to plants and aquatic life.

11.7.4 ENVIRONMENTAL CONCERNS OF CHEMICAL DEICERS

Corrosivity, not adverse environmental impact, has been the main problem associated with the use of chemical deicers. While each of the common deicers has potential environmental effects, studies show that none pose strong adverse threats (Michigan Department of Transportation 1993; Watershed Research 1995). Most deicing residues are highly soluble and have low toxicity; they flush quickly through soils and waterways and, in general, rapidly become diluted to levels that pose no environmental problems. Even under a worst-case scenario, undesirable effects are likely to be observed only close to the points of application, where concentrations are the highest. Studies by the Michigan Department of Transportation show that the greatest impacts have been to sensitive vegetation adjoining treated roadways; stream and lake concentrations of chloride and other deicing chemicals seldom reach levels that are detrimental to aquatic life. The state of Michigan has found that, although little surface water and groundwater contamination has been observed that is directly attributable to deicing practices, much of it has resulted from spillage and poor storage practices (Michigan Department of Transportation 1993).

11.7.5 DEICER COMPONENTS AND THEIR POTENTIAL ENVIRONMENTAL EFFECTS

11.7.5.1 Chloride Ion

There is no stream standard for chloride ion; it is generally regarded as a nondetrimental chemical component of state waters. Tests on fish showed no effect for concentrations of NaCl between 5,000 and 30,000 mg/L, depending on the species, exposure time, and water quality. Concentrations of chloride ion required to immobilize *Daphnia* in natural waters ranged between 2100 and 6143 mg/L (McKee and Wolf 1963). It has been recommended that concentrations above 3000 mg/L can be considered deleterious to both fish-food organisms and fish fry and that a permissible limit of 2000 mg/L can be established for fresh waters. However, these recommendations have not been acted upon at the federal level. The EPA secondary drinking water standard of 250 mg/L for chloride, based on average taste thresholds, was seldom exceeded in a Michigan study of road deicing impacts (Michigan Department of Transportation 1993).

Chloride in road splash can burn sensitive vegetation adjacent to treated roads; plant uptake may cause osmotic stress. Theoretically, chloride can form complexes that increase the mobility of metals in soils, but very few such cases have been confirmed (Michigan Department of Transportation 1993). Spring thaw surges may temporarily create surface water chloride levels detrimental to aquatic biota. However, dilution quickly occurs and flowing streams have not been significantly impacted. Spring thaw surges may temporarily cause high chloride levels in groundwater, although reports of excessive levels (>250 mg/L) are rare (Michigan Department of Transportation 1993).

11.7.5.2 Sodium Ion

Sodium is even less toxic to aquatic biota than chloride. There are no water quality standards for sodium ion. The main problems associated with sodium ion are its effects on agricultural soil permeability (see Section 6.11.5) and the necessity for restricted sodium intake by hypertensive people. Na^+ levels in groundwater can increase temporarily during spring thaws, posing a health threat to persons requiring low sodium intake.

11.7.5.3 Calcium, Magnesium, and Potassium Ions

Calcium, magnesium, and potassium are plant, animal, and human nutrients and there are no stream or drinking water standards for these substances. Calcium and magnesium improve soil aeration and permeability by decreasing the SAR. Calcium and magnesium also increase water hardness beneficially, reducing the toxic effects of dissolved heavy metals on aquatic life. Theoretically, these cations could increase heavy metal mobility in soils by exchange processes, but there is little documentation of such behavior.

11.7.5.4 Acetate

Acetate has no drinking water standards and has lower toxicity than NaCl. It biodegrades rapidly and does not accumulate in the environment. The only reported potential environmental problem with acetate is that, in large concentrations, it can deplete oxygen levels in surface waters by increasing BOD during biodegradation.

11.7.5.5 Impurities Present in Deicing Materials

Deicers contain trace amounts of heavy metals and sometimes phosphorus and nitrogen. These can be released with snowmelt, especially during spring thaw. Because the heavy metal impurities become mostly associated with solids, they are best controlled by sediment containment. Phosphorus and nitrogen will be controlled by infiltration of snowmelt into pervious areas, where they encourage vegetative growth.

REFERENCES

Aronheim, J. S. 1992. Virus transport in groundwater: Modeling of bacteriophage PRD1 transport through one-dimensional columns and a two-dimensional aquifer tank. Masters thesis, Department of Civil, Environmental, and Architectural Engineering, University of Colorado, Boulder, CO.

Asano, T., L. Y. C. Leong, M. G. Rigby, and R. H. Sakaji. 1992. "Evaluation of the California Wastewater Reclamation Criteria Using Enteric Virus Monitoring Data." *Water Science & Technology* 26 (7–8): 1513–24.

Bitton, G. 1975. "Adsorption of Viruses onto Surfaces in Soil and Water." *Water Research* 9: 473.

Bitton, G. 1980. *Introduction to Environmental Virology*. New York: John Wiley.

Bitton, G., J. M. Davidson, and S. R. Farra. 1979. "On the Value of Soil Columns for Assessing the Transport Pattern of Viruses through Soils: A Critical Outlook." *Water, Air, and Soil Pollution* 12: 449–57.

Burge, W. D., and N. K. Enkiri. 1978. "Adsorption Kinetics of Bacteriophage φχ-174 on Soil." *Journal of Environmental Quality* 7 (4): 536–41.

Chang, N.-Y., W. Pearson, J. I. J. Chang, A. Gross, M. Meyer, M. Jolly, B. Vang, and H. Samour. 25 November 1994. *Final Report on Environmentally Sensitive Sanding and Deicing Practices*. ESSD Research Group, Department of Civil Engineering, University of Colorado at Denver, Colorado Transportation Institute, and Colorado Department of Transportation CDOT-CTI-95-5.

Chu, Y., J. Yan, B. Thomas, and V. Y. Marylynn. 2003. "Effect of Soil Properties on Saturated and Unsaturated Virus Transport Through Columns." *Journal of Environmental Quality* 32: 2017–25.

Clement, R. 1997, "Femtograms or phantomgrams? An analytical view of the organochlorine issue," Ray Clement, Ontario Ministry of Environment and Energy, Laboratory Services Branch, Chemical Institute of Canada, 1997, www.thefreelibrary.com/Femtograms+or+phantomgrams%3F+an+analytical+view+of+the+ organochlorine...-a020029520.

Eaton, A. D., L. S. Clesceri, and A. E. Greenberg (Eds.). 1995. *Standard Methods for the Examination of Water and Wastewater*, 19th ed. Washington, DC: American Public Health Association, American Water Works Association, and Water Environment Federation.

Foster, D. M., M. A. Emerson, C. F. Buck, D. S. Walsh, and O. J. Sproul. 1980. "Ozone Inactivation of Cell- and Fecal-Associated Viruses and Bacteria." *Journal of the Water Pollution Control Federation* 52: 2174.

Gerba, C. P. 1975. "Fate of Wastewater Bacteria and Viruses in Soil." *Journal of the Irrigation and Drainage Division* 101 (IR3): 157–73.

Gerba, C. P., and J. E. Smith, Jr. 2005. "Sources of Pathogenic Microorganisms and Their Fate During Land Application of Wastes." *Journal of Environmental Quality* 34 (1): 42–48.

Hurst, C. J. 1989. "Fate of Viruses during Wastewater Sludge Treatment Processes." *CRC Critical Reviews in Environmental Control* 18 (4): 317–43.

Hurst, C. J., C. P. Gerba, and I. Cech. 1980. "Effects of Environmental Variables and Soil Characteristics on Virus Survival in Soil." *Applied and Environmental Microbiology* 40 (6): 1067–79.

Lefler, E., and Y. Kott. 1974. "Enteric Virus Behavior in Sand Dunes." *Israel Journal of Technology* 12: 298–304.

Lo, S. H., and O. J. Sproul. 1977. "Poliovirus Adsorption from Water onto Silicate Minerals." *Water Research* 11: 653–58.

McKee, J. E., and H. W. Wolf. 1963. *Water Quality Criteria, State of California*, 2nd ed. Sacramento, CA: Publication No. 3-A, The Resources Agency of California, State Water Quality Control Board.

Michigan Department of Transportation. December 1993. *The Use of Selected Deicing Materials on Michigan Roads: Environmental and Economic Impacts*. Lansing, MI: Public Sector Consultants, Michigan Department of Transportation.

National Research Council (NRC). 1994. *Groundwater Recharge: Using Waters of Impaired Quality*. Washington, DC: NRC.

Pinholster, G. 1995. "Drinking Recycled Wastewater." *Environmental Science & Technology* 29 (4): 174A–79A.

Rose, J. B., and C. P. Gerba. 1991. "Assessing Potential Health Risks from Viruses and Parasites in Reclaimed Water in Arizona and Florida, USA." *Water Science & Technology* 23: 2091–98.

Scheuerman, P. R., G. Bitton, A. R. Overman, and G. E. Gifford. 1979. "Retention of Viruses by Organic Soils and Sediments." *Journal of the Environmental Engineering Division* ASCE, EE4: 629–40.

Sobsey, M. D., C. H. Dean, M. E. Knuckles, and R. A. Wagner. 1980. "Interactions and Adsorption of Enteric Viruses in Soil Materials." *Applied and Environmental Microbiology* 40: 92.

Sobsey, M. D., and A. R. Hickey. 1985. "Effects of Humic and Fulvic Acids on Poliovirus Concentration from Water by Microporous Filtration." *Applied and Environmental Microbiology* 49 (2): 259–64.

Stotsky, G., M. Schiffenbauer, S. M. Lipson, and B. H. Yu. 1980. "Surface Interactions between Viruses and Clay Minerals and Microbes: Mechanisms and Implications." Presented at *The International Symposium on Viruses and Wastewater Treatment*, University of Surrey, Guilford, United Kingdom.

Stumm, W., and J. J. Morgan. 1981. *Aquatic Chemistry*, 2nd ed., Chapter 5, p. 780. New York: John Wiley.

USEPA. October 1981. *Process Design Manual for Land Treatment of Municipal Wastewater*. Washington, DC: EPA 625/1–81–013.

USEPA. 1986. *Bacteriological Ambient Water Quality Criteria for Marine and Fresh Recreational Waters*. Washington, DC: EPA 440/5-84-002, NTIS PB-86-158-045.

USEPA. September 1992a. *Manual: Guidelines for Water Reuse*. Washington, DC: EPA/625/R-92/004.

USEPA. July 1992b. *Test Methods for Evaluating Solid Waste*, 3rd ed., Vol. 1A, *Laboratory Manual Physical/Chemical Methods, SW 846*. Washington, DC: EPA.

USEPA. November 1994. *Drinking Water Standard Setting: Question and Answer Primer*. Washington, DC: EPA/811-K-94–001.

USEPA. August 1997. *Small Systems Compliance Technology List for the Surface Water Treatment Rule*. Washington, DC: EPA 815-R-97–002.

USEPA. April 1999. *Alternative Disinfectants and Oxidants Guidance Manual*. Washington, DC: EPA 815-R-99–014.

USEPA. 26 March 2007. "Guidelines Establishing Test Procedures for the Analysis of Pollutants." 40 CFR Part 136, Appendix B. *Federal Register* 72 (57): 14220–33.

USGS. 1997. "Cleaning a Cone Splitter." Office of Water Quality Technical Memorandum 97.03, Subject: Protocols for Cleaning a Teflon Cone Splitter to Produce Contaminant-Free Subsamples for Subsequent Determinations of Trace Elements. http://water.usgs.gov/admin/memo/QW/qw97.03.html

USGS. 2004a. "National Field Manual for the Collection of Water-Quality Data, Chapter A3, Cleaning of Equipment for Water Sampling." In Book 9 of *Handbooks for Water-Resources Investigations*, edited by F. D. Wilde, Revised 2004. U.S. Geological Survey, Techniques of Water-Resources Investigations. http://water.usgs.gov/owq/FieldManual/chapter3/final508Chap3book.pdf

USGS. 2004b. "National Field Manual for the Collection of Water-Quality Data, Chapter A4, Collection of Water Samples." In Book 9 of *Handbooks for Water-Resources Investigations*, edited by F. D. Wilde, D. B. Radtke, J. Gibs, and R. T. Iwatsubo, U.S. Geological Survey, Techniques of Water-Resources Investigations. http://water.usgs.gov/owq/FieldManual/chapter4/pdf/Chap4_v2.pdf

USGS. 2004c. "National Field Manual for the Collection of Water-Quality Data: Chapter A2, Selection of Equipment for Water Sampling." In Book 9 of *Handbooks for Water-Resources Investigations*, Version 2.0, revised by S. L. Lane, S. Flanagan, and F. D. Wilde, edited by F. D. Wilde, D. B. Radtke, J. Gibs, and R. T. Iwatsubo, U.S. Geological Survey, 3/2003 TWRI. http://water.usgs.gov/owq/FieldManual/Chapter2-Archive/Archive/Ch2.pdf

Voast, W. A. April 2003. "Geochemical Signature of Formation Waters Associated with Coal-Bed Methane." *AAPG Bulletin* 87 (4): 667–76.

Wallis, C., M. Henderson, and J. L. Melnick. 1972. "Enterovirus Concentration on Cellulose Membranes." *Applied Microbiology* 23 (3): 476–80.

Watershed Research. Summer 1995. "Rating Deicing Agents—Road Salt Stands Firm. Technical Note 55." *Watershed Protection Techniques* 1 (4): 217–20.

Even if all sampling procedures are performed correctly, most water sources, such as lakes, streams, wetlands, or stormwater runoff, are inherently heterogeneous and successive samples collected from the same location will generally have some degree of variability in composition.

Chapter 11, Section 11.2.1

Appendix A
A Selective Dictionary of Water Quality Parameters and Pollutants

A.1 INTRODUCTION

This section is a concise guide to useful information about some frequently measured water quality chemical species and chemical/physical parameters arranged alphabetically. The entries are not intended to be exhaustive, but to offer a brief summary of background data and information and, wherever appropriate, drinking water standards and health advisories (updated in January 2012). Most of the entries can have both natural and human origins. Wherever relevant, other parts of this book are referenced for further information.*

Where CAS identification numbers have been assigned, they are included for each entry. CAS stands for Chemical Abstracts Service Registry, a division of the

---

* Information in Appendix A has been compiled from many different sources, but particularly from Environmental Protection Agency (EPA) website pages.

American Chemical Society, which assigns unique identification numbers to each chemical compound and uses these numbers to facilitate literature and computer database searches for chemical information.

A.1.1 WATER QUALITY CHEMICAL SPECIES CLASSIFIED bY ABUNDANCE AND IMPORTANT CHEMICAL AND PHYSICAL PARAMETERS

Although chemicals that affect water quality are listed alphabetically in the dictionary section, it is sometimes useful to classify them in other ways, such as based on their typical abundance in natural waters.

Major chemical species are those most often present in natural waters in concentrations greater than 1.0 mg/L. These are the cations—calcium, magnesium, potassium, and sodium, and the anions—bicarbonate/carbonate, chloride, fluoride, nitrate, and sulfate. Silicon is usually present as a nonionic species and is reported by analytical laboratories as the equivalent concentration of silica (SiO_2).

Several additional chemical species are sometimes included with the major constituents because of their importance in determining water quality and because some of them sometimes attain concentrations comparable to the parameters mentioned above. These are aluminum, boron, iron, manganese, nitrogen in forms other than nitrate (such as ammonia and nitrite), organic carbon, phosphate, and the dissolved gases such as oxygen, carbon dioxide, and hydrogen sulfide.

Minor chemical species are those most often present in natural waters in concentrations less than 1.0 mg/L. These include the so-called trace elements and naturally occurring radioisotopes: antimony, arsenic, barium, beryllium, bromide, cadmium, cesium, chromium, cobalt, copper, iodine, lead, lithium, mercury, molybdenum, nickel, radium, radon, rubidium, selenium, silver, strontium, thorium, titanium, uranium, vanadium, and zinc.

In addition to chemical species, water quality measurements frequently include physical and chemical properties that do not identify particular chemical species, but are used as indicators of how water quality may affect water uses. These are acidity, alkalinity, hardness, hydrogen ion concentration (measured as pH), redox potential, biochemical oxygen demand (BOD), chemical oxygen demand (COD), color, corrosivity, gross alpha and beta emitters, odor, sodium adsorption ratio (SAR), Langelier index of corrosion, specific conductance (conductivity), specific gravity, temperature, total dissolved solids (TDS), total suspended solids (TSS), and turbidity.

DEFINITIONS AND ACRONYMS

Action level: The concentration of a contaminant, which, if exceeded, triggers treatment or other requirements that a water system must follow. For example, it could be the level of lead or copper, which, if exceeded in over 10% of the homes tested, triggers treatment for corrosion control.

10^{-4} cancer risk: The concentration of a chemical in drinking water corresponding to an estimated lifetime cancer risk of 1 in 10,000.

DWEL: Drinking water equivalent level. A lifetime exposure concentration, protective of adverse, noncancer health effects, that assumes all the exposure to a contaminant is from drinking water.

Health advisory (HA): An estimate of acceptable drinking water levels for a chemical substance based on health effects information; a health advisory is not a legally enforceable federal standard, but serves as technical guidance to assist federal, state, and local officials.

> **One-day HA:** The concentration of a chemical in drinking water that is not expected to cause any adverse noncarcinogenic effects for up to 1 day of exposure. The one-day HA is normally designed to protect a 10-kg child consuming 1 L of water per day.

> **Ten-day HA:** The concentration of a chemical in drinking water that is not expected to cause any adverse noncarcinogenic effects for up to 10 days of exposure. The ten-day HA is also normally designed to protect a 10-kg child consuming 1 L of water per day.

> **Lifetime HA:** The concentration of a chemical in drinking water that is not expected to cause any adverse noncarcinogenic effects for a lifetime of exposure. The lifetime HA is based on exposure of a 70-kg adult consuming 2 L of water per day. The lifetime HA for Group C carcinogens (see Table 3.10) includes an adjustment for possible carcinogenicity.

MCLG: Maximum contaminant level goal. A nonenforceable health goal, which is set at a level at which no known or anticipated adverse health effects occur and which allows an adequate margin of safety.

MCL: Maximum contaminant level. The highest level of a contaminant allowed in drinking water. MCLs are set as close to the MCLG as feasible, when using the best available analytical and treatment technologies and taking cost into consideration. MCLs are enforceable standards.

RfD: Reference dose. An estimate (with an uncertainty of up to a factor of 10) of a daily oral exposure to the human population (including sensitive subgroups) that is likely to be without an appreciable risk of deleterious effects during a lifetime. The units of the RfD are weight of pollutant consumed per day, per kilogram of body weight, typically estimated to be 70 kg for an average adult.

SDWR: Secondary Drinking Water Regulations. Nonenforceable federal drinking water guidelines concerning cosmetic effects (such as tooth or skin discoloration) or aesthetic effects (such as taste, odor, or color).

TT: Treatment technique. A particular method of water treatment that is legally required and enforceable, for reducing the level of a contaminant in drinking water. Used instead of an MCL to ensure best available treatment technology or when conditions beyond the control of a treatment plant operator (e.g., for lead and copper in a water distribution system) may affect contaminant concentrations at the required monitoring point.

A.2   ALPHABETICAL LISTING OF CHEMICAL AND PHYSICAL WATER QUALITY PARAMETERS AND POLLUTANTS

EPA drinking water standards and health advisories are updated to January 2012 (USEPA 2011).

A.2.1   2,4-D (2,4-DICHLOROPHENOXYACETIC ACID), CAS No. 94-75-7

2,4-D is a colorless, odorless powder used as a herbicide for the selective control of broad-leaved weeds in agriculture, and for the control of woody plants along roadsides, railways, and utilities rights of way. It is one of the most widely used herbicides in the world and is commonly used on crops such as wheat and corn, and on pasture and rangelands. It is also used to control broad-leaved aquatic weeds.

A.2.1.1   Environmental Behavior

2,4-D is rapidly degraded by microbes in soil and water, with a half-life of 3–22 days in different soils. 2,4-D is weakly sorbed by soil with sorption generally increasing with increasing soil organic carbon content. Leaching to groundwater is most likely in coarse-grained sandy soils with low organic content or with very basic soils. In general, little runoff occurs with 2,4-D or its amine salts.

A.2.1.2   Ecological Concerns

2,4-D is moderately toxic to birds and fish. Ester formulations are especially toxic to fish, and, in the case of a spill of a 2,4-D ester, some fish mortality is likely. The presence of emulsifiers with 2,4-D amine enhances the toxicity of the 2,4-D amine to aquatic species. 2,4-D exposures through spray drift and runoff are considered the greatest potential risks to terrestrial plants, mammals, and birds, whereas exposures to 2,4-D through direct application to water for aquatic weed control present the greatest potential risk to aquatic plants and animals. There is no evidence that bioconcentration of 2,4-D occurs through the food chain.

A.2.1.3   Health Concerns

Short-term exposure to 2,4-D at levels above the MCL can irritate the eyes, skin, and breathing passages and cause nervous system damage. Long-term chronic exposure at levels above the MCL can cause damage to the nervous system, kidneys, and liver. Some 2,4-D-based products contain surfactants, which, if inhaled, may cause neurological damage; skin contact may cause dermatitis if prolonged or repeated, and when ingested or inhaled may cause respiratory swelling and excess fluid in tissue.

2,4-D

EPA Primary Drinking Water Standard (mg/L)		Health Advisories (mg/L)	
MCL	0.07	1-day (10-kg child)	1
MCLG	0.07	2-day (10-kg child)	0.3
		RfD (lifetime + sensitive groups)	0.005 mg/kg/day
		DWEL	0.2
		Lifetime (adult)	—
		10^{-4} cancer risk	—

A.2.2   ACRYLAMIDE (C_3H_5NO), CAS No. 79-06-1

Acrylamide is an organic solid of white, odorless, flake-like crystals. Its major use is in the production of polyacrylamides, which are used in coagulants for water treatment, pulp and paper production, and mineral processing. Polyacrylamides are also used as grouting agents in the construction of drinking-water reservoirs and wells. Other uses include the synthesis of dyes, adhesives, contact lenses, soil conditioners, and permanent press fabrics.

A.2.2.1   Environmental Behavior

Acrylamide does not bind to soil and will move into soil rapidly, but it is degraded by microbes within a few days in soil and water. When released into the soil, acrylamide is expected to leach into groundwater, where it is likely to biodegrade to a moderate extent.

A.2.2.2   Ecological Concerns

Acrylamide is not expected to be toxic to aquatic life. The LC50/96-hour values for fish are over 100 mg/L. It is not expected to significantly bioaccumulate.

A.2.2.3   Health Concerns

Acrylamide is a neurotoxicant and has been shown to be a carcinogen, germ cell mutagen, and reproductive toxicant in animal studies. Short-term exposure at levels above the MCL may cause damage to the nervous system, weakness, and loss of coordination in the legs. Long-term exposure at levels above the MCL may cause damage to the nervous system, paralysis, and cancer.

Acrylamide

EPA Primary Drinking Water Standard (mg/L)		Health Advisories (mg/L)	
MCL	TT	1-day (10-kg child)	1.5
MCLG	Zero	2-day (10-kg child)	0.3
The EPA requires a water supplier to show that when acrylamide is added to water, the amount of uncoagulated acrylamide is less than 0.5 ppb		RfD (lifetime + sensitive groups)	0.002 mg/kg/day
		DWEL	0.07
		Lifetime (adult)	—
		10^{-4} cancer risk	—

A.2.2.4 Other Comments

Treatment/best available technologies: Conventional treatment processes do not remove acrylamide. Acrylamide concentrations in drinking water are controlled by limiting the acrylamide content of polyacrylamide flocculants and the dose used.

A.2.3 ALACHLOR ($C_{14}H_{20}ClNO_2$), CAS No. 15972-60-8

Alachlor is a pre- and postemergence herbicide used to control annual grasses and many broad-leaved weeds in corn and in many other crops. There are liquid, dry flowable, microencapsulated, and granular formulations. Alachlor is applied by ground, aerial, and chemigation equipment. It can also be mixed with dry bulk fertilizer. It is lost from soil mainly through volatilization, photodegradation, and biodegradation. Alachlor and its degradation products may be found in soil, groundwater, and surface water.

A.2.3.1 Environmental Behavior

Alachlor is highly mobile and moderately persistent, with a low affinity to adsorb to soils. It dissipates primarily by aerobic soil metabolism processes with a half-life of 2–3 weeks. Even when used according to label directions, it has a high probability of entering groundwater and surface water.

A.2.3.2 Ecological Concerns

Alachlor is slightly to practically nontoxic to birds, mammals, and honey bees; highly to moderately toxic to freshwater fish; and highly toxic to aquatic plants.

A.2.3.3 Health Concerns

Alachlor is considered to be a carcinogen and a metabolite of alachlor, 2,6-diethylaniline, and has been shown to be mutagenic.

Alachlor

EPA Primary Drinking Water Standard (mg/L)		Health Advisories (mg/L)	
MCL	0.002	1-day (10-kg child)	0.1
MCLG	Zero	2-day (10-kg child)	0.1
		RfD (lifetime + sensitive groups)	0.01 mg/kg/day
		DWEL	0.4
		Lifetime (adult)	—
		10^{-4} cancer risk	0.04

A.2.3.4 Other Comments

Treatment/best available technologies: Granulated activated carbon (GAC).

A.2.4 ALDICARB ($C_7H_{14}N_2O_2S$), CAS No. 116-06-3

Aldicarb is a broad-spectrum, systemic carbamate insecticide used to control nematodes in soil and a variety of insects and mites on citrus crops, dry beans, grain, sorghum, ornamentals, pecans, peanuts, potatoes, seed alfalfa, soybeans, sugar beets, sugarcane, sweet potatoes, and tobacco.

A.2.4.1 Environmental Behavior

Aldicarb and its degradation products are generally very soluble in water and mobile in soil. Adsorption in soil is primarily to organic matter, so leaching is most extensive in sandy or sandy loam soils. Aldicarb is very persistent in groundwater, typically degrading to nontoxic products with a half-life between a few weeks to as long as several years. The primary mode of degradation is chemical hydrolysis, with some microbial decay in shallow groundwater. In soils, the primary mode of degradation is oxidation by soil microorganisms and hydrolysis, depending on soil conditions.

A.2.4.2 Ecological Concerns

Aldicarb is one of the most acutely toxic pesticides in use, due to acetylcholinesterase inhibition. It is metabolized to the sulfoxide and sulfone. Aldicarb sulfoxide is a more potent inhibitor of acetylcholinesterase than aldicarb itself, whereas aldicarb sulfone is considerably less toxic than either aldicarb or the sulfoxide. The weight of evidence indicates that aldicarb, aldicarb sulfoxide, and aldicarb sulfone are not genotoxic or carcinogenic.

A.2.4.3 Health Concerns

Ingestion of aldicarb can cause dizziness, weakness, diarrhea, nausea, vomiting, abdominal pain, excessive perspiration, blurred vision, headache, convulsions, and temporary paralysis of the extremities. Recovery is rapid, usually within 6 hours. Aldicarb is considered an acutely toxic pesticide, with several incidents of accidental or intentional poisoning being reported. It is classified as probably not carcinogenic to humans.

Aldicarb

EPA Primary Drinking Water Standard (mg/L)		Health Advisories (mg/L)	
MCL (Aldicarb)	0.003	1-day (10-kg child)	0.01
(Aldicarb sulfone)	0.002	2-day (10-kg child)	0.01
(Aldicarb sulfoxide)	0.004	RfD (lifetime + sensitive groups)	0.001 mg/kg/day
MCLG (aldicarb)	0.001	DWEL	0.035
(Aldicarb sulfone)	0.001	Lifetime (adult)	0.07
(Aldicarb sulfoxide)	0.001	10^{-4} cancer risk	—

(Health advisory values are the same for each of the three chemicals, aldicarb + its two toxic metabolites, aldicarb sulfone and aldicarb sulfoxide)
The MCL for any two of the three chemicals (aldicarb, aldicarb sulfone, and aldicarb sulfoxide) should not exceed 0.007 mg/L because of similar modes of action

---

A.2.4.4 Other Comments

Treatment/best available technologies: Granulated activated carbon (GAC); ozonation.

A.2.5   ALDRIN ($C_{12}H_8Cl_6$), CAS No. 309-00-2 AND DIELDRIN ($C_{12}H_8Cl_6O$), CAS No. 60-57-1

Both aldrin and dieldrin are nonsystemic chlorinated pesticides used to control soil-dwelling pests, for wood protection, and, in the case of dieldrin, against insects of public health importance. The two compounds are closely related with respect to their toxicology and mode of action. Since the early 1970s, many countries have either severely restricted or banned the use of both compounds, particularly in agriculture. The last known manufacturer of aldrin and dieldrin, Shell International Chemical Co. (U.K.), ceased production of the pesticides in 1989.

A.2.5.1   Environmental Behavior

Aldrin and dieldrin have low water solubilities (0.027 mg/L for aldrin at 27°C and 0.186 mg/L for dieldrin at 20°C). Aldrin is rapidly oxidized to dieldrin under most environmental conditions and in the body. Dieldrin is a highly persistent organochlorine compound that has low mobility in soil, but readily volatizes to the atmosphere. Long-range atmospheric transport tends to be from warmer to colder regions where the pesticides are sometimes found where they have never been used. The high mobility of aldrin and dieldrin has led to widespread banning of their use because of concern about the difficulty of controlling the movement of these persistent chemicals in the environment.

A.2.5.2   Health Concerns

Aldrin and dieldrin are highly toxic to humans and animals, affecting the central nervous system and the liver. They are not considered carcinogenic. Exposure to aldrin and dieldrin occurs by oral ingestion, inhalation, and dermal absorption. In the body, aldrin is quickly metabolized to dieldrin. The biological half-life of dieldrin in humans is about 266 days. Signs and symptoms related to ingestion of or dermal contact with toxic doses of aldrin and dieldrin include headache, dizziness, nausea, general malaise, and vomiting, followed by muscle twitching, myoclonic jerks, and convulsions.

Aldrin and Dieldrin

EPA Primary Drinking Water Standard (mg/L)	Health Advisories (mg/L)	
Aldrin and dieldrin are not registered in the United States or Canada, and therefore should not be available for general use. No EPA MCL or MCLG is listed. Canada has set a maximum acceptable concentration (MAC) for combined aldrin + dieldrin in drinking water of 0.0007 mg/L	1-day (10-kg child)	0.0003 (Aldrin) 0.0005 (Dieldrin)
	2-day (10-kg child)	0.0003 (Aldrin) 0.0005 (Dieldrin)
	RfD (lifetime + sensitive groups)	0.00003 mg/kg/day (Aldrin) 0.00005 mg/kg/day (Dieldrin)
	DWEL	0.001 (Aldrin) 0.002 (Dieldrin)
	Lifetime (adult)	— (Dieldrin and Aldrin)
	10^{-4} cancer risk	0.0002 (Aldrin) 0.0002 (Dieldrin)

A.2.5.3   Other Comments

Treatment/best available technologies: Granulated activated carbon (GAC), coagulation, ozonation.

A.2.6   ALUMINUM (AL), CAS No. 7429-90-5

Aluminum is the most abundant metal and the third most abundant element in the earth's lithosphere (after oxygen and silicon) and its compounds are often found in natural waters. Aluminum is mobilized naturally in the environment by the weathering of rocks and minerals, particularly bauxite clays. It is a normal constituent of all soils and is found in low concentrations in all plant and animal tissues. Most naturally occurring aluminum compounds have very low solubility between pH 6 and 9. Therefore, dissolved forms rarely occur in natural waters in concentrations greater than about 0.01 mg/L. Concentrations in water greater than this usually indicate the presence of solid forms of aluminum, such as suspended solids and colloids.

Aluminum salts are widely used in water treatment as coagulants to remove organic matter, color, turbidity, and microorganisms. Concentrations above the SDWS can lead to undesirable color and turbidity. The concentration of Al^{3+} in water is controlled by the solubility of aluminum hydroxide, $Al(OH)_3$, which increases by a factor of about 10^3 for every unit decrease in pH. Thus, the concentration of dissolved aluminum, Al^{3+}, is about 3×10^{-5} mg/L at pH 6, 0.03 mg/L at pH 5, and 30 mg/L at pH 4.

Also, at lower water pH (<5) in the presence of clays and organic-rich soils, dissolved aluminum concentrations increase because of the release of Al^{3+} from the soil. Low pH means high H^+ concentrations. At pH < 5, the concentration of H^+ is high enough for H^+ to partially ion exchange with other, more strongly bound metals at ion-exchange sites on soil particles. Since Al^{3+} is bound more strongly than divalent and monovalent cations, it is among the last cations to be displaced by H^+ and requires a continued low pH to reach elevated dissolved levels.

RULE OF THUMB

The presence of an elevated concentration of Al^{3+}, often exceeding the concentrations of Ca^{2+} and Mg^{2+}, is a common characteristic of acidic waters (pH < 5), including mine drainage and waters affected by acid rain.

A.2.6.1   Health Concerns

Naturally occurring aluminum has a very low toxicity to humans and animals. Only a few industrially important aluminum compounds, such as the fumigant aluminum phosphide, are considered acutely hazardous. Exposure and ingestion of aluminum and its compounds is usually not considered to be harmful. Aluminum compounds are used in water treatment to remove color and turbidity, food packaging, medicines, soaps, dental cements, drug store items such as antacids and antiperspirants, and are present in many foods grown in soils containing aluminum.

Daily exposure to aluminum is inevitable due to its ubiquitous occurrence in nature and its many commercial uses. Estimated human consumption is about 88 mg/person of aluminum per day, mostly from food. Consumption of 2 L of water per day containing 1.5 mg/L of aluminum (well above the secondary drinking water standard, see below) only contributes 3.0 mg of aluminum per day, or less than 4% of the normal daily intake. Aluminum is not believed to be an essential nutrient. There is no human or animal evidence of carcinogenicity. There is some indication that ingestion of large doses of aluminum in medicines may cause skeletal problems.

Aluminum

EPA Primary Drinking Water Standard (mg/L)

		Health Advisories (mg/L)	
MCL	None	1-day (10-kg child)	—
MCLG	None	2-day (10-kg child)	—
SDWS	0.05–0.2	RfD (lifetime + sensitive groups)	—
		DWEL	—
		Lifetime (adult)	—
		10^{-4} cancer risk	—

The EPA has no primary drinking water standard for aluminum. The EPA secondary drinking water standard (nonenforceable) is expressed as a range: 0.05–0.2 mg/L. The EPA recommends that 0.05 mg/L be met wherever possible, but allows states to determine the required level on a case-by-case basis because water treatment technologies often use aluminum salts to remove color and turbidity and cannot always achieve the lower value. The EPA recommends that aluminum in drinking water should not exceed 0.2 mg/L because of taste and odor problems. In the presence of microorganisms, aluminum can react with iron, manganese, silica, and organic material to form fine sediments that can appear at the consumer's tap. If dissolved aluminum exceeds 0.1 mg/L, levels of iron normally acceptable may produce discoloration and staining. No lifetime health advisory has been established.

A.2.6.2 Other Comments

Because of its normally low dissolved concentrations, aluminum is generally of no concern in irrigation waters. However, a limit of 5.0 mg/L is recommended (National Academy of Sciences 1982) where irrigation waters are used regularly. Swimming pools treated with commercial grade aluminum sulfate compounds, known as alum, may cause eye irritation at concentrations greater than 0.1 mg/L.

Treatment/best available technologies: Careful control of pH and dosage in the coagulation process and filtration of the aluminum floc.

A.2.7 Ammonia/Ammonium Ion (NH_3/NH_4^+), CAS No. 7664-41-7

See Chapter 4 for a more detailed discussion.

In the biological decay of nitrogenous organic compounds, ammonia is the first nitrogenous product formed that does not contain carbon. It is among the waste products produced during decay that follows the death of plants, animals, and other life forms. Thus, ammonia is present naturally in surface and groundwaters. It also is discharged in many waste streams, particularly from municipal waste treatment. Ammonia in water is an indicator of possible bacterial, sewage, and animal waste pollution. Unpolluted waters have very low ammonia concentrations, generally less than 0.2 mg/L as nitrogen.

Continued oxidation of ammonia sequentially leads to nitrite and nitrate. Ammonia gas is very soluble in water, where it reacts as a base, raising the pH and forming an ammonium cation and a hydroxyl anion:

$$NH_3 + H_2O \rightleftarrows NH_4^+ + OH^- \qquad (A.1)$$

The unionized form (NH_3) is of greatest environmental concern because of its greater toxicity to aquatic life. However, because NH_4^+ readily converts to NH_3 by its pH-dependent equilibrium (Equation A.1), total ammonia is normally regulated in discharges. Concentrations of unionized (NH_3) or total ($NH_3 + NH_4^+$) ammonia are often reported in terms of the nitrogen content only, for example, $NH_3 = 10$ mg/L–N, or NH_3–N = 10 mg/L. This means that the sample contains unionized ammonia and the nitrogen portion of the unionized ammonia weighs 10 mg/L of the sample. The weight of hydrogen in the ammonia molecules is ignored. To convert milligrams per liter of NH_3 to milligrams per liter of NH_3–N, multiply by 0.822.

The equilibrium between the unionized (NH_3) and the ionized (NH_4^+) forms depends on pH, temperature, and, to a much lesser degree, on ionic strength (salinity or concentration of TDS; see Chapter 4).

- At 15°C and pH > 9.6, the fraction of NH_3 is greater than 0.5.
- At 15°C and pH < 9.6, the fraction of NH_4^+ is greater than 0.5.
- A temperature increase shifts the equilibrium of Equation A.1 to the left, increasing the NH_3 concentration.
- A temperature decrease shifts the equilibrium of Equation A.1 to the right, increasing the NH_4^+ concentration.
- An increase in ionic strength shifts the equilibrium of Equation A.1 to the right, increasing the NH_4^+ concentration slightly. In waters with very high TDS (>10,000 mg/L), there will be a small but measurable decrease in the percentage of NH_3.
- Because pH and temperature can vary considerably along a stream or within a lake, the fraction of total ammonia that is unionized is also variable at different locations. Therefore, the amount of total ammonia is usually of regulatory concern, rather than only the unionized form.

A.2.7.1 Health Concerns

Total ammonia ($NH_3 + NH_4^+$) in drinking water is more an esthetic than a health concern. The odor and taste of ammonia makes drinking water unpalatable at concentrations well below the appearance of any toxic effects to humans. The main health concern with ammonia is its potential oxidation to nitrite (NO_2^-) and nitrate (NO_3^-).

Ingested nitrate and nitrite react with iron in blood hemoglobin to cause a blood oxygen deficiency disease called methemoglobinemia, which is especially dangerous in infants (blue baby syndrome) because of their small total blood volume. There is no human or animal evidence of carcinogenicity.

Ammonia

EPA Primary Drinking Water Standard (mg/L)	Health Advisories (mg/L)	
The EPA has no primary or secondary drinking water standards for ammonia. However, the presence of NH_3 greater than 0.1 mg/L may raise the suspicion of recent pollution.	1-day (10-kg child)	—
	2-day (10-kg child)	—
	RfD (lifetime + sensitive groups)	—
Some states have adopted ammonia limits for water that will receive treatment to produce drinking water. For example, Colorado's ammonia standard for water classified as domestic water supply is 0.05 mg/L-N, 30-day average for total ammonia ($NH_3 + NH_4^+$)	DWEL	—
	Lifetime (adult)	30
	10–4 cancer risk	—

RULES OF THUMB

1. Only NH_3, the unionized form, has significant toxicity for aquatic life.
2. To convert milligrams per liter of unionized or total ammonia to milligrams per liter as nitrogen, multiply by 0.822. Example: 17.4 mg/L $NH_3 = 0.822 \times 17.4 = 14.3$ mg/L–N.
3. Since pH 9.6 is higher than the pH of most natural waters, NH_3–N in natural waters usually is mostly in the less toxic ionized ammonium form (NH_4^+).
4. In high-pH waters (>9), the NH_3 fraction can reach levels toxic to aquatic life.
5. The ionized form is not volatile and cannot be removed by air stripping. The unionized form, NH_3, is volatile and can be removed by air stripping.

A.2.8   ANTIMONY (Sb), CAS No. 1440-36-0

Antimony is a metalloid (having properties intermediate between metals and nonmetals). It is in the same chemical group (group 5A) as arsenic, with which it has some chemical similarities, including toxicity; however, it is only about one-tenth as abundant in the earth's crust and soils. The symbol Sb for the element is from stibium, the Latin name for antimony. In the environmental literature, antimony is often included with the metals because it is usually analyzed, along with other metals, by inductively coupled plasma (ICP) or atomic absorption techniques. Common sources of antimony in drinking water are discharges from petroleum refineries, fire retardants, ceramics, electronics, and solder. It is also found in batteries, pigments, ceramics, and glass.

Antimony is usually adsorbed strongly to iron, manganese, and aluminum compounds in soils and sediments. Soil concentrations normally range between 1 and 9 mg/L. The amount commonly dissolved in rivers is small, less than 0.005 mg/L. There is no evidence of bioconcentration of most antimony compounds.

A.2.8.1   Health Concerns

Antimony is used in medicines for treating parasite infections. It is present in meats, vegetables, and seafood in an average concentration of about 0.2–1.1 ppb (µg/L). An average person ingests about 5 µg of antimony every day in food and drink. Short-term exposures above the MCL may cause nausea, vomiting, and diarrhea. Potential health effects from long-term exposure above the MCL are an increase in blood cholesterol and a decrease in blood glucose. There is insufficient evidence to state whether antimony has the potential to cause cancer.

Antimony

EPA Primary Drinking Water Standard (mg/L)		Health Advisories (mg/L)	
MCL	0.006	1-day (10-kg child)	0.01
MCLG	0.006	2-day (10-kg child)	0.01
		RfD (lifetime + sensitive groups)	0.0004 mg/kg/day
		DWEL	0.01
		Lifetime (adult)	0.006
		10^{-4} cancer risk	—

A.2.8.2   Other Comments

Treatment/best available technologies: Coagulation and filtration, reverse osmosis.

A.2.9   ARSENIC (AS), CAS No. 7440-38-2

Chemically, arsenic is classified as a metalloid, having properties intermediate between metals and nonmetals. In the environmental literature, it is often included with the metals because it is usually analyzed, along with other metals, by ICP or atomic absorption techniques.

Arsenic is widely distributed throughout the earth's crust, most often as arsenic sulfide or as metal arsenates and arsenides. Inorganic arsenic occurs naturally in many minerals, especially in ores of copper and lead. Smelting of these ores introduces arsenic to the atmosphere as dust particles.

In minerals, arsenic is combined mostly with oxygen, chlorine, and sulfur. Inorganic arsenic compounds are used mainly as wood preservatives, insecticides, and herbicides. Organic forms of arsenic found in plants and animals are combined with carbon and hydrogen. Organic arsenic is generally less toxic than inorganic arsenic. Arsenic is not abundant, with an average concentration in the lithosphere of about 1.5 mg/kg (ppm). Background levels in soils typically range from 1 to 95 mg/kg. Average levels in U.S. soils are around 5–7 mg/kg. It is widely distributed and is found naturally in many foods at levels of 20–140 ppb, exposing most Americans to a constant low exposure, perhaps around 50 µg/day. Normal human blood contains 0.2–1.0 mg/L of arsenic; however, there is no evidence that arsenic is an essential nutrient.

Many arsenic compounds are water-soluble and may be found in groundwater, especially in the western United States. However, most arsenic compounds sorb strongly to soils and are therefore transported only over short distances in groundwater and surface water. Sorption and coprecipitation with hydrous iron oxides are the most important removal mechanisms under most environmental conditions. The average concentration for U.S. surface water is around 3 ppb. Groundwater levels average about 1–2 ppb, except in some western states where groundwater is in contact with volcanic rock and sulfide minerals high in arsenic. In western mining regions, groundwater arsenic levels as high as 48,000 ppb (48 ppm) have been observed. Many persons dependent on well water in the West ingest higher than average levels of inorganic arsenic through their drinking water supplies. The EPA reports arsenic as the second most common contaminant at Superfund sites.

A.2.9.1 Health Concerns

High levels (>60 ppm) of arsenic in food or water can be fatal. Arsenic damages tissues in the nervous system, stomach, intestine, and skin. Breathing high levels can irritate lungs and throat. Lower levels can cause nausea, diarrhea, irregular heartbeat, blood vessel damage, reduction of red and white blood cells, and tingling sensations in hands and feet. Long-term exposure to inorganic arsenic may cause darkening of the skin and the appearance of small warts on the palms, soles, and torso.

Inorganic arsenic was recognized as a possible carcinogen as early as 1879, when it was suggested that high rates of lung cancer in German miners might have been caused by inhaled arsenic. Arsenic is currently considered a carcinogen. Breathing inorganic arsenic increases the risk of lung cancer and ingesting inorganic arsenic increases the risk of skin cancer and tumors of the bladder, kidney, liver, and lung.

A crisis of well-water contamination by arsenic was discovered in Bangladesh in 1992. The crisis was created by a well-intended effort by the United Nation Children's Fund (UNICEF) to provide Bangladesh with reliable water sources free of cholera and dysentery organisms. Millions of water wells were installed and the water was tested for microbial contaminants, but not for arsenic and other toxic metals. It was estimated that 85% of Bangladesh's geographical area contained wells contaminated with inorganic arsenic. In recent years, deeper wells were sunk 500 ft or more to purer waters.

Arsenic

EPA Primary Drinking Water Standard (mg/L)		Health Advisories (mg/L)	
MCL	0.01	1-day (10-kg child)	—
MCLG	Zero	2-day (10-kg child)	—
		RfD (lifetime + sensitive groups)	0.0003 mg/kg/day
		DWEL	0.01
		Lifetime (adult)	—
		10^{-4} cancer risk	0.002

A.2.9.2 Other Comments

A maximum concentration of 0.1 mg/L is recommended for irrigation water and for the protection of aquatic plants.

Treatment/best available technologies: Iron coprecipitation, activated alumina or carbon sorption, ion exchange, reverse osmosis.

A.2.10 Asbestos, CAS No. 1332-21-4

Asbestos is a generic term for different naturally formed fibrous silicate minerals that are classified into two groups, serpentine and amphibole, based on structure. Six minerals have been characterized as asbestos: chrysotile, crosidolite, anthophyllite, tremolite, actinolite, and andamosite. The most common form is chrysotile, which has the serpentine structure; the others have the amphibole structure. These different forms of asbestos are composed of 40–60% silica, the remainder being oxides of iron, magnesium, and other metals.

Although asbestos may be introduced into the environment by the dissolution of asbestos-containing minerals and from industrial effluents, the primary source is through the wear or breakdown of asbestos-containing materials. Because asbestos fibers are resistant to heat and most chemicals, they have been mined for use in over 3000 different products in the United States, such as roofing materials, brake linings asbestos-reinforced pipe, packing seals, gaskets, fire-resistant textiles, and floor tiles. The EPA banned most uses of asbestos in the United States on July 12, 1989, because of potential adverse health effects in exposed persons. The remaining, currently allowed uses of asbestos include battery separators, sealant tape, asbestos thread, packing materials, and certain industrial uses of gaskets.

Typical background levels in lakes and streams range from 1 to 10 million fibers per liter. Asbestos is nonvolatile, insoluble, nonbiodegradable, and does not tend to adsorb to stream sediments. Asbestos fibers do not chemically decompose to other compounds in the environment and, therefore, can remain in the environment for decades or longer. Small asbestos fibers and fiber-containing particles may be carried long distances by water currents before settling out; larger fibers and particles tend to settle more quickly. Asbestos fibers do not pass through soils to groundwater.

There are no data regarding the bioaccumulation of asbestos in aquatic organisms, but asbestos is not expected to bioaccumulate. Ordinary sand filtration removes about 90% of the fibers.

A.2.10.1   Health Concerns

All types of asbestos fibers are known to cause serious health hazards in humans. Although asbestos is not known to cause any health problems when people are exposed to it at levels above the MCL for relatively short periods of time, long-term inhalation has the potential to cause cancer of the lung and other internal organs. Long-term ingestion above the MCL increases the risk of developing benign intestinal polyps.

Asbestos

EPA Primary Drinking Water Standard (million fibers > 10 μm length per L = MFL)		Health Advisories (mg/L)	
MCL	7 MFL	1-day (10-kg child)	—
MCLG	7 MFL	2-day (10-kg child)	—
		RfD (lifetime + sensitive groups)	—
		DWEL	—
		Lifetime (adult)	—
		10^{-4} cancer risk	700 MFL

A.2.10.2   Other Comments

Treatment/best available technologies: Coagulation and filtration, direct and diatomite filtration, corrosion control.

A.2.11   Atrazine ($C_8H_{14}ClN_5$), CAS No. 1912-24-9

Atrazine is a selective triazine herbicide used to control broad-leaved and grassy weeds in corn, sorghum, sugarcane, pineapple, Christmas trees, and other crops, and in conifer reforestation plantings. It is also used as a nonselective herbicide on noncropped industrial lands and on fallow lands.

A.2.11.1   Environmental Behavior

Atrazine is highly persistent in soil. In soil and water, atrazine degrades by hydrolysis, followed by biodegradation by soil microorganisms. Hydrolysis is rapid in acidic or basic environments, but is slower around neutral pH. Sunlight and evaporation do not affect its removal rate. Atrazine can persist for longer than 1 year under dry or cold conditions. It is moderately to highly mobile in soils with low clay or low organic matter content. Because it does not adsorb strongly to soil particles and has a lengthy half-life (60 to >100 days), it has a high potential for groundwater contamination despite being only moderately soluble in water. It is frequently detected in drinking water wells.

A.2.11.2 Ecological Concerns

Atrazine is practically nontoxic to birds, slightly toxic to fish and other aquatic life, and nontoxic to bees. It has a low level of bioaccumulation in fish.

A.2.11.3 Health Concerns

Atrazine is slightly to moderately toxic to humans and other animals. It can be absorbed orally, dermally, and by inhalation. The most likely source is through drinking water. Short exposures at levels above the MCL may cause nausea and dizziness.

Atrazine

EPA Primary Drinking Water Standard (mg/L)		Health Advisories (mg/L)	
MCL	0.003	1-day (10-kg child)	—
MCLG	0.003	2-day (10-kg child)	—
		RfD (lifetime + sensitive groups)	0.02 mg/kg/day
		DWEL	0.7
		Lifetime (adult)	—
		10^{-4} cancer risk	—

A.2.11.4 Other Comments

Treatment/best available technologies: Granulated activated carbon (GAC).

A.2.12 BARIUM (BA), CAS NO. 7440-39-3

Barium is the sixth most abundant element in the lithosphere, averaging about 500 mg/kg. It exists mainly as the sulfate ($BaSO_4$, barite) and, to a lesser extent, the carbonate ($BaCO_3$, witherite). Traces of barium are found in most soils, natural waters, and foods. Although most groundwaters contain only a trace of barium, some geothermal groundwaters may contain as much as 10 mg/L.

Barium is released to water and soil in the disposal of drilling wastes, from copper smelting, and industrial waste streams. It is not very mobile in most soil systems. In water, the more toxic soluble salts are likely to precipitate as the less toxic insoluble sulfate and carbonate compounds. Background levels for soil range from 100 to 3000 ppm. Barium occurs naturally in almost all surface waters in concentrations of 2–340 μg/L, with an average of 43 μg/L. However, some wells may contain barium levels 10 times higher than the drinking water standard. In seawater, marine animals concentrate the element 7–100 times, and marine plants 1000 times. Soybeans and tomatoes also accumulate soil barium 2–20 times.

A.2.12.1 Health Concerns

There is no evidence that barium is an essential nutrient. All soluble barium salts are considered toxic. Short-term exposure at levels above the MCL may cause gastrointestinal disturbances, muscular weakness, and liver, kidney, heart, and spleen damage. Long-term exposure above the MCL may cause hypertension. There is no evidence that barium can cause cancer. No health advisories have been established for short-term exposures.

Barium

EPA Primary Drinking Water Standard (mg/L)		Health Advisories (mg/L)	
MCL	2	1-day (10-kg child)	0.7
MCLG	2	2-day (10-kg child)	0.7
		RfD (lifetime + sensitive groups)	0.2 mg/kg/day
		DWEL	7
		Lifetime (adult)	—
		10^{-4} cancer risk	—

A.2.12.4 Other Comments

Treatment/best available technologies: Ion exchange, reverse osmosis, lime softening, electrodialysis.

A.2.13 Benzene (C_6H_6), CAS No. 71-43-2

Benzene is the simplest aromatic hydrocarbon. Its molecule consists of a single planar, conjugated ring with six carbon atoms arranged in a regular hexagon. See Chapter 7, Section 7.2.2 for its structural formula. Pure benzene is soluble in water to about 1700 mg/L and is miscible with many organic solvents. Benzene is a highly flammable, volatile, clear, colorless aromatic liquid. A major use of benzene is as a building block for making plastics, rubber, resins, and synthetic fabrics such as nylon and polyester. Other important uses include as a gasoline additive and a solvent in printing, paints, dry cleaning, etc. In the United States, gasolines are regulated to 1% benzene on a regional average. Canadian and European gasoline specifications contain a similar 1% limit on benzene content. Emissions from gasoline vehicles constitute the main source of benzene in the environment. Benzene, along with other LNAPLs, is treated in more detail in Chapter 7.

A.2.13.1 Environmental Behavior

Benzene is released to air primarily by vaporization and combustion emissions associated with its use in gasoline. Other sources are vapors from its production and use in manufacturing other chemicals. In addition, benzene may be in industrial effluents discharged into water and accidental releases from gas and oil production, refining and distribution industries. Benzene released to soil will either evaporate very quickly or leach to groundwater. It can be biodegraded by soil and groundwater microbes. Benzene released to surface water should mostly evaporate within a few hours to a few days, depending on quantity, temperature, water turbulence, etc. Although benzene does not degrade by hydrolysis, it may be biodegraded by microbes.

A.2.13.2 Ecological Concerns

Benzene has high acute and chronic toxicity to aquatic life. It bioaccumulates only slightly in aquatic organisms. It can cause death in plants and roots and membrane damage in leaves of various agricultural crops.

A.2.13.3 Health Concerns

Most exposure is by inhalation of benzene vapor and by ingestion of contaminated water. Short exposures above the MCL may cause temporary nervous system disorders, immune system depression, and anemia. Exposure to high atmospheric concentrations can affect the nervous system and may cause dizziness, nausea and vomiting, headache, and drowsiness and, at concentrations in the order of 20,000 ppm (65,000 mg/m^3), narcosis, coma and, sometimes, death. Long-term exposure above the MCL may cause chromosome aberrations and cancer.

Benzene			
EPA Primary Drinking Water Standard (mg/L)		**Health Advisories (mg/L)**	
MCL	0.005	1-day (10-kg child)	0.2
MCLG	Zero	2-day (10-kg child)	0.2
		RfD (lifetime + sensitive groups)	0.004 mg/kg/day
		DWEL	0.1
		Lifetime (adult)	—
		10^{-4} cancer risk	0.1

A.2.13.4 Other Comments

Treatment/best available technologies: Granulated activated carbon (GAC), air stripping

A.2.14 BENZO[A]PYRENE ($C_{20}H_{12}$, PAH), CAS No. 50-32-8

Benzo(a)pyrene (BaP) is a polycyclic aromatic hydrocarbon (PAH)*. PAHs have a molecular structure that consists of two or more fused aromatic rings, where adjacent rings share two or more carbon atoms. PAHs are not produced or used commercially but are formed as a result of incomplete combustion of organic materials. All PAHs have certain similar properties, see Chapter 8 and Table 8.3. If BaP is found in a sample, it is likely that other PAHs are present. They are of concern as a group because many have been identified as carcinogenic, mutagenic, and teratogenic. PAHs, including benzo(a)-pyrene, are found in exhaust from motor vehicles and other gasoline and diesel engines; emissions from coal-, oil-, and wood-burning stoves and furnaces; cigarette smoke; general soot and smoke of industrial, municipal, and domestic origin; cooked foods, especially charcoal broiled; and in incinerators, coke ovens, and asphalt processing. BaP is a solid, melting at 179°C, freely soluble in aromatic hydrocarbon solvents but only slightly in water.

---

* Other PAHs include acenaphthene, acenaphthylene, anthracene, benz[a]anthracene, benzo[e]pyrene, benzo[b]fluoranthene, benzo[ghi]perylene, benzo[j]fluoranthene, benzo[k]fluoranthene, chrysene, coronene, dibenz(a,h)anthracene, fluoranthene, fluorene, indeno(1,2,3-cd)pyrene, phenanthrene, pyrene, and naphthalene. Benzo(a)pyrene is the only one presently listed in the EPA Primary Drinking Water Standards.

A.2.14.1 Environmental Behavior

The main natural sources of BaP are forest fires and erupting volcanoes. Anthropogenic sources include the combustion of fossil fuels, coke oven emissions, and vehicle exhausts. In surface waters, direct deposition from the atmosphere appears to be the major source of BaP. Benzo(a)pyrene is moderately persistent in the environment. It readily binds to soils and does not readily leach to groundwater, though it has been detected in some groundwater. If released to water, it sorbs very strongly to sediments and particulate matter. In most waters and sediments, it resists breakdown by microbes or reactive chemicals, but it may evaporate or be degraded by sunlight. In water supply systems, it tends to sorb to any particulate matter and be removed by filtration before reaching the tap. In tap water, its source is mainly from PAH-containing materials in water storage and distribution systems.

A.2.14.2 Ecological Concerns

Benzo(a)pyrene is expected to bioconcentrate in aquatic organisms that cannot metabolize it, including plankton, oysters, and some fish.

A.2.14.3 Health Concerns

Short-term exposure at levels above the MCL can suppress the immune system and cause red blood cell damage, leading to anemia. Long-term exposure at levels above the MCL can cause developmental and reproductive effects and cancer.

Benzo[a]pyrene

EPA Primary Drinking Water Standard (mg/L)		Health Advisories (mg/L)	
MCL	0.0002	1-day (10-kg child)	—
MCLG	Zero	2-day (10-kg child)	—
		RfD (lifetime + sensitive groups)	—
		DWEL	—
		Lifetime (adult)	—
		10^{-4} cancer risk	0.0005

A.2.15 BERYLLIUM (BE), CAS NO. 7440-41-7

Beryllium is a metal found in natural deposits as ores containing other elements, and in some precious stones such as emeralds and aquamarine. It is not likely to be found in natural waters above trace levels due to the insolubility of oxides and hydroxides at normal environmental pHs. It has been reported to occur in U.S. drinking water at 0.01–0.7 µg/L.

A major use of beryllium is as an alloy hardener. Its greatest use is in making metal alloys for nuclear reactors and the aerospace industry. It is also used as an alloy and oxide in electrical equipment and electronic components, and in military vehicle armor. The chloride is used as a catalyst and intermediate in chemical manufacture. The oxide is used in glass and ceramic manufacture. Beryllium enters the environment principally as dust from burning coal and oil and from the slag and ash dumps of coal combustion. Some tobacco leaves contain significant levels of beryllium, which can enter the lungs of those exposed to tobacco smoke. It is also found in discharges from other industrial and municipal operations. Rocket exhausts contain beryllium oxide, fluoride, and chloride.

Very little is known about what happens to beryllium compounds when released to the environment. Beryllium compounds of very low water solubility appear to predominate in soils. Leaching and transport through soils to groundwater is unlikely to be of concern. Erosion or runoff of beryllium compounds into surface waters is not likely and it appears unlikely to leach to groundwater when released to land. Erosion and bulk transport of soil may carry beryllium sorbed to soils into surface waters, but most likely in particulate rather than dissolved form.

A.2.15.1 Health Concerns

Beryllium is more toxic when inhaled as fine particles than when ingested orally. Short-term air exposure can cause inflammation (chemical pneumonitis) of the lungs when inhaled. Some people develop a sensitivity, or allergy, to inhaled beryllium, leading to chronic beryllium disease. Long-term ingestion in water above the MCL may lead to intestinal lesions. There is some evidence that beryllium may cause cancer from lifetime exposures at levels above the MCL.

Beryllium			
EPA Primary Drinking Water Standard (mg/L)		**Health Advisories (mg/L)**	
MCL	0.004	1-day (10-kg child)	30
MCLG	0.004	2-day (10-kg child)	30
		RfD (lifetime + sensitive groups)	0.002 mg/kg/day
		DWEL	0.07
		Lifetime (adult)	—
		10^{-4} cancer risk	—

A.2.15.2 Other Comments

Treatment/best available technologies: Activated alumina, coagulation and filtration, ion exchange, lime softening, reverse osmosis

A.2.16 BORON (B), CAS NO. 7440-42-8

Boron is usually found in nature as the hydrated sodium borate salt kernite ($Na_2B_4O_7 \cdot 4H_2O$) or the calcium borate salt colemanite ($Ca_2B_6O_{11} \cdot 5H_2O$). Most environmentally important boron compounds are highly water-soluble. Natural weathering of boron-containing minerals is a major source of boron in certain geographical locations. In the United States, the minerals richest in boron are found in the Mojave Desert region of California, where concentrations above 300 mg/L have been observed in boron-rich lakes. In other U.S. surface waters, an average boron level is around 100 µg/L, but concentrations vary widely (from around 0.02 to 0.3 mg/L), depending on local geologic and industrial conditions. Background soil levels in the United States range up to 300 mg/kg, with an average of around 26 mg/kg.

Sodium tetraborate (kernite) is also known as borax and finds use as an additive in detergents and other cleaning agents. A major use of boron is in the manufacture of borosilicate glass, which, because of its low coefficient of thermal expansion, is used in ovenware, laboratory glassware, piping, and sealed-beam headlights. Boric acid (H_3BO_3) is used as a weak antiseptic and eyewash and as a "natural" insecticide. Other uses for boron compounds include fire retardants, leather tanning, pulp and paper whitening agents, and high-energy rocket fuels. Elemental boron is used for neutron absorption in nuclear reactors and in alloys with copper, aluminum, and steel. For these reasons, boron is common in sewage and industrial wastes. Effluent from municipal sewage treatment plants may contain up to 7 mg/L of boron, with an average (in California) of 1 mg/L.

Boron is essential to plant growth in very small amounts but may become toxic at higher amounts. For boron-sensitive plants, the toxic level may be as low as 1 mg/L. The optimum concentration depends on the plant type, but is generally around 0.3–0.4 mg/L to a maximum level of 0.75 mg/L in soil and irrigation water is generally accepted as protective for sensitive plants under long-term irrigation.

Boron is not known to be an essential nutrient for animals or humans. Boron mobility in water is greatest at pH < 7.5. Adsorption to soils and sediments is the main mechanism for removal from environmental waters. Sorption to oxide and hydroxide solids, particularly aluminum species, is enhanced above pH 7.5 and in the presence of Ca and Mg. There is no evidence that boron is bioconcentrated significantly by aquatic organisms and naturally occurring levels of boron do not appear to have an adverse effect on aquatic life. It is sometimes suggested that boron concentrations in discharges to fresh waters be limited to 10 mg/L.

A.2.16.1 Health Concerns

Moderately high doses of boron compounds appear to have little detrimental health effects. The lethal dose of boric acid for adults varies from 15 to 20 g. Chronic ingestion may cause dry skin, skin eruptions, and gastric disturbances.

Boron

EPA Primary Drinking Water Standard (mg/L)		Health Advisories (mg/L)	
MCL	None	1-day (10-kg child)	3
MCLG	None	2-day (10-kg child)	3
Boron in drinking water		RfD (lifetime + sensitive groups)	0.2 mg/kg/day
is not considered		DWEL	7
hazardous to human		Lifetime (adult)	6
health		10^{-4} cancer risk	—

A.2.16.2   Other Comments

Treatment/best available technologies: Because most boron compounds are highly water-soluble, boron is not significantly removed by conventional wastewater treatment. Boron may be coprecipitated with aluminum, silicon, or iron solids.

A.2.17   CADMIUM (CD), CAS NO. 7440-43-9

Cadmium is usually present in all soils and rocks, usually as CdS (in reducing conditions with sulfur present), $Cd(OH)_2$, and $CdCO_3$ (in oxidizing conditions at pH > 8). It is relatively mobile in surface water and groundwater systems, primarily as hydrated ions, $Cd^{2+}(aq)$, $CdSO_4$, or as complexes with humic acids and other organic ligands. It is also found in zinc, lead, and copper ores, in coal and other fossil fuels, and shales. It is often released during volcanic action. These deposits can serve as sources to ground and surface waters, especially when in contact with soft, acidic waters. Cadmium metal is used in the steel industry and in plastics. Cadmium compounds are widely used in batteries.

Cadmium is removed from natural waters by precipitation and sorption to soil surfaces, especially silicon and aluminum oxides. Removal by precipitation and sorption is strongly pH dependent, starting around pH = 6–7 and increasing as pH increases. Under reducing conditions, the concentration of dissolved cadmium species is limited by precipitation as CdS. When pH < 6–7, soluble cadmium is mobilized by desorbing from soil surfaces. The oxide and sulfide are relatively insoluble, whereas the chloride and sulfate salts are soluble. Soluble cadmium compounds have the potential to leach through soils to groundwater.

Average concentration in U.S. waters is about 0.001 mg/L. Cadmium concentrations in bed sediments are generally at least 10 times higher than in overlying water. Cadmium for industrial use is extracted during the production of other metals, chiefly zinc, lead, and copper. It is used for batteries, alloys, pigments, metal protective coatings, and as a stabilizer in plastics. It enters the environment mostly from industrial and domestic wastes, especially those associated with nonferrous mining, smelting, and municipal waste dumps. Because cadmium is chemically similar to zinc, an essential nutrient for plants and animals, it is readily assimilated into the food chain. Plants absorb cadmium from irrigation water and low levels are present in all foods, highest in shellfish, liver, and kidney meats. Smoking can double the average daily intake; one cigarette typically contains 1–2 μg of cadmium. The recommended upper limit in irrigation water is 0.01 mg/L.

A.2.17.1   Health Concerns

Cadmium is acutely toxic; a lethal dose is about 1 g. Acute exposure can cause nausea, vomiting, diarrhea, muscle cramps, salivation, sensory disturbances, liver injury, convulsions, shock, and renal failure. It is eliminated from the body slowly and can bioaccumulate over many years of low exposure. Long-term exposure to low levels of cadmium in air, food, and water leads to a buildup of cadmium in the kidneys and possible kidney disease. Other potential long-term effects are fragile bones and damage to lungs, liver, and blood. There is inadequate evidence to state whether or not cadmium has the potential to cause cancer from lifetime exposures in drinking water.

Cadmium

EPA Primary Drinking Water Standard (mg/L)

		Health Advisories (mg/L)	
MCL	0.005	1-day (10-kg child)	0.04
MCLG	0.005	2-day (10-kg child)	0.04
		RfD (lifetime + sensitive groups)	0.0005 mg/kg/day
		DWEL	0.02
		Lifetime (adult)	0.005
		10^{-4} cancer risk	—

A.2.17.2   Other Comments

Treatment/best available technologies: Coagulation and filtration, ion exchange, lime softening, reverse osmosis.

A.2.18   CALCIUM (CA), CAS No. 7440-70-2

Calcium cations (Ca^{2+}) and calcium salts are among the most commonly encountered substances in water, arising mostly from dissolution of minerals. Calcium often is the most abundant cation in river water. Among the most common calcium minerals are the two crystalline forms of calcium carbonate—calcite and aragonite ($CaCO_3$, limestone is primarily calcite), calcium sulfate (the dehydrated form, $CaSO_4$, is anhydrite; the hydrated form, $CaSO_4 \cdot 2H_2O$, is gypsum), calcium magnesium carbonate ($CaMg(CO_3)_2$, dolomite), and, less often, calcium fluoride (CaF_2, fluorite). Water hardness is caused by the presence of dissolved calcium, magnesium, and sometimes iron (Fe^{2+}), all of which form insoluble precipitates with soap and are prone to precipitating in water pipes and fixtures as carbonates (see Chapter 3). Limestone ($CaCO_3$), lime (CaO), and hydrated lime ($Ca(OH)_2$) are heavily used in the treatment of wastewater and water supplies to raise the pH and precipitate metal pollutants. Very low concentrations of calcium can enhance the deleterious effects of sodium in irrigation water by increasing the value of the sodium absorption ratio (SAR).

A.2.18.1   Health Concerns

Calcium is an essential nutrient for plants and animals, essential for bone, nervous system, and cell development. Recommended daily intakes for adults are between 800 and 1200 mg/day. Most of this is obtained in food; drinking water typically accounts for 50–300 mg/day, depending on the water hardness and assuming ingestion of 2 L/day. Calcium in food and water is essentially nontoxic. A number of studies suggest that water hardness protects against cardiovascular disease. One possible adverse effect from ingestion of high concentrations of calcium for long periods of time may be a greater risk of kidney stones. The presence of calcium in water decreases the toxicity of many metals to aquatic life. Stream standards for these metals are expressed as a function of hardness and pH. Thus, the presence of calcium in water is beneficial and no limits on calcium have been established for protection of human or aquatic health.

A.2.18.2   EPA Primary Drinking Water Standard

Since calcium is generally regarded as beneficial to health in all environmentally encountered concentrations, there are no health-based upper limits.

A.2.19   CARBOFURAN ($C_{12}H_{15}NO_3$), CAS No. 1563-66-2

Carbofuran is a white crystalline solid in the carbamate family, with a slightly phenolic odor. It is a broad-spectrum insecticide that is sprayed directly onto soil and plants just after emergence to control beetles, nematodes, and rootworm. The greatest use of carbofuran was on alfalfa and rice, with turf and grapes making up most of the remainder. Earlier uses were primarily on corn crops. Use of carbofuran is currently banned in the Unites States, Canada and the European Union.

A.2.19.1   Environmental Behavior

Carbofuran is soluble in water and is moderately persistent in soil. Its half-life is 30–120 days. It enters surface water as a result of runoff from treated fields and enters groundwater by leaching of treated crops. If released to soil, degradation occurs by chemical hydrolysis and biodegradation. The persistence of carbofuran in the soil increases as the clay and organic matter content of the soil increase, and as the pH and moisture content of soil decrease. Chemical hydrolysis occurs more rapidly in alkaline soil as compared to neutral or acidic soils. Carbofuran is likely to leach to groundwater in soils with low organic content. Volatilization from soil is not expected to be significant, although some evaporation from plants may occur. If released to water, carbofuran degrades by hydrolysis under alkaline conditions and by biodegradation. Aquatic volatilization, adsorption, and bioconcentration are not expected to be important.

A.2.19.2   Ecological Concerns

Carbofuran is very highly toxic to freshwater and estuarine/marine fish and to birds and mammals. It is reported to have been responsible for more bird deaths than any other pesticide. Carbofuran is also highly toxic to freshwater and estuarine/marine invertebrates. It is toxic to bees except in the granular formulation. It is not expected to accumulate in aquatic organisms.

A.2.19.3   Health Concerns

Occupational and general population exposure to carbofuran may occur by inhalation and dermal routes, particularly in the vicinity of aerial spraying of carbofuran as an insecticide. As with other N-methyl carbamate pesticides, the critical effect of carbofuran is cholinesterase inhibition, that is, it can overstimulate the nervous system, causing nausea, dizziness, and confusion. As with other carbamate compounds, carbofuran's cholinesterase-inhibiting effect is short term and reversible. Very high exposures (e.g., accidents or major spills) may cause respiratory paralysis and death. Long-term exposure at concentrations above the MCL can cause damage to the nervous and reproductive systems. Carbofuran is classified as "not likely" to be a human carcinogen.

Carbofuran

EPA Primary Drinking Water Standard (mg/L)		Health Advisories (mg/L)	
MCL	0.04	1-day (10-kg child)	—
MCLG	0.04	2-day (10-kg child)	—
		RfD (lifetime + sensitive groups)	0.00006 mg/kg/day
		DWEL	—
		Lifetime (adult)	—
		10^{-4} cancer risk	—

A.2.19.4 Other Comments

Treatment/best available technologies: Granulated activated carbon (GAC).

A.2.20 CHLORIDE (Cl^-), CAS No. 7440-39-3

Chlorides are widely distributed in nature, usually in the form of sodium, potassium, and calcium salts ($NaCl$, KCl, and $CaCl_2$), although many minerals contain small amounts of chloride as an impurity. Chloride in natural waters arises from weathering of chloride minerals, salting of roads for snow and ice control (see Chapter 11, Section 11.7.3), seawater intrusion in coastal regions, irrigation drainage, ancient groundwater brines, geothermal waters, and industrial wastewater.

Concentrations in unpolluted surface waters and nongeothermal groundwaters are generally low, usually below 100 mg/L. Thus, chloride concentrations in the absence of pollution are normally less than those of sulfate or bicarbonate.

A.2.20.1 Environmental Behavior

Chloride ion is extremely mobile; all chloride salts are very soluble except for lead chloride ($PbCl_2$), silver chloride ($AgCl$), and mercury chlorides (Hg_2Cl_2, $HgCl_2$). Chloride is not sorbed to soils and generally moves with water with little or no retardation. Consequently, it eventually moves to closed basins (e.g., the Great Salt Lake in Utah) or to the oceans.

Chloride ion is almost chemically and biologically inert when compared with the other major environmental ions. Under environmental conditions, chloride ions do not significantly enter into redox reactions, form no important solute complexes with other ions except at very high chloride ion concentrations (tens of thousands of milligrams per liter), form few metal precipitates (except with silver, mercury, and lead), are not significantly adsorbed to mineral surfaces, participate in few important biological processes, and have extremely low toxicities for mammalian and aquatic species.

Chloride circulates through the hydrologic cycle mainly by physical processes. Its lack of environmental reactivity is attested by the common use of chloride as a conservative tracer for groundwater movement and by its absence from EPA Drinking Water, CERCLA, and RCRA priority pollutant lists. Chloride in its most common form, as sodium chloride, is a "generally regarded as safe" (GRAS) substance. Chloride ion has no federal or state health-based treatment standards. The main environmental problems associated with chloride are reactivity with concrete, metal corrosion, adverse taste effects in drinking water, and toxicity to irrigated crops (less than 100 mg/L chloride recommended for most crops not classified as chloride sensitive, see Chapter 6, Section 6.11.2).

A.2.20.2 Ecological Concerns

To protect freshwater aquatic organisms and their uses, chloride levels should not exceed the acute and chronic levels given below more than once in 3 years (USEPA 1988):

- 860 mg/L for the criteria maximum concentration (CMC, 1 hour average acute)
- 230 mg/L for the criteria continuous concentration (CCC, 4 days average chronic)

A.2.20.3 Health Concerns

Chloride is the most abundant anion in the human body and is essential to normal electrolyte balance of body fluids. For adults, a daily dietary intake of about 9 mg chloride per kilogram of body weight is considered essential for good health. Chlorides in water are more of a taste than a health concern, although high concentrations may be harmful to people with heart or kidney problems.

A.2.20.4 EPA Primary Drinking Water Standard

There are no primary drinking water standards for chloride. The EPA secondary standard for chloride is 250 mg/L, based on adverse effect on taste.

A.2.20.5 Other Comments

Treatment/best available technologies: Conventional water treatment does not remove chloride ion. Reverse osmosis or nanofiltration is required.

A.2.21 CHROMIUM (CR), CAS No. 7440-47-3

Chromium is one of the less common elements and does not occur as a pure element, but only in compounds. Chromium occurs in either Cr(III) or Cr(VI) oxidation state, depending on pH and redox conditions. Cr(VI) can be reduced to Cr(III) by soil organic matter or the presence of S^{2-} and Fe^{2+} ions under the anaerobic conditions often encountered in deeper groundwater. Major Cr(VI) species include chromate (CrO_4^{2-}) and dichromate ($Cr_2O_7^{2-}$), which precipitate readily in the presence of metal cations (especially Ba^{2+}, Pb^{2+}, and Ag^+). Chromate and dichromate also adsorb on soil surfaces, especially iron and aluminum oxides.

Cr(III) is the dominant form of chromium at pH < 4. Cr^{3+} forms solution complexes with NH_3, OH^-, Cl^-, F^-, CN^-, SO_4^{2-}, and soluble organic ligands. Chromium mobility depends on the sorption characteristics of the soil, which depends on clay content, iron oxide content, and the amount of organic matter present. As a positively charged ion, trivalent chromium (Cr^{3+}) readily sorbs to negatively charged soils and minerals. Unsorbed Cr^{3+} forms insoluble colloidal hydroxides in the pH range of natural surface waters (6.5–9). Thus, Cr(III) mobility is decreased by adsorption to clays and oxide minerals below pH 5 and low solubility above pH 5 due to the formation of $Cr(OH)_3(s)$. Thus, it is unlikely that dissolved trivalent chromium will be present in surface waters at levels of concern. Trivalent chromium is also not likely to migrate to groundwater, most of it being retained in the upper 5–10 cm of soil.

Cr(VI) is the more toxic form of chromium. Cr(VI) exists mainly as the negatively charged complex chromate anion (CrO_4^{2-}), is not sorbed to any extent by soil or particulate matter, and is much more mobile than Cr(III). Cr(VI) is the form of chromium most commonly found at contaminated sites and is the dominant form of chromium in surface waters and shallow aquifers where aerobic conditions exist. The leachability of Cr(VI) increases as soil pH increases. However, Cr(VI) is a strong oxidant and readily oxidizes any oxidizable organic material present, itself being reduced to Cr(III). In the absence of organic matter, Cr(VI) can be stable for long periods of time, particularly under aerobic conditions.

Chromium occurs in minerals mostly as chrome iron ore, or chromite ($FeCr_2O_2$), in which it is present as Cr(III). Chromium in soils occurs mostly as insoluble chromium oxide (CrO_3), where it is present as Cr(VI). In natural waters, dissolved chromium exists either as Cr^{3+} cations or in anions such as chromate (CrO_4^{2-}) and dichromate ($Cr_2O_7^{2-}$). Although widely distributed in soils and plants, it is generally present at low concentrations in natural waters. Background levels in water typically range between 0.2 and 20 µg/L, with an average of 1 µg/L.

The main natural environmental source is weathering of rocks and soil. Major anthropomorphic sources include metal alloy production, metal plating, cement manufacturing, and incineration of municipal refuse and sewage sludge. Chromium has many industrial uses. Some major applications are in metal alloys, protective coatings on metal, magnetic tapes, paint pigments, cement, paper, rubber, and composition floor covering.

A.2.21.1   Health Concerns

Trivalent chromium is an essential trace nutrient, and plays a role in prevention of diabetes and atherosclerosis. Trivalent chromium is essentially nontoxic; the harmful effects of chromium to human health are caused by hexavalent chromium. Since oxidants such as chlorine or ozone readily oxidize trivalent chromium to the toxic hexavalent form, water quality limits are usually written for total chromium concentrations.

The EPA has found hexavalent chromium to potentially cause skin irritation or ulceration from acute exposures at levels above the MCL. Long-term exposures to chromium at levels above the MCL have the potential to cause dermatitis and damage to liver, kidney, and circulatory and nerve tissues. There is no evidence that chromium in drinking water has the potential to cause cancer from lifetime exposures in drinking water.

Chromium

EPA Primary Drinking Water Standard (mg/L; Values Are for Total Chromium)		Health Advisories (mg/L)	
MCL	0.1	1-day (10-kg child)	1
MCLG	0.1	2-day (10-kg child)	1
		RfD (lifetime + sensitive groups)	0.003 mg/kg/day Cr(VI)
		DWEL	0.1
		Lifetime (adult)	—
		10^{-4} cancer risk	—

These standards and criteria are based on the total concentration of dissolved chromium, Cr(III) + Cr(VI).

A.2.21.2 Other Comments

Treatment/best available technologies: Coagulation and filtration, ion exchange, reverse osmosis, lime softening (for Cr(III) only)

A.2.22 COPPER (CU), CAS NO. 7440-50-8

In nature, copper sometimes occurs as the pure metal, but more often, in the form of mineral ores of sulfide, oxide, arsenite, chloride, and carbonate that often contain 2% or less of the metal. Chalcopyrite ($CuFeS_2$) is the most abundant of the copper ores, accounting for about 50% of the world's copper deposits. Mining activities are the major human-caused source of copper contamination in groundwater and surface waters and the weathering of copper deposits is the main natural source of copper in environmental waters. However, dissolved copper rarely occurs in unpolluted waters above 10 µg/L, being limited by several processes: the low solubility of copper hydroxide ($Cu(OH)_2$), coprecipitation with less soluble metal hydroxides, and adsorption to sediments and soil surfaces.

In aerobic, alkaline water, cupric ion (Cu^{2+}), copper carbonate ($CuCO_3$), and copper hydroxide complexes ($CuOH^+$ and $Cu(OH)_2$) are the main dissolved copper species. Copper also forms strong dissolved complexes with humic acids. The affinity of Cu for humic acids increases as pH increases and TDS decreases. In anaerobic environments with sulfur present, Cu^{2+} precipitates as copper sulfide (CuS). The mobility of dissolved copper species is decreased by sorption to mineral surfaces. Cu^{2+} sorbs strongly to mineral surfaces over a wide range of pH values.

Other sources of copper include algaecides and lumber pressure-treated with chromated copper arsenate (CCA). In some cases, copper salts are added to reservoirs for the control of algae. Copper concentrations in acid mine drainage may reach several hundred milligrams per liter but if the pH is raised to 7 or higher, most of the copper precipitates. Smelting operations and municipal incineration may also generate airborne copper that is precipitated with rain and snow into surface waters.

Copper in drinking water is primarily caused by corrosion of copper pipes and fittings in plumbing systems. This is the reason for an EPA action level based on samples taken from distribution system taps, rather than an MCL. Corrosivity toward copper metal increases with decreasing pH, especially below pH 6.5.

A.2.22.1 Health Concerns

Copper is an essential nutrient, but at high doses, it has been shown to cause stomach and intestinal distress, liver and kidney damage, and anemia. The most toxic species of copper is cupric ion (Cu^{2+}). Mild copper toxicity has also been demonstrated for $CuOH^+$ and $Cu_2(OH)_2^{2+}$.

Persons with Wilson's disease may be at a higher risk of health effects due to copper than the general public. There is inadequate evidence to state whether or not copper has the potential to cause cancer from a lifetime exposure in drinking water.

Copper

EPA Primary Drinking Water Standard (mg/L; Values Are for Copper at the Water Tap)		Health Advisories (mg/L)	
MCL	TT	1-day (10-kg child)	—
Action Level*	1.3 (in 10% or more tap water samples)	2-day (10-kg child)	—
		RfD (lifetime + sensitive groups)	—
		DWEL	—
		Lifetime (adult)	—
MCLG	1.3	10^{-4} cancer risk	—

* Because plumbing in homes and commercial buildings is the main source of copper in drinking water supplies, the EPA has established a tap water action level rather than an MCL.

A.2.22.2 Other Comments

Treatment/best available technologies: For treating source water: Ion exchange, lime softening, reverse osmosis, coagulation and filtration. For corrosion control: pH and alkalinity adjustment, calcium adjustment, silica- or phosphate-based corrosion inhibition.

A.2.23 Cyanide (CN⁻), CAS No. 57-12-5 and Hydrogen Cyanide (HCN), CAS No. 74-90-8

Cyanide is a product of natural animal and vegetative decay processes and also is a component in many industrial waste streams. It is used extensively in mining to separate metals, particularly gold, from ores. In water, equilibrium exists between the ionized (CN^-, also called free cyanide) and unionized (HCN) forms, the fraction of each depending on pH (Equation A.2 and Figure A.1).

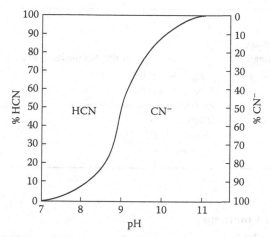

FIGURE A.1 Distribution of cyanide between the HCN and CN⁻ forms as a function of pH.

$$CN^- + H^+ \rightleftarrows HCN \tag{A.2}$$

Below pH 9, the predominant form is HCN. HCN is more toxic than CN⁻ and is the dominant form in most natural waters. HCN is volatile, whereas CN⁻ is nonvolatile.

The most common industrially used form, HCN, is used in the production of nylon and other synthetic fibers and resins. Some cyanide compounds are used as herbicides. The major sources of cyanide releases to water are discharges from metal finishing industries, iron and steel mills, and organic chemical industries. Disposal of cyanide wastes in landfills is a major source of releases to soil.

Cyanides are generally not persistent when released to water or soil, and are not likely to accumulate in aquatic life. They rapidly evaporate and are broken down by microbes. They do not bind to soils and may leach to groundwater. Cyanide-containing herbicides, such as Tabun, have moderate potential for leaching, but again are readily biodegraded so that they are not expected to bioconcentrate.

Soluble cyanide compounds, such as hydrogen and potassium cyanide, have low adsorption to soils with high pH, high carbonate, and low clay content. However, at pH less than 9.2, most free cyanide converts to HCN, which is highly volatile. Soluble cyanides are not expected to bioconcentrate.

Insoluble cyanide compounds such as the copper and silver salts adsorb to soils and sediments, and generally have the potential to bioconcentrate. Insoluble forms do not biodegrade to HCN.

A.2.23.1 Health Concerns

Short-term exposure to cyanide compounds above the MCL may cause rapid breathing, tremors, and other neurological effects. Long-term exposure at levels above the MCL may cause weight loss, thyroid effects, and nerve damage. There is inadequate evidence for carcinogenicity from lifetime exposures in drinking water.

Cyanide

EPA Primary Drinking Water Standard (mg/L)		Health Advisories (mg/L)	
MCL (CN⁻)	0.2	1-day (10-kg child)	0.2
MCLG (CN⁻)	0.2	2-day (10-kg child)	0.2
		RfD (lifetime + sensitive groups)	0.0006 mg/kg/day (HCN)
		DWEL	—
		Lifetime (adult)	—
		10⁻⁴ cancer risk	—

A.2.23.2   Other Comments

Treatment/best available technologies: Ion exchange, reverse osmosis, chlorination.

A.2.24   FLUORIDE (F⁻), CAS No. 16984-48-8

Most soils and rocks contain trace amounts of fluoride. Much higher concentrations are found in areas of active or dormant volcanic activity. Common fluoride-containing minerals include fluorite (or fluorspar, CaF_2), cryolite (Na_3AlF_6), and fluorapatite ($Ca_5F(PO_4)_3$). Weathering of minerals is the main source of fluoride in unpolluted waters, where concentrations are generally less than 1 mg/L but may sometimes exceed 50 mg/L. Where dissolved calcium is present, the formation of fluorite may limit fluoride concentrations. High fluoride concentrations are more likely in thermal waters and water with low calcium concentrations. Groundwater usually contains higher concentrations than surface water and groundwater concentrations as high as 10 mg/L are common.

The formation of fluoride complexes may be important in solubilizing beryllium, aluminum, tin, and iron in natural waters. Addition of fluoride to drinking water and toothpaste for reducing dental caries, and its subsequent discharge in sewage, also contributes to aquatic fluoride. Discharges from aluminum, steel, and phosphate production are important industrial sources of fluoride in water.

A.2.24.1   Health Concerns

Small amounts of fluoride appear to be an essential nutrient. People in the United States ingest about 2 mg/day in water and food. A concentration of about 1 mg/L in drinking water effectively reduces dental caries without harmful effects on health. Dental fluorosis can result from exposure to concentrations above 2 mg/L in children up to about 8 years of age. In its mild form, fluorosis is characterized by white opaque mottled areas on tooth surfaces. Severe fluorosis causes brown to black stains and pitting. Although the matter is controversial, the EPA has determined that dental fluorosis is a cosmetic and not a toxic or an adverse health effect. Water hardness limits fluoride toxicity to humans and fish. The severity of fluorosis decreases in harder drinking water. Crippling skeletal fluorosis in adults requires the consumption of about 20 mg or more of fluoride per day over a 20-year period. No cases of crippling skeletal fluorosis have been observed in the United States from the long-term consumption of 2 L/day of water containing 4 mg/L of fluoride. The EPA has concluded that 0.12 mg/kg/day of fluoride is protective of crippling skeletal fluorosis. Fluoride therapy, where 20 mg/day is ingested for medical purposes, is sometimes used to strengthen bone, particularly spinal bones.

Fluoride

EPA Primary Drinking Water Standard (mg/L)		Health Advisories (mg/L)	
MCL	4.0	1-day (10-kg child)	—
MCLG	4.0	2-day (10-kg child)	—
		RfD (lifetime + sensitive groups)	0.06 mg/kg/day
		DWEL	—
		Lifetime (adult)	—
		10^{-4} cancer risk	—

A.2.24.2 Other Comments

Treatment/best available technologies: anion exchange, nanofiltration, reverse osmosis.

A.2.25 Hydrogen Sulfide (H₂S), CAS No. 7783-06-4

See Chapter 4, Section 4.5 for more information.

Natural waters acquire sulfur compounds mainly from geochemical weathering of sulfide and sulfate minerals, fertilizers, decomposition of organic matter, and atmospheric deposition from industrial fuel combustion. During the decomposition of organic matter, sulfur is released largely as hydrogen sulfide (H_2S), which oxidizes to sulfate under aerobic conditions. Therefore, under aerobic conditions in aquatic systems, sulfate is the predominant form of sulfur and concentrations of hydrogen sulfide are very low.

Under anaerobic conditions, sulfate and organic sulfur compounds are reduced to the sulfide anion (S^{2-}) by bacterial reduction. The presence of sulfate-reducing bacteria in drinking water distribution systems can be a major cause of taste and odor problems. Sulfide anion is commonly found under aquatic anaerobic conditions wherever sulfur is present, for example, domestic and industrial wastewater and sludges, the hypolimnion of stratified lakes, and the bottoms of wetlands. Sulfide anion reacts with water to form hydrogen sulfide (H_2S). H_2S is a flammable, poisonous gas with a characteristic strong odor of rotten eggs. Hydrogen sulfide and the sulfide salts of the alkali and alkaline metals (groups 1A and 2A of the periodic table) are soluble in water. Soluble sulfide salts dissociate in water, forming the sulfide anion, which then reacts with water to form H_2S.

If transition metal cations are present, particularly iron and manganese, metal sulfides of low solubility are formed and precipitated at neutral to alkaline pH values. Thus, anaerobic zones of lakes and wetlands, where high levels of dissolved metals are present along with sulfide, are likely to contain metal sulfides in the sediments and very little dissolved sulfide anion. Only after most of the dissolved metals have been precipitated can H_2S accumulate in the water. Black sediments of eutrophic lakes and wetlands consist largely of precipitated iron and manganese sulfides, S^{2-} dissolved in interstitial water, acid-soluble sulfide compounds, elemental sulfur, organic sulfur, and sulfates. H_2S also is added in large quantities to the atmosphere from volcanic gases, industrial sources, and biochemical activity in water and soil.

The human nose is very sensitive to the rotten-egg odor of hydrogen sulfide. Most people can detect the smell of H_2S in water containing as little as 1 μg/L. Typical concentrations of H_2S in unpolluted surface water are <0.25 μg/L. H_2S > 2.0 μg/L constitutes a chronic hazard to aquatic life. Groundwater usually contains little or no sulfide, because contact with metal-bearing minerals results in the formation of metal sulfides with low solubility. Note, however, that sulfides may be found in wells, arising from sulfate-reducing bacteria present in the well. Some brines, especially those associated with petroleum deposits, may contain several hundred milligrams per liter of dissolved hydrogen sulfide.

A.2.25.1 Health Concerns

Hydrogen sulfide is acutely toxic to humans. In the workplace, most causes of accidental death are by accidental inhalation of high concentrations (>1000 ppm). However, the EPA has no drinking water standards for hydrogen sulfide because its disagreeable taste and odor make water unpalatable at concentrations much lower than the toxic levels. A guideline value is that the presence of H_2S should be not detectable by consumers.

Low levels of hydrogen sulfide are detoxified by enzymes in the body that oxidize H_2S to sulfate. However, these natural body defenses are overcome at air concentrations around 300–350 ppm. Industrial safety gas detectors, such as those used by utility, sewage, and petrochemical workers, are typically set to warn of exposure at 5–10 ppm and to go into emergency alarm at 15 ppm.

A.2.25.2 EPA Primary Drinking Water Standard

The EPA does not regulate hydrogen sulfide in drinking water because a concentration high enough to be a drinking water health hazard also makes the water unpalatable.

A.2.25.3 Other Comments

Treatment/best available technologies: See Chapter 3.

Typical state stream standards for undissociated H_2S:

- Aquatic life (cold and warm water biota): 0.002 mg/L (30-day average)
- Domestic water supply: 0.05 mg/L (30-day average).

A.2.26 IRON (FE), CAS NO. 7439-89-6

See also Chapter 5.

Iron is naturally released into waters by weathering of pyritic ores containing iron sulfide (FeS_2) and other iron-bearing minerals in igneous, sedimentary, and metamorphic rocks. In water, it commonly occurs with manganese,* but usually iron is in greater abundance. It also results from many human sources: mineral processing, coke and coal burning, acid-mine drainage, iron and steel industry wastes, and corrosion of iron and steel. Because iron is an essential nutrient for animals and plants, it is present in organic matter in soil and in sewage and landfill leachate. Many microorganisms use iron as an energy source and play an important part in iron oxidation and reduction processes.

In the aquatic environment, iron is present in two oxidation states: ferrous (Fe^{2+}) and ferric (Fe^{3+}). The reduced ferrous state is highly soluble in the pH range of unpolluted surface waters, whereas the oxidized ferric state is associated with compounds of low solubility at pH values above 5. For example, Fe^{3+} reacts with water to form low-solubility iron oxyhydroxides, which form yellow to red-brown precipitates often seen on rocks and sediments in surface waters with high iron concentrations. Iron oxyhydroxides often form as colloidal suspensions of gels or flocs. These have large surface areas and a strong absorptive capacity for other dissolved ionic species. Coprecipitation with iron at elevated pH has been developed as a treatment process for removing other dissolved metals. For example, removal of dissolved zinc by lime precipitation to a concentration below 0.1 mg/L generally requires coprecipitation with iron.

Since the ferrous state is easily oxidized to the ferric state (a process often enhanced by aerobic iron bacteria, which leaves slimy deposits of ferric iron), dissolved iron is mainly found under reducing conditions, in groundwater or anaerobic surface waters. In well-aerated water above pH 5, dissolved iron concentrations are generally less than 30 µg/L, whereas dissolved concentrations in groundwater with low DO may be as high as 50 mg/L. Concentrations in the milligram per liter range for aerated surface waters generally indicate that most of the measured iron is associated with sediments.

When sampling groundwater for dissolved iron, it is important to purge the well adequately, filter the water immediately onsite to remove iron sediments, acidify the filtrate to prevent further precipitation, and store in a container without any air headspace. It is common for dissolved Fe^{2+} in groundwater to become oxidized to Fe^{3+} after exposure to oxygen in a well or water distribution system; if the well becomes infected with iron bacteria, oxidation of ferrous to ferric iron is very rapid and the well may deliver "red" water unsuitable for domestic use. Oxidation converts dissolved ferrous iron to solid ferric iron particles, which initially are white, changing over time to yellow and finally to red-brown; the precipitated ferric iron particles range in size from large enough to settle out of the water to small enough to remain suspended as colloidal iron. Colloidal iron imparts a red tint color to water.† The result is rust-colored water that stains plumbing fixtures, dishes, and laundry. Such water often has objectionable taste and eventually may clog plumbing fixtures, reducing the flow of water.

---

* Because iron and manganese are chemically similar, they are commonly found in the same ore formations and water bodies.
† Colloidal manganese, if present, gives water a black tint.

The most common approach to controlling iron (or manganese) bacteria in wells is shock chlorination.* It is almost impossible to kill all the iron (or manganese) bacteria in a well because they also exist in the subsurface surrounding the well. They frequently come back, eventually requiring a repeat of the shock treatment. Automatic continuous application of low levels of chlorine may be more effective. In addition to killing the bacteria, chlorination also will oxidize ferrous to ferric iron. Thus, continuous chlorination may have to be followed by filtration to remove iron particulates.

A.2.26.1 Health Concerns

Iron is an essential nutrient in animal and plant metabolism. It is not normally considered a toxic substance. It is not regulated in drinking water except as a secondary standard for aesthetic reasons. Adults require between 10 and 20 mg of iron per day. Excessive iron ingestion may result in hemochromatosis, a condition of tissue damage from iron accumulation. This condition rarely occurs from dietary intake alone, but can result from prolonged consumption of acidic foods cooked in iron utensils and from the ingestion of large quantities of iron tablets.

Iron can be toxic to freshwater aquatic life above 1 mg/L and may interfere with fish uptake of oxygen through their gills above 0.3 mg/L.

A.2.26.2 EPA Primary Drinking Water Standard

The EPA has no primary drinking water standard for iron. The EPA secondary drinking water standard (nonenforceable) is 0.3 mg/L as total iron.

A.2.26.3 Other Comments

Treatment/best available technologies: Lime precipitation, aeration or chemical oxidation followed by filtration, cation exchange, microfiltration, reverse osmosis.

In surface waters, a high concentration of total iron (>1 mg/L) generally means that both dissolved and particulate forms are present. This requires sequential treatment steps:

1. Oxidation with chlorine, chlorine dioxide, permanganate, or ozone to oxidize dissolved Fe^{2+} to particulate Fe^{3+} and, also, to kill iron bacteria
2. Filtration to remove particles
3. Sorption with activated carbon to remove colloidal and dissolved iron, as well as excess chlorine

A.2.27 LEAD (PB), CAS NO. 7439-92-1

Lead minerals are found mostly in igneous, metamorphic, and sedimentary rocks. The most abundant lead mineral is galena (PbS). Oxide, carbonate, and sulfate minerals of lead are lanarkite (PbO), cerrusite ($PbCO_3$), and anglesite ($Pb(SO_4)$), respectively. Commercial ores have concentrations of lead in the range of 30–80 g/kg.

---

* The procedure for shock chlorination of iron (or manganese) bacteria is the same as that for treating odorous wells for sulfate-reducing bacteria, discussed in Chapter 4, Section 4.5.6.

Metallic lead and the common lead minerals have very low solubility. Most of the lead in aquatic environments, as much as 85%, is associated with sediments, the rest being in dissolved form. Although some lead enters the environment from natural sources by weathering of minerals, particularly galena, anthropogenic sources are about 100 times greater. Mining, milling, and smelting of lead and metals associated with lead, such as zinc, copper, silver, arsenic, and antimony, are major sources, as are combustion of fossil fuels and municipal sewage. Commercial products that are major sources of lead pollution include acid-storage batteries, electroplating, construction materials, ceramics and dyes, radiation shielding, ammunition, paints, glassware, solder, piping, cable sheathing, roofing, and, up to about 1980, gasoline additives such as tetramethyl- and tetraethyl-lead. Lead released to groundwater, surface water, and land is usually in the form of elemental lead, lead oxides and hydroxides, and lead–metal oxyanion complexes.

In areas away from mining and smelters, the use of leaded gasoline exceeded all other sources of lead contamination between about 1940 and 1980. In 1970, new regulations in the Clean Air Act led to a reduction in lead additives to gasoline as well as tighter restrictions on industrial emissions. By 1985, these measures had resulted in an overall decrease in lead emissions of around 20%. Organic lead additives to automotive gasoline were completely eliminated in 1996, but soils and water bodies still carry the lead legacy from earlier years.

Levels of dissolved lead in natural surface waters are generally low. Lead sulfides, sulfates, oxides, carbonates, and hydroxides are almost insoluble. Because of their greater abundance, the carbonates and hydroxides are considered to impose an upper limit on the concentrations of lead that can occur in lakes, rivers, and groundwaters. The global mean lead concentration in lakes and rivers is estimated to be between 1.0 and 10.0 $\mu g/L$.

Pb^{2+} is the stable ionic species in most of the natural environment. Sorption is the dominant mechanism controlling the distribution of lead in the aquatic environment, where it forms complexes with organic ligands to yield soluble, colloidal, and particulate compounds that sorb to humic materials. At the low lead concentrations typically found in the aquatic environment, most of the lead in the dissolved phase is likely to be in the form of organic ligand complexes. In the presence of clay suspensions at a pH range of 5–7, most lead is precipitated and sorbed as sparingly soluble hydroxides. Soluble lead is removed from natural waters mainly by association with sediments and suspended particulates. Lead solubility is very low (<1 $\mu g/L$ at pH range of 8.5–11) in water containing carbon dioxide and sulfate. At constant pH, the solubility of lead decreases with increasing alkalinity. Lead is bioaccumulated by aquatic organisms, including benthic bacteria, freshwater plants, invertebrates, and fish.

Lead in drinking water results primarily from corrosion of materials located throughout a distribution system containing lead and copper and from lead and copper plumbing materials used to plumb publicly and privately owned structures connected to the distribution system. Very little lead enters public distribution systems in water from treatment plants. Most public water systems serve at least some buildings with lead solder or lead service lines.

All water is corrosive to metal plumbing materials to some degree, even water termed "noncorrosive" or water treated to make it less corrosive. The corrosivity of water to lead is influenced by water pH, total alkalinity, dissolved inorganic carbonate, calcium, and hardness. Galvanic corrosion of lead into water also occurs with lead-soldered copper pipes, due to differences in the electrochemical potential of the two metals. Grounding of household electrical systems to plumbing can accelerate galvanic corrosion.

Lead is not very mobile under normal environmental conditions. It is generally retained in the upper 2–5 cm of soil, especially soils with at least 5% organic matter or a pH 5 or above. Leaching is not important under normal conditions. It is expected to slowly undergo speciation to the more insoluble sulfate, sulfide, oxide, and phosphate salts.

Metallic lead can be dissolved by pure water in the presence of oxygen, but if the water contains carbonates and silicates, protective films are formed, preventing further attack. The solubility of Pb is 10 µg/L above pH 8, whereas near pH 6.5, the solubility can exceed 100 µg/L. Lead is effectively removed from the water column by adsorption to organic matter and clay minerals, precipitation as insoluble salts, and reaction with hydrous iron and manganese oxides. Under appropriate conditions, dissolution due to anaerobic microbial action may be significant in subsurface environments. In an oxidizing environment, the least soluble common forms of lead are the carbonate, hydroxide, and hydroxycarbonate. In reducing conditions where sulfur is present, lead sulfide (PbS) is formed as an insoluble solid.

A.2.27.1 Health Concerns

Short-term exposure to lead at relatively low concentrations can cause interference with red blood cell chemistry, delays in normal physical and mental development in babies and young children, slight deficits in the attention span, hearing, and learning abilities of children, and slight increases in the blood pressure of some adults. It appears that some of these effects, particularly changes in the levels of certain blood enzymes and in aspects of children's neurobehavioral development, may occur at blood lead levels so low as to be essentially without a threshold. Long-term exposure to lead has been linked to cerebrovascular and kidney disease in humans. Lead has the potential to cause cancer from a lifetime exposure at levels above the action level.

Lead

EPA Primary Drinking Water Standard (mg/L; Values Are for Lead at the Water Tap)		Health Advisories (mg/L)	
MCL	TT	1-day (10-kg child)	—
Action Level*	0.015 (in 10% or more tap water samples)	2-day (10-kg child)	—
		RfD (lifetime + sensitive	—
MCLG	Zero	groups)	
		DWEL	—
		Lifetime (adult)	—
		10^{-4} cancer risk	—

* Because plumbing in homes and commercial buildings is the main source of lead in drinking water supplies, the EPA has established a tap water action level rather than an MCL.

A.2.27.2   Other Comments

Treatment/best available technologies: Ion exchange, lime softening, reverse osmosis, coagulation and filtration.

Corrosion control: pH and alkalinity adjustment, calcium adjustment, silica- or phosphate-based corrosion inhibition.

A.2.28   MAGNESIUM (MG), CAS No. 7439-95-4

Magnesium is used in the textile, tanning, and paper industries. Lightweight alloys of magnesium are used extensively in molds, die castings, extrusions, rolled sheets and plate forgings, mechanical handling equipment, portable tools, luggage, and general household goods. The carbonates, chlorides, hydroxides, oxides, and sulfates of magnesium are used in the production of magnesium metal, refractories, fertilizers, ceramics, explosives, and medicines. Magnesium hydroxide is applied as a flocculant in water purification.

Magnesium is abundant in the earth's crust and is a common constituent of natural water. Along with calcium, it is one of the main contributors to water hardness. The aqueous chemistry of magnesium is similar to that of calcium in the formation of carbonates and oxides. Magnesium compounds are, in general, more soluble than their calcium counterparts. As a result, large amounts of magnesium are rarely precipitated. Magnesium carbonates and hydroxides precipitate at high pH (>10). Magnesium concentrations can be extremely high in certain closed saline lakes. Natural sources contribute more magnesium to the environment than do all anthropogenic sources. Magnesium is commonly found in magnesite, dolomite, olivine, serpentine, talc, and asbestos minerals. The principal sources of magnesium in natural water are ferromagnesium minerals in igneous rocks and magnesium carbonates in sedimentary rocks. Water in watersheds with magnesium-containing rocks may contain magnesium in the concentration range of 1–100 mg/L. The sulfates and chlorides of magnesium are very soluble, and water in contact with such deposits may contain several hundred milligrams of magnesium per liter.

Magnesium contributes to water hardness (Section 3.6) and the removal of magnesium with water softening methods may have some negative environmental effects.

- Water softened by ion exchange has calcium and magnesium removed and replaced by sodium or hydrogen ion. This has two potentially harmful consequences:
  - The treated water is more corrosive to metals and may be damaging to metal plumbing, heat exchangers, and so on.
  - The SAR of the water is increased. This has the effect of making soils with clay less permeable, which, in turn, can make treated water unfit for irrigating vegetation (Chapter 6, Section 6.11). Where softened water is eventually discharged to septic onsite wastewater systems (OWS), a decrease in soil permeability may cause OWS overflow and potential pollution of nearby water sources.

- Softening additives in detergents and soaps, included to improve hard water washing, may also be environmentally harmful.
  - Even though the use of phosphate additives has been very much reduced, some municipal discharges are still significant sources of nutrient phosphorus.
  - The detergent softening replacements for phosphates are mainly complexing agents (sodium citrate, EDTA, NTA, etc.) or ion exchangers (e.g., zeolite A). Although these do not cause eutrophication, NTA may be mutagenic and EDTA is difficult to remove in wastewater treatment plants and may be released to waterways where it can solubilize sorbed metals from sediments and increase stream metal concentrations.

A.2.28.1   Health Concerns

There are no health guidelines for magnesium in drinking water because negative human and animal health effects are not expected. Magnesium is an essential nutrient for plants and animals, for bone and cell development. It accumulates in calcareous tissues, and is found in edible vegetables (700–5600 mg/kg), marine algae (6400–20,000 mg/kg), marine fish (1200 mg/kg), and mammalian muscle (900 mg/kg) and bone (700–1800 mg/kg). Magnesium is one of the principal cations of soft tissue. It is an essential part of the chlorophyll molecule. Recommended daily intake for adults is 400–450 mg/day, of which drinking water can supply from 12 to 250 mg/day, depending on the magnesium concentration and assuming ingestion of 2 L/day. Magnesium salts are used medicinally as cathartics and anticonvulsants. In general, the presence of magnesium in water is beneficial and no limits on magnesium have been established for protection of human or aquatic health.

A.2.28.2   EPA Primary Drinking Water Standard

There are no primary or secondary drinking water standards for magnesium. Magnesium in drinking water may provide nutritional benefits for persons with magnesium deficient diets.

A.2.29   MANGANESE (MN), CAS NO. 7439-96-5

Manganese is an abundant, widely distributed metal. It does not occur in nature as the elemental metal, but is found in various salts and minerals, frequently along with iron compounds. Soils, sediments, and metamorphic and sedimentary rocks are significant natural sources of manganese. The most important manganese mineral is pyrolusite (MnO_2). Other manganese minerals are manganese carbonate ($MnCO_3$, rhodocrosite) and manganese silicate ($MnSiO_3$, rhodonite). Ferromanganese minerals, such as biotite mica ($K(Mg,Fe)_3(AlSi_3O_{10})(OH)_2$) and amphibole ($(Mg, Fe)_7Si_8O_{22}(OH)_2$), contain large amounts of manganese. The weathering of manganese deposits contributes small amounts of manganese to natural waters.

Manganese, its alloys, and manganese compounds are commonly used in the steel industry for manufacturing metal alloys and dry-cell batteries, and in the chemical industry for making paints, varnishes, inks, dyes, glass, ceramics, matches, fireworks, and fertilizers. The iron and steel industry and acid mine drainage release a large portion of the manganese found in the environment. Iron and steel plants also release manganese into the atmosphere, from which it is redistributed by atmospheric deposition.

A.2.29.1 Environmental Behavior

Manganese seldom reaches concentrations of 1.0 mg/L in natural surface waters, and is usually present in quantities of 0.2 mg/L or less. Concentrations higher than 0.2 mg/L may occur in groundwaters and deep stratified lakes and reservoirs under reducing conditions. Subsurface and acid mine waters may contain 10 mg/L. Manganese is similar to iron in its chemical behavior, and is frequently found in association with iron. In the absence of dissolved oxygen, manganese normally is in the reduced manganous (Mn^{2+}) form, but it is readily oxidized to the manganic (Mn^{4+}) form. Permanganates (Mn^{7+}) are not persistent because they are strong oxidizers and are rapidly reduced in the process of oxidizing organic materials. Nitrate, sulfate, and chloride salts of manganese are quite soluble in water, whereas oxides, carbonates, phosphates, sulfides, and hydroxides are only sparingly soluble. In natural waters, a substantial fraction of manganese is present in suspended form. In surface waters, divalent manganese (Mn^{2+}) is rapidly oxidized to insoluble manganese dioxide (MnO_2), which then precipitates as a black solid often observed as black stains on rocks. In drinking water distribution systems, precipitation of MnO_2 may cause unsightly black staining of fixtures and laundry.

A.2.29.2 Health Concerns

Manganese is an essential trace element for microorganisms, plants, and animals, and, hence, is contained in all or nearly all organisms. Manganese is a ubiquitous element that is essential for normal physiologic functioning in all animal species. The total body load of manganese in an average adult is about 12 mg. Health problems in humans may arise from both deficient and excess intakes of manganese. Thus, any quantitative risk assessment for manganese must consider that, although manganese is an essential nutrient, excessive intake causes toxic symptoms. An average dietary intake in the United States ranges between 2 and 10 mg/day, with an average around 4 mg/day. Grains and cereals are the richest dietary sources of manganese, followed by fruits and vegetables. Meat, fish, and poultry contain little manganese. Drinking water supplies almost always contain less than the secondary standard of 0.05 mg/L and drinking water generally contributes no more than about 0.07 mg/day to an adult diet. A maximum adult dietary intake of 20 mg/day is recommended to avoid manganese toxicity. Manganese is not considered to be a cancer risk.

Manganese

EPA Primary Drinking Water Standard (mg/L)		Health Advisories (mg/L)	
MCL	—	1-day (10-kg child)	1.0
MCLG	—	2-day (10-kg child)	1.0
SDWS	0.05	RfD (lifetime + sensitive groups)	0.14 mg/kg/day
		DWEL	1.6
		Lifetime (adult)	0.3
		10^{-4} cancer risk	—

A.2.29.3 Other Comments

Treatment/best available technologies: Lime precipitation, aeration, cation exchange, microfiltration, reverse osmosis.

A.2.30 MERCURY (HG), CAS NO. 7439-97-6

Mercury is a liquid metal found in natural deposits of ores containing other elements. Mercury deposits occur in all types of rocks: igneous, sedimentary, and metamorphic. Although cinnabar (HgS) is the most common mercury ore, mercury is present in more than 30 common ore and gangue minerals. Mercury exists in the environment as the elemental metal, as monovalent and divalent salts, and as organic mercury compounds, the most important of which are methyl mercury ($HgCH_3^+$) and dimethyl mercury ($Hg(CH_3)_2$). Methyl and dimethyl mercury are formed from inorganic mercury by microorganisms found in bottom sediments and sewage sludge. There are other microorganisms that can demethylate mercury back to the inorganic form.

A.2.30.1 Environmental Behavior

Mercury is noteworthy among environmental metal pollutants by virtue of its volatility and the ease by which inorganic mercury can be converted to organic forms by microbial processes. Its volatility accounts for the fact that mercury is present in the atmosphere as metallic mercury and as volatilized organic mercury compounds. Terrestrial environments appear to be major sources of atmospheric mercury, with contributions from evapotranspiration of leaves, decaying vegetation, and degassing of soils. The major source of mercury movement in the environment is the natural degassing of the earth's crust, which may introduce between 25,000 and 150,000 t of mercury per year into the atmosphere. It is not unusual for atmospheric concentrations of mercury in an area to be up to four times the level in contaminated soils. Atmospheric mercury can enter terrestrial and aquatic habitats via particle deposition and precipitation. Measuring atmospheric concentrations of mercury at the earth surface or even from aircraft is a form of prospecting for mineral formations of other metals associated with elemental mercury. Inorganic forms of mercury (Hg) can be converted to soluble organic forms by anaerobic microbial action in the biosphere. In the atmosphere, 50% of volatile mercury is metallic mercury (Hg). As high as 25–50% of Hg in water is organic. Mercury in the environment is deposited and volatilized many times, with a residence time in the atmosphere of several days. In the volatile phase, it can be transported hundreds of kilometers.

Twenty thousand tons of mercury per year also is released into the environment by human activities such as combustion of fossil fuels, operation of metal smelters, cement manufacture, and other industrial releases. Mercury is used in the chloralkali industry, where mercury is used as an electrode to produce chlorine, caustic soda (sodium hydroxide), and hydrogen by electrolysis of molten sodium chloride. It is also used to produce electrical products such as dry-cell batteries, fluorescent light bulbs, switches, and other control equipment. Electrical products account for 50% of mercury used. Aquatic pollution originates in sewage, metal refining operations, chloralkali plant wastes, industrial and domestic products such as thermometers and batteries, and from solid wastes in major urban areas, where electrical mercury switches account for a significant release of mercury to the environment.

After release to the environment, mercury usually exists in mercuric (Hg2+), mercurous (Hg22+), elemental (HgO), or alkyllated form (methyl/ethyl mercury). The redox potential and pH of the system determine the stable forms of mercury that will be present. Mercurous and mercuric mercury are more stable under oxidizing conditions. When mildly reducing conditions exist, organic or inorganic mercury may be reduced to elemental mercury, which may then be converted to alkyllated forms by biotic or abiotic processes. Mercury is most toxic in its alkyllated forms which are soluble in water and volatile in air.

Hg(II) forms strong complexes with a variety of both inorganic and organic ligands, making it very soluble in oxidized aquatic systems. Sorption to soils, sediments, and humic materials is an important mechanism for removal of mercury from solution. Sorption is pH-dependent and increases as pH increases. Mercury may also be removed from solution by coprecipitation with sulfides.

Under anaerobic conditions, both organic and inorganic forms of mercury may be converted to alkyllated forms by microbial activity, such as by sulfur-reducing bacteria. Elemental mercury may also be formed under anaerobic conditions by demethylation of methyl mercury, or by reduction of Hg(II). Acidic conditions (pH < 4) also favor the formation of methyl mercury, whereas higher pH values favor precipitation of HgS(s).

In most unpolluted surface waters, mercuric hydroxide ($Hg(OH)_2$) and mercuric chloride ($HgCl_2$) are the predominant mercury species, with concentrations generally less than 0.001 mg/L. In polluted waters, concentrations up to 0.03 mg/L may occur. In aquatic systems, mercury binds to dissolved matter or fine particulates. In freshwater habitats, it is common for mercury compounds to be sorbed to particulate matter and sediments. Sediment binding capacity is related to organic content, and is little affected by pH. Mercury tends to combine with sulfur in anaerobic bottom sediments. Organic methyl mercury bioconcentrates along aquatic food chains to an extent that fish in mildly polluted waters may become unsafe for food use.

A.2.30.2 Health Concerns

Mercury is highly toxic. Organic alkyl mercury compounds, such as ethylmercuric chloride (C_2H_5HgCl) commonly used as fungicides, produce illness or death from the ingestion of only a few milligrams. Because inorganic forms of mercury can be converted to very toxic methyl and dimethyl mercury by anaerobic microorganisms, any form of mercury must be considered potentially hazardous to the environment. Most human mercury exposure is due to consumption of fish. The EPA has found that short-term and long-term exposure to mercury at levels in drinking water above the MCL may cause kidney damage. There is inadequate evidence to state whether mercury has the potential to cause cancer from lifetime exposures in drinking water.

Mercury

EPA Primary Drinking Water
Standard (mg/L; Values Are for
Inorganic Hg)   **Health Advisories (mg/L)**

MCL	0.002	1-day (10-kg child)	0.002
MCLG	0.002	2-day (10-kg child)	0.002
		RfD (lifetime + sensitive groups)	0.0003 mg/kg/day
		DWEL	0.01
		Lifetime (adult)	0.002
		10^{-4} cancer risk	—

A.2.30.3   Other Comments

Treatment/best available technologies: Granular activated carbon for influent mercury concentrations above 10 µg/L; coagulation and filtration, lime softening, and reverse osmosis for influent mercury concentrations less than 10 µg/L.

A.2.31   MOLYBDENUM (MO), CAS NO. 7439-98-7

Molybdenum is widely distributed in trace amounts in nature, occurring chiefly as insoluble molybdenite (MoS_2) and soluble molybdates (MoO_4^{2-}). Molybdenum is relatively mobile in the environment because soluble compounds predominate at pH > 5. The solubility of molybdenum increases as redox potential is lowered. Below pH 5, adsorption and coprecipitation of the molybdate anion by hydrous oxides of iron and aluminum are effective at removing dissolved molybdenum.

The weathering of igneous and sedimentary rocks (especially shale) is the main natural source of molybdenum in the aquatic environment.

Molybdenum metal is used in the manufacture of special steel alloys and electronic apparatus. Molybdenum salts are used in the manufacture of glass, ceramics, pigments, and fertilizers. The use of fertilizers containing molybdenum is the single most important anthropogenic input to the aquatic environment. Other contributions to the aquatic environment come from mining and milling of molybdenum, the use of molybdenum products, the mining and milling of some uranium and copper ores, and the burning of fossil fuels. Fresh water usually contains less than 1 mg/L molybdenum. Concentrations ranging between 0.03 and 10 µg/L are typical of unpolluted waters. Levels as high as 1500 µg/L have been observed in rivers in industrial areas. The average concentration of molybdenum in finished drinking water is about 1–4 µg/L.

A.2.31.1   Health Concerns

Molybdenum is an essential trace nutrient for all plants and animals. It is considered nontoxic to humans, but excessive levels (0.14 mg/kg body weight; 10 mg/day for a 70-kg adult) may cause high uric acid levels and an increased chance of gout. The recommended daily intake is 70–250 µg/day for adults. Local concentrations may vary by a factor of 10 or more depending on regional geology, causing both deficient and excessive intake of molybdenum by plants and ruminants. Average adults contain about 5 mg of molybdenum in their body, and ingest about 100–300 µg/day. Twenty enzymes in plants and animals are known to be built around molybdenum, including xanthine oxidase, which helps to produce uric acid, essential for eliminating excess nitrogen from the body.

Molybdenum

EPA Primary Drinking Water Standard (mg/L)

Health Advisories (mg/L)

EPA Primary Drinking Water Standard (mg/L)		Health Advisories (mg/L)	
MCL	—	1-day (10-kg child)	0.08
MCLG	—	2-day (10-kg child)	0.08
There are no primary or		RfD (lifetime + sensitive groups)	0.005 mg/kg/day
secondary drinking		DWEL	0.2
water standards for		Lifetime (adult)	0.04
molybdenum		10^{-4} cancer risk	—

EPA Primary Drinking Water Standard.

A.2.32 NICKEL (NI), CAS No. 7440-02-0

Nickel is found in many ores such as sulfides, arsenides, antimonides, silicates, and oxides. Its average crustal concentration is about 75 mg/kg. Because nickel is an important industrial metal, industrial waste streams can be a major source of environmental nickel. Inadvertent formation of volatile and poisonous nickel carbonyl (C_4NiO_4) can occur in various industrial processes that use nickel catalysts, such as coal gasification, petroleum refining, and hydrogenation of fats and oils. Nickel oxide is present in residual fuel oil and in atmospheric emissions from nickel refineries. The atmosphere is a major conduit for nickel as particulate matter. Contributions to atmospheric loading come from both natural sources and anthropogenic activity, with input from both stationary and mobile sources. Nickel particulates eventually precipitate from the atmosphere to soils and waters. Soil-borne nickel enters waters with surface runoff or by percolation of dissolved nickel into groundwater.

Nickel is one of the most mobile heavy metals in the aquatic environment. Its concentration in unpolluted water is controlled largely by coprecipitation and sorption with hydrous oxides of iron and manganese. In polluted environments, nickel forms soluble complexes with organic material. In reducing environments where sulfides are present, insoluble nickel sulfide is formed. Average concentrations in U.S. surface waters are typically between 10 and 100 µg/L, with concentrations as high as 11,000 µg/L in streams receiving mine discharges. In surface waters, sediments generally contain more nickel than the overlying water.

A.2.32.1 Health Concerns

Nickel appears to be an essential trace nutrient in animals and humans, but its exact role is not yet understood. Daily intake of nickel, mainly from food, is around 150 µg/day; the adult requirement appears to be between 5 and 50 µg/day. Although a few nickel compounds, such as nickel carbonyl, are poisonous and carcinogenic, most nickel compounds are nontoxic. The toxicity of dissolved nickel ingested orally is low, comparable to zinc, chromium, and manganese, perhaps because only 2–3% of ingested nickel is absorbed.

The EPA has not found nickel to cause adverse human health effects from short-term exposures at levels above the MCL. Long-term exposure above the MCL can cause decreased body weight, heart and liver damage, and dermatitis. There is no evidence that nickel has the potential to cause cancer from lifetime exposures in drinking water.

Nickel

EPA Primary Drinking Water Standard (mg/L)		Health Advisories (mg/L)	
MCL	—	1-day (10-kg child)	1.0
MCLG	—	2-day (10-kg child)	1.0
The EPA remanded the drinking water		RfD (lifetime + sensitive	0.02 mg/kg/day
standard for nickel in 1995. Prior to		groups)	0.7
1995, the MCLG and MCL both		DWEL	0.1
were 0.1 mg/L. Currently, there are		Lifetime (adult)	—
no drinking water standards for		10^{-4} cancer risk	
nickel.			

A.2.32.2 Other Comments

Treatment/best available technologies: Ion exchange, lime softening, and reverse osmosis.

A.2.33 Nitrate (NO_3^-), CAS No. 14797-55-8 and Nitrite (NO_2^-), CAS No. 14797-65-0

See Chapter 4, Section 4.2.5 for a more detailed discussion.

Nitrate and nitrite anions are highly soluble in water. Owing to their high solubility and weak retention by soil, nitrate and nitrite are very mobile, moving through soil at approximately the same rate as water. Thus, nitrate has a high potential to migrate to groundwater. Because they are not volatile, nitrate and nitrite are likely to remain in water until consumed by plants or other organisms.

Nitrate is the oxidized form and nitrite is the reduced form. Aerated surface waters will contain mainly nitrate and groundwaters, with lower levels of dissolved oxygen, will contain mostly nitrite. They readily convert between the oxidized and reduced forms depending on the redox potential. Nitrite in groundwater is converted to nitrate when brought to the surface or exposed to air in wells. Nitrate in surface water is converted to nitrite when it percolates through soil to oxygen-depleted groundwater.

The main inorganic sources of contamination of drinking water by nitrate are potassium nitrate and ammonium nitrate. Both salts are used mainly as fertilizers. Ammonium nitrate is also used in explosives and blasting agents. Because nitrogenous materials in natural waters tend to be converted to nitrate, all environmental nitrogen compounds, particularly organic nitrogen and ammonia, should be considered potential nitrate sources. Primary sources of organic nitrates include human sewage and livestock manure, especially from feedlots.

A.2.33.1   Health Concerns

Nitrate is a normal dietary component. A typical adult ingests around 75 mg/day, mostly from the natural nitrate content of vegetables, particularly beets, celery, lettuce, and spinach. Short-term exposure to levels of nitrate in drinking water higher than the MCL can cause serious illness or death, particularly in infants. Nitrate is converted to nitrite in the body, and nitrite oxidizes Fe^{2+} in blood hemoglobin to Fe^{3+}, rendering the blood unable to transport oxygen, a condition called methemoglobinemia. Infants are much more sensitive than adults to this problem because of their small total blood supply. Symptoms include shortness of breath and blueness of the skin. This can be an acute condition in which health deteriorates rapidly over a period of days.

Long-term exposure to levels of nitrate or nitrite in excess of the MCL may cause diuresis, increased starchy deposits, and hemorrhaging of the spleen. There is inadequate evidence to state whether or not nitrates or nitrites have the potential to cause cancer from lifetime exposures in drinking water.

Nitrate and Nitrite

EPA Primary Drinking Water Standard (mg/L)

		Health Advisories (mg/L)	
MCL (nitrate as N)	10	1-day (10-kg child)	10 (nitrate)
MCLG (nitrate as N)	10		1.0 (nitrite)
MCL (nitrite as N)	1.0	2-day (10-kg child)	10 (nitrate)
MCLG (nitrite as N)	1.0		1.0 (nitrite)
MCL (nitrite + nitrite as N)	10	RfD (lifetime + sensitive	1.6 mg/kg/day (nitrate)
MCLG (nitrite + nitrite as N)	10	groups)	0.16 mg/kg/day (nitrite)
		DWEL	—
		Lifetime (adult)	—
		10^{-4} cancer risk	—

A.2.34   PERCHLORATE (CIO_4^-), CAS No. 014797-73-0

Perchlorate salts:

Ammonium perchlorate, CAS No.: 7790-98-9
Lithium perchlorate, CAS No.: 7791-03-9
Potassium perchlorate, CAS No.: 7778-74-7
Sodium perchlorate, CAS No.: 7601-89-0

Perchlorate is a soluble oxychloro anion most commonly used as a solid salt in the form of ammonium perchlorate, potassium perchlorate, lithium perchlorate, or sodium perchlorate, all of which are highly soluble. Ammonium perchlorate is the most widely used perchlorate compound. In their pure forms, these salts are white or colorless crystals or powders. Perchlorate salts dissolve in water and readily move from surface to groundwater. Perchlorate is known to originate from both natural and man-made sources.

The most common uses for ammonium perchlorate are in explosives, military munitions, and rocket propellants. In addition, perchlorate salts are used in a wide range of nonmilitary applications, including pyrotechnics and fireworks, blasting agents, solid rocket fuel, matches, lubricating oils, nuclear reactors, air bags, and certain types of fertilizers. Improper storage and disposal related to these uses is the most typical route for perchlorate to enter into the environment. Monitoring data show that more than 4% of public water systems, serving between 5 million and 17 million people, have detected perchlorate in their finished water.

A.2.34.1 Environmental Behavior

Perchlorate is extremely soluble in water and, because it adheres poorly to mineral surfaces and organic material, it is very mobile in aquifer systems. It is relatively inert in typical groundwater and surface water conditions and can persist in the environment for many decades.

A.2.34.2 Ecological Concerns

Available information about the ecological effects of the perchlorate anion is very limited, but is receiving increased attention. The current judgment is that perchlorate may have deleterious effects on other species throughout the environment. Field studies have demonstrated an overall lack of bioconcentration. Detectable concentrations of perchlorate are found only in a limited number of terrestrial mammals, birds, fish, amphibians, and insects exposed to elevated levels of perchlorate in the environment.

A.2.34.3 Health Concerns

Perchlorate interferes with thyroid functions because it is similar in size to the iodide anion and can mistakenly be taken up in place of iodide by the thyroid gland. Even at low concentrations, perchlorate impairs normal thyroid function and may contribute to thyroid cancer, although perchlorate has not positively been linked to cancer in

---

Perchlorate

EPA Primary Drinking Water Standard (mg/L)	Health Advisories	
Proposed rule expected in 2013.	1-day (10-kg child)	—
	2-day (10-kg child)	—
	RfD (lifetime + sensitive groups)	0.0007 mg/kg/day
	DWEL	0.0245
	Lifetime (adult)	—
	10^{-4} cancer risk	—

humans. Recent studies also show possible adverse effects on normal growth and development of both fetuses and children. Perchlorate has been detected in some food products, but the primary route of exposure is through the consumption of water containing perchlorate.

A.2.34.4 EPA Primary Drinking Water Standard

No national drinking water standards are currently (2012) established for perchlorate, although a few states, including California, Arizona and Texas, have set state standards. EPA is in the process of developing a national drinking water standard for perchlorate and expects to publish a proposed rule for public review and comment in 2013.

A.2.35 POLYCHLORINATED BIPHENYLS

There are 209 different polychlorinated biphenyls (PCBs), each with its own CAS identification number. (See Chapter 8, Section 8.4 for more information about PCBs.)

PCBs are a family of stable man-made organic compounds produced commercially by direct chlorination of biphenyl. PCBs have very high chemical, thermal, and biological stability; low water solubility; low vapor pressure; high dielectric constant; and high flame resistance. PCBs were manufactured and sold under various trade names (Aroclor, Pyranol, Phenoclor, Pyralene, Clophen, and Kaneclor) as complex mixtures differ in their average chlorination level. PCB mixtures range from oily liquids to waxy solids. Owing to their nonflammability, chemical stability, high boiling point, and electrical insulating properties, PCBs were used in hundreds of industrial and commercial applications including electrical, heat transfer, and hydraulic equipment; as plasticizers in paints, plastics, and rubber products; in pigments, dyes, and carbonless copy paper; and in many other applications. More than 1.5 billion pounds of PCBs were manufactured in the United States until production was stopped in 1977.

A.2.35.1 Environmental Behavior

PCBs are very stable and do not degrade readily in the environment. They even survive ordinary incineration and can escape as vapors up the smokestack. The wide use of PCBs has resulted in their common presence in soil, water, and air. PCB dispersion from source regions to global distribution occurs mainly through atmospheric transport and subsequent deposition. They sorb strongly to soil and sediments. In aquatic systems, sediments are an important reservoir.

A.2.35.2 Ecological Concerns

Environmental contamination was first reported in 1966, when high levels of PCBs were found in fish due to bioaccumulation. Fish consumption remains the major route of exposure to PCBs and there are health consequences associated with these exposures.

PCBs do not easily biodegrade. Although they are no longer manufactured, they still leak from old electrical devices, including power transformers, capacitors, television sets, and fluorescent lights, and can be released from hazardous waste sites and historic and illegal refuse dumps. They also persist in fatty foods, such as certain fish, meat, and dairy products.

A.2.35.3 Health Concerns

The toxicity of PCBs is a complicated issue because each congener, of which there are 209, differs in its toxicity. Furthermore, commercial PCBs are blends of the different pure congeners. In general, PCBs cause a wide variety of health effects, often at very low exposure levels. PCBs alter major systems in the body (immune, hormone, nervous, and enzyme systems), and therefore they can affect most of the body organs and functions. All PCBs are listed by the EPA as known carcinogens and priority pollutants. When they are incinerated, they can produce dioxins, which are rated by the EPA among the most toxic substances.

Polychlorinated Biphenyls

EPA Primary Drinking Water Standard (mg/L)		Health Advisories (mg/L)	
MCL	0.0005	1-day (10-kg child)	—
MCLG	Zero	2-day (10-kg child)	—
		RfD (lifetime + sensitive groups)	—
		DWEL	—
		Lifetime (adult)	—
		10^{-4} cancer risk	0.01

A.2.36 SELENIUM (SE), CAS No. 7782-49-2

Selenium is widely distributed in the earth's crust at concentrations averaging 0.09 mg/kg. It occurs in igneous rocks, with sulfides in volcanic sulfur deposits, in hydrothermal deposits, and in porphyry copper deposits. The major source of selenium in the environment is the weathering of rocks and soils. In addition, volcanic activity contributes to its natural occurrence in waters in trace amounts. Volcanic activity is an important source of selenium in regions with high soil concentrations.

Most selenium for industrial and commercial purposes is produced from electrolytic copper-refining shines and from flue dusts from copper and lead smelters. Anthropogenic sources of selenium in water bodies include effluents from copper and lead refineries, municipal sewage, and fallout of emissions from fossil fuel combustion. Selenium in surface waters can range between 0.1 and 2700 µg/L, with most values between 0.2 and 20 µg/L.

Dissolved selenium exists mostly as the selenite (SeO_3^{2-}) and selenate (SeO_4^{2-}) anions. Selenate is more mobile under oxidizing conditions than under reducing conditions and can be reduced by bacteria in anaerobic environments to form methylated selenium compounds, which are volatile. Ferric selenite, however, is insoluble and offers a treatment for removing dissolved selenium. Alkaline and oxidizing conditions favor the formation of soluble selenates, which also are the biologically available forms for plants and animals. Acidic and reducing conditions readily reduce selenates and selenites to insoluble elemental selenium, which precipitates from the water column.

A.2.36.1 Health Concerns

Selenium is a nutritionally essential trace element for all vertebrates and most plants. Human blood contains about 0.2 mg/L, 1000 times higher than typical surface waters, demonstrating that selenium is bioaccumulated. The adult daily requirement is between 20 and 200 µg/day. Drinking water seldom contains more than a few micrograms per liter of selenium, not enough for it to be a significant dietary source. Depending on the food eaten, daily intake is between 6 and 200, with levels near 150 µg/day being typical. Cereals, nuts, and seafood are good sources of selenium. Ingestion of quantities above the recommended maximum daily intake of 450 µg/day increases the risk of selenium poisoning, the most obvious symptom of which is bad breath and body odor caused by volatile methyl selenium produced by the body to eliminate excess selenium.

The EPA has found that short-term exposure to selenium at levels above the MCL may cause hair and fingernail changes, damage to the peripheral nervous system, fatigue, and irritability. Long-term exposure above the MCL may cause hair and fingernail loss, damage to kidney and liver tissue, and damage to the nervous and circulatory systems. There is no evidence that selenium has the potential to cause cancer from lifetime exposures in drinking water.

Selenium

EPA Primary Drinking Water Standard (mg/L)		Health Advisories (mg/L)	
MCL	0.05	1-day (10-kg child)	—
MCLG	0.05	2-day (10-kg child)	—
		RfD (lifetime + sensitive groups)	0.005 mg/kg/day
		DWEL	0.2
		Lifetime (adult)	0.05
		10^{-4} cancer risk	—

A.2.36.2 Other Comments

Treatment/best available technologies: Activated alumina, coagulation and filtration, lime softening, reverse osmosis, electrodialysis.

A.2.37 SILVER (AG), CAS NO. 7440-22-4

Silver is a white, lustrous, ductile metal that occurs naturally in its pure, elemental form and in ores, mostly as argentite (Ag_2S). Other silver ores include cerargyrite ($AgCl$), proustite ($3AgS \cdot As_2S_3$), and pyrargyrite ($(Ag_2S)_3 \cdot Sb_2S_3$). Silver is also found associated with lead, gold, copper, and zinc ores. Silver is among the less common but most widely distributed elements in the earth's crust. Its concentration in normal soil averages around 0.3 mg/kg.

Before the recent widespread adoption of digital photography, a large portion of silver consumption was for photographic materials. Also, because silver has the highest known electrical and thermal conductivities of all metals, it finds extensive use in electrical and electronic products such as batteries, switch contacts, and conductors. Other major uses include sterling and plated metalwork, jewelry, coins and medallions, brazing alloys and solders, catalysts, mirrors, fungicides, and dental and medical supplies.

Natural processes, such as weathering and volcanic activity, release silver to the environment. Silver has been found associated with sulfides, sulfates, chlorides, and ammonia salts in deposits and discharges of hot springs and volcanic materials. Anthropogenic sources of silver include discharges from landfills and waste lagoons, fallout from incineration and industrial emissions, and direct waste discharge to water. Some home water treatment devices use silver as an antibacterial agent and may represent a contamination source. Surface waters in nonindustrial regions average around 0.2–0.3 µg/L of silver, while streambed sediments range between 140 and 600 µg/kg of silver. In industrial areas, silver concentrations in surface waters may reach 40 µg/L and stream sediment concentrations 1500 µg/kg. Finished drinking water seldom contains more than 1 µ/L of silver.

Metallic silver is stable over much of the pH and redox range found in natural waters, but has very low water solubility. Insoluble silver compounds, such as $AgCl$, Ag_2S, Ag_2Se, and Ag_3AsS3, may be present in aquatic systems in colloidal form, adsorbed to various humic substances, or incorporated with sediments. At pH < 7.5 under aerobic conditions, the Ag^+ cation is soluble and mobile. Around pH 7.5–8.0, aquatic Ag^+ reacts with water to form the insoluble oxide. Silver is dispersed through the aquatic environment as dissolved and colloidal species, but it eventually resides in the bottom sediments. Sorption, particularly by manganese dioxide, and precipitation of silver halides, particularly silver chloride, are the main processes that remove dissolved silver from the water column. These processes, along with the low crustal abundance of silver, account for its low observed concentrations in the aqueous phase.

A.2.37.1   Health Concerns

There is no evidence that silver is an essential nutrient. Metallic silver is not considered toxic to humans, but most of its salts exhibit toxic properties. Silver in all forms is acutely toxic to aquatic life. Large oral doses of silver nitrate can cause severe gastrointestinal irritation and ingestion of 10 g is likely to be fatal. Chronic human exposure to silver in drinking water seems to cause only argyria, a discoloration of the skin resulting from the deposition of metallic silver in tissues. The EPA considers this condition to be mainly cosmetic. The minimum adult cumulative dose of silver for inducing argyria is about 1000 mg, an amount likely to be encountered only in industrial environments. The accumulation of 1000 mg of silver over a lifetime (70 years) would require the retention of 40 µg/day. Since around 90% of the silver intake is excreted, the required daily intake for inducing argyria would be more like 400 µg/day. An average daily diet may contain 20–80 µg of silver and drinking 2 L of water may contribute an additional 2 µg. Thus, food and drinking water are not likely to deliver toxic quantities of silver.

Silver

EPA Primary Drinking Water Standard (mg/L)		Health Advisories (mg/L)	
MCL	—	1-day (10-kg child)	0.2
MCLG	—	2-day (10-kg child)	0.2
SDWS	0.10	RfD (lifetime + sensitive groups)	0.005 mg/kg/day
		DWEL	0.2
		Lifetime (adult)	0.1
		10^{-4} cancer risk	—

A.2.37.2 Other Comments

Treatment/best available technologies: Lime softening at pH 11, sand filtration followed by activated carbon, ion exchange, nano- and ultrafiltration.

A.2.38 Sulfate (SO_4^{2-}), CAS No. 14808-79-8

The sulfate anion (SO_4^{2-}) is the stable, oxidized form of sulfur. Sulfate minerals are widely distributed in nature, and most sulfate compounds are readily soluble in water. All sulfate salts are very soluble except for calcium and silver sulfates, which are moderately soluble, and barium, mercury, lead, and strontium sulfates, which are insoluble.

It is estimated that about one-half of the river sulfate load arises from mineral weathering and volcanism, and the other half from biochemical and anthropogenic sources. Industrial discharges are another significant source of sulfates. Mine and tailings drainage, smelter emissions, agricultural runoff from fertilized lands, pulp and paper mills, textile mills, tanneries, sulfuric acid production, and metalworking industries are all sources of sulfate-polluted water. Aluminum sulfate (alum) is used as a sedimentation agent for treating drinking water. Copper sulfate is used for controlling algae in raw and public water supplies.

Air emissions from industrial fuel combustion and the roasting of sulfur-containing ores carry large amounts of sulfur dioxide and sulfur trioxide into the atmosphere, adding sulfates to surface waters through precipitation. Sulfate concentrations normally vary between 10 and 80 mg/L in most surface waters, although they may reach several thousand milligrams per liter near industrial discharges. High sulfate concentrations are also present in areas of acid mine drainage and in well waters and surface waters in arid regions where sulfate minerals are present.

Sulfate minerals can cause scale buildup in water pipes similar to carbonates. Elevated sulfate levels in combination with chlorine bleach can make cleaning clothes difficult. In aerobic water, sulfur-oxidizing bacteria produce effects similar to those of iron bacteria. They convert sulfide into sulfate, producing a dark slime that can clog plumbing and/or stain clothing. Blackening of water or dark slime coating the inside of toilet tanks may indicate a sulfur-oxidizing bacteria problem. Sulfur-oxidizing bacteria are less common than sulfur-reducing bacteria. In anaerobic water, sulfate-reducing bacteria convert sulfate into sulfide, which leads to the generation of malodorous hydrogen sulfide gas and precipitation of dissolved metals (Chapter 4, Section 4.5 and Chapter 5, Section 5.4.3).

A.2.38.1 Environmental Behavior

Nearly all natural surface waters and shallow groundwaters contain sulfate anions. Sulfate is commonly found as a prominent component of unpolluted waters and is included among the six major surface and shallow groundwater ions (Na^+, Ca^+, Mg^+ Cl^-, HCO_3^{2-}, and SO_4^{2-}), second to bicarbonate as the most abundant anion in most freshwaters. Sulfur is an essential plant and animal nutrient, and sulfate is the most common inorganic form of sulfur in aerobic environments. Sulfate water concentrations that are too low have a detrimental effect on both land and aquatic plant growth.

Sulfate is redox sensitive and is bacterially reduced to sulfide ion under anaerobic conditions. Sulfide may be released to the atmosphere as H_2S gas or precipitated as insoluble metal sulfides. Oxidation of sulfides returns sulfur to the sulfate form.

Sulfates may be leached from most sedimentary rocks, including shales, with the most appreciable contributions from such sulfate deposits as gypsum ($CaSO_4 \cdot 2H_2O$) and anhydrite ($CaSO_4$). The oxidation of sulfur-bearing organic materials can contribute sulfates to waters.

A.2.38.2 Ecological Concerns

Sulfates serve as an oxygen source for bacteria. Under anaerobic conditions, sulfate-reducing bacteria reduce dissolved sulfate to sulfide, which then may be volatilized to the atmosphere as H_2S, precipitated as insoluble salts, or incorporated in living organisms. These processes are common in the anaerobic regions of wetlands and lakes fed by surface and groundwaters with high sulfate levels. Oxidation of sulfides returns the sulfur to the sulfate form.

Sulfate is a major contributor to salinity in many irrigation waters. However, toxicity usually is not an issue, except at very high concentrations, where high sulfate can interfere with uptake of other nutrients. As with boron, sulfate in irrigation water is a plant nutrient, and irrigation water often carries enough sulfate for maximum production of most crops. Seawater contains about 2700 mg/L of sulfate.

A.2.38.3 Health Concerns

The sulfate anion is generally considered nontoxic to animal, aquatic, and plant life. It is an important source of sulfur, an essential nutrient for plants and animals. Sulfates are used as additives in the food industry, and the average daily intake of sulfate from drinking water, air, and food is approximately 500 mg. As examples, some measured sulfate concentrations in beverages are 100–500 mg/L in drinking water, 500 mg/L in coconut milk, 260 mg/L in beer (bitter), 250 mg/L in tomato juice, and 300 mg/L in red wine (FNB 2004). Available data suggest that people acclimate rapidly to the presence of sulfates in their drinking water.

No upper limit likely to cause detrimental human health effects has been determined for sulfate in drinking water. However, concentrations of 500–750 mg/L may cause a temporary mild laxative effect, although doses of several thousand milligrams per liter generally do not cause any long-term ill effects. Because of the laxative effects resulting from ingestion of drinking water containing high sulfate levels, the EPA recommends that health authorities be notified of sources of drinking water that contain sulfate concentrations in excess of 500 mg/L.

The presence of sulfate can adversely affect the taste of drinking water, imparting a bitter taste. The lowest taste threshold concentration for sulfate is approximately 250 mg/L as sodium salt, but higher as calcium or magnesium salts (up to 1000 mg/L).

A.2.38.4   EPA Primary Drinking Water Standard

The EPA has no primary drinking water standard for sulfate. The EPA secondary drinking water standard (nonenforceable) is 250 mg/L, based on aesthetic effects.

A.2.38.5   Other Comments

Treatment/best available technologies: Anion exchange, reverse osmosis, nanofiltration.

A.2.39   SULFIDE (S^{2-}, SEE HYDROGEN SULFIDE)

A.2.40   THALLIUM (Tl), CAS No. 7440-28-0

Thallium is a soft, lead-like metal that is widely distributed in trace amounts. It may be found naturally as pure metal or associated with potassium and rubidium in copper, gold, zinc, and cadmium ores. Thallium compounds are nonvolatile, although many are water-soluble. Thallium is generally present in trace amounts in fresh water. Unpolluted soil levels range between 0.1 and 0.8 mg/kg, with an average around 0.2 mg/kg.

Thallium compounds are used mainly in electronics industry, and to a limited extent in the manufacture of pharmaceuticals, alloys, and high refractive-index glass. Man-made sources of thallium pollution are gaseous emissions from cement factories, coal-burning power plants, and metal sewers. Small amounts of thallium in fallout from these sources frequently contaminates nearby farms and gardens, where it is readily absorbed by the plant roots to contaminate food crops. The leaching of thallium from ore-processing operations is the major source of elevated thallium concentrations in water. Thallium is a trace metal associated with copper, gold, zinc, and cadmium.

Most thallium is released into the environment by weathering of minerals. Human sources of thallium are wastes from the production of other metals, for example, from the roasting of pyrite during the production of sulfuric acid, and in mining and smelting operations of copper, gold, zinc, lead, and cadmium. Waste streams of these industries may contain as much as 90 µg/L of thallium.

In the aquatic environment, thallium is transported as soluble complexes with humic materials (above pH 7), sorption to clay minerals, and bioaccumulation. In reducing environments, thallium may be precipitated as elemental metal or, in the presence of sulfur, as the insoluble sulfide. In waters of high oxygen content, Tl^+ is the dominant oxidation state, forming soluble chloride, carbonate, and hydroxy salts. Thallium sorption to sediments is pH dependent. Thallium is strongly sorbed by montmorillonite clay at pH 8, but only slightly at pH 4. In a study of heavy metal cycling in a lake in southwestern Michigan, thallium was detected only in the sediments. Since thallium is soluble in most aquatic systems, it is readily available to aquatic organisms and is quickly bioaccumulated by fish and plants.

A.2.40.1 Health Concerns

Thallium is a toxic metal with no known nutritional value. On the contrary, it is notorious for its use by murderers as a poison; the lethal dose for an adult is around 800 mg. Environmental exposure is mainly through contaminated foods, which are estimated to contain, on average, about 2 ppb of thallium. The adult total body burden of thallium is 0.1–0.5 mg, the greatest part of which is carried by muscle tissue.

Short-term exposures to thallium at levels above the MCL can cause gastrointestinal irritation, numbness of toes and fingers, the sensation of burning feet, and muscle cramps. Long-term exposure to thallium at levels above the MCL can cause damage to liver, kidney, intestinal and testicular tissues, as well as changes in blood chemistry and hair loss. There is no evidence that thallium has the potential to cause cancer from lifetime exposures in drinking water.

Thallium

EPA Primary Drinking Water Standard (mg/L)		Health Advisories (mg/L)	
MCL	0.002	1-day (10-kg child)	0.007
MCLG	0.0005	2-day (10-kg child)	0.007
		RfD (lifetime + sensitive groups)	—
		DWEL	—
		Lifetime (adult)	—
		10^{-4} cancer risk	—

A.2.40.2 Other Comments

Treatment/best available technologies: Activated alumina, ion exchange, reverse osmosis, nanofiltration.

A.2.41 Vanadium (V), CAS No. 7440-62-2

Vanadium is widely dispersed in the earth's crust at an average concentration of about 150 ppm. Deposits of ore-grade mineable vanadium are rare. There are more than 65 known vanadium-bearing minerals. In the United States, vanadium occurs in uranium-bearing ores and in phosphate shales and rocks in parts of the western states. Vanadium is also present in coal and crude oil. The bulk of commercial vanadium is obtained as a by-product or coproduct from the processing of iron, titanium, and uranium ores, and to a lesser extent, from phosphate, bauxite, and chromium ores and the ash or coke from burning or refining petroleum. Vanadium is mainly used as an alloy additive.

Vanadium enters the aquatic environment mainly by surface erosion and natural seepage from carbon-rich deposits such as tar and oil sands, atmospheric deposition, weathering of vanadium-rich ores and clays, and leaching of coal mine wastes. Vanadium concentrations in fresh water typically range between <0.3 µg/L and around 200 µg/L. Groundwater concentrations of vanadium are typically <1 µg/L.

A.2.41.1 Health Concerns

There are no particular health or safety hazards associated with vanadium and its compounds. Dust and fine powders present a moderate fire hazard. Vanadium compounds, of which the most common is vanadium pentoxide, may irritate the conjunctivae and respiratory tract. Toxic effects have been observed from airborne concentrations of vanadium compounds of several milligrams or more per cubic meter of air. OSHA threshold limits in the workplace are 0.5 mg/m^3 for dust and 0.05 mg/m^3 for fumes. Oral toxicity in humans is minimal.

As an environmental pollutant, vanadium is of concern mainly because of its high levels in residual fuel oils and its subsequent contribution to atmospheric particulate levels from the combustion of these fuels in urban areas.

A.2.41.2 EPA Primary Drinking Water Standard

The EPA has no drinking water standards for vanadium.

A.2.42 Zinc (Zn), CAS No. 7440-66-6

Zinc is a common contaminant in surface and groundwaters, stormwater runoff, and industrial waste streams. Its average concentration in the earth's crust is around 70 mg/kg. Because it is very chemically reactive, pure zinc is not found free in nature, but is always associated with one of the 55 known zinc minerals (mainly sulfides, oxides, carbonates, and silicates). Ninety percent of industrial zinc production is from the minerals sphalerite (also called zincblende) and wurtzite, which are different crystalline forms of zinc sulfide (ZnS). Zinc minerals tend to be associated with minerals of other metals, particularly lead, copper, cadmium, mercury, and silver. The most industrially important chemical property of zinc is its high oxidation–reduction potential, which places it above iron in the electromotive series. Because of this, zinc coatings are used to protect iron and steel from corrosion. When iron or steel coated with zinc is exposed to corrosive conditions, zinc displaces iron from solution and prevents dissolution of the iron. In the process of dissolving, zinc reduces dissolved iron, and many other dissolved metals, to their elemental metallic state. Its use as a corrosion-resistant coating for iron and steel is the most important industrial use for zinc.

Zinc occurs in natural waters in both suspended and dissolved forms. It is one of the most mobile heavy metals in surface waters and groundwater because it forms soluble compounds at neutral and acidic pH values. At higher pH values, zinc can form carbonate and hydroxide complexes, which control zinc solubility. The dissolved form is the divalent cation, Zn^{2+}. Dissolved zinc is readily sorbed to or occluded in mineral clays and humic colloids. In water of low alkalinity and below pH 7, Zn^{2+} is the dominant form. Sorption to sediments or suspended solids, including hydrous iron and manganese oxides, clay minerals, and organic matter, is the primary fate of zinc in aquatic environments. As pH and alkalinity increase, the potential for sorption and humic complex formation also increase. Above pH 7, sorption to metal oxides, clays, and apatite (calcium phosphate minerals, $Ca_5(F,Cl,OH,0.5CO_3)(PO_4)_3$) can bind over 90% of dissolved zinc. Sorption is generally not significant below pH

6. At low pH or under anaerobic conditions, metal oxides can redissolve and release bound zinc cation into solution. However, if the redox potential rises and aerobic conditions are established, zinc sulfide is oxidized to soluble zinc sulfate ($ZnSO_4$), also releasing Zn^{2+} into solution.

Zinc concentrations in unpolluted surface waters typically range between about 5 and 50 µg/L. Streams draining mined areas often exceed 100 µg/L. Industries with waste streams containing significant levels of zinc include steel works with galvanizing operations, zinc and brass metal works, zinc and brass plating, and production of viscose rayon yarn, ground wood pulp, and newsprint paper. Reported concentrations of zinc in industrial waste streams reach 48,000 mg/L. More representative values are between 10 and 200 mg/L.

A.2.42.1 Health Concerns

Zinc is an essential nutrient and is not toxic to humans. About 1 g/day may be ingested

---

Zinc

EPA Primary Drinking Water Standard (mg/L)		Health Advisories (mg/L)	
MCL	—	1-day (10-kg child)	6.0
MCLG	—	2-day (10-kg child)	6.0
SDWS	5.0	RfD (lifetime + sensitive groups)	0.3 mg/kg/day
		DWEL	10
		Lifetime (adult)	2.0
		10^{-4} cancer risk	—

---

without ill effects. Recommended dietary allowance is 15 mg/day for adults.

A.2.42.2 Other Comments

Treatment/best available technologies: Chemical precipitation, ion exchange, activated carbon filtration, reverse osmosis, electrolytic plating.

REFERENCES

FNB, 2004, *Dietary Reference Intakes for Water, Potassium, Sodium, Chloride, and Sulfate*, Panel on dietary reference intakes for electrolytes and water, food and nutrition board, National Academies Press, Washington, DC, 640 p.

National Academy of Sciences, 1982, *Drinking Water and Health*, vol. 4, National Academies Press, Washington, DC, 299 p.

USEPA, 1988, United States Environmental Protection Agency, *Ambient Water Quality Criteria for Chloride—1988*. Office of Research and Development, Environmental Research Laboratory, Duluth, MN. EPA 440/5-88-001.

USEPA, 2011, *2011 Edition of the Drinking Water Standards and Health Advisories*, EPA 820-R-11-002, Office of Water, U.S. Environmental Protection Agency, Washington, DC. http://water.epa.gov/action/advisories/drinking/upload/dwstandards2011.pdf

When a solid salt is dissolved in water, the attractive forces that had held the solid together continue to act, working to reverse the dissolving process and bring the dissolving molecules back together again into crystalline form. However, new attractive forces between the dissolved ions and water molecules act to keep the solutes dissolved. Because of this "tug-of-war," salt dissolution reactions always are reversible.

Appendix B, Section B.1

Appendix B
Solubility of Slightly Soluble Metal Salts

B.1   THE MECHANISM OF DISSOLVING

Metal salts are metal ionic compounds that dissolve in water by breaking ionic bonds in the solid to form dissolved ionic species. Most crystalline metal compounds are classified as salts and are common in subsurface mineral formations. Many important water quality parameters are dependent on the solubility behavior of metal salts. For example,

- Alkalinity depends, in large part, on the concentrations of dissolved metal carbonates (e.g., $CaCO_3$).
- Hardness is calculated using the cation concentrations of dissolved calcium and magnesium salts.
- Total dissolved solids (TDS), an important parameter in many water quality assessments is mainly a measure of the total dissolved metal salts in a sample.
- Certain dissolved metal ions are toxic to aquatic life and require close control in municipal and industrial wastewater discharges.
- The suitability of water for irrigation depends on, among other factors, TDS and the balance of dissolved sodium, calcium, and magnesium ions (see "sodium absorption ratio" in Chapter 6, Section 6.11.5).

The following discussion assumes that only one pure salt is being dissolved in pure water and that the temperature of the system is constant. When a solid salt is dissolved in water, the attractive forces that held the solid together before it dissolved continue to act, working to reverse the dissolving process and bring the dissolving molecules back together again into crystalline form. However, new attractive forces between the dissolved ions and water molecules act to keep the solutes dissolved. Because of this "tug-of-war," salt dissolution reactions always are reversible (Equation B.1):

$$\text{Solid salt (the } \textit{solute}\text{)} + \text{water} \underset{R_b}{\overset{R_f}{\rightleftharpoons}} \text{dissolved salt ions (the } \textit{solution}\text{)} \quad \text{(B.1)}$$

In Equation B.1, removal of solute ions from the crystalline solid into solution is the forward reaction (R_f), and the recombining of dissolved solute ions,* resulting in the precipitation of solid solute out of solution, is the backward reaction (R_b).

- If $R_f > R_b$, the total mass of solid salt decreases as solid dissolves and the concentration of dissolved ions increases.
- If $R_f < R_b$, the concentration of dissolved ions decreases as they recombine and precipitate as solid salt or enter the crystal structure of already solid salt, and the mass of solid salt increases.
- If $R_f = R_b$, the system is at equilibrium and no changes in mass of solid salt or concentration of dissolved ions are observable, although the forward and backward reactions continue to occur. The magnitude of R_f and R_b at equilibrium is proportional to the salt solubility; low solubility means low reaction rates and higher solubility means correspondingly higher reaction rates.

When solid solute is first added to water, initially no dissolved salt ions are present to collide and recombine, so the rate of the reverse reaction is zero; the overall reaction can only progress in the forward reaction. As more and more solid solute dissolves, the concentration of dissolved ions increases, the frequency of recombination collisions (between dissolved ions and between dissolved ions and solid salt surfaces) increases, and the rate of the backward reaction increases. While the backward reaction is increasing, the rate of the forward reaction is either constant or decreases (if the exposed crystal surface area decreases). Eventually, the rate of the backward reaction increases to be exactly equal to that of the forward reaction, at which point the system is said to have reached equilibrium and no further change in dissolved concentrations can be observed. Depending on the initial value of R_f, there is a limit to how much salt can be dissolved in a given volume of water; for example, if R_f is initially relatively small (indicating strong attractive forces within the solid), R_b increases to match R_f and establish equilibrium rather quickly. Thus, a salt with strong internal attractive forces will have relatively low solubility in water.

---

* There are two ways by which dissolved salt ions can recombine to the solid form:
  1. Dissolved cations collide with dissolved anions and precipitate as the solid salt.
  2. Dissolved cations and anions collide with the surface of the solid salt and become incorporated into the salt crystal structure.

- R_f mainly depends on the relative strengths of attractive forces between ions in the solid salt crystal and between dissolved ions and polar water molecules. It also depends on the solid surface area exposed to water, which may or may not change significantly during the dissolution process.
  - The stronger the ion–ion attractions within the crystal, as compared to the ion–water molecule attractions, the slower the rate of R_f.
  - The greater the solid surface area exposed to water, the greater the rate of R_f; if the exposed surface area does not change significantly, R_f is essentially constant.
- R_b also depends on the relative strengths of attractive forces between ions in the solid salt crystal and between dissolved ions and polar water molecules, but in addition it depends on the collision frequency between dissolved ions of opposite charge or dissolved ions and solid salt surfaces. The collision frequency is dependent on the concentrations of dissolved ions.
  - As solid salt dissolves, R_b continues to increase as the concentrations of dissolved species rise, causing the frequency of recombination collisions to increase, producing precipitated solid.
- R_f and R_b typically depend differently on temperature, which means that the state of equilibrium and the solubility of different salts are also temperature dependent. For most metal salts (about 95%), water solubility increases significantly with increasing temperature. A few metal salts have solubilities that change very little with temperature and very few have solubilities that decrease with increasing temperature.*

RULE OF THUMB

The most important factors determining the relative forward and backward reaction rates are the strengths of ion–ion bond strengths within the solid salt crystal, the strengths of ion–water molecule attractions in solution, concentrations of dissolved salt ions, temperature, and, to a lesser extent, surface area of solid salt exposed to water.

B.2 SOLUBILITY PRODUCT OF SLIGHTLY SOLUBLE METAL SALTS

Because the dissolution of metal salts is a reversible reaction, there is always a limit to how much salt can dissolve in a given volume of water at a given temperature. This limit determines the solubility of the salt and is reached when $R_f = R_b$ and the system reaches equilibrium. The equilibrium condition is expressed by an equilibrium constant called the solubility product, K_{sp}. The mathematical form and value of K_{sp}

---

* Salts that dissolve exothermically (release heat when dissolving) become less soluble with increasing temperature. This is the case when the water–ion attractive forces are stronger than the internal solid salt cation–anion attractive forces. Salts that dissolve endothermically (absorb heat and cool the solution when dissolving) become more soluble with increasing temperature.

depends on the chemical formula of the salt, the molar concentrations of dissolved salt ions at equilibrium, and the temperature.

For a generalized salt, A_aB_b, the dissolution reaction (Equation B.1) is written as

$$A_aB_b(s) \underset{R_b}{\overset{R_f}{\rightleftharpoons}} aA^m(aq) + bB^n(aq) \tag{B.2}$$

Its generalized solubility product is written as

$$K_{sp} = [A^{m+}]^a [B^{n+}]^b \tag{B.3}$$

where square brackets indicate molar concentrations of the indicated ions.

B.2.1 LIMITATIONS OF EQUATION B.3

It is important to understand why the emphasis on "slightly soluble" salts is necessary. The water solubility of any solid depends as much on attractive forces between water and solid (i.e., between ions in the salt crystal at the surface of the dissolving solid and polar water molecules adjacent to the solid surface) as it does on attractive forces within the solid (i.e., those that act to hold the solid together and prevent dissolution). If too many dissolved ions are present in a water solution, their electrical charges influence the net forces that attract additional ions from the solid surface into solution.

When a slightly soluble salt dissolves to the point of equilibrium in pure water, the dissolved salt concentration is very small and the solid surface still "sees" mainly water molecules; then, solid-to-water forces remain essentially constant. Under these dilute salt solution conditions, calculations using the solubility product, K_{sp}, with measured ion concentrations are reasonably accurate.

However, when a moderately or highly soluble salt dissolves to the point of equilibrium in pure water, a larger number of solvated ions and fewer water molecules are adjacent to the solid surface. Under these conditions of a more concentrated salt solution, solid-to-liquid forces are different from solid-to-pure water forces, and calculations using the solubility product with measured ion concentrations have significant error. This difficulty is handled by determining concentration correction factors called activity coefficients, which yield "effective ion concentrations" for use in solubility product calculations. However, it turns out that many metal salts of environmental interest (as opposed to industrial interest) involve slightly soluble salts, for which calculations using the solubility product with measured ion concentrations are satisfactory.

RULES OF THUMB

Calculations using the solubility product, K_{sp}, should be used only with "slightly soluble" salts, and they become less accurate as salt solubility increases.

1. Practical definition of a slightly soluble salt: A salt that dissolves in pure water to make a solution containing less than 0.001 mol/L of dissolved ions.

    a. Ion concentrations are small enough that ionic attractions between solid and dissolved salt ions are negligible (i.e., ions in the solid surface are attracted into solution by essentially the same concentration of water molecules initially as at equilibrium). Then measured ion concentrations can normally be used to calculate K_{sp} with adequate accuracy.

    b. Solutions that meet this condition are said to be essentially "ideal." Solutions approach an "ideal" condition as dissolved ion concentrations approach zero.

2. Practical definition of a moderately or highly soluble salt: A salt that dissolves in pure water to make a solution containing more than 0.001 mol/L of dissolved ions.

    a. Ionic attractions between solid and dissolved salt ions may be significant (i.e., ions in the solid surface are influenced by both dissolved salt ions and water molecules), and "effective" dissolved ion concentrations, also called activities, must be used to calculate K_{sp}.

    b. The activity (a) is calculated by adjusting the measured concentration with an experimentally determined correction factor called the activity coefficient.

    c. Solutions that require the use of activity coefficients for calculating K_{sp} are called nonideal solutions.

3. Estimating the accuracy of K_{sp} calculations:

    a. If dissolved ion concentrations $<10^{-3}$ mol/L, the accuracy of Equation B.3 is usually sufficient for most purposes.

    b. If dissolved ion concentrations $>10^{-1}$ mol/L, the accuracy of Equation B.3 is usually too poor for most purposes.

    c. If dissolved ion concentrations are between 10^{-3} and 10^{-1} mol/L, the accuracy of Equation B.3 should be considered as possibly useful for estimating "ballpark" values.

B.3 FINDING THE SOLUBILITY PRODUCT OF A SALT FROM ITS MEASURED SOLUBILITY

The solubility product, K_{sp}, and the solubility, S, of a salt are different, but related, quantities. In other words, the value of a solubility product does not directly indicate the solubility of the salt.

- The solubility product depends on a measurement of dissolved ion concentrations after a solid salt has dissolved in water to the point of attaining a solid–water equilibrium. The units of the solubility product are moles per liter raised to some exponential power, x, that is, $(mol/L)^x$. The value of the exponent depends on the salt type (e.g., AB, AB_2, A_2B_3, etc., see Table B.1).
- The solubility is a measurement of the maximum amount of solid salt that will dissolve into a water solution. The units of solubility are weight per unit volume, for example, grams of solid salt per 100 mL of water.

To illustrate the difference between solubility and solubility product, consider the case of calcium carbonate, $CaCO_3$ (MW = 100.1 g/mol), for which a measured solubility of 6.2 mg/L at 20°C has been reported.

Determine the molar solubility, S, of $CaCO_3$:

$$S = 6.2 \text{ mg/L} = 6.2 \times 10^{-3} \text{ g/L} = \frac{6.2 \times 10^{-3} \text{ g/L}}{100.1 \text{ g/mol}} = 6.2 \times 10^{-5} \text{ mol/L}$$

The dissolution reaction for $CaCO_3$ is $CaCO_3(s) \rightleftarrows Ca^{2+}(aq) + CO_3^{2-}(aq)$.

The dissolution reaction shows that the concentrations of Ca^{2+} and CO_3^{2-} are each equal to the moles of $CaCO_3$ that have dissolved into pure water, which, at equilibrium, is the measured solubility, $S = 6.2 \times 10^{-5}$ mol/L. From Equations B.2 and B.3, $K_{sp} = [Ca^{2+}][CO_3^{2-}]$, where the dissolved ion concentration units are moles per liter.

$$K_{sp} = [Ca^{2+}][CO_3^{2-}] = [S][S] = S^2 = (6.2 \times 10^{-5} \text{mol/L})^2 = 3.8 \times 10^{-9} (\text{mol/L})^2$$

Not only are the magnitudes of S and K_{sp} different, but also the units.

Table B.1 shows how the relation between K_{sp} and S depends on salt type. Note that two salts of different types (e.g., AB and A_2B_3) will have different solubility products even if they chanced to have the same water solubility.* The examples below illustrate this relation for different types of salts.

1. **Silver chloride, AgCl**
   Dissolution reaction: $AgCl(s) \rightleftarrows Ag^+(aq) + Cl^-(aq)$
   Silver chloride is called an AB salt (see Table B.1) because the coefficients a and b in Equation B.2 are both equal to 1, which makes the exponents a and b in Equation B.3 also equal to 1. Therefore, $K_{sp} = [Ag^+][Cl^-] = (S)(S) = S^2$.

2. **Lead chloride, $PbCl_2$**
   Dissolution reaction: $PbCl_2(s) \rightleftarrows Pb^{2+}(aq) + 2Cl^-(aq)$
   Lead chloride is an AB_2 salt because the coefficients $a = 1$ and $b = 2$ in Equation B.2, which makes the exponents $a = 1$ and $b = 2$ in Equation B.3. Therefore, $K_{sp} = [Pb^{2+}][Cl^-]^2 = (S)(2S)^2 = 4S^3$.

3. **Aluminum chloride, $AlCl_3$**
   Dissolution reaction: $AlCl_3(s) \rightleftarrows Al^{3+}(aq) + 3\,Cl^-(aq)$
   Aluminum chloride is an AB_3 salt. Therefore, $K_{sp} = [Al^{3+}][Cl^-]^3 = (S)(2S)^2 = 9S^4$.

4. **Magnesium phosphate, $Mg_3(PO_4)_2$**
   Dissolution reaction: $Mg_3(PO_4)_2(s) \rightleftarrows 3\,Mg^{2+}(aq) + 2\,PO_4^{3-}(aq)$
   Magnesium phosphate is an A_3B_2 salt. Therefore, $K_{sp} = [Mg^{2+}]^3[PO_4^{3-}]^2 = (3S)^3(2S)^2 = 108S^5$.

The above examples are generalized in Table B.1. Table B.2 lists measured solubility products for many salts, along with their dissolution reactions.

---

* By the same reasoning, two salts of different types (e.g., AB and A_2B_3; see Table B.1) will have different water solubilities even if they chance to have the same solubility products.

TABLE B.1
Relationship between Solubility Product, K_{sp}, and Solubility, S, for Several Generalized Salts of Composition A_aB_b in Pure Water

A	B	Salt Type	Equations for K_{sp} and S in Pure Water	Example Salts and Dissolution Equations
1	1	AB	$K_{sp} = [A^{m+}][B^{n+}] = (S)(S) = S^2$ $S = K_{sp}^{1/2}$	AgCl, NaBr, CaS, AlPO$_4$, CaCO$_3$, AgNO$_3$, BaCrO$_4$ BaCrO$_4$(s) $\rightleftarrows$ Ba^{2+}(aq) + CrO$_4^{2-}$(aq)
1	2	AB$_2$	$K_{sp} = [A^{m+}][B^{n+}]^2 = (S)(2S)^2 = 4S^3$ $S = \left(\dfrac{K_{sp}}{4}\right)^{1/3}$	BaF$_2$, Ca(OH)$_2$, PbI$_2$, Cu(OH)$_2$, MgCl$_2$ BaF$_2$(s) $\rightleftarrows$ Ba^{2+}(aq) + 2 F$^-$(aq)
1	3	AB$_3$	$K_{sp} = [A^{m+}][B^{n+}]^3 = (S)(3S)^3 = 27S^4$ $S = \left(\dfrac{K_{sp}}{27}\right)^{1/4}$	Al(OH)$_3$, AlCl$_3$, Cr(OH)$_3$, Al(ClO$_4$)$_3$, SbF$_3$, Fe(ClO$_4$)$_3$, Cr(NO$_3$)$_3$ Cr(OH)$_3$(s) $\rightleftarrows$ Cr^{3+}(aq) + 3 OH$^-$(aq)
2	1	A$_2$B	$K_{sp} = [A^{m+}]^2[B^{n+}] = (2S)^2(S) = 4S^3$ $S = \left(\dfrac{K_{sp}}{4}\right)^{1/3}$	Ag$_2$SO$_4$, Cs$_2$SO$_4$, Li$_2$MoO$_4$, Na$_2$CO$_3$, K$_2$CrO$_4$, (NH$_4$)$_2$CrO$_4$ Na$_2$CO$_3$(s) $\rightleftarrows$ 2 Na$^+$(aq) + CO$_3^{2-}$(aq)
2	3	A$_2$B$_3$	$K_{sp} = [A^{m+}]^2[B^{n+}]^3 = (2S)^2(3S)^3 = 108S^5$ $S = \left(\dfrac{K_{sp}}{108}\right)^{1/5}$	Al$_2$(SO$_4$)$_3$, Pr$_2$(SeO$_3$)$_3$ Al$_2$(SO$_4$)$_3$(s) $\rightleftarrows$ 2 Al^{3+}(aq) + 3 SO$_4^{2-}$(aq)
3	1	A$_3$B	$K_{sp} = [A^{m+}]^3[B^{n+}] = (3S)^3(S) = 27S^4$ $S = \left(\dfrac{K_{sp}}{27}\right)^{1/4}$	Na$_3$PO$_4$, K$_3$CrO$_4$, Ag$_3$AsO$_4$, Li$_3$VO$_4$ Na$_3$PO$_4$(s) $\rightleftarrows$ 3 Na$^+$(aq) + 3 PO$_4^{3-}$(aq)
3	2	A$_3$B$_2$	$K_{sp} = [A^{m+}]^3[B^{n+}]^2 = (3S)^3(2S)^2 = 108S^5$ $S = \left(\dfrac{K_{sp}}{108}\right)^{1/5}$	Mg$_3$(PO$_4$)$_2$(s), Ca$_3$(PO$_4$)$_2$(s), Sr$_3$(PO$_4$)$_2$(s) Mg$_3$(PO$_4$)$_2$(s) $\rightleftarrows$ 3 Mg^{2+}(aq) + 2 PO$_4^{3-}$(aq)

TABLE B.2
Solubility Products,* Solubilities, and Dissolution Reactions of Some Metal Salts and Minerals

Metal Salt	Formula	Dissolution Reaction	K_{sp}
Aluminum hydroxide	$Al(OH)_3$	$Al(OH)_3(s) \rightleftarrows Al^{3+}(aq) + 3\,OH^-(aq)$	1.3×10^{-33}
Aluminum phosphate	$AlPO_4$	$AlPO_4(s) \rightleftarrows Al^{3+}(aq) + PO_4^{3-}(aq)$	6.3×10^{-19}
Aluminum phosphate diihydrate	$AlPO_4 \cdot 2H_2O$	$AlPO_4 \cdot 2H_2O(s) \rightleftarrows Al^{3+}(aq) + PO_4^{3-}(aq) + 2H_2O$	1.4×10^{-25}
Barium carbonate	$BaCO_3$	$BaCO_3(s) \rightleftarrows Ba^{2+}(aq) + CO_3^{2-}(aq)$	5.1×10^{-9}
Barium chromate	$BaCrO_4$	$BaCrO_4(s) \rightleftarrows Ba^{2+}(aq) + CrO_4^{2-}(aq)$	1.2×10^{-10}
Barium fluoride	BaF_2	$BaF_2(s) \rightleftarrows Ba^{2+}(aq) + F_2^-(aq)$	1.0×10^{-6}
Barium hydroxide	$Ba(OH)_2$	$Ba(OH)_2(s) \rightleftarrows Ba^{2+}(aq) + 2\,OH^-(aq)$	5×10^{-3}
Barium iodate	$Ba(IO_3)_2$	$Ba(IO_3)_2(s) \rightleftarrows Ba^{2+}(aq) + 2\,IO_3^-(aq)$	3.1×10^{-8}
Barium nitrate	$Ba(NO_3)_2$	$Ba(NO_3)_2(s) \rightleftarrows Ba^{2+}(aq) + 2\,NO_3^-(aq)$	4.64×10^{-3}
Barium sulfate	$BaSO_4$	$BaSO_4(s) \rightleftarrows Ba^{2+}(aq) + SO_4^{2-}(aq)$	1.1×10^{-10}
Barium sulfite	$BaSO_3$	$BaSO_3(s) \rightleftarrows Ba^{2+}(aq) + SO_3^{2-}(aq)$	8×10^{-7}
Barium thiosulfate	BaS_2O_3	$BaS_2O_3(s) \rightleftarrows Ba^{2+}(aq) + S_2O_3^{2-}(aq)$	1.6×10^{-5}
Cadmium carbonate	$CdCO_3$	$CdCO_3(s) \rightleftarrows Cd^{2+}(aq) + CO_3^{2-}(aq)$	1.8×10^{-14}
Cadmium fluoride	CdF_2	$CdF_2(s) \rightleftarrows Cd^{2+}(aq) + 2\,F_2^-(aq)$	6.44×10^{-3}
Cadmium hydroxide	$Cd(OH)_2$	$Cd(OH)_2(s) \rightleftarrows Cd^{2+}(aq) + 2\,OH^-(aq)$	2.5×10^{-14}
Cadmium iodate	$Cd(IO_3)_2$	$Cd(IO_3)_2(s) \rightleftarrows Cd^{2+}(aq) + 2\,IO_3^-(aq)$	2.3×10^{-8}
Cadmium phosphate	$Cd_3(PO_4)_2$	$Cd_3(PO_4)_2(s) \rightleftarrows 3\,Cd^{2+}(aq) + 2\,PO_4^{3+}(aq)$	2.53×10^{-33}
Cadmium sulfide	CdS	$CdS(s) + H_2O \rightleftarrows Cd^{2+}(aq) + HS^-(aq) + OH^-(aq)$	8×10^{-27}
Calcium carbonate	$CaCO_3$	$CaCO_3(s) \rightleftarrows Ca^{2+}(aq) + CO_3^{2-}(aq)$	2.8×10^{-9}

Calcium chromate	$CaCrO_4$	$CaCrO_4(s) \rightleftarrows Ca^{2+}(aq) + CrO_4^{2-}(aq)$	7.1×10^{-4}
Calcium fluoride	CaF_2	$CaF_2(s) \rightleftarrows Ca^{2+}(aq) + 2\,F^-(aq)$	5.3×10^{-9}
Calcium fluorophosphate	$Ca_5F(PO_4)_3$	$Ca_5(PO_4)_3F(s) \rightleftarrows 5\,Ca^{2+}(aq) + 3PO_4^{3-}(aq) + F^-$	1×10^{-60}
Calcium hydrogen phosphate	$CaHPO_4$	$CaHPO_4(s) \rightleftarrows Ca^{2+}(aq) + HPO_4^{2-}(aq)$	1×10^{-7}
Calcium hydroxide	$Ca(OH)_2$	$Ca(OH)_2(s) \rightleftarrows Ca^{2+}(aq) + 2\,OH^-(aq)$	5.5×10^{-6}
Calcium hydroxyphosphate	$Ca_5OH(PO_4)_3$	$Ca_5(PO_4)_3OH(s) \rightleftarrows 5\,Ca^{2+}(aq) + 3\,PO_4^{3-}(aq) + OH^-$	1×10^{-36}
Calcium iodate	$Ca(IO_3)_2$	$Ca(IO_3)_2(s) \rightleftarrows Ca^{2+}(aq) + 2\,IO_3^-(aq)$	7.1×10^{-7}
Calcium magnesium carbonate	$CaMg(CO_3)_2$	$CaMg(CO_3)_2(s) \rightleftarrows Ca^{2+}(aq) + Mg^{2+}(aq) + 2CO_3^{2-}(aq)$	2.9×10^{-17}
Calcium oxalate	CaC_2O_4	$CaC_2O_4\ (s) \rightleftarrows Ca^{2+}(aq) + C_2O_4(aq)$	4×10^{-9}
Calcium phosphate	$Ca_3(PO_4)_2$	$Ca_3(PO_4)_2 \rightleftarrows Ca^{2+}(aq) + SO_4^{2-}(aq) + 2H_2O$	2.0×10^{-29}
Calcium sulfate (anhydrous)	$CaSO_4$	$CaSO_4 \rightleftarrows Ca^{2+}(aq) + SO_4^{2-}(aq) + 2H_2O$	9.1×10^{-6}
Calcium sulfate dihydrate	$CaSO_4 \cdot 2H_2O$	$CaSO_4 \cdot 2H_2O(s) \rightleftarrows Ca^{2+}(aq) + SO_4^{2-}(aq) + 2H_2O$	4.93×10^{-5}
Calcium sulfite	$CaSO_3$	$CaSO_3(s) \rightleftarrows Ca^{2+}(aq) + SO_3^{2-}(aq) + 2H_2O$	6.8×10^{-8}
Chromium(II) hydroxide	$Cr(OH)_2$	$Cr(OH)_2(s) \rightleftarrows Cr^{2+}(aq) + 2\,OH^-(aq)$	2×10^{-16}
Chromium(III) hydroxide	$Cr(OH)_3$	$Cr(OH)_3(s) \rightleftarrows Cr^{3+}(aq) + 3\,OH^-(aq)$	6.3×10^{-31}
Cobalt(II) carbonate	$CoCO_3$	$CoCO_3(s) \rightleftarrows Co^{2+}(aq) + CO_3^{2-}(aq)$	1.0×10^{-10}
Cobalt(II) hydroxide	$Co(OH)_2$	$Co(OH)_2(s) \rightleftarrows Co^{2+}(aq) + 2\,OH^-(aq)$	1.3×10^{-15}
Cobalt(II) sulfide	CoS	$CdS(s) + H_2O \rightleftarrows Cd^{2+}(aq) + HS^-(aq) + OH^-(aq)$	5×10^{-22}
Cobalt(III) hydroxide	$Co(OH)_3$	$Co(OH)_3(s) \rightleftarrows Co^{2+}(aq) + 3\,OH^-(aq)$	1.6×10^{-44}
Copper(I) bromide	$CuBr$	$CuBr(s) \rightleftarrows Cu^+(aq) + Br^-(aq)$	5.3×10^{-9}
Copper(I) chloride	$CuCl$	$CuCl(s) \rightleftarrows Cu^+(aq) + Cl^-(aq)$	1.2×10^{-6}
Copper(I) cyanide	$CuCN$	$CuCN(s) \rightleftarrows Cu^+(aq) + CN^-(aq)$	3.2×10^{-20}
Copper(I) iodide	CuI	$CuI(s) \rightleftarrows Cu^+(aq) + I^-(aq)$	1.1×10^{-12}

(Continued)

TABLE B.2 *(Continued)*

Solubility Products, Solubilities, and Dissolution Reactions of Some Metal Salts and Minerals

Metal Salt	Formula	Dissolution Reaction	K_{sp}
Copper(II) arsenate	$Cu_3(AsO_4)_2$	$Cu_3(AsO_4)_2(s) \rightleftarrows 3\ Cu^{2+}(aq) + 2\ AsO_4^{3-}(aq)$	7.6×10^{-36}
Copper(II) bromide	$CuBr_2$	$CuBr_2(s) \rightleftarrows Cu^{2+}(aq) + 2\ Br^-(aq)$	5.3×10^{-9}
Copper(II) carbonate	$CuCO_3$	$CuCO_3(s) \rightleftarrows Cu^{2+}(aq) + CO_3^{2-}(aq)$	1.4×10^{-10}
Copper(II) chloride	$CuCl_2$	$CuCl_2(s) \rightleftarrows Cu^{2+}(aq) + 2\ Cl^-(aq)$	1.2×10^{-6}
Copper(II) chromate	$CuCrO_4$	$CuCrO_4(s) \rightleftarrows Cu^{2+}(aq) + CrO_4^{2-}(aq)$	3.6×10^{-6}
Copper(II) iodate	$Cu(IO_3)_2$	$Cu(IO_3)_2(s) \rightleftarrows Cu^{2+}(aq) + 2\ IO_3^-(aq)$	7.4×10^{-8}
Copper(II) hydroxide	$Cu(OH)_2$	$Cu(OH)_2(s) \rightleftarrows Cu^{2+}(aq) + 2\ OH^-(aq)$	2.2×10^{-20}
Copper(II) sulfide	CuS	$CuS(s) + H_2O \rightleftarrows Cu^{2+}(aq) + HS^-(aq) + OH^-(aq)$	6×10^{-36}
Copper(II) thiocyanate	$Cu(SCN)_2$	$Cu(SCN)_2(s) \rightleftarrows Cu^{2+}(aq) + 2\ SCN^-(aq)$	1.1×10^{-14}
Iron(II) carbonate	$FeCO_3$	$FeCO_3(s) \rightleftarrows Fe^{2+}(aq) + CO_3^{2-}(aq)$	3.2×10^{-11}
Iron(II) hydroxide	$Fe(OH)_2$	$Fe(OH)_2(s) \rightleftarrows Fe^{2+}(aq) + 2\ OH^-(aq)$	8.0×10^{-16}
Iron(II) sulfide	FeS	$FeS(s) + H_2O \rightleftarrows Fe^{2+}(aq) + HS^-(aq) + OH^-(aq)$	6.3×10^{-18}
Iron(III) arsenate	$FeAsO_4$	$FeAsO_4(s) \rightleftarrows Fe^{3+}(aq) + AsO_4^{3-}(aq)$	5.7×10^{-21}
Iron(III) hydroxide	$Fe(OH)_3$	$Fe(OH)_3(s) \rightleftarrows Fe^{3+}(aq) + 3\ OH^-(aq)$	1.0×10^{-38}
Iron(III) phosphate	$FePO_4$	$FePO_4(s) \rightleftarrows Fe^{3+}(aq) + PO_4^{3-}(aq)$	1.3×10^{-22}
Lead(II) arsenate	$Pb_3(AsO_4)_2$	$Pb_3(AsO_4)_2(s) \rightleftarrows 3\ Pb^{2+}(aq) + 2\ AsO_4^{3-}(aq)$	4×10^{-36}
Lead(II) bromate	$Pb(BrO_3)_2$	$Pb(BrO_3)_2(s) \rightleftarrows Pb^{2+}(aq) + 2\ BrO_3^-(aq)$	7.9×10^{-6}
Lead(II) bromide	$PbBr_2$	$PbBr_2(s) \rightleftarrows Pb^{2+}(aq) + 2\ Br^-(aq)$	4.0×10^{-5}
Lead(II) carbonate	$PbCO_3$	$PbCO_3(s) \rightleftarrows Pb^{2+}(aq) + CO_3^{2-}(aq)$	7.4×10^{-14}

Lead(II) chloride	$PbCl_2$	$PbCl_2(s) \rightleftarrows Pb^{2+}(aq) + 2\ Cl^-(aq)$	1.6×10^{-5}
Lead(II) chromate	$PbCrO_4$	$PbCrO_4(s) \rightleftarrows Pb^{2+}(aq) + CrO_4^{2-}(aq)$	2.8×10^{-13}
Lead(II) fluoride	PbF_2	$PbF_2(s) \rightleftarrows Pb^{2+}(aq) + 2\ F^-(aq)$	7.7×10^{-8}
Lead(II) hydroxide	$Pb(OH)_2$	$Pb(OH)_2(s) \rightleftarrows Pb^{2+}(aq) + 2\ OH^-(aq)$	1.2×10^{-15}
Lead(II) iodate	$Pb(IO_3)_2$	$Pb(IO_3)_2(s) \rightleftarrows Pb^{2+}(aq) + 2\ IO_3^-(aq)$	2.5×10^{-13}
Lead(II) iodide	$PbBr_2$	$PbI_2(s) \rightleftarrows Pb^{2+}(aq) + 2\ I^-(aq)$	7.1×10^{-9}
Lead(II) sulfate	$PbSO_4$	$PbSO_4(s) \rightleftarrows Pb^{2+}(aq) + SO_4^{2-}(aq)$	1.6×10^{-8}
Lead(II) sulfide	PbS	$PbS(s) + H_2O \rightleftarrows Pb^{2+}(aq) + HS^-(aq) + OH^-(aq)$	8.0×10^{-28}
Lithium phosphate	Li_3PO_4	$Li_3PO_4(s) \rightleftarrows 3\ Li^+(aq) + PO_4^{3-}(aq)$	3.2×10^{-9}
Magnesium carbonate	$MgCO_3$	$MgCO_3(s) \rightleftarrows Mg^{2+}(aq) + CO_3^{2-}(aq)$	3.5×10^{-8}
Magnesium fluoride	MgF_2	$MgF_2(s) \rightleftarrows Mg^{2+}(aq) + 2\ F^-(aq)$	6.5×10^{-9}
Magnesium hydroxide	$Mg(OH)_2$	$Mg(OH)_2(s) \rightleftarrows Mg^{2+}(aq) + 2\ OH^-(aq)$	1.8×10^{-11}
Magnesium oxalate	MgC_2O_4	$MgC_2O_4(s) \rightleftarrows Mg^{2+}(aq) + C_2O_4^{2-}(aq)$	7×10^{-7}
Magnesium phosphate	$Mg_3(PO_4)_2$	$Mg_3(PO_4)_2(s) \rightleftarrows 3\ Mg^{2+}(aq) + 2\ PO_4^{3-}(aq)$	1×10^{-25}
Manganese(II) carbonate	$MnCO_3$	$MnCO_3(s) \rightleftarrows Mn^{2+}(aq) + CO_3^{2-}(aq)$	1.8×10^{-11}
Manganese(II) hydroxide	$Mn(OH)_2$	$Mn(OH)_2(s) \rightleftarrows Mn^{2+}(aq) + 2\ OH^-(aq)$	1.9×10^{-13}
Manganese(II) sulfide	MnS	$MnS(s) + H_2O \rightleftarrows Mn^{2+}(aq) + HS^-(aq) + OH^-(aq)$	2.5×10^{-13}
Mercury(I) bromide	Hg_2Br_2	$Hg_2Br_2(s) \rightleftarrows Hg_2^{2+}(aq) + 2\ Br^-(aq)$	5.6×10^{-23}
Mercury(I) chloride	Hg_2Cl_2	$Hg_2Cl_2(s) \rightleftarrows Hg_2^{2+}(aq) + 2\ Cl^-(aq)$	1.3×10^{-18}
Mercury(I) chromate	Hg_2CrO_4	$Hg_2CrO_4(s) \rightleftarrows Hg_2^{2+}(aq) + CrO_4^{2-}(aq)$	2.0×10^{-9}
Mercury(I) cyanide	$Hg_2(CN)_2$	$Hg_2(CN)_2(s) \rightleftarrows Hg_2^{2+}(aq) + 2\ CN^-(aq)$	5×10^{-40}
Mercury(I) iodide	Hg_2I_2	$Hg_2I_2(s) \rightleftarrows Hg_2^{2+}(aq) + 2\ I^-(aq)$	4.5×10^{-29}
Mercury(I) sulfate	Hg_2SO_4	$Hg_2SO_4(s) \rightleftarrows Hg_2^{2+}(aq) + SO_4^{2-}(aq)$	7.4×10^{-7}
Mercury(I) thiocyanate	$Hg_2(SCN)_2$	$Hg_2(SCN)_2(s) \rightleftarrows Hg_2^{2+}(aq) + 2\ SCN^-(aq)$	3.0×10^{-20}

(Continued)

TABLE B.2 (Continued)
Solubility Products, Solubilities, and Dissolution Reactions of Some Metal Salts and Minerals

Metal Salt	Formula	Dissolution Reaction	K_{sp}
Mercury(II) thiocyanate	$Hg(SCN)_2$	$Hg(SCN)_2(s) \rightleftarrows Hg^{2+}(aq) + 2\ SCN^-(aq)$	2.8×10^{-20}
Mercury(II) bromide	$HgBr_2$	$HgBr_2(s) \rightleftarrows Hg^{2+}(aq) + 2\ Br^-(aq)$	1.3×10^{-19}
Mercury(II) sulfide	HgS	$HgS(s) + H_2O \rightleftarrows Hg^{2+}(aq) + HS^-(aq) + OH^-(aq)$	4×10^{-53}
Nickel(II) carbonate	$NiCO_3$	$NiCO_3(s) \rightleftarrows Ni^{2+}(aq) + CO_3^{2-}(aq)$	1.3×10^{-11}
Nickel(II) hydroxide	$Ni(OH)_2$	$Ni(OH)_2(s) \rightleftarrows Ni^{2+}(aq) + 2\ OH^-(aq)$	2.0×10^{-15}
Nickel(II) sulfide	NiS	$NiS(s) + H_2O \rightleftarrows Mn^{2+}(aq) + HS^-(aq) + OH^-(aq)$	1.3×10^{-25}
Scandium fluoride	ScF_3	$ScF_3(s) \rightleftarrows Sc^{3+}(aq) + 3\ F^-(aq)$	4.2×10^{-18}
Scandium hydroxide	$Sc(OH)_3$	$Sc(OH)_3(s) \rightleftarrows Sc^{3+}(aq) + 3\ OH^-(aq)$	8×10^{-31}
Silver arsenate	Ag_3AsO_4	$Ag_3AsO_4(s) \rightleftarrows 3\ Ag^+(aq) + AsO_4^{3-}(aq)$	1×10^{-22}
Silver bromate	$AgBrO_3$	$AgBrO_3(s) \rightleftarrows Ag^+(aq) + BrO_3^-(aq)$	5.8×10^{-5}
Silver bromide	$AgBr$	$AgBr(s) \rightleftarrows Ag^+(aq) + Br^-(aq)$	5×10^{-13}
Silver carbonate	Ag_2CO_3	$Ag_2CO_3(s) \rightleftarrows 2\ Ag^+(aq) + CO_3^{2-}(aq)$	8.1×10^{-12}
Silver chloride	$AgCl$	$AgCl(s) \rightleftarrows Ag^+(aq) + Cl^-(aq)$	1.8×10^{-10}
Silver cyanide	$AgCN$	$AgCN(s) \rightleftarrows Ag^+(aq) + CN^-(aq)$	2.2×10^{-16}
Silver chromate	Ag_2CrO_4	$Ag_2CrO_4(s) \rightleftarrows 2\ Ag^+(aq) + CrO_4^{2-}(aq)$	1.1×10^{-12}
Silver iodate	$AgIO_3$	$AgIO_3(s) \rightleftarrows Ag^+(aq) + IO_3^-(aq)$	6.0×10^{-4}
Silver iodide	AgI	$AgI(s) \rightleftarrows Ag^+(aq) + I^-(aq)$	8.3×10^{-17}
Silver nitrite	$AgNO_2$	$AgNO_2(s) \rightleftarrows Ag^+(aq) + NO_2^-(aq)$	6.0×10^{-4}
Silver oxalate	$Ag_2C_2O_4$	$Ag_2C_2O_4(s) \rightleftarrows 2\ Ag^+(aq) + C_2O_4^{2-}(aq)$	3.4×10^{-11}
Silver sulfate	Ag_2SO_4	$Ag_2SO_4(s) \rightleftarrows 2\ Ag^+(aq) + SO_4^{2-}(aq)$	1.4×10^{-5}
Silver sulfide	Ag_2S	$Ag_2S(s) + H_2O \rightleftarrows 2\ Ag^+(aq) + HS^-(aq) + OH^-(aq)$	8×10^{-51}

Name	Formula	Equation	K_{sp}
Silver sulfite	Ag_2SO_3	$Ag_2SO_3(s) \rightleftarrows 2\,Ag^+(aq) + SO_3^{2-}(aq)$	1.5×10^{-14}
Silver thiocyanate	$AgSCN$	$AgSCN(s) \rightleftarrows Ag^+(aq) + SCN^-(aq)$	1.1×10^{-12}
Strontium carbonate	$SrCO_3$	$SrCO_3(s) \rightleftarrows Sr^{2+}(aq) + CO_3^{2-}(aq)$	1.1×10^{-10}
Strontium chromate	$SrCrO_4$	$SrCrO_4(s) \rightleftarrows Sr^{2+}(aq) + CrO_4^{2-}(aq)$	2.2×10^{-5}
Strontium fluoride	SrF_2	$SrF_2(s) \rightleftarrows Sr^{2+}(aq) + 2\,F^-(aq)$	2.5×10^{-9}
Strontium oxalate	SrC_2O_4	$SrC_2O_4(s) \rightleftarrows Mg^{2+}(aq) + C_2O_4^{2-}(aq)$	4×10^{-7}
Strontium sulfate	$SrSO_4$	$SrSO_4(s) \rightleftarrows Sr^{2+}(aq) + SO_4^{2-}(aq)$	3.2×10^{-7}
Thallium(I) bromate	$TlBrO_3$	$TlBrO_3(s) \rightleftarrows Tl^+(aq) + BrO_3^-(aq)$	1.7×10^{-4}
Thallium(I) bromide	$TlBr$	$TlBr(s) \rightleftarrows Tl^+(aq) + Br^-(aq)$	3.4×10^{-6}
Thallium(I) chloride	$TlCl$	$TlCl(s) \rightleftarrows Tl^+(aq) + Cl^-(aq)$	1.7×10^{-4}
Thallium(I) chromate	Tl_2CrO_4	$Tl_2CrO_4(s) \rightleftarrows 2\,Tl^+(aq) + CrO_4^{2-}(aq)$	9.8×10^{-15}
Thallium(I) iodate	$TlIO_3$	$TlIO_3(s) \rightleftarrows Tl^+(aq) + IO_3^-(aq)$	3.1×10^{-6}
Thallium(I) iodide	TlI	$TlI(s) \rightleftarrows Tl^+(aq) + I^-(aq)$	1.6×10^{-4}
Thallium(I) sulfide	Tl_2S	$Tl_2S(s) + H_2O \rightleftarrows 2\,Tl^+(aq) + HS^-(aq) + OH^-(aq)$	6×10^{-22}
Thallium(I) thiocyanate	$TlSCN$	$TlSCN(s) \rightleftarrows Tl^+(aq) + SCN^-(aq)$	3.0×10^{-20}
Thallium(III) hydroxide	$Tl(OH)_3$	$Tl(OH)_3(s) \rightleftarrows Tl^{3+}(aq) + 3\,OH^-(aq)$	6.3×10^{-16}
Tin(II) hydroxide	$Sn(OH)_2$	$Sn(OH)_2(s) \rightleftarrows Sn^{2+}(aq) + 2\,OH^-(aq)$	1.4×10^{-28}
Tin(II) sulfide	SnS	$SnS(s) + H_2O \rightleftarrows Sn^{2+}(aq) + HS^-(aq) + OH^-(aq)$	1.0×10^{-25}
Zinc carbonate	$ZnCO_3$	$ZnCO_3(s) \rightleftarrows Zn^{2+}(aq) + CO_3^{2-}(aq)$	1.4×10^{-11}
Zinc cyanide	$Zn(CN)_2$	$Zn(CN)_2(s) \rightleftarrows Zn^{2+}(aq) + 2\,CN^-(aq)$	3×10^{-16}
Zinc hydroxide	$Zn(OH)_2$	$Zn(OH)_2(s) \rightleftarrows Zn^{2+}(aq) + 2\,OH^-(aq)$	1.0×10^{-15}
Zinc iodate	$Zn(IO_3)_2$	$Zn(IO_3)_2(s) \rightleftarrows Zn^{2+}(aq) + 2\,IO_3^-(aq)$	3.9×10^{-6}
Zinc oxalate	ZnC_2O_4	$ZnC_2O_4(s) \rightleftarrows Zn^{2+}(aq) + C_2O_4^{2-}(aq)$	2.7×10^{-8}
Zinc phosphate	$Zn_3(PO_4)_2$	$Zn_3(PO_4)_2(s) \rightleftarrows 3\,Zn^{2+}(aq) + 2\,PO_4^{3-}(aq)$	9.0×10^{-33}
Zinc sulfide	ZnS	$ZnS(s) + H_2O \rightleftarrows Zn^{2+}(aq) + HS^-(aq) + OH^-(aq)$	1.6×10^{-24}

* Solubility products in this table are taken from several sources, but mostly from Lange's *Handbook of Chemistry*, 14th edition. They are valid for temperatures around "room temperature," between 18 and 25°C.

B.4 USES OF THE EQUILIBRIUM CONSTANT

There are many kinds of calculations involving water–salt equilibria. Five of the most common calculations are listed below, followed by examples.

1. Use experimental equilibrium data to determine a value for the solubility product, K_{sp}, of a salt.
2. Use K_{sp} to calculate the solubility of a salt and the equilibrium ion concentrations after a salt is dissolved into initially pure water.
3. Use K_{sp} to calculate the solubility of a salt and the equilibrium ion concentrations when the salt is dissolved into water that already contains ions of the type present in the dissolving salt.
4. Predict whether a precipitate will form when two (or more) soluble salts are dissolved in the same solution.
5. Determine the pH dependence of the solubility of salts that change the concentrations of hydroxyl (OH^-) or hydrogen (H^+) ions when dissolved.

Example 1: Use Experimental Equilibrium Data to Determine a Value for a Solubility Product

A. A handbook table lists the solubility of lead sulfate, $PbSO_4$, as 0.0041 g of salt in 100 mL of saturated water solution at 20°C. Determine the value of the solubility product.

Step 1: Write the dissolution reaction: $PbSO_4(s) \rightleftarrows Pb^{2+}(aq) + SO_4^{2-}(aq)$. Note that $PbSO_4$ is a type AB salt.

Step 2: Write the solubility product expression: $K_{sp} = [Pb^{2+}][SO_4^{2-}]$.

Step 3: Because concentration units in Equation B.3 are mol/L, the solubility, S, of $PbSO_4$ must be expressed in mol/L. The molecular weight of $PbSO_4$ is

$$MW_{PbSO_4} = 207.2 + 32.07 + 4(16.00) = 303.27 \text{ g/mol}$$

$$S_{PbSO_4} = \frac{0.0041 \text{ g}}{100 \text{ mL}} \times \frac{1000 \text{ mL}}{L} \times \frac{1 \text{ mol}}{303.3 \text{ g}} = 1.352 \times 10^{-4} \text{ mol/L}$$

Step 4: From Step 1, we see that the moles of Pb^{2+} and SO_4^{2-} on the product side of the reaction are each equal to the moles of $PbSO_4$ that dissolve. Therefore, at equilibrium,

$$[Pb^{2+}] = 1.352 \times 10^{-4} \text{ mol/L and } [SO_4^{2-}] = 1.352 \times 10^{-4} \text{ mol/L}$$

Step 5: The solubility product at 20°C is

$$K_{sp}(PbSO_4) = (1.352 \times 10^{-4} \text{ mol/L})(1.352 \times 10^{-4} \text{ mol/L}) = 1.827 \times 10^{-8} \text{ mol}^2/L^2$$

Since the measured solubility was reported to only two significant figures, the calculated solubility product is known only to two significant figures. Therefore, the experimental data indicate that $K_{sp}(PbSO_4) = 1.8 \times 10^{-8}$ mol²/L². Note that

K_{sp} has units, mol^2/L^2 in this case, that are dependent on the type of salt (AB, AB_2, A_2B_3, etc.) being measured (see Table B.1).

B. A laboratory technician measured the solubility of magnesium hydroxide, $Mg(OH)_2$, to be 0.0012 g of salt in 100 mL of saturated water solution at 20°C. Use this measurement to calculate a value for the solubility product of $Mg(OH)_2$. Follow the procedure in part A above.

Step 1: Write the dissolution reaction: $Mg(OH)_2(s) \rightleftarrows Mg^{2+}(aq) + 2\, OH^-(aq)$. Note that $Mg(OH)_2$ is a type AB_2 salt.

Step 2: Write the solubility product expression: $K_{sp} = [Mg^{2+}][OH^-]^2$.

Step 3: The molecular weight of $Mg(OH)_2$ is

$$MW_{Mg(OH)_2} = 24.31 + 2(16.00 + 1.008) = 58.33 \text{ g/mol}$$

$$S_{Mg(OH)_2} = \frac{0.0012 \text{ g}}{100 \text{ mL}} \times \frac{1000 \text{ mL}}{L} \times \frac{1 \text{ mol}}{58.33 \text{ g}} = 2.057 \times 10^{-4} \text{ mol/L}$$

Step 4: From Step 1, we see that the moles of Mg^{2+} on the product side of the reaction are equal to the moles of $Mg(OH)_2$ that dissolve. However, the moles of OH^- are equal to two times the moles of $Mg(OH)_2$ that dissolve. Therefore, at equilibrium,

$$[Mg^{2+}] = 2.057 \times 10^{-4} \text{ mol/L and } [OH^-] = 4.114 \times 10^{-4} \text{ mol/L}$$

Step 5: The solubility product at 20°C is

$$K_{sp}(Mg(OH)_2) = (2.057 \times 10^{-4})(4.114 \times 10^{-4})^2 = 3.4824 \times 10^{-11} \text{ mol}^3/L^3$$

To two significant figures, $K_{sp}(Mg(OH)_2) = 3.5 \times 10^{-11} \text{ mol}^3/L^3$.

Example 2: Use K_{sp} to Calculate the Solubility of a Salt and the Equilibrium Ion Concentrations after a Salt Is Dissolved into Initially Pure Water

A. The dissolution reaction of silver chloride, AgCl, is

$$AgCl(s) \rightleftarrows Ag^+(aq) + Cl^-(aq)$$

A handbook table gives $K_{sp}(AgCl) = 1.8 \times 10^{-10}$. If AgCl is dissolved into pure water, find the equilibrium concentrations of Ag^+ and Cl^- in mg/L.

Step 1: From the dissolution reaction, AgCl is an AB salt. Therefore,

$$K_{sp} = [Ag^+][Cl^-] = 1.8 \times 10^{-10} \text{ mol}^2/L^2$$

Step 2: When AgCl is dissolved into pure water, no other salt ions are initially present. Since AgCl is an AB salt, the molar concentrations of Ag^+ and Cl^- at saturation are equal to each other and are equal in value to the molar solubility, S, of AgCl(s) that will dissolve into 1 L of pure water.

Let $S = [Ag^+] = [Cl^-]$, then $K_{sp} = [S][S] = S^2 = 1.8 \times 10^{-10}$ mol^2/l^2

The solubility of AgCl is

$$S = \sqrt{1.8 \times 10^{-10} \text{ mol}^2 L^{-2}} = 1.3 \times 10^{-5} \text{ mol/L}$$

The concentrations of dissolved salt ions in mg/L are (to two significant figures):

$$[Ag^+] = [Cl^-] = 1.3 \times 10^{-5} \text{ mol/L}$$

Convert to mg/L using the atomic weights of Ag and Cl, which are 107.9 and 35.45, respectively.

$$C_{Ag^+} = \left(1.3 \times 10^{-5} \frac{\text{mol}}{\text{L}}\right)\left(107.9 \frac{\text{g}}{\text{mol}}\right)\left(1000 \frac{\text{mg}}{\text{g}}\right) = 1.4 \text{ mg/L}$$

$$C_{Cl^-} = \left(1.3 \times 10^{-5} \frac{\text{mol}}{\text{L}}\right)\left(35.45 \frac{\text{g}}{\text{mol}}\right)\left(1000 \frac{\text{mg}}{\text{g}}\right) = 0.46 \text{ mg/L}$$

B. The dissolution reaction of copper arsenate, $Cu_3(AsO_4)_2$, is

$$Cu_3(AsO_4)_2(s) \rightleftarrows 3\ Cu^{2+}(aq) + 2\ AsO_4^{3-}(aq)$$

A handbook table gives $K_{sp}(Cu_3(AsO_4)_2) = 7.6 \times 10^{-36}$. If $Cu_3(AsO_4)_2$ is dissolved into pure water, find the equilibrium concentrations of Cu^{2+} and AsO_4^{3-} in mg/L.

Step 1: From the dissolution reaction, $Mg_3(PO_4)_2$ is an A_3B_2 salt so that

$$K_{sp} = [Mg^{2+}]^3[PO_4^{3-}]^2 = 1.0 \times 10^{-25} \text{ mol}^5/L^5$$

Step 2: When $Mg_3(PO_4)_2$ is dissolved into pure water, no other salt ions are initially present. Let the molar solubility (mol/L of $Mg_3(PO_4)_2$(s) that will dissolve into 1 L of pure water) be designated as S. Then, the dissolution reaction shows that, at saturation, the molar concentration of $Mg^{2+} = 3S$ and the molar concentration of $PO_4^{3-} = 2S$:

$$K_{sp} = [3S]^3[2S]^2 = (27S^3)(4S^2) = 108S^5 = 1.0 \times 10^{-25} \text{ mol}^5/L^5$$

The solubility of $Mg_3(PO_4)_2$ is

$$S = \left(\frac{1.0 \times 10^{-25} \text{ mol}^5/\text{L}^5}{108} \right)^{1/5} = \left(9.26 \times 10^{-28} \text{ mol}^5/\text{L}^5 \right)^{1/5} = 3.92 \times 10^{-6} \text{ mol/L}$$

The concentrations of dissolved salt ions in mg/L are

$$[Mg^{2+}] = 3S = 1.2 \times 10^{-5} \text{ mol/L} \quad \text{and} \quad [PO_4^{3-}] = 2S = 7.8 \times 10^{-6} \text{ mol/L}$$

Convert to mg/L using the molecular weights of Mg^{2+} and PO_4^{-3}, which are 24.3 and 95.0, respectively.

$$C_{Mg^{2+}} = \left(1.2 \times 10^{-5} \frac{\text{mol}}{\text{L}} \right) \left(24.3 \frac{\text{g}}{\text{mol}} \right) \left(1000 \frac{\text{mg}}{\text{g}} \right) = 0.292 \text{ mg/L}$$

$$C_{PO_4^{-3}} = \left(7.8 \times 10^{-6} \frac{\text{mol}}{\text{L}} \right) \left(95.0 \frac{\text{g}}{\text{mol}} \right) \left(1000 \frac{\text{mg}}{\text{g}} \right) = 0.741 \text{ mg/L}$$

Example 3: Use K_{sp} to Calculate the Solubility of a Salt and the Equilibrium Ion Concentrations When the Salt is Dissolved into Water That Already Contains Ions of the Type Present in the Dissolving Salt

This example illustrates the "common-ion effect": The solubility of a salt in the presence of a common ion, that is, an ion that is the same as one formed by the salt but is supplied from another source, is always lower than the solubility of the salt in pure water.* The common ion might already be present before the salt is dissolved, as in this example, or added later after the salt has dissolved.

A. Determine the solubility of AgCl in water already containing 50 mg/L of NaCl, a highly soluble salt.

---

* This statement always is true when the common ion is in low concentration. However, it sometimes happens that a common ion in high concentration will form complexes with the salt, actually increasing its solubility. For example, the solubility of AgCl in water decreases, as expected, if additional Cl^- is present from another source, such as HCl, as long as $[Cl^-]$ remains less than about 3×10^{-3} mol/L. If $[Cl^-]$ is increased above this value, the solubility of AgCl increases. In a solution where $[Cl^-] = 1.0$ mol/L, the solubility of AgCl is nearly the same as in pure water. This behavior is attributed to reaction that form additional soluble species such as

$$AgCl(s) + Cl^- \rightleftarrows AgCl_2^-(aq)$$

$$AgCl(s) + 2 Cl^- \rightleftarrows AgCl_3^{2-}(aq)$$

$$AgCl(s) + 3 Cl^- \rightleftarrows AgCl_4^{3-}(aq), \text{ etc.}$$

Chapter 5 discusses in more detail how complex formation influences metal removal treatment from wastewaters.

The dissolution reactions of AgCl and NaCl are

$$AgCl(s) \rightleftarrows Ag^+(aq) + Cl^-(aq)$$

$$NaCl(s) \rightleftarrows Na^+(aq) + Cl^-(aq)$$

From Example 1, $K_{sp}(AgCl) = [Ag^+][Cl^-] = 1.8 \times 10^{-10}$. NaCl is highly soluble, and its solubility is unchanged after the addition of AgCl. After adding AgCl, the solution contains the ions: Na^+ from NaCl, Ag^+ from AgCl, and Cl^- from both AgCl and NaCl.

1. The concentration of Na is not relevant to the solubility of AgCl.
2. Because NaCl is an AB salt, the molar concentration of Cl^- from NaCl is equal to the molar concentration of dissolved NaCl. The molecular weight of NaCl = 58.5 g/mol. Therefore, 50 mg/L of dissolved NaCl = $50 \times 10^{-3}/58.5 = 0.85 \times 10^{-3}$ mol/L and $[Cl^-_{NaCl}] = 0.85 \times 10^{-3}$ mol/L.
3. $K_{sp} = [Ag^+][Cl^-] = 1.8 \times 10^{-10}$ mol^2/L^2 is always valid. Note that the ion concentrations equal the total dissolved concentrations in the solution, no matter their source.
4. Let S^*_{AgCl} equal the new molar solubility of AgCl in the solution with the common ion Cl^-. Then $[Ag^+] = S^*_{AgCl}$. However, there are two sources of Cl^-, AgCl and NaCl. Therefore, $[Cl^-_{NaCl}] = 0.85 \times 10^{-3}$ mol/L and $[Cl^-_{AgCl}] = S^*_{AgCl}$.
5. $K_{sp} = [Ag^+][Cl^-] = (S^*_{AgCl})(S^*_{AgCl} + 0.85 \times 10^{-3}$ mol/L$) = 1.8 \times 10^{-10}$ mol^2/L^2.
6. This gives a second-order equation to solve for S^*:

$$S^{*2} + 0.85 \quad 10^{-3} S^* - 1.8 \quad 10^{-10} = 0$$

Fortunately, it is unnecessary to solve this equation exactly. We saw in Example 1 that when AgCl is dissolved in pure water, its chloride equilibrium concentration is $[Cl^-] = S_{AgCl} = 1.3 \times 10^{-5}$ mol/L, quite small compared to $[Cl^-_{NaCl}] = 0.85 \times 10^{-3}$ mol/L. Because of the common-ion effect, $[Cl^-_{AgCl}]$ will be even smaller in the NaCl solution. Therefore, $[Cl^-_{AgCl}] = S^*_{AgCl} \ll 0.85 \times 10^{-3}$ mol/L and can be neglected, to a first approximation, in the calculation of S^*_{AgCl}. This type of approximation is generally valid for determining the common-ion effect for slightly soluble salts in solutions containing relatively high concentrations of a common ion. With this approximation, the equation in Step 5 becomes

$$K_{sp} = [Ag^+][Cl^-] = (S^*_{AgCl})(S^*_{AgCl} + 0.85 \times 10^{-3}) \cong (S^*_{AgCl})(0.85 \times 10^{-3})$$
$$= 1.8 \times 10^{-10} \ mol^2/L^2$$

The solubility of AgCl in water already containing 50 mg/L of NaCl is

$$S^*_{AgCl} \cong \frac{1.8 \times 10^{-10} \ mol^2/L^2}{0.85 \times 10^{-3} \ mol/L} = 2.1 \times 10^{-7} \ mol/L$$

S^*_{AgCl} is reduced from the solubility of AgCl in pure water ($S_{AgCl} = 1.3 \times 10^{-5}$ mol/L) because of the common-ion effect.

The concentrations of dissolved salt ions in mg/L are

$$[Ag^+] = S^*_{AgCl} = 2.1 \times 10^{-7} \, mol/L$$

$$[Cl^-] = S^*_{AgCl} + 0.85 \times 10^{-3} \, mol/L = (2.1 \times 10^{-7} + 0.85 \times 10^{-3}) \, mol/L$$
$$= 0.85021 \times 10^{-3} \, mol/L$$

$$[Cl^-] \cong 0.85 \times 10^{-3} \, mol/L$$

$$[Na^+] = 0.85 \times 10^{-3} \, mol/L$$

Convert to mg/L using the molecular weights of Ag^+, Cl^-, and Na^+, which are 107.9, 35.45, and 23.0, respectively.

$$C_{Ag^+} = \left(2.1 \times 10^{-7} \frac{mol}{L}\right)\left(107.9 \frac{g}{mol}\right)\left(1000 \frac{mg}{g}\right)$$
$$= 2.27 \times 10^{-2} \, mg/L = 0.0227 \, mg/L$$

$$C_{Cl^-} \cong \left(0.85 \times 10^{-3} \frac{mol}{L}\right)\left(35.45 \frac{g}{mol}\right)\left(1000 \frac{mg}{g}\right) = 30.1 \, mg/L$$

$$C_{Na^+} = \left(0.85 \times 10^{-3} \frac{mol}{L}\right)\left(23.0 \frac{g}{mol}\right)\left(1000 \frac{mg}{g}\right) = 19.5 \, mg/L$$

Example 4: Predict Whether a Precipitate Will Form When Two (or More) Soluble Salts Are Dissolved in the Same Solution

Chemical precipitation can also occur if two soluble salts react to form a different salt of low solubility. For example, silver nitrate ($AgNO_3$) and sodium chloride (NaCl) are both highly soluble. However, they react in solution to form the slightly soluble salt silver chloride (AgCl, see Example 2) and the soluble salt sodium nitrate ($NaNO_3$). The slightly soluble silver chloride may precipitate as a solid, while the soluble silver nitrate remains dissolved.

Will a precipitate form if 20 mg of $AgNO_3$ and 20 mg of NaCl are dissolved in 1 L of pure water?

1. The dissolution reactions are

$$AgNO_3(s) \xrightarrow{H_2O} Ag^+ (aq) + NO_3^-(aq) \tag{B.4}$$

$$NaCl(s) \xrightarrow{H_2O} Na^+ (aq) + Cl^-(aq) \tag{B.5}$$

Because both salts are highly soluble, after the salts dissolve and before any further reaction, the solution would contain 20 mg/L each of Ag^+, Na^+, Cl^-, and NO_3^- ions.

While in solution, all ions move about freely. Ions with charges having opposite signs (positive/negative) are attracted to one another, while ions with charges having the same sign (positive/positive and negative/negative) are repelled from one another. Charged ions with opposite signs tend to pair up randomly, regardless of their chemical identity.

Therefore, the ions can combine in all possible ways that pair a positive ion with a negative ion. Besides the original Ag^+/NO^-_3 and Na^+/Cl^- pairs, both of which are highly soluble, Ag^+/Cl^- and Na^+/NO_3^- also are possible pairs. Since $NaNO_3$ is a highly soluble salt, which dissolves to form Na^+ and NO_3^-, the Na^+ and NO_3^- ions simply remain in solution. However, $AgCl$ is only slightly soluble and can precipitate as a solid. The overall reaction is written as

$$AgNO_3(aq) + NaCl(aq) \xrightarrow{H_2O} Na^+(aq) + NO_3^-(aq) + AgCl(s) \qquad (B.6)$$

Whether or not the precipitation reaction B.7 occurs depends on the concentrations of Ag^+ and Cl^-.

$$Ag^+(aq) + Cl^-(aq) \xrightarrow{H_2O} AgCl(s) \qquad (B.7)$$

2. The molar concentrations of Ag^+ and Cl^- are

$$[Ag^+] = \frac{20\ mg/L}{107.9\ g/mol} \times \frac{1\ g}{1000\ mg} = 0.185\ mol/L$$

$$[Cl^-] = \frac{20\ mg/L}{35.5\ g/mol} \times \frac{1\ g}{1000\ mg} = 0.563\ mol/L$$

3. From Example 2, the solubility product for $AgCl$ is $K_{sp} = [Ag^+][Cl^-] = 1.8 \times 10^{-10}\ mol^2/L^2$. If the product of the equilibrium molar concentrations, $[Ag^+][Cl^-]$, exceeds $1.8 \times 10^{-10}\ mol^2/L^2$, solid $AgCl$ must precipitate until the dissolved ion molar concentrations are equal to the value of K_{sp}.

$$[Ag^+][Cl^-] = (0.185\ mol/L)(0.563\ mol/L) = 0.104\ mol^2/L^2 \gg 1.8 \times 10^{-10}\ mol^2/L^2$$

$AgCl$ solid will precipitate. By Equation B.7, the moles of Ag^+ and Cl^- that precipitate must be equal. Therefore, essentially all the Ag^+ and 0.185 mol/L of the Cl^- will precipitate. The final equilibrium solution will contain $(0.563 - 0.185) = 0.378$ mol/L of Cl^-.

Use $K_{sp} = [Ag^+][Cl^-] = 1.8 \times 10^{-10}\ mol^2/L^2$ to determine the amount of Ag^+ remaining in solution.

$$K_{sp} = [Ag^+][0.378] = 1.8 \times 10^{-10}\ mol^2/L^2$$

$$[Ag^+] = \frac{1.8 \times 10^{-10}}{0.378} = 4.16 \times 10^{-10}\ mol/L = (4.16 \times 10^{-10}\ mol/L)(107.9\ g/mol)$$

$$[Ag^+] = 5.14 \times 10^{-8}\ g/L = 5.14 \times 10^{-5}\ mg/L\ \text{remaining in solution}$$

Example 5: Determine the pH Dependence of the Solubility of Salts that Change the Concentrations of Hydroxyl (OH⁻) or Hydrogen (H⁺) Ions when Dissolved

When dissolving a salt into water changes the hydroxyl or hydrogen ion concentrations,* the solubility of that salt will be pH dependent.

A. Show that the solubility of the salt $Mg(OH)_2$ is pH dependent.
 1. The dissolution reactions from Equations B.4 and B.5 are repeated below:

$$AgNO_3(s) \xrightarrow{H_2O} Ag^+(aq) + NO_3^-(aq)$$

$$NaCl(s) \xrightarrow{H_2O} Na^+(aq) + Cl^-(aq)$$

Because both salts are highly soluble, after the salts dissolve and before any further reaction, the solution would contain 20 mg/L each of Ag^+, Na^+, Cl^-, and NO_3^- ions.

While in solution, all ions move about freely. Ions with charges having opposite signs (positive/negative) are attracted to one another while ions with charges having the same sign (positive/positive and negative/negative) are repelled from one another. Charged ions with opposite signs tend to pair up randomly, regardless of their chemical identity.

Therefore, the ions can combine in all possible ways that pair a positive ion with a negative ion. Besides the original Ag^+/NO_3^- and Na^+/Cl^- pairs, both of which are highly soluble, Ag^+/Cl^- and Na^+/NO_3^- also are possible pairs. Since $NaNO_3$ is a highly soluble salt, which dissolves to form Na^+ and NO_3^-, the Na^+ and NO_3^- ions simply remain in solution. However, AgCl is only slightly soluble and can precipitate as a solid. The overall reaction is written as was shown in Equation B.6:

$$AgNO_3(aq) + NaCl(aq) \xrightarrow{H_2O} Na^+(aq) + NO_3^-(aq) + AgCl(s)$$

Whether or not the precipitation reaction from Equation B.7 occurs depends on the concentrations of Ag^+ and Cl^-.

$$Ag^+(aq) + Cl^-(aq) \xrightarrow{H_2O} AgCl(s)$$

 2. The molar concentrations of Ag^+ and Cl^- are

$$[Ag^+] = \frac{20 \text{ mg/L}}{107.9 \text{ g/mol}} \times \frac{1 \text{ g}}{1000 \text{ mg}} = 0.185 \text{ mol/L}$$

$$[Cl^-] = \frac{20 \text{ mg/L}}{35.5 \text{ g/mol}} \times \frac{1 \text{ g}}{1000 \text{ mg}} = 0.563 \text{ mol/L}$$

---

* Such salts act as acids or bases (see Chapter 3, Section 3.2).

3. From Example 2, the solubility product for AgCl is $K_{sp} = [Ag^+][Cl^-] = 1.8 \times 10^{-10}$ mol^2/L^2. If the product of the equilibrium molar concentrations, $[Ag^+][Cl^-]$, exceeds 1.8×10^{-10} mol^2/L^2, solid AgCl must precipitate until the dissolved ion molar concentrations are equal to the value of K_{sp}.

$$[Ag^+][Cl^-] = (0.185 \text{ mol/L})(0.563 \text{ mol/L}) = 0.104 \text{ mol}^2/L^2 >> 1.8 \times 10^{-10} \text{ mol}^2/L^2$$

AgCl solid will precipitate. By Equation B.7, the moles of Ag^+ and Cl^- that precipitate must be equal. Therefore, essentially all the Ag^+ and 0.185 mol/L of the Cl^- will precipitate. The final equilibrium solution will contain $(0.563 - 0.185) = 0.378$ mol/L of Cl^-.

Use $K_{sp} = [Ag^+][Cl^-] = 1.8 \times 10^{-10}$ mol^2/L^2 to determine the amount of Ag^+ remaining in solution.

$$K_{sp} = [Ag^+][0.378] = 1.8 \times 10^{-10} \text{ mol}^2/L^2$$

$$[Ag^+] = \frac{1.8 \times 10^{-10}}{0.378} = 4.16 \times 10^{-10} \text{ mol/L} = (4.16 \times 10^{-10} \text{ mol/L})(107.9 \text{ g/mol})$$

$$[Ag^+] = 5.14 \times 10^{-8} \text{ g/L} = 5.14 \times 10^{-5} \text{ mg/L remaining in solution}$$

Every remediation project is unique and site specific. The reason for this is that, although each pollutant has its predictable and, frequently, tabulated chemical and physical properties, each project site has properties of water, soil, and environment that are always different from other sites to some extent, depending on climate, long-term geologic history, and more recent anthropomorphic disturbances.

Chapter 2, Section 2.1

Appendix C
Glossary of Acronyms and Abbreviations

(See inside front cover for chemical element abbreviations.)

Acronym	Definition	Chapter Section with First or Main Entry
10^{-4} cancer risk	the concentration of a chemical in drinking water corresponding to an estimated lifetime cancer risk of 1 in 10,000	Appendix A
2,3,7,8-TCDD	2,3,7,8-tetrachlorodibenzo-p-dioxin	11.6.5.1
2,4-D	2,4-dichlorophenoxyacetic acid, synthetic systemic pesticide/herbicide	2.4.2
A	mass number of an element; equals the sum of neutrons plus protons in an atom's nucleus	10.2.2.1
AA	activated alumina	10.6
ABP	acid–base potential	4.9.1.3
AEC	anion exchange capacity	6.10.2
AMD	acid mine drainage	5.9.1
ANP	acid neutralizing potential	5.9.1.3
AOC	assimilable organic carbon	3.11.5
AOP	advanced oxidation process	3.11.6.1
aq	aqueous	2.3.3
ARD	acid rock drainage	5.9.1
atm	atmosphere	1.3.4
AW	atomic weight	1.3.4
BAP	bioavailable phosphorus	4.3.4.2
BAT	best available technology	10.6
BDOC	biodegradable organic carbon	3.11.5
BOD	biological oxygen demand	3.8
BOD_5	biological oxygen demand analysis with a 5-day incubation period	3.8.2
bp	boiling point	2.7.1; 2.8.7
Bq	Becquerel; SI unit of radiation activity	10.4
BTEX	the aromatic hydrocarbons benzene, toluene, ethylbenzene, and the xylene isomers	7.2.2
°C	temperature in degrees centigrade (also called degrees Celcius)	1.3.4
C	concentration	1.3.4
C_a	concentration of a volatile species in air	6.6
C_w	concentration of a chemical species sorbed to soil surfaces	6.6
CBOD	carbonaceous biological oxygen demand	3.8.2
$CBOD_5$	carbonaceous biological oxygen demand analysis with a 5-day incubation period	3.8.4
CCL	Drinking Water Contaminant Candidate List	1.1.2
CEC	cation exchange capacity	6.10.2
Ci	curie; unit of radionuclide activity	10.4
cm	centimeter; unit of length	3.9.4
CMA	calcium magnesium acetate	11.7.3
CNDC	carbon number distribution curve	7.8

Acronym	Definition	Chapter Section with First or Main Entry
COC	chain of custody	11.6.2.1
COD	chemical oxygen demand	3.8
C_{soil}	concentration of a dissolved species	6.7.1
CWA	Clean Water Act	1.1
d	density	3.10.1
Da	Dalton; 1 Da = 1 mol. wt. in g/mol; formerly atomic mass unit (AMU)	1.3.4
DBP	disinfection by-product	3.11.5
DCA	dichloroethane	6.7.2
DCE	dichloroethene, dichloroethylene	9.6.1
DDT	dichlorodiphenyltrichloroethane, synthetic pesticide	2.4.2
DIC	dissolved inorganic carbon	3.4.3
DNA	deoxyribonucleic acid, nucleic acid containing the genetic code	4.2.2
DNAPL	dense nonaqueous phase liquid	7.1
DO	dissolved oxygen	3.7
DOC	dissolved organic carbon	3.11.5
dps	radioisotope disintegrations per second; unit of radiation activity	10.4
dS/m	decisiemens per meter; unit of conductivity	3.9.4
DWEL	drinking water equivalent level; 70-year lifetime exposure concentration protective of adverse, noncancer health effects, which assumes that all of the exposure to a contaminant is from drinking water	Appendix A
EC	electrical conductivity, also called specific conductivity or just conductivity	3.9.4
E. coli	Escherichia coli; bacterial species within the fecal coliform group	11.3
EDB	ethylene dibromide	1.3.5
EDC	ethylene dichloride (also, endocrine disrupter chemicals)	1.3.5 (8.4.3)
EDTA	ethylenediaminetetraacetate, common metal chelating agent	2.3.3
EN	electronegativity; ΔEN = difference in electronegativity between two adjacent atoms	2.8.1
EPA	U.S. Environmental Protection Agency	1.1.2
eq. wt.	equivalent weight	1.3.6
erg	unit of energy	10.4.2
fc	fecal coliforms	11.3
fg	femtogram; unit of mass or weight, 10^{-15} g	1.3.4
f_{oc}	weight fraction of organic carbon in soil	6.6.4
f_{oc}^*	critical lower value for the fraction of organic carbon in soil below which sorption to inorganic matter becomes dominant	6.6.4

(Continued)

(*Continued*)

Acronym	Definition	Chapter Section with First or Main Entry
f_{om}	weight fraction of organic matter in soil	6.6.4
fp	freezing point	2.7.1; 2.8.7
g	gram; unit of mass or weight	1.3.4
GC	gas chromatograph	7.72
GC/MS	gas chromatograph–mass spectrometer combination instrument	7.72
g/eq	grams per equivalent	1.3.6
g/L	grams per liter	1.3.4
GAC	granulated activated carbon	4.6.1.4
Gy	gray; unit of absorbed radiation dose	10.4
HA	Health Advisory; an estimate of acceptable drinking water levels for a chemical substance based on health effects information	Appendix A
HAAs	haloacetic acids; regulated disinfection by-products (DBP) with a cancer risk	3.11.5
HMO	hydrous manganese oxide	10.6
ICP	ion-coupled plasma spectroscopy	5.1
IX	ion exchange	6.10.2; 10.6
k	reaction rate constant	6.7
K_d	soil–water partition coefficient	6.6
kg	kilogram; unit of mass or weight, 10^3 g	1.3.4
K_H	Henry's law constant	6.6
kJ	kilojoules; unit of energy	2.8.6
K_{oc}	soil organic carbon–water partition coefficient	6.6.4
K_{ow}	octanol–water partition coefficient	6.6.5
K_p	free product–water partition coefficient	6.6
K_{sp}	solubility product	Appendix B
L	liter	1.1.1
LRL or RL	laboratory reporting limit	11.6.4.2
LLE	liquid–liquid extraction	11.5
LNAPL	light nonaqueous phase liquid	7.1
LUFT	leaking underground fuel tank	9.3.1
MCL	maximum contaminant level; highest level of a contaminant allowed in drinking water. MCLs are enforceable standards	Appendix A
MCLG	maximum contaminant level goal; a nonenforceable health goal, which is set at a level at which no known or anticipated adverse effect on the health of persons occurs and which allows an adequate margin of safety	Appendix A
MDL	method detection limit	11.6.4.2
meq	milliequivalent	1.3.6
MeV	million electron volts	10.3
MF	membrane microfiltration	3.11.6.8
mg	milligram; unit of mass or weight, 10^{-3} g	1.3.4

Acronym	Definition	Chapter Section with First or Main Entry
mg/L	milligrams per liter	1.3.4
mmho/cm	millimhos per centimeter; unit of conductivity	3.9.4
mol	mol = mole; unit of quantity = 6.022×10^{23} items (usually atoms or molecules)	1.3.4
mp	melting point	2.8.7
MPN or mpn	most probable number	11.3.2
MQL	method quantitation limit	11.6.4.2
MRDL	maximum residual disinfection level	3.11.4.3
MTBE	methyl-*tert*-butyl ether, a gasoline additive	7.2.2
MW	molecular weight	1.3.4
N	neutron number; the number of neutrons in an element's nucleus	10.2.2.1
NAPL	nonaqueous phase liquid	7.1
NBOD	nitrogenous biochemical oxygen demand	3.8.3
NF	membrane nanofiltration	3.11.6.8
ng	nanogram; unit of mass or weight, 10^{-9} g	1.3.4
ng/L	nanogram per liter	1.3.4
NOAEL	no observed adverse effect level	Appendix A
NOM	naturally occurring organic matter	3.11.5
NPDES	National Pollutant Discharge Elimination System	1.1.1
NTA	nitrilotriacetate, metal chelating agent	2.3.3
NTU	nephelometric turbidity units	3.9.2
O&G	oil and grease analysis	11.5
OR	oxidation–reduction reaction, also called redox reaction	3.5
ORP	oxidation–reduction potential, also called redox potential	2.3.3
OSHA	Occupational Safety and Health Administration	3.11.6.1
OSWER	EPA Office of Solid Waste and Emergency Response	2.3.1
P	pressure	1.3.4
PAC	polyaluminum chloride	4.3.6.2
PAH	polycyclic aromatic hydrocarbon	2.4
PCB	polychlorinated biphenyl	2.4; 8.4
PCE	tetrachloroethene; also called perchloroethylene or PERC	9.6
pCi	picocurie; unit of radionuclide activity; 10^{-12} curies	10.4
PERC	perchloroethylene; also called tetrachloroethene or PCE	9.6
pg	picogram; unit of mass or weight, unit of mass or weight, 10^{-12} g.	1.3.4
pg/L	picogram per liter	1.3.4
pH	logarithmic measure of hydrogen ion concentration acidity	3.2
PNA	percent natural abundance of a particular isotope in a given sample of an element	10.2.2.1

(Continued)

(*Continued*)

Acronym	Definition	Chapter Section with First or Main Entry
PP	particulate phosphorus	4.3.4.2
ppb	parts per billion	1.3.4
ppm	parts per million	1.3.4
ppmv	parts per million by volume	1.3.4
ppmw	parts per million by weight	1.3.4
ppq	parts per quadrillion	1.3.4
ppt	parts per trillion	1.3.4
PQL	practical quantitation limit	11.6.4.2
PRP	potentially responsible party	7.7
psi	pounds per square inch	1.3.4; 11.6.9
PWS	public water system	10.5.1
PZC	point of zero charge	6.8.3
QA/QC	quality assurance and quality control	11.6
QC	quality control	11.6
R or r	roentgen; unit of photon (as opposed to particle) radiation exposure dose	10.4
R	retardation factor for contaminant movement in the subsurface	6.7.1
rad	radiation absorbed dose; unit of radiation energy absorbed by a receiving body	10.4
RBE	relative biological effectiveness; a quality factor to convert absorbed radiation dose to rems or sieverts	10.4.3
RCRA	Resource Conservation Recovery Act; law governing disposal of solid and hazardous wastes	1.3.4
rem	roentgen equivalent man; unit of radiation dose equivalent	10.4
R_fD	reference dose; an estimate of a daily oral exposure to the human population (including sensitive subgroups) that is likely to be without appreciable risk of deleterious effects during a 70-year lifetime	Appendix A
RO	reverse osmosis	3.6.3
RS	residual saturation	8.2.1
s	second; unit of time	10.4
S	siemens; unit of conductivity	3.9.4
S	solubility	Appendix B
SAR	sodium adsorption ratio	6.11.5.1
SAR-adj.	adjusted sodium absorption ratio, also called adjusted sodium hazard	6.11.5.2
SDC	simulated distillation curve	7.8
SDWR	Secondary Drinking Water Regulations; nonenforceable federal guidelines regarding cosmetic effects or aesthetic effects of drinking water	Appendix A
SHMP	sodium hexametaphosphate, $Na_3(PO_4)_6$	4.3.4.1

Acronym	Definition	Chapter Section with First or Main Entry
SI	International System of Units	10.4
SP	soluble phosphorus	4.3.4.2
SPE	solid phase extraction	11.5
SRB	sulfate-reducing bacteria	4.5.2
SRP	soluble reactive phosphorus	4.3.4.2
SRPRB	sulfate-reducing permeable reactive barrier	5.9.3
SSCT	small system compliance technology	10.6
STP	standard temperature and pressure; 0°C and 1 atm	10.5.4
STPP	sodium tripolyphosphate, $Na_5P_3O_{10}$	4.3.4.1
SUP	soluble unreactive phosphorus	4.3.4.2
S_v	sievert; unit of radiation dose equivalent	10.4
$t_{1/2}$	half-life of a radioisotope	10.2.8.1
TCA	trichloroethene	8.2.3; 9.6.2
TCE	trichloroethylene, trichloroethene	8.2.3
TDS	total dissolved solids, sometimes called nonfilterable solids	3.9.1
TEH	total extractable hydrocarbons, also called semivolatiles	9.5.2
TEL	tetraethyl lead; a gasoline additive	7.71
THMs	trihalomethanes	3.11.5.1
TMDL	total maximum daily loads	1.3.3
TML	tetramethyl lead; a gasoline additive	7.71
TN	total nitrogen	4.4.1.2
TOC	total organic carbon	6.7.2
TP	total phosphorus	4.3.4.2
TPH	total petroleum hydrocarbons	9.1
T_{rec}	total recoverable metals; metals analysis specified in permits or monitoring plan	5.8
TSPP	tetrasodium pyrophosphate, $Na_4P_2O_7$	4.3.4.1
TSS	total suspended solids, sometimes called filterable solids	3.9.1
TT	treatment technique; a particular method of water treatment that is legally required and enforceable, for reducing the level of a contaminant in drinking water	Appendix A
TVH	total volatile hydrocarbons	9.5.2
UF	membrane ultrafiltration	3.11.6.8
USEPA	U.S. Environmental Protection Agency	1.1.2
USGS	U.S. Geological Survey	4.4
UST	underground storage tank	7.1
UV	ultraviolet radiation	3.11.4; 3.11.6.6
VOC	volatile organic compound	9.1
WET	whole effluent toxicity	1.1.3
WE_{well}	water elevation at a water–LNAPL interface in a well	7.5
WHO	World Health Organization	3.9.2

(Continued)

(Continued)

Acronym	Definition	Chapter Section with First or Main Entry
WTE	water table elevation in subsurface adjacent to a well	7.5
X_i	mole fraction of ith component in a mixture	7.3.8
y	year; unit of time	10.2.8.1
Z	proton number; the number of protons in an element's nucleus; also called the atomic number	10.2.2.1
μg	microgram; unit of mass or weight, 10^{-6} g	1.3.4
μg/L	microgram per liter	1.3.4
μmhos/cm	micromhos per centimeter; unit of conductivity	3.9.4

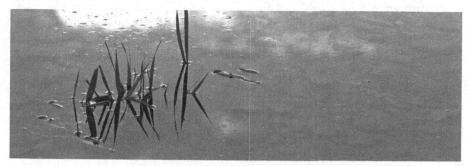

Unpolluted rainwater is acidic, around pH 5.6, because of dissolved CO_2 from the atmosphere. Rainwater acidity appears to be increasing as atmospheric CO_2 levels increase, mainly due to increased burning of fossil fuels

Chapter 3, Section 3.3.3.

Answers to Selected Chapter Exercises

CHAPTER 1

1. To determine the amount of dissolved salts contained in it, 475 mL of a water sample was evaporated. After evaporation, the dried precipitated salts weighed 1475 mg. What was the concentration in parts per million (ppm) of dissolved salts (also called total dissolved solids [TDS])?
Answer: 3105 ppm

2. The annual arithmetic mean ambient air quality standard for sulfur dioxide (SO_2) is 0.03 ppmv. What is this standard in $\mu g/m^3$?
Answer: 1300 $\mu g/m^3$

3. The primary drinking water maximum contaminant level (MCL) for barium (Ba) is 2.0 mg/L. If the sole source of barium is barium sulfate ($BaSO_4$), how much $BaSO_4$ salt is present in 1 L of water that contains 2.0 mg/L of Ba? (Hint: The moles of Ba in 2.0 mg equal the moles of $BaSO_4$ in 1 L of sample.)
Answer: 3.40 mg

4. Most people can detect the odor of ozone in concentrations as low as 10 ppb. Could they detect the odor of ozone in samples with an ozone level of (a) 0.118 ppm, (b) 25 ppm, and (c) 0.001 ppm?

Answer:
a. 0.118 ppm = 0.118 ppm × 1000 ppb/ppm = 118 ppb ozone. Since this is much higher than 10 ppb, you probably can smell the sample.
b. 25 ppm = 25 ppm × 1000 ppb/ppm = 25,000 ppb ozone. This also is much higher than 10 ppb. You probably can smell it.
c. 0.001 ppm × 1000 ppb/ppm = 1 ppb ozone. You probably cannot smell it.

5. Determine the percentage by volume of the different gases in a mixture containing 0.3 L of O_2, 1.6 L of N_2, and 0.1 L of CO_2. Assume that each separate gas is at 1 atm pressure before mixing and that the pressure of the combined gases after mixing is also at 1 atm.

Answer:
When the pressures of the individual gases before mixing and the pressure of the combined gases after mixing are the same, the individual volumes are additive in the mixture and the percentage by volume of any gas in the mixture will be

$$\text{Percent of given gas} = \frac{\text{volume of given gas}}{\text{total volume of all gases}} \times 100$$

Thus,

$$\text{Percentage of } O_2 = \frac{0.3L}{0.3+1.6+0.1L} \times 100$$
$$= 15\% \text{ (normal atmosphere \% = 21\%)}$$

$$\text{Percentage of } N_2 = \frac{1.6L}{0.3+1.6+0.1L} \times 100$$
$$= 80\% \text{ (normal atmosphere \% = 78\%)}$$

$$\text{Percentage of } CO_2 = \frac{0.1L}{0.3+1.6+0.1L} \times 100$$
$$= 5\% \text{ (normal atmosphere \% = 0.03\%)}$$

As a check on the accuracy of the answers, note that all the percentages add up to 100%. When figures are rounded to maintain the correct significant figures, percentages might not always add to exactly 100% because of cumulative rounding errors.

6. What is the significance of the fact that the percentage of oxygen in air is 21% by numbers of molecules and 23% by mass?

Answer:
The mass of O_2 molecules must be greater than the average mass of all other gases in air. The average mass of air molecules depends mainly on the two most abundant gases: nitrogen and oxygen. The predominant gas in air is N_2, with a number (same as volume) percentage of 78% and a molecular mass of 28. Oxygen is heavier, having a molecular mass of 32, so its mass percentage is greater than its number percentage. The mass percentage of N_2 will be a little lower than its number percentage. If all the gas molecules in air have the same mass, then the percentages by number and mass would be the same.

7. Express the 0.9% argon content of air in ppm.

Answer:

Percentage is the same as parts per hundred (pph). One percent is the same as 1 part per 100 parts, or $\dfrac{1 \text{ part}}{100 \text{ parts}}$ and 1 ppm is the same as $\dfrac{1 \text{ part}}{10^6 \text{ parts}}$. To express percent as ppm, you must convert parts per hundred to parts per million. Therefore,

$$0.9\% = \frac{0.9 \text{ parts}}{100 \text{ parts}} \times \frac{10^2 \times 10^4}{10^6} = \frac{0.9 \times 10^4}{10^6} = 9000 \text{ ppm}$$

8. Express 400 ppm of CO_2 in cigarette smoke as a percentage of the smoke inhaled.

Answer:

Since percent is parts per hundred, the question is how many pph will be equal to 400 ppm?

$$400 \text{ ppm} = \frac{400}{1,000,000} = \frac{0.04}{100} = 0.04\%$$

9. The permissible limit for ozone for a 1 h average is 0.12 ppm. If Little Rock, Arkansas, registers a reading of 0.15 ppm for 1 h, by what percent does Little Rock exceed the limit for atmospheric ozone?

Answer:

In this case, you want to compare the standard with the amount that the Little Rock measured value exceeds the standard value. The percent exceedance will be the exceedance divided by the standard, times 100. The difference between the Little Rock value and the standard is $0.15 - 0.12 = 0.03$. The percentage that the Little Rock value exceeds the standard is

$$\frac{0.03}{0.12} \times 100 = 25\% \text{ above the standard}$$

10. A certain water-soluble pesticide is fatal to fish at 0.5 mg/L (ppm). Five kilograms of the pesticide are spilled into a stream. The stream flow rate was 10 L/s at 1 km/h. For what approximate distance downstream could fish potentially be killed?

Answer:

First, find the volume of water that 5 kg of the pesticide will contaminate to a lethal concentration. Then, find what length of stream contains that concentration.

a. Total volume of water contaminated to lethal concentration is

$$\frac{5000 \text{ g}}{0.5 \times 10^{-3} \text{ g/L}} = 1 \times 10^7 \text{L}$$

b. Find the time required until the stream water is diluted by new flow to less than lethal concentration.

At a flow of 10 L/s: $\dfrac{10^7 \text{L of contaminated water}}{10 \text{ L/s}} = 10^6 \text{s}$

$\dfrac{1 \times 10^6 \text{ s}}{3600 \text{ s/h}} = 278$ h of flow until pesticide is diluted

Potential distance of fish kill is 278 h × 1 km/h = 278 km or 173 miles. This assumes that the pesticide does not move downstream by plug flow but is distributed uniformly throughout the contaminated length of the stream.

11. Chromium(III) in a water sample is reported as 0.15 mg/L. Express the concentration as eq/L. (The Roman numeral III indicates that the oxidation number of chromium in the sample is +3. It also indicates that the dissolved ionic form would have a charge of +3.)

Answer:

The atomic weight of chromium is 52.0 g/mol. Chromium(III) ionizes as Cr^{3+}, so its concentration in mol/L is multiplied by 3 to obtain its equivalent weight.

$$\text{mol/L of } Cr^{3+} = [Cr^3] = \frac{0.15 \text{ mg/L}}{(52.0 \text{ g/mol})(1000 \text{ mg/g})}$$

$$= 2.88 \times 16^{-6} \text{ mol/L (or 2.88 μmol/L)}$$

Concentration of Cr^{3+} in equivalents = $[Cr^3]_{eq}$

$$[Cr^{3+}]_{eq} = 2.88 \times 10^{-6} \text{ mol/L} \times 3 \text{ eq/mol}$$

$$= 8.64 \times 10^{-6} \text{ eq/L} = 8.64 \text{ μeq/L}$$

12. Alkalinity in a water sample is reported as 450 mg/L of $CaCO_3$. Convert this result to eq/L of $CaCO_3$. Alkalinity is a water quality parameter that results from more than one constituent. It is expressed as the amount of $CaCO_3$ that would produce the same analytical result as the actual sample (see Chapter 2).

Answer:

The molecular weight of $CaCO_3$ is $(1 \times 40 + 1 \times 12 + 3 \times 16) = 100$ g/mol. The dissolution reaction of $CaCO_3$ is

$$CaCO_3 \xrightarrow{H_2O} Ca^{2+} + CO_3^{2-}$$

Since the absolute value of charge for either the positive or negative species equals 2, eq/L = mol/L × 2.

$$450 \text{ mg/L} = \frac{450 \text{ mg/L}}{(100 \text{ g/mol})(1000 \text{ mg/g})}$$

$$= 4.5 \times 10^{-3} \text{ mol/L or 4.5 mmol/L}$$

4.5×10^{-3} mol/L × 2 eq/mol = 9.0×10^{-3} eq/L or 9.0 meq/L

CHAPTER 2

5. Various compounds and some of their properties are tabulated as follows:

Compound	Boiling Point (°C)	Water Solubility (g/100 mL)	Dipole Moment (D)
H_2	−253	2×10^{-4}	0
HCl	−84.9	82	1.08
HBr	−67	221	0.82
CO_2	−78	0.15	0
CH_4	−164	2×10^{-3}	0
NH_3	−33.5	90	1.3
H_2O	100	∞	1.85
HF	19.5	∞	1.82
LiF	1676	0.27	6.33

   a. Which compounds are nonpolar?

     Answer:

     All compounds with a dipole moment equal to zero are nonpolar, for example, H_2, CO_2, and CH_4.

   b. Which compounds are polar?

     Answer:

     All compounds with a dipole moment greater than zero are polar, for example, HCl, HBr, NH_3, H_2O, HF, and LiF.

   c. Make a rough plot of boiling point versus dipole moment. What conclusions may be inferred from the graph?

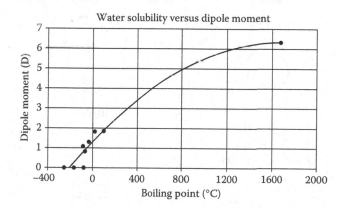

Water solubility versus dipole moment

     Answer:

     The plot shows a general increase of boiling point with dipole moment.

   d. Discuss briefly the trends in water solubility. Why is solubility not related to dipole moment in the same manner as boiling point? Zero dipole moment indicates low solubility and higher dipole moments up to at least about $D = 2$ tend to indicate higher solubility. But LiF, with a very high dipole moment of $D = 6.3$, has a low solubility characteristic of compounds with dipole moments around $D = 0$ (see part e for the reason).

Answer:

While the boiling point of a compound is related only to the strength of the attractive forces between molecules of the pure compound, its water solubility is related to that plus the strength of the attractive forces to water molecules.

e. The ionic compound LiF appears to be unique. Try to suggest a reason for its low solubility.

Answer:

The large dipole moment of LiF indicates that its bonding is ionic. Both the Li^+ and the F^- ions are small, allowing them to come very close together, which results in an unusually strong ionic bond. The very strong ionic attractions within the LiF solid compound are too strong for the attractive forces between LiF and water molecules to pull apart.

6. The chemical structures of several compounds are shown below. Fill in the table with estimated properties based only on their chemical structures. Do not look up reference data for the answers.

Answer:

Compound	Structure	Physical State at Room Temperature (Gas, Liquid, or Solid)	Water Solubility (Very Low, Low, Moderate, High)	Lipid Solubility (Very Low, Low, Moderate, High)
Ethane	H_3C-CH_3	Gas	Very low	High
Benzene		Liquid	Low	High
Citric acid		Solid	High	Low
Anthracene		Solid	Very low	High
Glyphosate		Solid	High	Low

CHAPTER 3

2. a. Water in a stream had $[H^+] = 6.1 \times 10^{-8}$ mol/L. What is the pH?
      Answer: pH = 7.21
   b. The pH of water in a stream is 9.3. What is the hydrogen ion concentration?
      Answer: $[H^+] = 5.0 \times 10^{-10}$ mol/L

3. An engineer requested a water sample analysis that included the following parameters: pH, carbonate ion, and bicarbonate ion. Explain why she probably is wasting money.
   Answer:
   If there is a charge for each analysis, the request for a carbonate analysis is unnecessary because the carbonate concentration can be found from Figure 3.2 and accompanying discussion when pH and bicarbonate are known.

6. A water sample from a lake has a measured alkalinity of 0.8 eq/L. In the early morning, a monitoring team measures the lake's pH as part of an acid rain study and finds pH 6.0. The survey team returns after lunch to recheck their data. By this time, algae and other aquatic plants have consumed enough dissolved CO_2 to reduce the lake's total carbonate (C_T) to one-half of its morning value.
   a. What was the morning value for C_T?
      Answer:
      Use the Deffeyes diagram, Figure 3.3. A horizontal line from alkalinity = 0.8 eq/L intercepts the pH 6.0 line at about $C_{T,morning} = 2.5 \times 10^{-3}$ mol/L.
   b. What was the pH after lunch?
      Answer:
      C_T after lunch was $(2.5 \times 10^{-3}$ mol/L$)/2 = 1.25 \times 10^{-3}$ mol/L. Since no acid was added, the alkalinity is essentially unchanged.
      Move vertically up from $C_{T,afternoon} = 1.25 \times 10^{-3}$ mol/L to alkalinity = 0.8 eq/L.
      The pH at that point is about 6.6.
      Depletion of CO_2 by photosynthesis caused pH to rise from 6.0 to 6.6.

7. A groundwater has the following analysis at pH 7.6:

Analyte	Concentration (mg/L)
Calcium	75
Magnesium	40
Sodium	10
Bicarbonate	300
Chloride	10
Sulfate	112

Calculate alkalinity, total hardness, carbonate (temporary) hardness, and noncarbonate (permanent) hardness.

Answer:

Note that the carbonate concentration will be 0 at pH 7.6 (see the distribution diagram, Figure 3.2).

$$\text{Alkalinity} = 0.820\,[HCO_3^-] + 1.668[CO_3^{-2}]$$
$$= (0.820)(300) + (1.668)(0) = 246 \text{ mg/L}$$
$$\text{Total hardness} = 2.497[\text{Ca, mg/L}] + 4.118\,[\text{Mg, mg/L}]$$
$$= (2.497)(75) + (4.118)(40) = 352 \text{ mg/L}$$
$$\text{Carbonate hardness} = 0.820\,[HCO_3^-] + 1.668\,[CO_3^{-2}]$$
$$= (0.820)(300) + (1.668)(0) = 246 \text{ mg/L}$$
$$\text{Noncarbonate hardness} = \text{total hardness} - \text{carbonate hardness}$$
$$= 352 - 246 = 106 \text{ mg/L}$$

Note: Noncarbonate hardness is the hardness from noncarbonate salts of calcium and magnesium, mainly the salts of sulfate, chloride, and silicate, all expressed as mg/L of $CaCO_3$.

8. A stream has a measured DO of 6.0 mg/L at 12°C. What can you say about its ability to support aquatic life?

Answer:

At 12°C, saturated DO concentration 10.76 mg/L (from Table 3.4) and

$$\% \text{ DO saturation} = \frac{6.0 \text{ mg/L}}{10.76 \text{ mg/L}} \times 100 = 55.8\%.$$

From Table 3.5, 55.8% saturation indicates moderately polluted water at 10°C, a condition made slightly worse at 12°C. The Rules of Thumb at the end of Section 3.7 says that DO concentrations of at least 5–6 mg/L are required for healthy fish. A reasonable conclusion would be that the stream probably is presently supportive of aquatic life, but cannot suffer much further degradation without becoming unable to support a healthy habitat for aquatic life.

CHAPTER 4

1. A wastewater treatment plant removed ammonia with a nitrification–denitrification process that also reduced alkalinity. For each gram of NH_3–N removed, the process also removed 7.14 g of alkalinity-$CaCO_3$. If the plant was designed to remove 25 mg/L of NH_3–N and the total throughput was 250,000 gal/day, how many pounds of caustic soda (sodium hydroxide, NaOH) must be added each day to restore the alkalinity to its original value before ammonia removal?

Answer:

According to Figure 3.3, the addition of NaOH, a strong base, represents a vertical increase on the alkalinity versus total carbonate graph. Therefore, the milliequivalents per liter of NaOH needed is equal to the milliequivalents per liter of alkalinity lost.

Alkalinity-$CaCO_3$ lost $= 7.14 \times 25$ mg/L NH_3–N $= 178.5$ mg/L alkalinity-$CaCO_3$

Eq. wt of $CaCO_3 = 50.04$ g/eq (Table 1.1)

Eq/L of alkalinity-$CaCO_3$ lost $= (178.5$ mg/L$)/(50.04$ mg/meq$) = 3.57$ meq/L

Equivalents of NaOH required $=$ eq/L of alkalinity-$CaCO_3$ lost $= 3.57$ meq/L

Eq. wt of NaOH $= 40.0$ g/eq

$$\text{Wt/d NaOH} = \frac{40.0 \text{ mg/ meq} \times 3.57 \text{ meq/L}}{4.54 \times 10^5 \text{ mg /lb} \times 0.264 \text{ gal/L}}$$
$$\times 2.50 \times 10^5 \text{ gal/day} = 298 \text{ lb/day}$$

2. A wastewater flow contains 30 g/L total ammonia nitrogen and has a 10 g/L discharge limit. An air-stripping tower is to be used. Its temperature ranges from 20°C to 30°C and the pH is normally about 9. At what pH must the stripper be operated?

Answer:

Total $NH_3 - N = [NH_3] + [NH_4^+]$. The pH must be adjusted so that enough ammonia is in the volatile form to meet the discharge requirement.

The ratio $\dfrac{[NH_3]}{[NH_3] + [NH_4^+]}$ must be at least as large as the fraction of NH_3–N to be removed.

Fraction of NH_3–N to be removed is $\dfrac{20}{30} = 0.67$.

Stripping efficiency is increased with increasing pH and higher temperatures. Figure 4.2, Chapter 4, shows that to get 0.67 (67%) of the total NH_3–N in the form of volatile NH_3, the pH must be raised to about 10.0, for the worst case of 20°C

CHAPTER 5

1. Is a solution of ferric sulfate, $Fe_2(SO)_3$, in water expected to be acidic, basic, or neutral? Explain your answer.

Answer:

The solution is expected to be acidic. Ferric sulfate dissolves to form ferric cations and sulfate anions:

$$Fe_2(SO)_3 = 2Fe^{3+} + 3SO_4^{2-}$$

Each ion attracts a hydration shell of water molecules. The triple positive charge of a Fe^{3+} ion can attract the electrons of adjacent water molecules so strongly that the O–H bond in a water molecule is broken, releasing H^+ into the solution and making the solution acidic (see Figure 5.3).

3. A groundwater sample contains dissolved Fe^{2+}. What difficulties might be encountered in using this sample for determining the in situ pH of the groundwater?

Answer:

The pH must be measured without exposing the sample to the atmosphere. Once the sample is exposed to the atmosphere, two sources of error can occur:

a. Carbon dioxide dissolved in the sample will equilibrate with CO_2 in the atmosphere. Typically, groundwater is enriched in CO_2 relative to the atmosphere because it is produced by biodegradation reactions in the subsurface. In this case, CO_2 will be lost from the sample and tend to cause an increase in pH over the in situ value.

b. Oxygen dissolved in the sample will equilibrate with O_2 in the atmosphere. Typically, groundwater is depleted in O_2 relative to the atmosphere because it is consumed by biodegradation reactions in the subsurface. This establishes the anaerobic redox conditions that reduce any iron present to the ferrous Fe^{2+} form. When the sample is exposed to the atmosphere, O_2 will enter the sample causing Fe^{2+} to be oxidized to Fe^{3+}. Fe^{3+} dissolved in water produces acidity as in Example 1, and will tend to lower the pH below the in situ value.

The net effect on pH will depend on the relative concentrations of CO_2 and Fe^{2+} in the sample.

5. An industrial discharge permit has limits for chromium, copper, and cadmium of 0.01 mg/L each. Assuming that the untreated discharge will always exceed these limits, what difficulties are encountered if you try to comply by using only pH adjustment? Use Figure 5.1 and assume that the only complexing anion present is OH^-. How might you meet the limit for all three metals?

Answer:

The most soluble of the metals is cadmium, which has a solubility for $Cd(OH)_2$ of 0.01 mg/L at a pH of 10.8. For a safety margin, you might want to go to a little higher pH, but not above pH 12 (see why in Figure 5.1). However, pH 10.8 is above the pH for the minimum solubilities of chromium and copper. At pH 10.8, the solubility of $Cu(OH)_2$ is about 0.017 mg/L and that of $Cr(OH)_3$ is about 1.3 mg/L. If you lower the pH to a value where chromium and copper will meet the discharge limit, around pH 8.1, cadmium will not; at pH = 8.1, the solubility of $Cd(OH)_2$ is greater than 100 mg/L, off the scale of Figure 5.1. To meet the limits for all three metals, first lower the pH to around 11, which lowers the solubility of $Cd(OH)_2$ well below the limit, and remove the precipitated metals by filtering. Then adjust the pH to around 8.1 to meet the limits for chromium and copper.

CHAPTER 6

1. What are soil horizons?

Answer:

Stratified soil layers formed by the cumulative effects of many rock-weathering reactions and the actions of bioorganisms on organic matter.

2. The pores of soil contain a significant amount of air. How does the composition of soil air is different from atmospheric air and why is it so?
Answer:
It tends to be depleted in oxygen and enhanced in CO_2 because of microbial respiration.

3. Many soils have ion-exchange properties. What does this mean?
Answer:
They can immobilize many metals by charge attraction. They contain materials such as clays or humus that have many polar and charged molecular sites to which dissolved ions are attracted and adsorbed. Typically, these sites are occupied by the abundant Na^+ and H_3O^+ ions, which are relatively weakly held. When multiply charged metal cations are available, they attach to the charged sites and displace (exchange) the Na^+ and H_3O^+ ions, which then enter the water as dissolved ions.

4. A chemical spill has contaminated a lake with carbon tetrachloride (CCl_4). For carbon tetrachloride, Table 6.5 shows that log $K_{ow} = 2.73$ (thus, $K_{ow} = 5.4 \times 10^2$). The maximum concentration of carbon tetrachloride that a certain species of fish can accumulate with a low probability of health risks is reported to be 10 ppm. Assuming that the fatty tissue of these fish behaves similarly to octanol with respect to carbon tetrachloride partitioning, at what value should the water quality standard for carbon tetrachloride be set in this lake to protect this fish?
Answer:

$$K_{ow} = 5.4 \times 10^2 = \frac{\text{concentration in octanol}}{\text{concentration in water}}$$

$$= \frac{\text{maximum concentration in fish}}{\text{maximum concentratin in water}}$$

$$\text{Maximum concentration in water} = \frac{10 \text{ ppm}}{5.4 \times 10^2}$$

$$= 1.85 \times 10^{-2} \text{ ppm}$$

$$= 18.5 \text{ ppb} = 18.5 \text{ µg/L}$$

5. A mountain stream is classified for Cold Water Aquatic Life use. The average atmospheric pressure at the stream's elevation of 6600 ft is about 0.8 atm. A very low BOD and turbulence in the shallow stream assures that oxygen in the stream water is in equilibrium with oxygen in the atmosphere. What is the DO (concentration of dissolved oxygen in mg/L) of the stream? Use Henry's law. The level of approximation allows you to neglect the vapor pressure of water and to assume a temperature of 20°C. Compare your answer with Table 3.5 in Chapter 3.
Answer:
From Equation 6.12: $K_H(O_2, 20°C) = \dfrac{C_a}{C_w} = 0.0198$ L · atm/mg

Percent O_2 in dry air = 21%: $C_a = (0.21)(0.8 \text{ atm}) = 0.17 \text{ atm}$

$$C_w = \frac{0.17 \text{ atm}}{0.0198 \text{ L} \cdot \text{atm/mg}} = 8.6 \text{ mg/L}$$

According to Table 3.5, water quality with respect to dissolved oxygen is good whenever DO > 8.0, making the water quality of this stream very good at the time of sampling.

CHAPTER 7

1. Discuss how fluctuations in the groundwater level can influence the distribution of light nonaqueous phase liquids (LNAPL) hydrocarbon contamination in the subsurface.

   Answer:

   LNAPL "float" on the surface of the groundwater. When the water table elevation fluctuates, it pushes the LNAPL up and down, spreading the LNAPL into a "smear zone," effectively widening the region where LNAPL is immobilized by sorption to soil particles. When the water table moves up, some LNAPL is trapped by capillary action in the saturated zone below the water table and the floating LNAPL layer is thinned. When the water table moves down again, some of the previously trapped LNAPL is released. This released LNAPL moves downward under gravity to the new water table surface and the floating LNAPL layer becomes thicker.

CHAPTER 10

1. Complete the following nuclear equations (in every case, the unknown is just one kind of particle):

   a. $^{218}_{84}\text{Po} \rightarrow {}^{4}_{-2}\text{He} + \underline{\quad}$

   Answer:

   $^{218}_{84}\text{Po} \rightarrow {}^{4}_{-2}\text{He} + {}^{214}_{86}\text{Rn}$

   c. $^{1}_{0}\text{n} + {}^{235}_{92}\text{U} \rightarrow {}^{90}_{38}\text{Sr} + 2{}^{1}_{0}\text{n} + \underline{\quad}$

   Answer:

   $^{1}_{0}\text{n} + {}^{235}_{92}\text{U} \rightarrow {}^{90}_{38}\text{Sr} + 2{}^{1}_{0}\text{n} + {}^{144}_{54}\text{Xe}$

5. A scientist is studying isotopes of the element thorium (Th), which has the atomic number 90. He isolates two different samples, each of which contains a single pure isotope of thorium. All the atoms in one sample have an atomic weight of 230 mass units and all the atoms in the other sample have an atomic weight of 234 mass units. Which is the correct explanation?

   Answer:

   c. Each atom of the heavier sample has four more neutrons in the nucleus than do the atoms in the lighter sample.

7. Use Figure 10.2 to write an equation for the probable mode of decay of each of the following radionuclides. Your equation should be in the form of the answers to Question 1.

a. $^{38}_{16}S$

Answer: The radioactive sulfur isotope with 16 protons and (38–16) = 22 neutrons is seen in Figure 10.2 to lie just above the zone of stability. In this region, radionuclides tend to decay by beta emission. Therefore, the decay reaction for S-38 will be:

$^{38}_{16}S \rightarrow {}^{0}_{-1}e + {}^{38}_{17}Cl$

9. Radon-222 has a half-life of 3.8 days. The source water for a drinking water treatment plant contains 1200 pCi/L of radon. Because aeration is not an option at this plant, how long must they store the water to meet the EPA proposed MCL of 300 pCi/L?

Answer:

$$k = \frac{0.693}{3.8d} = 0.182/day$$

$$\ln\frac{300 \ pCi/L}{1200 \ pCi/L} = -1.39 = -0.182/day \times t$$

$$t = 7.62 \ days$$

Index

Printed in the United States
by Baker & Taylor Publisher Services